Critical point. For example:
$t_{.025}$ leaves .025 probability
in the tail.

TABLE **V** *t* **Critical Points**

d.f.	$t_{.25}$	$t_{.10}$	$t_{.05}$	$t_{.025}$	$t_{.010}$	$t_{.005}$	$t_{.0025}$	$t_{.0010}$	$t_{.0005}$
1	1.00	3.08	6.31	12.7	31.8	63.7	127	318	637
2	.82	1.89	2.92	4.30	6.96	9.92	14.1	22.3	31.6
3	.76	1.64	2.35	3.18	4.54	5.84	7.45	10.2	12.9
4	.74	1.53	2.13	2.78	3.75	4.60	5.60	7.17	8.61
5	.73	1.48	2.02	2.57	3.36	4.03	4.77	5.89	6.87
6	.72	1.44	1.94	2.45	3.14	3.71	4.32	5.21	5.96
7	.71	1.41	1.89	2.36	3.00	3.50	4.03	4.79	5.41
8	.71	1.40	1.86	2.31	2.90	3.36	3.83	4.50	5.04
9	.70	1.38	1.83	2.26	2.82	3.25	3.69	4.30	4.78
10	.70	1.37	1.81	2.23	2.76	3.17	3.58	4.14	4.59
11	.70	1.36	1.80	2.20	2.72	3.11	3.50	4.02	4.44
12	.70	1.36	1.78	2.18	2.68	3.05	3.43	3.93	4.32
13	.69	1.35	1.77	2.16	2.65	3.01	3.37	3.85	4.22
14	.69	1.35	1.76	2.14	2.62	2.98	3.33	3.79	4.14
15	.69	1.34	1.75	2.13	2.60	2.95	3.29	3.73	4.07
16	.69	1.34	1.75	2.12	2.58	2.92	3.25	3.69	4.01
17	.69	1.33	1.74	2.11	2.57	2.90	3.22	3.65	3.97
18	.69	1.33	1.73	2.10	2.55	2.88	3.20	3.61	3.92
19	.69	1.33	1.73	2.09	2.54	2.86	3.17	3.58	3.88
20	.69	1.33	1.72	2.09	2.53	2.85	3.15	3.55	3.85
21	.69	1.32	1.72	2.08	2.52	2.83	3.14	3.53	3.82
22	.69	1.32	1.72	2.07	2.51	2.82	3.12	3.50	3.79
23	.69	1.32	1.71	2.07	2.50	2.81	3.10	3.48	3.77
24	.68	1.32	1.71	2.06	2.49	2.80	3.09	3.47	3.75
25	.68	1.32	1.71	2.06	2.49	2.79	3.08	3.45	3.73
26	.68	1.31	1.71	2.06	2.48	2.78	3.07	3.43	3.71
27	.68	1.31	1.70	2.05	2.47	2.77	3.06	3.42	3.69
28	.68	1.31	1.70	2.05	2.47	2.76	3.05	3.41	3.67
29	.68	1.31	1.70	2.05	2.46	2.76	3.04	3.40	3.66
30	.68	1.31	1.70	2.04	2.46	2.75	3.03	3.39	3.65
40	.68	1.30	1.68	2.02	2.42	2.70	2.97	3.31	3.55
60	.68	1.30	1.67	2.00	2.39	2.66	2.92	3.23	3.46
120	.68	1.29	1.66	1.98	2.36	2.62	2.86	3.16	3.37
∞	.67	1.28	1.64	1.96	2.33	2.58	2.81	3.09	3.29
	$= z_{.25}$	$= z_{.10}$	$= z_{.05}$	$= z_{.025}$	$= z_{.010}$	$= z_{.005}$	$= z_{.0025}$	$= z_{.0010}$	$= z_{.0005}$

Introductory Statistics for Business and Economics

WILEY SERIES IN PROBABILITY
AND MATHEMATICAL STATISTICS

ESTABLISHED BY WALTER A. SHEWHART AND SAMUEL S. WILKS

Editors

Vic Barnett, Ralph A. Bradley, J. Stuart Hunter,
David G. Kendall, Rupert G. Miller, Jr., Stephen M. Stigler,
Geoffrey S. Watson

Probability and Mathematical Statistics

ADLER • The Geometry of Random Fields
ANDERSON • The Statistical Analysis of Time Series
ANDERSON • An Introduction to Multivariate Statistical Analysis
ARAUJO and GINE • The Central Limit Theorem for Real and Banach Valued Random Variables
ARNOLD • The Theory of Linear Models and Multivariate Analysis
BARLOW, BARTHOLOMEW, BREMNER, and BRUNK • Statistical Inference Under Order Restrictions
BARNETT • Comparative Statistical Inference, *Second Edition*
BHATTACHARYYA and JOHNSON • Statistical Concepts and Methods
BILLINGSLEY • Probability and Measure
BOROVKOV • Asymptotic Methods in Queuing Theory
BOSE and MANVEL • Introduction to Combinatorial Theory
CASSEL, SARNDAL, and WRETMAN • Foundations of Inference in Survey Sampling
COCHRAN • Contributions to Statistics
COCHRAN • Planning and Analysis of Observational Studies
DE FINETTI • Theory of Probability, Volumes I and II
DOOB • Stochastic Processes
EATON • Multivariate Statistics: A Vector Space Approach
FELLER • An Introduction to Probability Theory and Its Applications, Volume I, *Third Edition,* Revised; Volume II, *Second Edition*
FULLER • Introduction to Statistical Time Series
GRENANDER • Abstract Inference
GUTTMAN • Linear Models: An Introduction
HANNAN • Multiple Time Series
HANSEN, HURWITZ, and MADOW • Sample Survey Methods and Theory, Volumes I and II
HARDING and KENDALL • Stochastic Geometry
HETTMANSPERGER • Statistical Inference Based on Ranks
HOEL • Introduction to Mathematical Statistics, *Fifth Edition*
HUBER • Robust Statistics
IMAN and CONOVER • A Modern Approach to Statistics
IOSIFESCU • Finite Markov Processes and Applications
ISAACSON and MADSEN • Markov Chains
LAHA and ROHATGI • Probability Theory
LARSON • Introduction to Probability Theory and Statistical Inference, *Third Edition*
LEHMANN • Testing Statistical Hypotheses
LEHMANN • Theory of Point Estimation
MATTHES, KERSTAN, and MECKE • Infinitely Divisible Point Processes
MUIRHEAD • Aspects of Multivariate Statistical Theory
PARZEN • Modern Probability Theory and Its Applications
PURI and SEN • Nonparametric Methods in Multivariate Analysis
RANDLES and WOLFE • Introduction to the Theory of Nonparametric Statistics
RAO • Linear Statistical Inference and Its Applications, *Second Edition*
RAO and SEDRANSK • W.G. Cochran's Impact on Statistics
ROHATGI • An Introduction to Probability Theory and Mathematical Statistics
ROHATGI • Statistical Inference
ROSS • Stochastic Processes
RUBINSTEIN • Simulation and The Monte Carlo Method
SCHEFFE • The Analysis of Variance

Introductory Statistics for Business and Economics

THIRD EDITION

Thomas H. Wonnacott
University of Western Ontario

Ronald J. Wonnacott
University of Western Ontario

JOHN WILEY & SONS
New York · Chichester · Brisbane · Toronto · Singapore

Library of Congress Cataloging in Publication Data

Wonnacott, Thomas H., 1935-
  Introductory statistics for business and economics.

  Bibliography: p.
  Includes index.
  1. Social sciences—Statistical methods.   2. Statistics.
3. Commercial statistics.   4. Economics—Statistical
methods.   I. Wonnacott, Ronald J.   II. Title.
HA29.W622   1984   519.5   84-3588
ISBN 0-471-09716-0

Printed in the United States of America

10 9 8 7 6 5 4 3

To Elizabeth and John Tukey

About the Authors

Thomas H. Wonnacott
Ronald J. Wonnacott

The authors both studied mathematics, statistics, and economics as undergraduates at the University of Western Ontario. Ron later received a Ph.D. in economics at Harvard, and Tom, a Ph.D. in statistics at Princeton. Between them, they have taught at Wesleyan University, the University of Minnesota, Duke University, and the University of California at Berkeley, and currently both teach at the University of Western Ontario. Both professors are prolific authors. Tom concentrates on statistics applied to business, social science, and medicine. Ron specializes in economic policy, especially Canada–U.S. trade. Together they have written several books for Wiley, including *Econometrics* and *Regression*.

The hobbies they share include skiing, tennis, Mozart and Verdi. Tom also enjoys touch football, and Ron plays golf. Most important, each is fortunate to have a wife and family who provide a large measure of moral support.

Preface

This book is a two-semester introduction to statistics for students in business and economics. It also is designed so that the first eight to twelve chapters can be used in a one-semester course. Our objective is to introduce students to most of the underlying logic normally available only in much more advanced texts. Yet we use the simplest possible mathematics consistent with a sound presentation.

To The Student

Statistics is the intriguing study of how you can describe an unknown world by opening a few windows on it. You will discover the excitement of thinking in a way you have never thought before.

This book is not a novel, and it cannot be read that way. Whenever you come to a numbered example in the text, try first to answer it yourself. Only after you have solved it, or at least have given it a lot of hard thought, should you consult the solution we provide. The same advice holds for the exercise problems at the end of each section. These problems have been kept computationally as simple as possible, so that you can concentrate on insight rather than arithmetic. At the same time, we have tried to make them realistic by the frequent use of real data—or at least small subsets of real data. The point of going through the hand calculations in the text is *not* to become an expert at calculating, but to develop a feeling for what the concepts mean. For this purpose, small sets of numbers will do (the much larger sets of real data are usually handled by computers anyway).

The more challenging problems and sections are indicated by a star (*). For example, in Chapters 13 and 14 we give some problems that are best answered by using a computer package: we want students who like computers to see their power. But at the same time we keep these exercises optional, so that other students can fully master the text without using a computer.

Brief answers to all odd-numbered problems are given in the back of the book. Their completely worked-out solutions are available in the student's workbook.

To The Instructor

A basic objective of this book is to explain statistical concepts as clearly and simply as possible. The only prerequisite is high-school algebra. (Calculus is also very helpful in providing greater mathematical maturity and an understanding of some optional proofs.) Problems and examples, rather than abstract theory, are used to introduce new material; the necessary theory is presented only after the student has gained a clear, intuitive idea of the concepts discussed.

This book also shows the logical relation between topics that often have appeared in texts as separate and isolated chapters; for example, the equivalence of interval estimation and hypothesis testing, the t test and the F test, analysis of variance and regression using dummy variables, and so on. In every case, our motivation has been to help students appreciate the underlying logic, so that they can arrive at answers to practical problems.

We have placed high priority on the regression model, not only because regression is widely regarded as the most powerful tool of the practicing statistician but also because it provides a good focal point for understanding such related techniques as correlation and analysis of variance. We give a great deal of coverage to nonlinear and multiple regression, and emphasize the value of multiple regression in reducing bias in observational studies.

This text is designed for maximum flexibility. Basic classical statistics are presented in the first fifteen chapters, while the last ten chapters include special but important topics, including nonparametric statistics, index numbers, decision trees, Bayesian inference, and time series. The instructor can choose any combination of topics in these last ten chapters to complete the course.

This New Edition

To make the coverage of material more complete, we have revised this edition more heavily than previous editions. We have introduced many modern techniques that applied statisticians are using, but which are often unavailable in introductory texts; for example:

- Exploratory data analysis
- The jacknife, for confidence intervals in difficult cases

- Path analysis, to illuminate regression
- Bayes shrinkage, to reduce estimation error
- Box-Jenkins ARIMA forecasting
- Decision trees
- Robust estimation, as well as nonparametric statistics
- Randomization to eliminate bias

As in earlier editions, the more advanced topics are left to the end of the chapter and marked with a star (*) to indicate that they are optional. To further improve the format, we have made two new changes: (1) To help the student keep the material in perspective, we give a summary at the end of each chapter. (2) We have moved many of the proofs to Appendixes at the end of the book. (Instructors who are used to teaching with the proofs can of course still do so. At the same time, instructors can also use this book for less demanding courses where there is little or no time for proofs.)

Finally, we have tried to keep the length of the book manageable by deleting a few less popular topics. However, to keep them accessible to instructors who feel they are important, we have retained them in the instructor's manual, where they can be easily photocopied for class use. The manual includes as well about a dozen new topics such as the Poisson and exponential distributions. Also in the instructor's manual are the completely worked out solutions to the even-numbered problems, to complement the odd-numbered solutions in the student's workbook.

Acknowledgments

So many people have contributed their help to this project that it is impossible to thank them all. But special thanks go to all those cited in the previous editions, and to the following for their generous help: David Bellhouse, Robin Carter, Lai Chan, Andy Grindlay, John Koval, Ian MacLeod, and Ian MacNeill.

We also warmly thank the many instructors and students who have forwarded suggestions to us based on their experience with earlier editions, especially the students of Business 306, Economics 135 and 255, and Statistics 135 here at the University of Western Ontario. Finally, we thank the reviewers of the manuscript, whose comments have greatly improved the final product:

Dr. I. E. Allen
The Wharton School
University of Pennsylvania

Professor William Burrell
Wayne State University

Professor Ramona K. First
San Francisco State University

Professor Richard G. Flood
College of William and Mary

Professor Michael Klima
DePaul University

Dr. Julia A. Norton
California State University-
Hayward

Professor Andrew Russakoff
St. John's University

Dr. Leonard W. Swanson
Northwestern University

Professor H. F. Williamson, Jr.
University of Illinois, Urbana

THOMAS H. WONNACOTT
RONALD J. WONNACOTT

Contents

Part

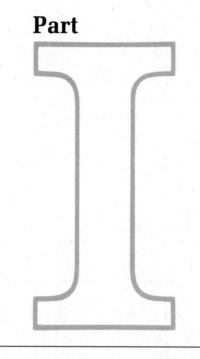

BASIC PROBABILITY
AND STATISTICS

The Nature of Statistics

He uses statistics as a drunken man uses lampposts—for support rather than for illumination.   ANDREW LANG

People often think of statistics as simply collecting numbers. Indeed, this was its original meaning: State-istics was the collection of population and economic information vital to the state. But statistics is now much more than this. It has developed into a scientific method of analysis widely applied in business and all the social and natural sciences. To illustrate the power of modern statistics, we will examine two typical applications—first, a pre-election poll, and second, the evaluation of a routine of hospital administration.

1-1 Random Sampling: A Political Poll

A—THE RELIABILITY OF A RANDOM SAMPLE

Before every presidential election, the Gallup poll estimates the percentage of the population that will vote for each candidate, and picks the winner. Clearly, canvassing the entire population would be an unrealistic task. Instead, a sample of voters is taken in the hope that the percentage of, say, Democrats that turns up in the sample will provide a good estimate of the percentage of Democrats in the population.

Just how should one select the individual Americans who will make up this sample? Some interesting lessons can be learned from past mistakes. For example, when polling was in its infancy in 1936, the editors of *Literary Digest* tried to predict the U.S. presidential election. But they used a sample of voters chosen from lists such as telephone books and club memberships—lists that tended to be more heavily Republican than the voting population at large. Even worse, only a quarter responded to their questionnaire and, as it turned out, they also tended to be more Republican than the nonrespondents. Altogether, this sample was so *biased* (i.e., lopsided, and not fairly representative of the population) that it led to a misleading prediction of a Republican victory. Election day produced a rude surprise: Less than 40% voted Republican, and the Democratic incumbent, Roosevelt, was elected by a landslide. The Republican candidate who woke up that morning expecting to become President—Alf Landon—is remembered now only by a few historians.

Other examples of biased samples are easy to find. Informal polls of people on the street are often biased because the interviewer may select people who seem civil and well dressed; a surly worker or harrassed mother is overlooked. Members of congress cannot rely on unsolicited mail as an unbiased sample of their constituency, since it over-represents organized pressure groups and people with strong opinions.

From such bitter experience important lessons have been learned: To avoid bias, *every voter must have a chance to be counted*. And to avoid slighting any voter, even unintentionally, the sample should be selected *randomly*. There are various ways of doing this, but the simplest to visualize is the following: Put each voter's name on a chip, stir the chips thoroughly in a large bowl, and draw out a sample of, say, a thousand chips. This gives the names of the thousand voters who make up what is called a *simple random sample* of size $n = 1000$.

Unfortunately, in practice simple random sampling is often very slow and expensive. For example, a random sample of American voters would include many in isolated areas who would be very difficult to track down. Much more efficient is *multistage sampling:* From the nation as a whole, take a random sample of a few cities (and counties); within each of these, take a random sample of a few blocks; finally, within each block, take a random sample of several individuals. While methods like this are frequently used, we will start with simple random sampling (as in drawing chips from a bowl).

A simple random sample will not reflect the population perfectly, of course. For example, if the sample includes only a few voters, the luck of the draw may be a big factor. To see why, suppose the population of voters is split 50-50 Democrat and Republican. How might a small sample of ten voters turn out? The likeliest result would be 5 Democrats (and 5 Republicans), but the luck of the draw might produce 8 or 9 Democrats—just as 10 flips of a fair coin might produce 8 or 9 heads. In other words, in such a small sample, the proportion of Democrats might be 80% or 90%—a far cry from the population proportion of 50%.

In a larger sample, the sample proportion of Democrats (which we call

P) will be a more reliable estimate of the population proportion (which we call π, the Greek equivalent of P. A list of Greek letters is given just before the Index.) In fact, the easiest way to show how well π is estimated by P is to construct a so-called *confidence interval*:

$$\pi = P \pm \text{sampling allowance} \qquad (1\text{-}1)$$

with crucial questions being, "How small is this sampling allowance?" and "How sure are we that we are right?" Since this typifies the very core of the book, we state the answer more precisely, in the language of Chapter 8 (where you will find it fully derived):

> **For simple random sampling, we can state with approximately 95% confidence that**
>
> $$\pi = P \pm 1.96\sqrt{\frac{P(1 - P)}{n}} \qquad (1\text{-}2)$$
>
> **where π and P are the population and sample proportions, and n is the sample size.**

We shall illustrate this confidence interval in Example 1-1 below. But first, we repeat a warning in the Preface: Every numbered example of this kind is an exercise you should first work out yourself, rather than just read. We therefore have put each example in the form of a question for you to answer; if you get stuck, then you can read the solution. But in all cases remember that *statistics is not a spectator sport.* You cannot learn it by watching, any more than you can learn to ride a bike by watching. You have to jump on and take a few spills.

EXAMPLE **1-1** Just before the 1980 presidential election, a Gallup poll of about 1500 voters showed 780 for Reagan and the remaining 720 for Carter. Calculate the 95% confidence interval for the population proportion π of Reagan supporters. (Since the Gallup poll combined multistage and other kinds of random sampling that together provided about the same accuracy as simple random sampling, equation (1-2) gives a good approximation.)

Solution The sample size is $n = 1500$ and the sample proportion is

$$P = \frac{780}{1500} = .52$$

Substitute these into equation (1-2):

$$\pi = .52 \pm 1.96 \sqrt{\frac{.52(.48)}{1500}}$$

$$\pi = .52 \pm .03 \tag{1-3}$$

That is, with 95% confidence, the proportion for Reagan in the whole population of voters was between 49% and 55%.

We must always remember that a confidence interval is an uncertain business. In equation (1-3) for example, we were only 95% confident. We must concede the 5% possibility that the "luck of the draw" turned up a misleading sample—just as flipping a coin 10 times may yield 8 or 9 heads.

A confidence interval can be made more precise, of course, by increasing sample size: As n increases in equation (1-2), the sampling allowance shrinks. For example, if we increased our sample to 15,000 voters, and continued to observe a proportion of .52 for Reagan, the 95% confidence interval would shrink to the more precise value:

$$\pi = .52 \pm .01 \tag{1-4}$$

This is also intuitively correct: a larger sample contains more information about the population, and hence allows a more precise conclusion.

B—INDUCTION AND DEDUCTION

One of the major objectives of this book will be to construct confidence intervals like (1-3). Another related objective is to *test hypotheses*. For example, suppose a claim is made that 60% of the population supports Reagan. In mathematical terms, this hypothesis may be written $\pi = .60$. It seems reasonable to reject this hypothesis, because it does not fall within the likely range (1-3). In general, there will always be this kind of relation: An hypothesis can be rejected if it lies outside the confidence interval.

Both confidence intervals and hypothesis tests are examples of *statistical inference* or *inductive* reasoning—using an observed sample to make a statement about the unknown population. In this text we will also study *probability theory* or *deductive* reasoning—arguing in the reverse direction, from a known population to the unknown sample. (In fact, this will be studied first because it is philosophically simpler and provides the necessary base for statistical inference later.) See Figure 1-1.

C—WHY SAMPLE?

Sampling was used in the Gallup poll—even though it involved some uncertainty—because polling the whole population is much too large a task. There are a variety of reasons why sampling is done in general, including:

1. *Limited* resources. Not only in political polls, but in many other cases such as market surveys, neither funds nor time are available to observe the whole population.
2. *Scarcity.* Sometimes only a small sample is available. For example, in heredity versus environment controversies, identical twins provide ideal data because they have identical heredity. Yet very few such twins are available.

There are many examples in business. An allegedly more efficient machine may be introduced for testing, with a view to the purchase

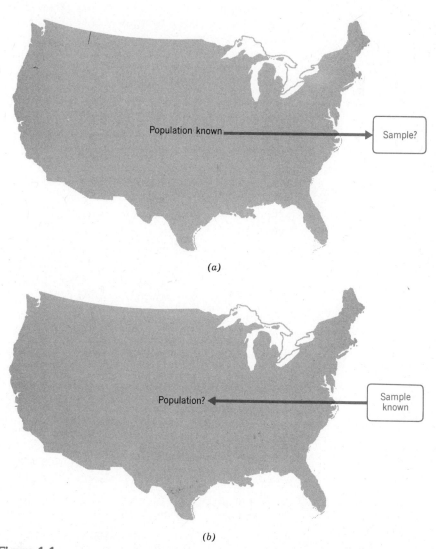

(a)

(b)

Figure 1-1
Deduction and Induction contrasted. (a) Deduction (probability). (b) Induction (statistical inference).

of additional similar units. The manager of quality control simply cannot wait around to observe the entire population that this machine will produce. Instead, a sample run is observed, and the decision on this machine is based on an inference from this sample.

3. *Destructive testing.* For example, suppose that we wish to know the average life of all the light bulbs produced by a certain factory. It would be absurd to insist on observing the whole population of bulbs until they burn out.

4. *Sampling may be more accurate.* How can a mere sample be more accurate than looking at the whole population? If the sample is carefully done, its error can be pretty well confined to the confidence allowance given in (1-2). On the other hand, a survey of the whole population might be such a gargantuan task that much worse errors could occur: A large force of inadequately trained personnel, for instance, may make measurement errors that a smaller and better trained force would avoid.

For example, the U.S. Census Bureau conducts a monthly sample of about 100,000 Americans called the *Current Population Survey* (to determine unemployment, among other things). This is a model of careful work, and in some ways is more accurate than the complete census of all 200,000,000 Americans taken every 10 years.

PROBLEMS

1-1 Project yourself back in time to six recent U.S. presidential elections. In parentheses we give the results of the Gallup pre-election poll of 1500 voters. (As we mentioned already, each sample has about the same accuracy as a simple random sample. And we continue to ignore third parties.)

YEAR	DEMOCRAT	REPUBLICAN
1960	Kennedy (51%)	Nixon (49%)
1964	Johnson (64%)	Goldwater (36%)
1968	Humphrey (50%)	Nixon (50%)
1972	McGovern (38%)	Nixon (62%)
1976	Carter (51%)	Ford (49%)
1980	Carter (48%)	Reagan (52%)

a. In each case, construct a 95% confidence interval for the proportion of Democratic supporters in the whole population.

b. Mark each case where the interval is wrong—that is, fails to include the true proportion π given in the following list of *actual* voting results:

1960	Kennedy	50.1%
1964	Johnson	61.3%

1968	Humphrey	49.7%
1972	McGovern	38.2%
1976	Carter	51.1%
1980	Carter	44.7%

1-2 In order to serve its advertisers better, a radio station specializing in fm music commissioned a market research survey of 500 listeners to determine their preference for classical or popular music. The survey results broken down by age and sex were as follows:

Numbers who prefer classical

AGE	SEX	
	MALE	FEMALE
under 25	19	26
25–50	38	34
over 50	48	60

Numbers who prefer popular

AGE	SEX	
	MALE	FEMALE
under 25	63	45
25–50	38	33
over 50	44	52

a. For each of the following, calculate the appropriate estimate and then a 95% confidence interval around it:

    i. The percentage of young males (under 25) who prefer popular.

    ii. The percentage of the young (under 25) who prefer popular.

    iii. The percentage of males who prefer popular.

    iv. The percentage of females who prefer popular.

    v. The percentage of people who prefer popular. How is this answer related to the previous two?

b. What assumption did you make in part **a**?

1-3 Criticize each of the following sampling plans, pointing out some possible biases and suggesting how to reduce them.

a. In order to estimate how many of her constituents support a gun-control bill, a Senator found from her mail that 132 supported it while 429 opposed it.

b. In order to predict the vote on a municipal subsidy to child day-care centers, a survey selected every corner house and asked whoever answered the door which way they intended to vote. Out of 2180 corner homes canvassed between 9 a.m. and 5 p.m., 960 replies were obtained.

c. To estimate the average income of its MBA graduates 10 years later, a university questioned all those who returned to their 10th reunion. Of the 281 graduates, 56 returned to their reunion and 14 were willing to provide information on their income.

1-2   Randomized Experiments: Testing a Hospital Routine

Thus far we have seen how randomizing ensures a sample against bias. In this section we will see how randomization similarly frees an experiment of bias.

Is the emotional bond of a mother to her infant weakened by the traditional hospital routine (allowing the woman only a glimpse of her newborn and then keeping them apart for about 8 hours)? To test this, a group of mothers were provided with a different treatment—*extended contact* with their infants. They were allowed a full hour with their baby just after birth, plus extra contact for the first three afternoons (Klaus and Kennel, 1972).

A—TREATMENT VS. CONTROL GROUPS

Rather than giving the extended contact treatment to all 28 women selected for this study, half were kept in the traditional routine as a *control group* for comparison. The question was: Who should be given the new treatment, and who should be kept as a control? The new treatment should not be given just to the women who requested it, since these women might inherently be the ones most interested in their children. Then, if the mother-child relationship thrived, how could one tell if it was because of the treatment, or because of this *extraneous influence*—that is, because the women who got the treatment were initially better mothers?

In fact, there are probably many other extraneous factors as well. For example, the woman's education, age, or marital status might also influence the bond with the infant. The most effective way to neutralize *all* these extraneous influences is to randomize.

B—RANDOMIZATION

To randomize, we could simply put the name of each of the 28 women on a chip, stir the chips in a bowl, and draw out half of them at random.[1] These will be given the new treatment, while the other half are put into the control group. Then the single women, for example, will on average[2] be equally spread into the treatment group and control group. Similarly, the treatment and control groups will on average be equal in terms of every other possible extraneous influence—such as education, age, and so on (the list may be almost endless).

---

[1] In the actual experiment, it was expedient to assign the patients to the treatment or control groups depending on the day of delivery —which strikes a woman pretty well at random.

[2] Suppose that, of the 28 women in this experiment, 10 are single. Then randomized assignment will *on average* put 5 of these single women in the treatment group (and 5 in the control group); but *in any single experiment*, the luck of the draw may overload the treatment group with 6 or 7 of these single women.   *(cont'd)*

Thus randomization tends to neutralize all extraneous influences. If we observe that the group getting the treatment has a better result, we can therefore conclude it is *caused* by the treatment, rather than some extraneous influence.

C—BLIND AND DOUBLE-BLIND

To ensure a fair test of a treatment, the treatment and control groups must not only be initially *created* equal by randomization, they must also be *kept* equal (except, of course, for the fact that one is getting the treatment and the other is not). To see how they might not be kept equal, suppose the doctor who finally evaluates the subjects (patients) knows who has received the new treatment and who has not; she might tend unconsciously to give a more favorable report to the treated subjects (especially if it is a treatment she herself has recommended). Consequently, the evaluator should be kept blind about who has been treated and, if possible, so should the nurses and anyone else who deals with the subjects.

Sometimes it is possible to keep even the subjects themselves blind[3], in a *double-blind* experiment. In drug trials, for example, the control subjects can be given sugar pills (a *placebo*) that they cannot distinguish from the treatment pill; then none of the participants know which group they are in.

In conclusion, an experiment should be rigorously controlled by being randomized and blind (double-blind, if possible). As shown in Figure 1-2, such experiments are the scientific ideal:

> **Randomized, blind experiments ensure that on average the two groups are initially equal, and continue to be treated equally. Thus a fair comparison is possible.** (1-5)

D—THE RESULTS OF THE EXPERIMENT

The treated women acted differently: They scored higher on care and interest in the baby during the doctor's physical examination a month later. Even two years later, there were still differences: For example, in talking to their child, the treated women tended to use questions rather than commands. (Kennel, Voos, and Klaus, 1979)

One way of preventing this from happening is to divide the 10 single women into 5 *matched pairs*. In each pair, one woman goes into the control group, the other into the treatment group. Then there will be the exact 5-5 split desired. (But one must be careful to *randomly* select the one woman within each pair to be assigned to the treatment group.)

[3]In the actual experiment, the evaluators could be kept blind, but the subjects and many of the personnel providing the treatment could not: Every woman knew of course whether or not she was getting her baby immediately after birth, and so did the attending nurses.

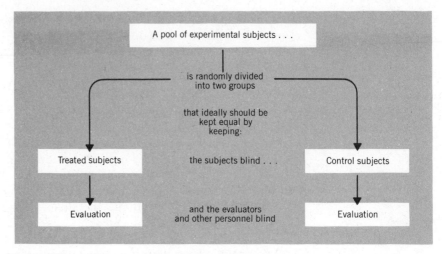

Figure 1-2
The logic of a randomized double-blind experiment

In conclusion, this experiment suggested the importance of early contact of mothers with their infants. (Interestingly, the subjects of this study were relatively poor women with few social supports. Some similar experiments on other women have not shown such strong bonding effects.)

PROBLEMS

1-4   To determine how well vitamin C prevents colds, a study in Toronto under T.W. Anderson divided volunteers randomly into two groups of about 400 each—a treatment group given vitamin C, and a control group given a placebo (a "dummy" pill that had no drug, but looked indistinguishable). The proportions who were free of colds during the winter were 26% and 18%, respectively. (Consumer Reports, 1976)

a.   What was the purpose of the placebo?

b.   It is customary to imagine that the volunteers represent a random sample from a large hypothetical population, for which we can construct 95% confidence intervals. Construct such a confidence interval for the proportion of people treated with vitamin C who would be free of colds.

c.   Repeat part **b** for the people treated with a placebo.

d.   Write a brief verbal conclusion about the effect of vitamin C.

1-5   Give an example of some "treatment" whose full effects are still unknown (for example, alcohol, compulsory education, certain forms of psychotherapy). Suggest how you might conduct an experiment to evaluate its effects. (A thorough answer to this question would qualify for a Ph.D. proposal, or a large grant application. You will have to be satisfied with a rough sketch.) Include in your discussion:

a. Who would be the subjects?

b. Who would get the treatment, and who would not?

c. Who would be kept blind?

d. How would the effects be evaluated?

1-6 Suppose the 28 women in the infant care experiment described in the text were lettered to preserve confidentiality, as follows (with single women denoted by capital letters):

> A a b B C   D c d e E   f F g h j
> k G m n p   H J q r s   t K u

One way to randomly assign them to treatment or control is to put all their names in a hat, and draw out half. Sometimes, however, the full list of subjects is not known initially; for example, they may just come in one by one until the researchers decide they have enough data. Then a different randomization has to be done. For example, as each subject enters the hospital, she could be assigned to treatment or control depending on whether a coin turns up heads or tails.

a. Actually flip a coin for each of the women lettered above to see how this randomization works. (Or use the table of random digits in Appendix Table I, letting an odd digit represent heads, and an even digit represent tails.) Circle the subjects who get "heads," and consequently go into the treatment group.

b. As already noted, the 10 capital letters represent the 10 women who are single. Does the randomization in part **a** leave them about evenly split into the treatment and control groups?

c. Suppose this experiment were repeated 5 times (or, equivalently, conducted on 5 times as large a scale). How evenly would you expect the single women to be split now?

d. To confirm your guess in part **c**, repeat the randomization 5 times and see whether the overall split is nearly 50-50. Does this show how randomization tends to "even out extraneous factors on average"?

1-7 If the list of 28 women in Problem 1-6 is known initially, we can exploit this information to get an exact 50-50 split on marital status in our one particular experiment, as follows:

a. Draw up a list of 14 *matched pairs*—5 pairs of single women (capital letters) and 9 pairs of married women (little letters).

b. In each pair, flip a coin for the first woman, and circle her if she gets "heads" and consequently goes into the treatment group (of course, the second woman should then be assigned to the other group).

How are the single women split? And the married women? Has this achieved the exact balance desired?

1-3 Observational Studies vs. Randomized Experiments

As we have seen, randomized assignment of subjects to treatment or control is what frees an experiment of bias. Sometimes, however, a treatment is studied less rigorously by simply observing how it has worked on those who happened to have got it: This is called an *historical* or *observational* study. Although such studies are less satisfactory than randomized experiments, they are sometimes all that is possible or practical, for a number of reasons:

A—RANDOMIZATION IS SOMETIMES NOT POSSIBLE

Especially in the social sciences, there are often examples where randomized assignment of treatment and control is just not possible. For example, suppose we wish to determine whether sex affects college faculty salaries. Specifically, do women earn less, simply because they are women?

In this case a randomized experiment would involve taking an initial pool of professors, and assigning a sex to each by some random process (heads you're a woman, tails you're a man), and then watching what happens to the salaries of women compared to men over the next 10 years. In this way we could remove the effect on salaries of extraneous influences, such as years an individual has been teaching. (Because of randomization, about half of the most experienced teachers would be assigned to the male group, and half to the female group.)

But, of course, we cannot randomly assign sex. We just have to take professors as they come in an observational study.

B—RANDOMIZATION IS SOMETIMES NOT PRACTICAL

Suppose we wish to examine whether the crime rate is affected by whether people live in the city or in the country. Could we randomly assign half the subjects to the city and half to the country (regardless of where they are now living)? Obviously, people would be unwilling to move and change their way of life, just to assist in a statistical study. Again, we have to take people as they come—some in the city, and some in the country—in an observational study.

C—RANDOMIZATION IS SOMETIMES NOT DONE, EVEN WHEN PRACTICAL

Suppose we wish to evaluate the benefits of a free educational program for prekindergarten children (such as Headstart). Since there are many more applicants than places available (due to limited funds), selecting individuals at random would be a fair way of deciding which children would be allowed to participate. At the same time, such randomization would admirably suit the needs of a valid scientific experiment. Unfortunately, randomization is seldom done, and so the value of many such programs remains in dispute.

It is interesting to speculate on why randomization isn't used more often, even when it costs relatively little. Is it because some investigators just don't appreciate its importance? Or is it because some administrators cannot admit that a mere coin or a bowl of chips does a better job of assignment than they do? Or is it politics?

Whatever the reason, randomization could be done more frequently; indeed, whenever practical, it should be undertaken. We cannot repeat this often enough, since it is one of the most important points we make: *Randomized assignment ensures that an experiment is free of bias.*

D—SOME ETHICAL ISSUES

Is it ethical to experiment with people? In our original example, for instance, was it ethical for the hospitals to experiment with mothers and infants? Since the new experimental treatment seemed to harm no one, it is hard to imagine any ethical objection to trying it out. In fact, a far more interesting question is this: Was it ethical to install the earlier routine measures (separation of mother and infant) *without* gathering careful experimental evidence?

In general, *every* time a new program is introduced—medical, educational, social, business, or whatever— at some stage it *has* to be tried on people for the first time; that is, an experiment has to be undertaken. So the real question is: Do we experiment carefully, or haphazardly? Do we experiment with randomized control and learn quickly, or do we run poor experiments that yield misinformation and hence result in policies and practices that cause unnecessary harm? The ethical problem has been nicely summarized by a surgeon (Peacock, 1972; via Tufte, 1974):

> One day when I was a junior medical student, a very important Boston surgeon visited the school and delivered a great treatise on a large number of patients who had undergone successful operations for vascular reconstruction. At the end of the lecture, a young student at the back of the room timidly asked, "Do you have any controls?" Well, the great surgeon drew himself up to his full height, hit the desk, and said, "Do you mean did I not operate on half of the pa-

tients?'' The hall grew very quiet then. The voice at the back of the room very hesitantly replied, ''Yes, that's what I had in mind.'' Then the visitor's fist really came down as he thundered, ''Of course not. That would have doomed half of them to their death.'' .. It was absolutely silent then, and one could scarcely hear the small voice ask, ''Which half?''

E—REDUCING BIAS IN OBSERVATIONAL STUDIES: REGRESSION

As already noted, the difficulty of randomized experiments in the social sciences often means that there is no alternative but to take individuals as they come, in an observational study. But in such a study how can we reduce the *bias* (or *confounding*) due to uncontrolled extraneous factors? To use our earlier example in estimating the effect of sex on professors' income, how do we control for such extraneous factors as years of experience? Fortunately, there is a general answer.

> **Along with the variables we are studying (such as sex and income) we also observe and record the extraneous factors (such as years of experience).**
>
> **When we cannot hold these extraneous factors constant by randomized assignment, we analyze our data in a compensating way that gives us, insofar as possible, the same answer *as if we had held the extraneous factors constant*. The technical method used to accomplish this is called *multiple regression analysis*, or just *regression* (which will be described in Chapter 13).**

(1-6)

Although it does the best job possible under the circumstances, regression still cannot do a perfect job—it is just not possible to record or even identify the endless list of extraneous influences. So we must recognize that no method of analyzing an observational study—not even regression—can *completely* compensate for a lack of randomized control.

Yet regression remains a very valuable tool because it powerfully describes, in a single equation, how one variable is related to several others. For example, it can show how a professor's income is related to sex, age, teaching performance, research record, and so on. In fact, regression is probably the single most important tool you will ever use, and so we will place heavy emphasis on it.

F—CONCLUSION

We have seen that when randomized experiments are not possible, we have to be satisfied with observational studies that passively observe how the treatment happened to be given, rather than actively assign it randomly and fairly. The problem with such observational studies is that the effect of the treatment may be confounded (biased) by extraneous influences.

Yet all is not lost in an observational study. The confounding effects of extraneous influences can be reduced by taking these influences into account in a regression analysis.

PROBLEMS

1-8 The Graduate School of Berkeley in 1975 admitted 3700 out of 8300 men applicants, and 1500 out of 4300 women applicants (Bickel and O'Connell, 1975). Assume men and women applicants to each faculty were equally qualified.

a. What is the difference in admission rates, between men and women? Do you think this provides evidence of sex discrimination?

b. To find out where this difference arises, the data was broken down by faculty, as follows (although this breakdown is hypothetical, it preserves the spirit of the problem while simplifying the computations):

	MEN		WOMEN	
FACULTY	**NUMBER OF APPLICANTS**	**NUMBER ADMITTED**	**NUMBER OF APPLICANTS**	**NUMBER ADMITTED**
Arts	2300	700	3200	900
Science	6000	3000	1100	600
Totals	8300	3700	4300	1500

Now what is the difference in admission rates between men and women, for Arts? And for Science? What evidence of sex discrimination is there?

c. Explain why your answers to parts **a** and **b** seem to be in conflict. (This is an example of *Simpson's paradox*: What is true for the parts may not be true for the whole.)

d. Answer True or False; if False, correct it.

i. If faculty is kept constant, men and women are admitted about equally. However, there is a tendency for women to apply to the tougher faculty, which explains why their overall admission rate is considerably lower.

ii. In part **a,** the effect of sex on admissions could not be properly understood because there was an extraneous influence (faculty) that wasn't being controlled. When appropriate control was introduced in part **b,** however, a more accurate picture emerged.

iii. With regression, the analysis could be improved further by taking into account other extraneous variables as well as faculty. For example, if we aren't sure of our initial assumption that men and women are equally qualified, we could allow for different qualifications by using their graduate record exam scores in a regression analysis.

1-9 Naive studies earlier in this century showed a positive relation between population density and social pathology such as crime. (Higher density areas of cities had higher crime rates.) However, more careful studies using regression (Choldin, 1978 or Simon, 1981) have shown that for extraneous factors kept constant, the relation often disappears or even becomes slightly negative.

a. Suggest some extraneous factors that often accompany high density, and might be responsible for some of the crime.

b. Underline the correct choice in each bracket:

i. This illustrates how [double-blindedness, observational studies, randomized experiments] can often be deceptive when analyzed naively.

ii. Specifically, the [extraneous factors, confidence intervals, sociological theories] may be what produce some—or all—of the effects.

iii. We therefore say the effect of population density is [confounded with, interchangeable with, multiplied by] the effect of the extraneous factors.

1-10 An historical study of crime in American urban cores (Simon, 1981) shows that since 1950, the density has tended to go down while the crime rate has increased. To what extent does this prove that low population density produces crime?

1-4   Brief Outline of the Book

In the first 10 chapters, we study how a random sample such as the Gallup poll can be used to make a statement about the underlying population from which the sample was drawn. The necessary foundation of deduction (probability theory) is laid in Chapters 3 to 5, so that induction (statistical inference) can be developed in Chapters 6 to 10. By the time we complete Chapter 10, we will be in a position to estimate a wide variety of things, such as the average air pollution in Cleveland, or the average quality of a product coming off an assembly line. Each of these is relatively simple to estimate, because just one variable is being considered.

In Chapters 11 to 15 we take up the even more interesting question of how one variable is related to several others. For example, how is the mortality rate in American cities related to air pollution, the average age of the population, and so on? It is in these chapters that we develop regression analysis, along with numerous applications.

Finally, from Chapter 16 to the end, we cover a wide variety of additional topics, including some of the most modern and challenging applications of statistics.

CHAPTER 1 SUMMARY

1-1 Statistics can estimate a whole population just by looking at a sample that is properly drawn from it. To avoid bias, the sample must be *randomly* drawn. Then a confidence interval can be constructed, with an error allowance that shows the sampling uncertainty.

1-2 In an experiment to evaluate a new treatment, how can we avoid bias? We must *randomly* determine who gets the treatment and who gets left as a control. And anyone who might prejudice the results must be kept blind about who has received the treatment and who has not.

1-3 When randomized assignment of treatment in an experiment is not possible, we have to be satisfied with observational studies that simply observe how the treatment happened to be given. Then the effect of the treatment may be biased by extraneous factors. Fortunately, regression can reduce this bias.

As a whole, this chapter has illustrated a truth as old as science itself: *The way we collect data is at least as important as how we analyze it.* There are many excellent books that develop this vital issue further, including some entertaining and inexpensive paperbacks—by Campbell, Huff, Moore, Slonim, Tanur and others, and Wallis and Roberts. Their titles are listed in the Reference Section at the back of the book.

REVIEW PROBLEMS

These review problems included at the end of each chapter give an overview of all the material. Because they do not fit neatly into a pigeonhole, they require more thought. Consequently, they provide the best preparation for meeting real life—and exams.

1-11 Give some historical examples of useless or even harmful "treatments" that persisted for many years, because they were not evaluated properly (e.g., bloodletting in medicine).

1-12 Give some present-day examples of "treatments" that are perhaps useless or harmful, but still persist because they have not been evaluated well and nobody really knows their true effect (e.g., life imprisonment versus capital punishment).

In which cases would it be relatively easy to evaluate the treatment properly? How?

In which cases would it be extremely difficult to evaluate the treatment properly? Why?

1-13   "The possession of such a degree as the MBA will double lifetime earnings as compared to those of a high school graduate." (Bostwick, 1977)

To back this up, suppose the author found out that MBAs' annual incomes are twice as high on average as high school graduates'. Is the author's conclusion accurate? Explain.

1-14   Do seat belts prevent injury and death? They definitely do, according to experiments that have been done, smashing up cars and measuring the damage to dummy passengers. It is also valuable to look for confirming evidence from real people, in observational studies. For example, in a study of accidents, we could compare injury rates of those who were wearing seat belts to those who were not. In order to keep extraneous factors to a minimum, which of the two alternatives in each of the cases below would be better, and why?

a.   Using data from all accidents, or from just those involving cars equipped with seat belts?

b.   Using data on all the occupants, or on just the drivers?

c.   Using data that lumps together all injuries, or that categorizes them into several levels of severity?

d.   Having the doctor who evaluates the injury be informed, or not informed, of whether the patient had been wearing a seat belt?

1-15   The following data on seat belt usage and injury rates was collected for selected years in the 1970s, from selected states that kept good records (U.S. Dept. of Transportation, 1981):

Belt Usage and Injury Rates for Accident-Involved Occupants With Safety Belts Available

| | | INJURY RATE | |
SEAT BELT USAGE	PERCENT OF ALL OCCUPANTS	MODERATE INJURY $(2 \leq AIS < 3)$[a]	SERIOUS OR GREATER INJURY $(AIS \geq 3)$
Unbelted	85.9	.023	.013
Lap Belt	3.9	.011	.009
Lap and Shoulder Belt	7.8	.005	.004
Unknown	2.4	—	—

[a]AIS: Abbreviated Injury Scale 1980 Revision, American Association for Automotive Medicine, Norton Grove, IL 60053.

a.   Regarding the data as a random sample from the population of accidents over the whole decade from the whole country, construct a confidence interval for the rate of moderate injury among those wearing lap and shoulder belts. Then do the same for those wearing no seat belts. Assume $n = 10,000$.

b.   Repeat part **a** for injury that is serious or worse.

c.   What can you conclude about the value of lap and shoulder belts?

1-16 **a.** Of the 18,000 deaths in Arizona in 1977, 1440 were from respiratory disease. Assuming these deaths can be regarded as a random sample from a hypothetical population of millions who might have lived and died in Arizona, calculate a 95% confidence interval for the population proportion (π_A) of deaths that are due to respiratory disease.

b. For the U.S. in 1977, calculate the corresponding proportion $\pi_{U.S.}$, given the information that out of 1,900,000 deaths, 110,000 were from respiratory disease.

c. To what extent does the data show that:

   i. Arizona has a higher proportion of deaths due to respiratory disease?

   ii. Arizona's climate worsens respiratory disease?

1-17 United States unemployment statistics are obtained through the Current Population Survey, a sample of about 100,000 adults conducted monthly by the U.S. Census Bureau. Like the Gallup poll, it is a combination of multistage and other kinds of random sampling that altogether provide about the same accuracy as simple random sampling, so that equation (1-2) gives a good approximation. In December 1977, the sample gave roughly the following figures for the noninstitutional population aged 16 and over:

Employed	58,770 ⎱ 62,500
Unemployed	3,730 ⎰
Outside the Labor Force	37,500 ← should be
Total	100,000 ignored

What was the unemployment rate in the entire U.S. population? Construct a 95% confidence interval.

Descriptive Statistics

The average statistician is married to 1.75 wives who try their level best to drag him out of the house 2¼ nights a week with only 50 percent success.

He has a sloping forehead with a 2 percent grade (denoting mental strength), ⅝ of a bank account, and 3.06 children who drive him ½ crazy; 1.65 of the children are male.

Only .07 percent of all statisticians are ¼ awake at the breakfast table where they consume 1.68 cups of coffee—the remaining .32 dribbling down their shirt fronts. . . . On Saturday nights, he engages ⅓ of a baby-sitter for his 3.06 kiddies, unless he happens to have ⅝ of a mother-in-law living with him who will sit for ½ the price. . . .   W. F. MIKSCH (1950)

We already have discussed the primary purpose of statistics—to make an inference from a sample to the whole population. As a preliminary step, the sample must be simplified and reduced to a few descriptive numbers, called sample *statistics*.

For instance, in the Gallup poll of Example 1-1, the polltaker would record the answers of the 1500 people in the sample, obtaining a sequence such as D,R,R,D,D,R, . . . , where D and R represent Democrats and Republicans. An appropriate summary of this sample is the statistic P, the sample proportion of Republicans; this can be used to make an inference about π, the population proportion. Admittedly, this statistic P is trivial to compute. It requires merely counting the number of R's and then dividing by the sample size, obtaining $P = 780/1500 = .52$.

Now we will take a look at some other samples, and calculate the appropriate statistics to summarize them.

2-1 Frequency Tables and Graphs

A—DISCRETE EXAMPLE

Suppose we take a sample of 50 U.S. families, and record the number of children in each family as X. We call X a *discrete* random variable because it can take on only a few values that can be counted (0,1,2,3, ... ; but X cannot take on the continuous values in between). Suppose the 50 values of X turn out to be

$$3, \ 2, \ 2, \ 0, \ 5, \ 1, \ 4, \ 0, \ \ldots, \ 7, \ 2$$

To simplify, we keep a running tally of the outcomes in Table 2-1. In the third column we record, for example, that a zero-child family was observed 9 times; the *relative* frequency is 9/50 = .18 (or 18%),[1] and is recorded in the final column.

TABLE **2-1** **Frequency and Relative Frequency of the Number of Children, in 50 Completed American Families**

NUMBER OF CHILDREN	TALLY OF 50 FAMILIES	FREQUENCY f	RELATIVE FREQUENCY $\dfrac{f}{n}$
0	●●●●● ●●●●	9	.18
1	●●●●● ●●	7	.14
2	●●●●● ●●●●● ●●	12	.24
3	●●●●● ●●●●	9	.18
4	●●●●●	5	.10
5	●●●●● ●	6	.12
6		0	0
7	●●	2	.04
		n = 50	1.00 √

---

[1] Throughout this book, relative frequencies (and probabilities) will be expressed as either decimals or percentages, whichever seems more convenient at the time.

Usually, mathematicians prefer decimals, while applied statisticians prefer percentages. Therefore, we usually do our calculations in decimal form and provide the verbal interpretations in percentage form.

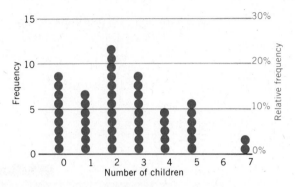

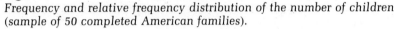

Figure 2-1

Frequency and relative frequency distribution of the number of children (sample of 50 completed American families).

In Figure 2-1 we graph the frequency distribution—essentially just the tallies of Table 2-1 turned 90°. The *relative* frequency distribution in the last column can be shown on the same graph, using a different vertical scale on the right. To keep clear the meaning of this graph, always remember that each dot represents a family. Thus the initial stack of 9 dots represents the 9 families having zero children, and so on.

B—CONTINUOUS EXAMPLE

Suppose we take a sample of 200 American men, and record each height in inches. We call height X a *continuous* variable, since its possible values vary continuously (X could be 64, 65, 66, . . . , or *anywhere in between,* such as 64.328 . . . inches[2]). It no longer makes sense to talk about the frequency of a specific value of X, since we will never again observe anyone exactly 64.328 . . . inches tall. Instead, we can tally the frequency of heights within a cell, as a Table 2-2. (For a change this time we tally each observation with a customary stroke, instead of a dot.)

The cells have been chosen somewhat arbitrarily, but with the following conveniences in mind:

1. The number of cells is a reasonable compromise between too much detail and too little. Usually 5 to 15 cells is appropriate.

---

[2]We shall overlook the fact that although height is conceptually continuous, in practice the measured height is rounded to a few decimal places at most, and is therefore discrete. In this case we sometimes find observations occurring right at the cell boundary. Where should they be placed—in the cell above or the cell below? (For example, if we observe a height of 61.5 inches, should it be recorded in the first or second row of Table 2-2?)

One of the best solutions is systematically to move the first such borderline observation up into the cell above, move the second such observation into the cell below, the third up, the fourth down, and so on. This procedure avoids the bias that would occur if, for example, all borderline observations were moved to the cell above.

TABLE **2-2**   **Frequency and Relative Frequency of the Heights of 200 Men**

CELL BOUNDARIES	CELL MIDPOINT	TALLY	FREQUENCY f	RELATIVE FREQUENCY f/n
58.5–61.5	60	‖‖	4	.02
61.5–64.5	63	‖‖ ‖‖ ‖	12	.06
.	66	.	44	.22
.	69	.	64	.32
.	72	.	56	.28
73.5–76.5	75		16	.08
76.5–79.5	78	‖‖	4	.02
			n = 200	1.00

2. Each cell midpoint, which hereafter will represent all observations in the cell, is a convenient whole number.

Figure 2-2 illustrates the grouping of the 200 heights into cells. In panel (*a*), the 200 men are represented by 200 tiny dots strung out along the X-axis. The grouped data can then be graphed in panel (*b*). (Note how this is simply a graph of the frequency distribution in the second-to-last column of Table 2-2.) Bars are used to represent frequencies as a reminder that the observations occurred throughout the cell, and not just at the midpoint. Such a graph is called a *bar graph* or *histogram*.

Once more, to emphasize the meaning of this graph, we have represented each observation (man's height) by a dot. Thus the initial 4 dots represent the 4 shortest men, and so on.

C—PERCENTILES

A bar graph can be extremely useful in showing relative position. For example, returning to Figure 2-2*b*, what can we say about a man 64.5 inches tall (at the right-hand boundary of the second cell)? He is relatively small, with only a few men smaller than he. To be specific, only 8% are smaller. (From the last column of Table 2-2, we add up the first two cells: 2% + 6% = 8%.) His height is therefore said to be the 8th percentile. At the other end, a height of 73.5 inches would be the 90th percentile (since only 8% + 2% = 10% of the observations are higher).

The percentiles that cut the data up into four quarters have special names: The 25th percentile and 75th percentiles are called the *lower* and *upper* quartiles. The 50th percentile is called the *median*, because it is the middle value that cuts the data in half. In Figure 2-2, for example, we see (and confirm in Appendix 2-2) that:

$$\left. \begin{array}{ll} \text{Lower quartile} & \simeq 67 \text{ inches} \\ \text{Median} & \simeq 69.5 \text{ inches} \\ \text{Upper quartile} & \simeq 72 \text{ inches} \end{array} \right\} \qquad (2\text{-}1)$$

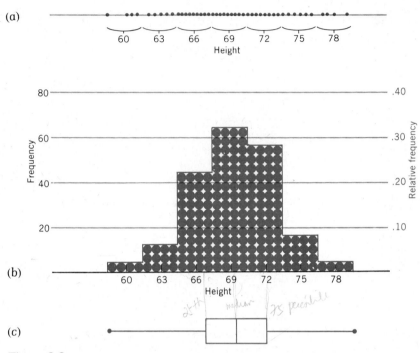

Figure 2-2
Different graphs of the same dots (heights of 200 American men). (a) The grouping of observations into cells, illustrating the first two columns of Table 2-2. (b) The bar graph for the grouped data. (c) The very brief box plot.

D—BOX PLOTS

To show the quartiles of a distribution very clearly, an alternative graph of the data has been recently proposed, called the *box plot* (Tukey, 1977). Figure 2-2c shows the box plot alternative for the distribution of men's heights. The two quartiles mark the ends of the box, while the median is shown as the vertical line near the middle of the box. The extent of the distribution on each side is indicated by a protruding line, so that the graph is sometimes called a *box and whisker* plot.

The box plot is not only easy to draw and understand; it also gives a good idea of where the distribution is centered, and how spread out it is. These two issues, center and spread, are important enough that we will spend the rest of the chapter describing how they can be measured.

PROBLEMS

In many of the exercises based on real data, such as Problem 2-1 or 2-2, we have tried to preserve the reality yet keep the arithmetic manageable by using the following device: From the available data (which usually is a whole population, or at least a very large sample) we construct a small sample that reflects the population as closely as possible.

2-1 In a large American university, a random sample of 5 women professors gave the following annual salaries (in thousands of dollars, rounded from Katz, 1973):

$$9, 12, 8, 10, 16$$

Without sorting into cells, graph the salaries as dots along an X-axis.

2-2 At the same university, a random sample of 25 men professors gave the following annual salaries (in thousands of dollars, rounded):

13	11	19	11	22	22	13	11	17	13
27	14	16	13	24	31	9	12	15	15
21	18	11	9	13					

Without sorting into cells, graph the salaries as dots along an X-axis. Be sure to use the same scale as in Problem 2-1, so you can compare the graphs of the men and women. In your view how good is the evidence that, over the whole university, men tended to earn more than women? (This issue will be answered more precisely in Chapter 8.)

2-3 Sort the data of Problem 2-2 into cells with midpoints of 10, 15, 20, 25, 30, and draw the bar graph.

2-4 Draw the box plot for the 5 women's salaries plotted in Problem 2-1. (*Hint:* The median is the middle observation. For the quartiles, cut off ¼ of 5 ≃ 1 observation from each end.)

2-5 Draw the box plot for the 25 men's salaries plotted in Problem 2-2. (*Hint:* The median is the middle observation, with 12 below and 12 above. For the quartiles, cut off ¼ of 25 ≃ 6 observations from each end.) Use the same scale as in Problem 2-4, so that you can compare.

2-6 Using the 25 men's salaries plotted in Problem 2-2 (where each observation represents 4% of the data), what percentile is a salary of 10 thousand? 20 thousand? 30 thousand?

2-2 Center of a Distribution

There are many different ways to define the center of a distribution. We will discuss the three most popular—the mode, the median, and the mean—starting with the simplest.

A—THE MODE

Since *mode* is the French word for fashion, the mode of a distribution is defined as the most frequent (fashionable) value. In the example of men's heights, the mode is 69 inches, since this cell has the greatest frequency

To generalize this, suppose a sample of n observations is denoted by $X_1, X_2, \ldots, X_n$. Then the average or mean is denoted by $\overline{X}$. It is found by summing and dividing by the sample size n:

$$\text{Mean } \overline{X} \equiv \frac{1}{n}(X_1 + X_2 + \cdots + X_n) \tag{2-3}$$

We customarily abbreviate the sum of all the X values by[3] ΣX (where Σ is sigma, the Greek equivalent of our S as in Sum. For a complete list of Greek symbols, see the glossary just before the index.) Thus (2-3) can be written briefly as

$$\overline{X} \equiv \frac{1}{n} \Sigma X \tag{2-4}$$

The average for our earlier sample of heights could be computed by summing all 200 observations and dividing by 200. However, this tedious calculation can be greatly simplified by using the grouped data in Table 2-3b. Let us denote the first cell midpoint by x_1 and use it to approximate all the observations in the first cell (f_1 in number). Similar approximations hold for all the other cells, too, so that

$$\overline{X} \simeq \frac{1}{n}\left[\underbrace{(x_1 + x_1 + \cdots + x_1)}_{f_1 \text{ times}} + \underbrace{(x_2 + x_2 + \cdots + x_2)}_{f_2 \text{ times}} + \cdots\right]$$

$$= \frac{1}{n}\left[x_1 f_1 + x_2 f_2 + \cdots\right] \tag{2-5}$$

where $\simeq$ means *approximately equals*.[4] In brief notation,

---

[3]The sum ΣX could be written more formally as

$$\sum_{i=1}^{n} X_i$$

That is, the typical observation X_i is being summed over all its n values. When all our sums are over the range from 1 to n, however, we need not explicitly state so every time. The informal notation ΣX is quite adequate.

[4]In approximating each observed value by the midpoint of its cell, we sometimes err positively, sometimes negatively; but these errors will tend to cancel out, so that finally (2-6) should be a good approximation. [For a discrete distribution such as family size, there is no approximation necessary; then (2-6) will be exactly true.]

Note that cell midpoints are denoted by small x, to distinguish them from observed values X. If there are c cells altogether, then the sum in (2-6) could be written more formally as

$$\overline{X} = \sum_{i=1}^{c} x_i f_i$$

$$\boxed{\text{For grouped data,} \quad \overline{X} \simeq \frac{1}{n} \Sigma \, xf} \qquad (2\text{-}6)$$

Using (2-6), we calculate the mean height in Table 2-3, part (b). The calculation is just like the ungrouped calculation in part (a), except that we carefully count each x with its frequency f.

D—THE MEAN INTERPRETED AS THE BALANCING POINT

The 200 heights appeared in Figure 2-2a as points along the X-axis. If we think of each observation as a one-pound mass, and the X-axis as a weightless supporting rod, we might ask where this rod balances. Our intuition suggests "the center."

The precise balancing point, also called the center of gravity, is given in physics by exactly the same formula as the mean. Thus we may think of the sample mean as the "balancing point" of the data, symbolized by ▲ in our graphs.

E—RELATIVE POSITIONS OF THE MODE, MEDIAN, AND MEAN

Figure 2-3 shows a distribution that has a single peak and is symmetric (i.e., one half is the mirror image of the other). In this case the mode, median, and mean all coincide.

But what if the distribution is skewed? For example, Figure 2-4a shows a long tail to the right. Then will the median, for example, coincide with the mode? With so many observations strung out in the right-hand tail, we have to move from the peak value toward the right in order to pick up half the observations. Thus the median is to the right of the mode.

TABLE **2-3** **Calculation of Mean (a) for ungrouped data of 5 shipping losses (b) for grouped data of 200 men's heights**

(a)		(b)	
		GIVEN DATA		CALCULATION OF $\overline{X}$
X	x	f	xf
10	60	4	240
20	63	12	756
30	66	44	2904
50	69	64	4416
90	72	56	4032
	75	16	1200
	78	4	312
$\overline{X} = \dfrac{200}{5}$		n = 200	$\overline{X} = \dfrac{13{,}860}{200}$
= 40			= 69.3

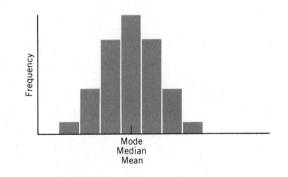

Figure 2-3
A symmetric distribution with a single peak. The mode, median, and mean
coincide at the point of symmetry.

Finally, where will the mean lie? Near the median perhaps? In Figure
2-4, panel (a) shows what happens if we try to balance the distribution
at the median. Half the observations lie on either side, but the observations
on the right are farther out and exert more downward leverage. To find
the actual balancing point (mean) we have to go farther to the right, as in
panel (b). Thus the mean lies to the right of the median.

What then are the conclusions for a skewed distribution? *Relative to
the mode, the median lies out in the direction of the long tail. And the
mean lies even farther out.*

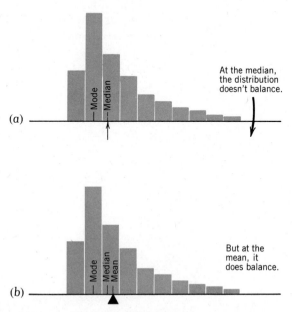

Figure 2-4
For a skewed distribution, the median is not the balancing point (mean). (a)
The distant observations on the right tip the attempted balance at the median.
(b) The actual balancing point (mean) is therefore to the right.

F—WHICH CENTER IS APPROPRIATE—THE MODE, MEDIAN, OR MEAN?

For some purposes, one measure of center may be appropriate; for other purposes, another may be. To illustrate, consider the distribution of incomes of the 78 million American men in 1975 shown in Figure 2-5.

The mode is about 0, which merely reflects the fact that there is a concentration of men with practically no income—some of the unemployed and retired. The vast majority of incomes are scattered more thinly throughout the range 2 to 40 thousand—and this important information is not picked up by the mode at all. So if we quoted the mode of 0, we could not distinguish the 1975 U.S. income from the 1875 income! In this case, the mode is useless.

The median income is about 8 thousand dollars, and this is the figure that in a sense is the most representative—50% above, and 50% below. It perhaps is the best measure of what the "typical" American man earns. Furthermore, it is *resistant*; that is, it is changed very little (or not at all) by wild changes in a few observations. For example, if the top few incomes were to increase dramatically, nothing would happen to the median.

Finally, the mean is about 10 thousand dollars, a figure that was obtained by counting up every dollar—the pauper's and the millionaire's equally. This is both its advantage and disadvantage: it is the figure of most use to the tax department, since it gives the total income (78 million men × 10 thousand dollars = 780 billion dollars); yet it is not as good a measure of typical income as the median because it can be inflated by a few very large incomes (that is, it lacks the resistance of the median).

To sum up, we can conclude that the mode is the easiest, but the most inadequate measure of center. The median is more useful because it rep-

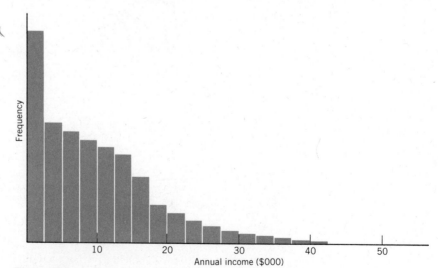

Figure 2-5
Incomes of American men, 1975.

resents the most typical value, as many people understand the term. Finally, the mean is the only central measure that takes into account the total of all the observations. For this reason, the mean is often the most useful value in such fields as engineering, economics, and business. Since it has other advantages as well (outlined later in Section 11-3), we will primarily use the mean for the rest of the book.

PROBLEMS

2-7   Calculate the mean salary for:

    a.   The 5 women (in Problem 2-1) whose salaries were 9, 12, 8, 10, 16.

    b.   The 25 men (in Problem 2-3) whose salaries were:

x	f	
10	7	70
15	10	150
20	5	100
25	2	50
30	1	30

$= 16.0$

$\frac{400}{25} = 80$

2-8   Sort the following daily profit figures of 20 newsstands into five cells, whose midpoints are $60, $65, $70, $75, and $80:

$81.32   58.47   64.90   67.88   76.02   75.06   76.73   64.21
74.92   77.56   58.01   68.05   73.37   75.41   59.41   65.43
74.76   76.51   65.10   76.02

    a.   Graph the relative frequency distribution.

    b.   Approximately what are the mean and mode? Mark the mean as the balancing point ▲.

2-9   Sort the data of Problem 2-8 into three cells, whose midpoints are $60, $70, and $80. Then answer the same questions.

2-10   Summarize the answers to the previous two problems by completing the table below.

GROUPING	MEAN	MODE
Original Data	$70.46	Not Defined
Fine Grouping (Problem 2-8)		
Coarse Grouping (Problem 2-9)		

    a.   Why is the mode not a good measure?

    b.   Which gives a closer approximation to the mean of the original data: the coarse or the fine grouping?

2-11   In the following distributions, the mean, median, and mode are among the five little marks on the X-axis. Pick them out—without resorting to calculation.

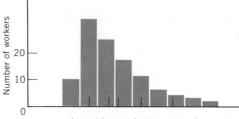

Annual income in thousands of dollars

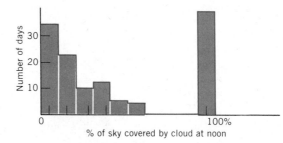

% of sky covered by cloud at noon

2-12   In Problem 2-11a, suppose 2 zeros were mistakenly added to one of the incomes, making it 100 times too large. How would this affect the mean and median? Underline the correct choice in each of the following square brackets:

a.   The [mean, median] would erroneously be increased substantially.

b.   The [mean, median] would be changed very little or not at all, depending on whether the one mutilated observation was originally below or above the [mean, median]. That is, the [mean, median] is *resistant*.

2-13   In Problem 2-11b suppose the rightmost bar was reduced slightly, to ¾ of its present height. Although the distribution isn't much changed, one of the measures of center is drastically changed (lacks resistance). Which one is it: the mean, median, or mode?

2-14   The annual tractor production of a multinational farm implement corporation in 7 different countries was as follows (in thousands):

6, 8, 6, 9, 11, 5, 60

a.   Graph the distribution, representing each figure as a dot on the X-axis.

b. What is the total production? What is the mean? The median? The mode? Mark these on the graph.

c. For another corporation operating in 10 countries, production (in thousands) has a mean of 7.8 per country, a median of 6.5, and a mode of 5.0. What is the total production of this corporation?

2-15 **a.** Calculate the mean of the following 5 numbers:

$$3, 7, 8, 12, 15$$

b. Calculate the five deviations from the mean, $X - \bar{X}$ (keeping the $+$ or $-$ sign). Then calculate the average of these deviations.

c. Write down any set of numbers. Calculate their mean, and then the average deviation from the mean.

d. Prove that, for every possible sample of n observations, the average deviation from the mean is exactly zero. Is this equally true for deviations from the median?

2-16 **a.** Two samples had means of $\bar{X}_1 = 100$ and $\bar{X}_2 = 110$. Then the samples were pooled into one large sample, for which the mean $\bar{X}$ was calculated. What is $\bar{X}$ if the sample sizes are:

    i. $n_1 = 30$, $n_2 = 70$?

    ii. $n_1 = 80$, $n_2 = 20$?

    iii. $n_1 = 50$, $n_2 = 50$?

    iv. $n_1 = 15$, $n_2 = 15$?

b. Answer true or false; if false, correct it:
To generalize part **a**, we can express the average as:

$$\bar{X} = \frac{n_1\bar{X}_1 + n_2\bar{X}_2}{n_1 + n_2} \tag{2-7}$$

2-3 Spread of a Distribution

Although the average may be the most important single statistic, it is also important to know how spread out or varied the observations are. As with measures of center, we find that there are several measures of spread. Two commonly used are the Inter-quartile range and the standard deviation. But we consider several others as well because of the way they illuminate these two.

A—THE RANGE

The range is simply the distance between the largest and smallest value:

$$\text{Range} \equiv \text{largest} - \text{smallest observation}$$

For the heights in Figure 2-2, the range is $79 - 59 = 20$ inches. The trouble with the range is that it lacks resistance, since it depends entirely upon two observations—the largest and smallest. If, for example, the tallest person happened to be 3 inches taller, the range would be 3 inches more.

B—THE INTER-QUARTILE RANGE (IQR)

As an alternative to the two end observations, let us take the two quartiles. In Figure 2-2c, for example, they were already shown in the box plot as about 67 inches and 72 inches. These are much more resistant (i.e., resistant to undue influence by a single observation. Because the distribution is so dense in the neighborhood of the quartiles, moving any single observation by 3 inches will change the quartiles very little.)

The distance between the quartiles then measures the spread of the middle half of the observations: It is therefore called the *inter-quartile range* (IQR), or *midspread*:

$$\boxed{\text{IQR} \equiv \text{upper quartile } - \text{ lower quartile}} \tag{2-8}$$

For the heights in Figure 2-2, the IQR is about $72 - 67 = 5$ inches. Note that the IQR is just the *length of the box* in panel (c).

The IQR is so easy to calculate and so resistant that our story could well stop here. Nevertheless, for some other purposes it is important to use *all* the observations to calculate a measure of spread. We will therefore embark next on a new course that will eventually take us to the standard measure of spread (the "standard deviation").

C—MEAN ABSOLUTE DEVIATION

To incorporate all the observations, we first list them in a table; for example, in Table 2-4 we show again the 5 shipment losses X, and their average $\overline{X} = 40$. In the next column, we calculate all 5 *deviations from the mean* $X - \overline{X}$. Some are positive, and some are negative; since these cancel out, their average[5] is exactly 0. The average deviation is therefore

---

[5]To prove the average deviation is exactly zero, we first calculate the sum:

$$\Sigma(X - \overline{X}) \equiv (X_1 - \overline{X}) + (X_2 - \overline{X}) + \cdots$$

$$= (X_1 + X_2 + \cdots) - (\overline{X} + \overline{X} + \cdots)$$

$$= \Sigma X - n\overline{X}$$

$$= \Sigma X - n\left(\frac{1}{n}\Sigma X\right)$$

$$\Sigma(X - \overline{X}) = 0 \tag{2-9}$$

$$\text{Average deviation} = \frac{0}{n} = 0$$

TABLE **2-4** **Various measures of deviation, leading up to the standard deviation _s_**

(1) DATA X	(2) DEVIATIONS $X - \overline{X}$	(3) ABSOLUTE DEVIATIONS $\lvert X - \overline{X} \rvert$	(4) SQUARED DEVIATIONS $(X - \overline{X})^2$	(5) SQUARED DEVIATIONS $(X - \overline{X})^2$
10	-30	30	900	900
20	-20	20	400	400
30	-10	10	100	100
50	10	10	100	100
90	50	50	2500	2500
$\overline{X} = \dfrac{200}{5}$	0	$\text{MAD} = \dfrac{120}{5}$	$\text{MSD} = \dfrac{4000}{5}$	$s^2 = \dfrac{4000}{4}$
$= 40$		$= 24$	$= 800$	$= 1000$
				$s = 32$

a useless measure of spread: It is _always_ zero no matter how spread out the distribution may be.

We can solve this problem of "cancelling signs" by ignoring all negative signs. In other words, take the average of the _absolute_ values of the deviations:

$$\text{Mean Absolute Deviation, MAD} \equiv \frac{1}{n}\Sigma\lvert X - \overline{X}\rvert$$

This is calculated in column (3) of Table 2-4.

D—MEAN SQUARED DEVIATION (MSD)

Although MAD intuitively is a good measure of spread, in many ways a better solution to the problem of "cancelling signs" is to _square_ each deviation[6]:

$$\text{Mean Squared Deviation, MSD} \equiv \frac{1}{n}\Sigma(X - \overline{X})^2 \qquad (2\text{-}10)$$

This is calculated in column (4) of Table 2-4.

---

*[6]Squares are mathematically more tractable than absolute values. (For example, the squaring function is everywhere differentiable, whereas the absolute value function is not differentiable at the origin.)

Squares also have the same important relationship in statistics as they do in geometry. (In geometry, the relationship between squares is called the Pythagorean theorem. In statistics, this same relationship is called the analysis of variance, as we shall see in Chapter 10.)

E—VARIANCE AND STANDARD DEVIATION

For certain technical reasons described later in part (f), it is customary in (2-10) to use the divisor $n - 1$ instead of n. This gives a slightly different measure of spread, the variance:

$$\text{Variance, } s^2 \equiv \frac{1}{n - 1} \Sigma (X - \overline{X})^2 \qquad (2\text{-}11)$$

For grouped data, we must as usual modify this formula, by counting each deviation with the frequency f that it occurs. Then the variance becomes:

$$\boxed{\text{Variance for grouped data, } s^2 \simeq \frac{1}{n - 1} \Sigma (x - \overline{X})^2 f} \qquad (2\text{-}12)$$

To compensate for having squared the deviations, let us finally take the square root. This will give us the standard way to measure the deviation from the mean—so it is called the *standard deviation, s*:

$$\text{Standard deviation, } s \equiv \sqrt{\text{variance}}$$

$$\boxed{\text{For grouped data, } s \simeq \sqrt{\frac{1}{n - 1} \Sigma (x - \overline{X})^2 f}} \qquad (2\text{-}13)$$

In the last column of Table 2-4, we calculate the variance and standard deviation. The variance is huge (1000)—because of the squaring. By taking the square root, we obtain $s = 32$, a more suitable measure of deviation. After all, if we look at the individual deviations in column (3), they range from 10 to 50. Since the standard deviation $s = 32$ lies in between, it may therefore be viewed as the typical deviation.

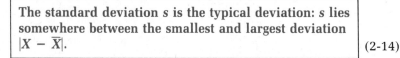

$$\boxed{\begin{array}{l}\text{The standard deviation } s \text{ is the typical deviation: } s \text{ lies} \\ \text{somewhere between the smallest and largest deviation} \\ |X - \overline{X}|.\end{array}} \qquad (2\text{-}14)$$

For grouped data, the standard deviation s is calculated in Table 2-5. (Note that here, or anywhere else we use grouped data, n is the *sample size*, not the number of rows in the table. And note that it is appropriate to round[7] $\overline{X}$.)

---

[7]This rounding makes good sense for hand calculations. And it produces an error in s^2 that is surprisingly small. Specifically, if we let e denote the error in rounding off the mean, then (as we will prove in Appendix 4-2):

$$\text{Variance is overestimated by approximately } e^2 \qquad (2\text{-}15)$$
(cont'd)

TABLE **2-5** **Calculation of Standard Deviation of Men's Heights**

GIVEN DATA		CALCULATION OF S		
x	f	$(x - \overline{X})$	$(x - \overline{X})^2$	$(x - \overline{X})^2 f$
60	4	−9	81	324
63	12	−6	36	432
66	44	−3	9	396
69	64	0	0	0
72	56	3	9	504
75	16	6	36	576
78	4	9	81	324

$$n = 200$$

Recall:

$$\overline{X} = 69.3$$
$$\simeq 69$$

$$s^2 = \frac{2556}{199}$$
$$= 12.84$$
$$s = \sqrt{12.84} = 3.6$$

The calculated value $s = 3.6$ is indeed the typical deviation, lying between the smallest and largest deviation (0 and 9). To emphasize this, in Figure 2-6 we mark the standard deviation s as a bar extending on either side of the mean.

In conclusion, Figure 2-6 has shown how a large sample can be summarized by the frequency distribution, and further condensed to a couple of numbers that measure the center and spread—the mean $\overline{X}$ and standard deviation s. (We call $\overline{X}$ and s^2 the first and second *moments* of the sample.)

F—DEGREES OF FREEDOM (d.f.)

The MSD was a good measure of spread, provided we only want to describe the sample. But typically we want to go one step further, and make a statistical inference about the underlying population. For this purpose the sample variance is better, as the following intuitive argument indicates.

If only $n = 1$ observation was available, this observation would be the sample mean and would give us some idea of the underlying population mean. Since there is no spread in the sample, however, we would have absolutely no idea about the underlying population spread. For example suppose we observe only one basketball player, and his height is 6'6". This provides us with an estimate of the *average* height of all players, but no information whatever about how spread out their heights may be (6'4" to 6'8"? Or 5' to 8'? From this single observation, we have no clue whatsoever.)

Only to the extent that n exceeds one can we get information about the

---

This overestimation is usually so small that we may ignore it. For example, in rounding $\overline{X}$ from 69.3 in Table 2-5, the rounding error e is .3 and e^2 is only .09. Thus the variance $s^2 = 12.84$ is overestimated by only .09, a relatively trivial amount. (And if it is ever required, the correct value is easily obtained by subtraction: $12.84 - .09 = 12.75$.)

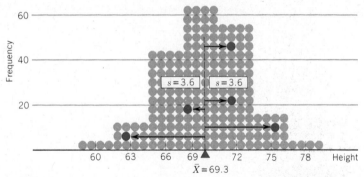

Figure 2-6

The standard deviation s is the typical deviation—some deviations are smaller and some are larger, as shown. (Same distribution as Figure 2-2).

spread. That is, there are essentially only $(n - 1)$ pieces of information for the spread, and this is the appropriate divisor for the variance. Customarily, pieces of information are called *degrees of freedom* (d.f.),[8] and our argument is summarized as:

> **For the variance, there are $n - 1$ d.f. (degrees of freedom, or pieces of information).**   (2-16)

PROBLEMS

2-17   Recall that the women's salaries in Problem 2-1 ranked in order were

$$8, 9, 10, 12, 16$$

    a.   Find the range and IQR. (*Hint:* Read them off the box plot in Problem 2-4.)

    b.   Calculate the MAD, MSD, variance, and standard deviation.

2-18   Recall that the 25 men's salaries in Problem 2-3 were:

x	f
10	7
15	10
20	5
25	2
30	1

---

[8]To see where the phrase "degrees of *freedom*" comes from, consider a sample of $n = 2$ observations, 21 and 15, say. Since $\overline{X} = 18$, the residuals are $+3$ and -3, the second residual necessarily being just the negative of the first. While the first residual is "free," the second is strictly determined. Hence there is only 1 degree of freedom in the residuals.

Generally, for a sample of size n, while the first $n - 1$ residuals are free, the last residual is strictly determined by the requirement that the sum of all residuals be zero—that is, $\Sigma(X - \overline{X}) = 0$, as shown in (2-9).

    a. Find the IQR. (*Hint:* Read it off the box plot in Problem 2-5.)

    b. Calculate the standard deviation s.

2-19 In a test of the reliability of his machine, a technician measured the viscosity of a specimen of crude oil 150 times. Broken down into the 3 time periods, the three frequency distributions were:

VISCOSITY	FREQUENCY		
X	FIRST 50	MIDDLE 50	LAST 50
60	0	1	0
65	2	7	5
70	15	22	38
75	19	18	6
80	11	2	1
85	3	0	0
	50	50	50

    a. Graph the 3 sets of data, side by side. Do you discern any trends?

    b. For each of the 3 time periods, calculate the mean and standard deviation. Do these calculations show the same trends you observed in part *a* ?

    c. For the complete set of 150 observations, calculate the mean and standard deviation. How are they related to the means and standard deviations found in part *b*?

2-4 Statistics by Computer

Most large-scale statistical studies are carried out on the computer, using a computing system that is either *batch* or *interactive*. In a batch system (such as SAS or SPSS), all data and computing instructions are typed up ahead of time, run through the computer, and then all the output is printed up for the user to sort out.

    On the other hand, in an interactive system, the computer carries on a running conversation with the operator, who gives one instruction at a time. Depending on the computer's reply, the operator can then choose the next instruction. If a mistake is made, it is easily corrected, and interactive systems are so forgiving that they are ideal for amateurs. One of the most popular interactive systems is MINITAB (Ryan, Joiner, Ryan, 1976), and we will use it throughout the text to illustrate how the computer works.

    For example, Table 2-6 shows how MINITAB handles the data on 200 men's heights. To understand who is typing—the operator or the computer—it is important to look for the symbol $>$ written by the computer, which means "Your turn, operator." For example, $>$ appears at the very beginning of Table 2-6. In response, the operator types READ 'X', which commands the computer to prepare to read the data into a column denoted by 'X'. The operator then types in the 200 heights.

TABLE **2-6** **Computer Summary of the 200 Heights in Table 2-2: The Histogram, Mean, and Standard Deviation.**

```
      > READ 'X'
DATA>     64.3
DATA>     71.7
DATA>     69.4
DATA>     70.8
              .
              .
              .
DATA>     68.5

      > HISTOGRAM OF 'X', FIRST MIDPOINT AT 60, CELLWIDTH  3

  EACH * REPRESENTS    2 OBSERVATIONS

  MIDDLE OF     NUMBER OF
  INTERVAL      OBSERVATIONS
     60.00         4      **
     63.00        12      ******
     66.00        44      **********************
     69.00        64      ********************************
     72.00        56      ****************************
     75.00        16      ********
     78.00         4      **

      > MEAN OF 'X'
  MEAN     =          69.302
      > STANDARD DEVIATION OF 'X'
  ST.DEV. =          3.4842
```

Next, using the command HISTOGRAM OF 'X' . . . the operator instructs the computer to make a histogram of the data.

Finally, using the commands MEAN OF 'X' and STANDARD DEVIA-TION OF 'X', the operator instructs the computer to calculate these two statistics. Note that they are slightly different from the approximate answers in Tables 2-3 and 2-5. (The computer finds it easy to compute the exact mean $\bar{X}$ based on the original 200 observations, whereas the approximate mean in Table 2-3 was based on the grouped data.)

2-5 Linear Transformations

A—CHANGE OF ORIGIN

Table 2-7(a) gives the elevation X for four midwestern states, in feet above sea level. Their mean $\bar{X}$ and standard deviation s_X also are calculated.

It often happens that a new reference level is more convenient—for example, Lake Superior, which is 600 feet above sea level. In Table 2-7(b), we therefore give the elevation in feet above Lake Superior, X'. Of course, each new elevation will be 600 feet less than the old (600 feet being the elevation of Lake Superior):

$$X' = X - 600$$

TABLE **2-7** Altitudes of Four Midwestern States

STATE	(a) ALTITUDE IN FEET ABOVE SEA LEVEL GIVEN			(b) ALTITUDE IN FEET ABOVE LAKE SUPERIOR			(c) ALTITUDE IN YARDS ABOVE SEA LEVEL		
	given X	$(X - \bar{X})$	$(X - \bar{X})^2$	$X' =$ $X - 600$	$(X' - \bar{X}')$	$(X' - \bar{X}')^2$	$X^* =$ $X/3$	$(X^* - \bar{X}^*)$	$(X^* - \bar{X}^*)^2$
Illinois	600	−330	108,900	0	−330	108,900	200	−110	12,100
Iowa	1,170	240	57,600	570	240	57,600	390	80	6,400
Michigan	900	−30	900	300	−30	900	300	−10	100
Wisconsin	1,050	120	14,400	450	120	14,400	350	40	1,600

$$\bar{X} = \frac{3,720}{4} \qquad s_X^2 = \frac{181,800}{3} \qquad \bar{X}' = \frac{1,320}{4} \qquad s_{X'}^2 = \frac{181,800}{3} \qquad \bar{X}^* = \frac{1,240}{4} \qquad s_{X^*}^2 = \frac{20,200}{3}$$

$$= 930 \qquad\qquad = 60,600 \qquad\qquad = 330 \qquad\qquad = 60,600 \qquad\qquad = 310 \qquad\qquad = 6,733$$

$$s_X = 246 \qquad\qquad\qquad\qquad s_{X'} = 246 \qquad\qquad\qquad\qquad s_{X^*} = 82$$

43

Now what will the new mean elevation $\overline{X}'$ be? It will, of course, be 600 feet less than the old mean $\overline{X}$:

$$\overline{X}' = \overline{X} - 600$$

On the other hand, the spread of the new elevations will be exactly the same as the old:

$$s_{X'} = s_X$$

These two equations are easy to verify by actual calculation: in Table 2-7(b), we find that $\overline{X}' = 330$, which indeed is 600 feet less than $\overline{X} = 930$; we also find that $s_{X'} = 246 = s_X$.

These issues are illustrated in Figure 2-7, and generalized in the following theorem:

$$\boxed{\begin{aligned} \text{If } X' &= X + a \\ \text{then } \overline{X}' &= \overline{X} + a \\ \text{and } s_{X'} &= s_X \end{aligned}}$$

(2-17)

B—CHANGE OF SCALE

Sometimes a new scale of measurement is more convenient—for example, yards instead of feet. In Table 2-7(c), we therefore give the elevations in yards above sea level, X^*. Of course, each new elevation will be 1/3 of the old (since there are three feet in every yard):

$$X^* = \tfrac{1}{3} X$$

What will the new mean elevation $\overline{X}^*$ be? It will, of course, be 1/3 of the old mean $\overline{X}$:

$$\overline{X}^* = \tfrac{1}{3} \overline{X}$$

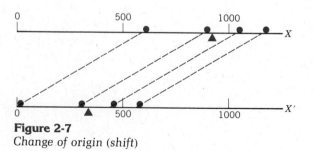

Figure 2-7
Change of origin (shift)

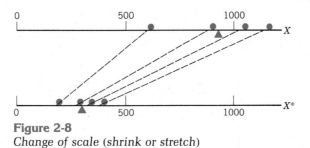

Figure 2-8
Change of scale (shrink or stretch)

Similarly, the spread of the new elevations will be 1/3 of the old:

$$s_{X^*} = \tfrac{1}{3} s_X$$

These two equations again are easy to verify by actual calculation: In Table 2-7(c), we find that $\overline{X}^* = 310$, which indeed is 1/3 of $\overline{X} = 930$. We also find that $s_{X^*} = 82$, which is 1/3 of $s_X = 246$.

These issues are illustrated in Figure 2-8, and generalized in the following theorem:[9]

$$\boxed{\begin{aligned} &\text{If } X^* = bX \\ &\text{then } \overline{X}^* = b\overline{X} \\ &\text{and } s_{X^*} = |b|s_X \end{aligned}}$$
(2-18)

C—GENERAL LINEAR TRANSFORMATION

It now is appropriate to combine the above two cases into one, by considering $Y = a + bX$. (Because the graph of Y is a straight line, this is called a *linear* transformation.) We find the mean and standard deviation are transformed just as we would expect:

$$\boxed{\begin{aligned} &\text{If } Y = a + bX \\ &\text{then } \overline{Y} = a + b\overline{X} \\ &\text{and } s_Y = |b|s_X \end{aligned}}$$
(2-19)

This theorem is proved in Appendix 2-5, and may be interpreted very simply: If the *individual* observations are linearly transformed, then the *mean* observation is transformed in exactly the same way, and the *standard deviation* is changed by the factor $|b|$, with no effect from a.

---

[9]Since b can be either positive or negative, we must use its absolute value $|b|$ in equation (2-18).

Incidentally, if $|b| < 1$, the transformation is a shrinking, as in Figure 2-8. If $|b| > 1$, then it is a stretching.

PROBLEMS

2-20 The altitudes of four mountain states were approximately as follows (feet above sea level):

Arizona	4100
Nevada	5500
Colorado	6900
Wyoming	6700

   a. Calculate the mean and standard deviation.

   b. What are the mean and standard deviation in *yards* above sea level?

   c. Compared to the midwestern states in Table 2-7, how much higher are these mountain states (by what factor)? Does your answer depend on whether you use feet or yards?

2-21 An agricultural experimental station had five square plots, whose lengths (in feet) were:

$$10, 20, 30, 50, 90$$

   a. What is the average length?

   b. What is the average area?

   c. What is the total area?

2-22 The following is the grouped frequency table for the actual weight (in ounces) of 50 "one-pound" bags of cashews that a supermarket clerk filled from bulk stock.

ACTUAL WEIGHT	NUMBER OF BAGS
15.9 ozs.	2
16.0	16
16.1	22
16.2	10

Handwritten annotations:
$\bar{x} = \frac{1}{n} \sum x f = \frac{1}{50}(31.8 + 256 + 354.2 + 162)$

$\bar{x} = 16.08$

$s = \sqrt{\frac{1}{49}(.0648 + .1024 + .0088 + .144)}$

$s = .08$

   a. Find the mean and the standard deviation of the weights.

   b. Each bag costs the supermarket 21 cents per ounce plus 12 cents (for the bag itself and labor to fill it). Find the mean and standard deviation of the costs.

2-23 Find the mean and standard deviation of the following 5 estimates of the population of a country:

24,580,000
24,460,000
24,510,000
24,480,000
24,520,000

Hint: It is natural to simply drop the first 2 digits and last 4 digits of every number, which leaves us with the numbers 58, 46, ... This is mathematically justified—it is just the linear transformation

$$Y = \frac{X - 24,000,000}{10,000}$$

2-6 Calculations Using Relative Frequencies

Sometimes original data is not available and only a summary is given in the form of the relative frequency distribution. (For example, suppose the original information in the middle columns of Table 2-1 has been discarded, and only the relative frequency distribution in the first and last columns is available.) From this, how can we calculate $\overline{X}$ and s?

To derive the appropriate formula for $\overline{X}$, recall that for grouped data,

$$\overline{X} \simeq \frac{1}{n} \left(x_1 f_1 + x_2 f_2 + \cdots \right) \qquad \text{(2-5) repeated}$$

$$= x_1 \left(\frac{f_1}{n} \right) + x_2 \left(\frac{f_2}{n} \right) + \cdots$$

$$\boxed{\overline{X} \simeq \Sigma x \left(\frac{f}{n} \right)} \qquad \begin{array}{l} \text{(2-20)} \\ \text{like (2-6)} \end{array}$$

That is, $\overline{X}$ is just the sum of the x values weighted according to their relative frequencies f/n. In the same way the MSD can also be calculated from the relative frequencies:

$$\boxed{\text{MSD} \simeq \Sigma (x - \overline{X})^2 \left(\frac{f}{n} \right)} \qquad \begin{array}{l} \text{(2-21)} \\ \text{like (2-10)} \end{array}$$

Finally, we can easily calculate s^2 from the MSD [by comparing (2-10) with (2-11)]:

$$\text{Variance, } s^2 = \left(\frac{n}{n - 1} \right) \text{MSD} \qquad \text{(2-22)}$$

For large n, the factor $n/(n-1)$ is practically 1. (For example, when $n = 1000$, it is 1.001.) Then

$$\text{Variance } s^2 \simeq \text{MSD} \qquad \text{(2-23)}$$

TABLE **2-8** **Calculation of Mean and Standard Deviation from the Relative Frequency Distribution ($n = 200$ observations).**

	GIVEN RELATIVE FREQUENCY	CALCULATION OF $\overline{X}$ USING (2-20)	CALCULATION OF MSD USING (2-21)		
x	$\dfrac{f}{n}$	$x\left(\dfrac{f}{n}\right)$	$(x - \overline{X})$	$(x - \overline{X})^2$	$(x - \overline{X})^2\left(\dfrac{f}{n}\right)$
60	.02	1.20	-9	81	1.62
63	.06	3.78	-6	36	2.16
66	.22	14.52	-3	9	1.98
69	.32	22.08	0	0	0
72	.28	20.16	3	9	2.52
75	.08	6.00	6	36	2.88
78	.02	1.56	9	81	1.62

$$\overline{X} = 69.30 \qquad\qquad\qquad\qquad \text{MSD} \simeq 12.78$$
$$\simeq 69$$
$$\text{by (2-22), } s^2 = \left(\frac{200}{199}\right) 12.78 = 12.84$$
$$s = \sqrt{12.84} = 3.6$$

To show how these formulas work, in Table 2-8 we use them to calculate $\overline{X}$ and s. Note that the answers agree, of course, with the previous answers in Tables 2-3 and 2-5.

PROBLEM

2-24 Compute the mean and standard deviation of the following sample of 25 salaries (same grouped data as in Problem 2-3):

SALARY (MIDPOINT OF CELL)	RELATIVE FREQUENCY
10	.28
15	.40
20	.20
25	.08
30	.04
	1.00

CHAPTER 2 SUMMARY

2-1 Data can be easily summarized by tabulation into a frequency distribution, and then graphed. Alternatively, data can be pictured with a box plot that shows the median and upper and lower quartiles.

2-2 To define the center of a distribution, the most common measures are the median $\overset{\downarrow}{X}$ (middle value) and the mean $\overline{X}$ (balancing point).

2-3 To define the spread of a distribution, the most common measures are the interquartile range IQR and the standard deviation s.

2-4 To illustrate how easily data can be analyzed on a computer (using an interactive computing system such as MINITAB), we summarized some data by computing the frequency distribution, and then $\overline{X}$ and s.

2-5 If a sample of observations are linearly transformed (shifted, and shrunk or stretched), then the mean is transformed exactly the same way. And the standard deviation is similarly shrunk or stretched, but not shifted.

2-6 The mean and standard deviation can be calculated from the *relative* frequency distribution as well as from the frequency distribution itself.

REVIEW PROBLEMS

2-25 A stand of 40 red pines gave the following diameters (feet):

1.6, 2.8, 2.2, 2.5, 2.8	1.9, 2.2, 1.7, 1.8, 2.8
1.8, 1.8, 2.1, 3.1, 2.2	1.8, 1.5, 1.4, 1.5, 1.9
1.4, 1.5, 2.1, 2.5, 2.6	1.4, 2.5, 1.9, 1.9, 1.8
2.4, 1.6, 1.5, 2.0, 1.5	1.4, 2.4, 2.0, 2.4, 1.8

a. Without sorting into cells, graph the diameters as dots along an X axis.

b. What is the median? The interquartile range?

c. Draw the box plot.

2-26 Group the data of Problem 2-25 into cells, with midpoints 1.5, 2.0, . . .

a. Graph the *relative* frequency distribution.

b. Calculate $\overline{X}$, s, and the relative standard deviation (or coefficient of variation) defined as $CV = s/\overline{X}$.

c. If the trees were measured in inches instead of feet, what would be $\overline{X}$, s, and CV?

*d.[10] The volume of lumber V (in cubic feet) that can be cut from a tree of diameter D (in feet) is estimated as

$$V = 8D^2$$

Estimate the total amount of lumber in the stand of 40 trees.

2-27 The 1980 U.S. population of 222 million was widely distributed among the 50 states—from Alaska (0.4 million) to California (23 million). A box graph nicely shows an appropriate amount of detail.

---

[10]As we mentioned in the Preface, a star indicates that it is more difficult.

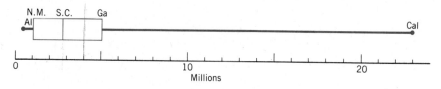

Roughly estimate:

a. The median population.

b. The mean population.

c. The midrange (IQR).

d. The percentile ranking of Louisiana (population 4.0 million).

2-28 Five states were randomly sampled, and gave the following areas (in thousands of square miles):

Montana	147	Utah	85
Minnesota	84	California	159
W. Virginia	24		

a. Calculate the mean $\overline{X}$, and standard deviation s.

b. On the basis of this sample of 5 states, estimate the area for the whole country (50 states).

2-29 Repeat Problem 2-28 for another random sample of 5 states, that turned out this time to be:

Maryland	11	Idaho	84
Ohio	41	New Jersey	8
Nebraska	77		

Note that the estimated area of the country is different—of course, since the sample is different. This illustrates the inherent variability in statistical estimation. (The true area of the country is 3620 thousand square miles.)

2-30 Suppose that the disposable annual incomes of the 5 million residents of a certain country had a mean of $4800 and a median of $3400.

a. What is the disposable income of the whole country?

b. What can you say about the shape of the distribution of incomes?

2-31 In the 1970s, the U.S. unemployment rate was as follows[11]:

YEAR	UNEMPLOYMENT	YEAR	UNEMPLOYMENT
1970	4.9%	1975	8.5%
1971	5.9	1976	7.7
1972	5.6	1977	7.0
1973	4.9	1978	6.0
1974	5.6	1979	5.7

---

[11]The data for Problems 2-31, 2-32, and 2-33 comes from the Statistical Abstract of the United States (Bureau of the Census, U.S. Dept. of Commerce). This is also the source for many other Problems throughout the book that quote a specific year.

Calculate the average and standard deviation:

a. For the first 5 years, 1970-1974.

b. For the last 5 years, 1975-1979.

c. For all 10 years. How is this answer related to the answers in parts **a** and **b**?

*2-32 In 1975 the world's population was 4 billion, with the following approximate breakdown. What was the growth rate for the world as a whole?

REGION	POPULATION (MILLIONS)	ANNUAL GROWTH RATE (%)
Europe	500	0.5%
N. America	240	0.8
Russia	260	0.9
Asia	2300	2.1
S. America	300	2.6
Africa	400	2.8
World	4000	?

*2-33 At the local university three Commerce professors offered a graduate course, whose sizes were 8, 12, and 120 students, respectively.

a. The average commerce professor who offers a graduate course faces a class of what size? And what is the standard deviation?

b. The average commerce student who is *taking* a graduate course has a different view. He faces a class of what size? And what is the standard deviation?

Chapter

Probability

The urge to gamble is so universal and its practice so
pleasurable that I assume it must be evil.   HEYWOOD BROUN

3-1 Introduction

In the next four chapters, we will make deductions about a sample from
a known population. For example, if the population of American voters
is 55% Democrat, we can hardly hope to draw exactly that same percentage
of Democrats in a random sample. Nevertheless, it is "likely" that "close
to" this percentage will turn up in our sample. Our objective is to define
"likely" and "close to" more precisely, so we can make useful predictions.
First, however, we must lay a good deal of groundwork. Predicting in the
face of uncertainty requires a knowledge of the laws of *probability*, and
this chapter is devoted to their development. We shall begin with the
simplest example—rolling dice—which was also the historical beginning
of probability theory, several hundred years ago.

EXAMPLE **3-1**

> Throw a single fair die 50 times. Or simulate this by consulting the
> random numbers in Appendix Table I at the back of the book, dis-
> regarding the digits 7, 8, 9, and 0. Since the remaining digits 1, 2,
> ..., 6 will of course still be equally likely, they will provide an
> accurate simulation of a die.
>   Graph the relative frequency distribution:
>
> (a)   After 10 throws.
>
> (b)   After 50 throws.
>
> (c)   After zillions of throws (guess).

Solution Since we don't have a die at hand, we will simulate (it's also faster). Although it would be best to start at a random spot in Table I, we will start at the beginning so that our work will be easy to follow. The first few digits in Table I are

$$3\cancel{9}65\cancel{7}645451\cancel{9}\cancel{9}\cancel{0}69 \ldots$$

with the irrelevant digits stroked out.

The first sample ($n = 10$ throws) is summarized in Table 3-1 in column (a). The second sample (with n increased to 50 throws) is shown in column (b). The emerging picture begins to confirm our expectations: In the long run, a fair die approaches a simple pattern; it will come up equally often on every face (in fact, this is the very definition of fairness), and so we don't have to actually carry out the zillions of throws for part (c). These three relative frequency distributions are displayed in graph form in Figure 3-1.

In Figure 3-1, we note in the first graph how relative frequency fluctuates wildly when there are few observations in the sample. Yet eventually, in the third graph relative frequency settles down. And the limiting value is called the probability:

$$\boxed{\textbf{Probability} = \textbf{limiting relative frequency}} \qquad (3\text{-}1)$$

Or, in symbols,

$$\Pr \equiv \lim\left(\frac{f}{n}\right) \qquad (3\text{-}2)$$

TABLE **3-1** **Relative Frequency Distributions for a Die, Using Various Sample Sizes**

X NUMBER OF DOTS	RELATIVE FREQUENCY $= \dfrac{f}{n}$		
	(a) $n = 10$	(b) $n = 50$	(c) $n = \infty$
1	.10	.22	$1/6 = .167$
2	0	.12	$1/6 = .167$
3	.10	.14	$1/6 = .167$
4	.20	.14	$1/6 = .167$
5	.30	.14	$1/6 = .167$
6	.30	.24	$1/6 = .167$
	1.00$\sqrt{}$	1.00$\sqrt{}$	1.00$\sqrt{}$

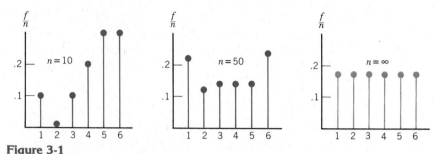

Figure 3-1
Relative frequency distributions for a die, using various sample sizes.

Just as all the relative frequencies must add to 1 (as in Table 3-1, for example) so must all the probabilities add to 1. That is,

$$Pr(e_1) + Pr(e_2) + \cdots = 1 \qquad (3\text{-}3)$$

where $e_1, e_2, \ldots$ represent the elementary outcomes of an experiment. For example, in our experiment of tossing a die, the probability of the first outcome (the die turning up with one dot), plus the probability of the second outcome (two dots), plus the probability of all other outcomes, must add to 1.

PROBLEMS

3-1 To see how random fluctuation settles down consistently in the long run, repeat Example 3-1. That is, throw a die 50 times (or simulate with random digits, starting at a random spot in Appendix Table I). Then graph the relative frequency distribution:

  a. After 10 throws.

  b. After 50 throws.

  c. After zillions of throws (guess).

3-2 The graphs of Problem 3-1 can alternatively be condensed onto a time axis. That is, in the following graph, above $n = 10$ plot all six relative frequencies found in Problem 3-1**a**. Similarly, show the relative frequencies for $n = 50$ and $n = \infty$.

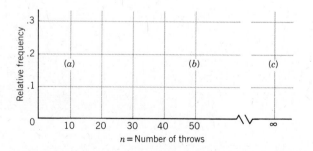

To make the graph more complete, also graph the relative frequencies at $n = 5$ and $n = 20$. Then note how the relative frequencies converge. (In Chapter 6 we will get a precise formula for this convergence.)

3-3   **a.**   Toss a coin 50 times, and record how often it comes up heads. What is your best guess for the *probability* of a head?

b.   Toss a thumb tack 50 times, and record how often it comes point up. What is your best guess for the *probability* of this?

c.   Roll a pair of dice 50 times (or simulate by drawing pairs of digits from Table II), and record how often you get a "total of 7 or 11" (an outright win in "shooting craps"). What is your best guess for the *probability*?

3-2   Probabilities of Events

A—PROBABILITY TREES

In the previous section, the die example was an experiment in which the 6 outcomes (faces) were extremely simple. Usually, an experiment has a more complex set of outcomes.

For example, suppose that the "experiment" consists of having a family of three children. As everyone knows, a typical outcome might be "a boy, then a girl, finally another boy," which we will abbreviate to (BGB). How can we find a systematic way to list *all* such outcomes, and at the same time, find their probabilities?

The simplest solution is to conduct a *thought experiment*: Imagine a million couples performing the experiment, and see what will happen. A systematic breakdown, birth by birth, is shown in Figure 3-2. At the left we imagine the million couples starting out, forming the trunk of what

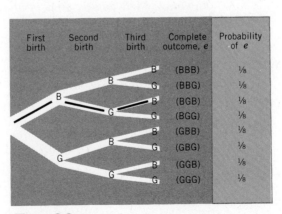

Figure 3-2
The probability tree for a family of 3 children.

will soon become a tree. At the first birth, half the couples have B and the other half have G—as indicated by the first branching in Figure 3-2.

At the second birth, one-half would again report B and the other half would report G. This is illustrated by the further branching of the tree. When all such possible branchings are shown, we obtain a tree with 8 twigs at the end ($2 \times 2 \times 2$ branchings).

Each path, from beginning to end, represents a complete outcome (family). The outcome (BGB), for example, is the path outlined in Figure 3-2. What is its probability? Of the millions of couples who start out, only one-half have a boy initially. Of these, only one half have a girl next, and of these, only one half finally have a boy. Thus the proportion of the couples who would report the complete outcome (BGB) is

$$\frac{1}{2} \text{ of } \frac{1}{2} \text{ of } \frac{1}{2} = \frac{1}{8}$$

This long-run proportion, or probability, is duly recorded in Figure 3-2 in the last column. Similarly, we calculate and record the probability for every one of the 8 possible outcomes. Since boys and girls are equally likely, each probability in this example is the same (1/8).

In general, trees provide a very powerful way of analyzing any step-by-step experiment—even when there are more than two branchings at each step, or the probabilities are unequal. Another example will illustrate.

EXAMPLE **3-2**   Suppose on every birth the probability of a boy is 52% and a girl is 48%. (According to U.S. birth statistics, this is a more realistic assumption than our earlier .50-.50 assumption.) Calculate the probability tree for a couple having 3 children.

Solution   The tree will be similar to Figure 3-2. But now the branchings are asymmetric, with 52% of the couples having B at each birth. This will affect the final probabilities. For example, what proportion of the couples will finally attain the outcome (BGB)?

$$\begin{aligned} \text{Pr(BGB)} &= 52\% \text{ of } 48\% \text{ of } 52\% \\ &= .52 \times .48 \times .52 = .13 \end{aligned} \quad (3\text{-}4)$$

Continuing in this way, we obtain the tree shown in Figure 3-3.

Example 3-2 provides a more realistic model: For example, the actual statistics on American 3-child families would show the proportion that are (BBB) is closer to 14% than to 12.5% (1/8). Yet even this model is not perfectly realistic. After all, the purpose of a mathematical model, whether in statistics, economics, or chemistry, is just to *provide approximate answers easily:* A model should give good predictions by capturing the

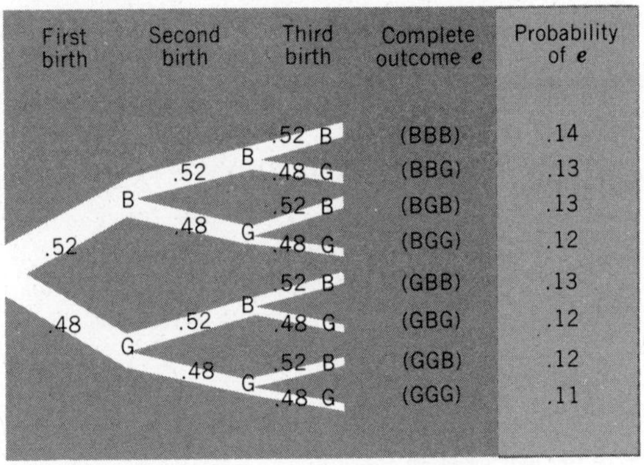

First birth	Second birth	Third birth	Complete outcome *e*	Probability of *e*
		.52 B	(BBB)	.14
	.52 B	.48 G	(BBG)	.13
.52 B	.48 G	.52 B	(BGB)	.13
		.48 G	(BGG)	.12
	.52 B	.52 B	(GBB)	.13
.48 G		.48 G	(GBG)	.12
	.48 G	.52 B	(GGB)	.12
		.48 G	(GGG)	.11

Figure 3-3
The probability tree for a family of 3 children, if the probability of a boy on each birth is 52%.

essential features of a problem while omitting trivial, needlessly complicated detail. We would therefore regard the model of Figure 3-2 as adequate for many purposes, although the refined model of Figure 3-3 would sometimes be required.

B—OUTCOME SETS AND EVENTS

We have seen how a probability tree is one of the most effective ways to get a complete list of outcomes and their probabilities. For example, Figure 3-4 repeats this information for the simple model of a 3-child family. Since we will often be referring to these eight outcomes $e_1, e_2, \ldots, e_8$, we represent each as a *point* in this figure. All of these points then make up the *outcome set* S (shown as the shaded area in Figure 3-4). This is

Outcome Set S		Probabilities
(BBB)	• e_1	1/8
(BBG)	• e_2	1/8
(BGB)	• e_3	1/8
(BGG)	• e_4	1/8
(GBB)	• e_5	1/8
(GBG)	• e_6	1/8
(GGB)	• e_7	1/8
(GGG)	• e_8	1/8

Figure 3-4
Outcome Set for 3-child family, with probabilities (assuming boys and girls are equally likely).

also called the sample space, since statisticians are mostly interested in outcome sets in sampling contexts.

In planning 3 children, suppose that a couple is hoping for the event

$$E: \text{ at least 2 girls} \tag{3-5}$$

This event includes outcomes e_4, e_6, e_7, and e_8, illustrated in the Venn diagram shown in Figure 3-5. This is customarily written:

$$E = \{e_4, e_6, e_7, e_8\}$$

In fact, this method provides a convenient way to define an event in general:

$$\boxed{\text{An event } E \text{ is a subset of the outcome set } S} \tag{3-6}$$

Now, the interesting question is: What is the probability of E? If we imagine many families carrying out this experiment, 1/8 of the time e_4 will occur; 1/8 of the time e_6 will occur; and so on. Thus, in one way or another, E will occur $1/8 + 1/8 + \cdots = 4/8$ of the time; that is,

$$\Pr(E) = \frac{1}{8} + \frac{1}{8} + \frac{1}{8} + \frac{1}{8} = \frac{4}{8} = .50$$

The obvious generalization is that the probability of an event is the sum of the probabilities of all the points (or outcomes) included in that event. That is,

$$\boxed{\Pr(E) = \sum \Pr(e)} \tag{3-7}$$

where we sum over just those outcomes e that are in E. Again, note an analogy between mass and probability: The mass of an object is the sum of the masses of all the atoms in that object; the probability of an event is the sum of the probabilities of all the outcomes included in that event.

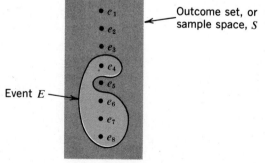

Figure 3-5
Venn diagram: An event as a subset (same outcome set as in Figure 3–4).

EXAMPLE **3-3** Refer again to the planning of 3 children in Figure 3-4.

a. Make a Venn diagram to show the event:
F = second child a girl, followed by a boy.
Then calculate its probability.

b. Calculate the probability for each of the following events:

G = less than 2 girls
H = all the same sex
K = less than 2 boys
I = no girls
I_1 = exactly one girl
I_2 = exactly 2 girls
I_3 = exactly 3 girls

Solution a. Using the sample space of Figure 3-4, we scan the points one by one. We find that there are two in the event F. And so its probability is 2/8.

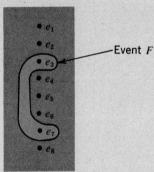

b. The probability of each event simply requires counting up the outcomes in it—because every outcome has the same probability, 1/8. Table 3-2 gives the answers (and for completeness, also lists the events E and F discussed earlier.)

TABLE **3-2** **Several Events in Planning 3 Children**

	THREE ALTERNATIVE WAYS OF NAMING AN EVENT		
(1) **ARBITRARY** **SYMBOL** **FOR EVENT**	**(2)** **VERBAL** **DESCRIPTION**	**(3)** **OUTCOME LIST**	**(4)** **PROBABILITY**
E	At least 2 girls	$\{e_4, e_6, e_7, e_8\}$	4/8
F	Second child a girl followed by a boy	$\{e_3, e_7\}$	2/8
G	Less than 2 girls	$\{e_1, e_2, e_3, e_5\}$	4/8
H	All the same sex	$\{e_1, e_8\}$	2/8
K	Less than 2 boys	$\{e_4, e_6, e_7, e_8\}$	4/8
I	No girls	$\{e_1\}$	1/8
I_1	Exactly 1 girl	$\{e_2, e_3, e_5\}$	3/8
I_2	Exactly 2 girls	$\{e_4, e_6, e_7\}$	3/8
I_3	Exactly 3 girls	$\{e_8\}$	1/8

The first 3 columns of Table 3-2 show 3 different ways to name or specify an event. The value of specifying an event the third way—by its outcomes list—is evident when we look at the event K (less than 2 boys). The list for K is the same as the list for the first event E (at least 2 girls). That is, $K = E$, an equality that is not so evident from the verbal description.

PROBLEMS

3-4   In a learning experiment, a subject attempts a certain task twice in a row. Each time his chance of failure is .40. Draw the probability tree, and then calculate the chance of:

    a. No failures.

    b. Exactly one failure.

3-5   Repeat Problem 3-4, if the subject attempts his task once more, for a total of three tries.

3-6   Repeat Problem 3-4, if the subject makes three tries and he learns from his previous trials, especially his previous successes, as follows: His chance of failure is still .40 at the first trial. However, for later trials his chance of failure drops to .30 if his previous trial was a failure, and drops way down to .20 if his previous trial was a success.

3-7   **a.** (Acceptance sampling) The manager of a small hardware store buys electric clocks in cartons of 12 clocks each. To see whether each carton is acceptable, 3 clocks are randomly selected and thoroughly tested. If all 3 are of acceptable quality, then the carton of 12 is accepted.

        Suppose in a certain carton, unknown to the hardware manager, only 8 of the 12 are of acceptable quality. What is the chance that the sampling scheme will inadvertently accept the carton?

    b. Repeat part **a**, if the given carton of 12 clocks has only 6 of acceptable quality.

3-8   When a penny and a nickel are tossed, the outcome set could be written as a tree or as a rectangular array:

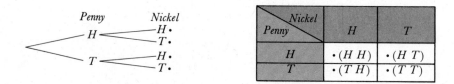

Draw up the same two versions of the outcome set when a pair of dice are thrown—one red, one white. Then, using whichever version is more convenient, calculate the probability of:

a. A total of 4 dots.

b. A total of 7 dots.

c. A total of 7 or 11 dots (as in Problem 3-3**c**).

d. A double (the same value on both dice).

e. A total of at least 8 dots.

f. A 1 on one die, 5 on the other.

g. A 1 on both dice ("snake eyes").

h. Would you get the same answers if both dice were painted white? In particular, would you get the same answers as before, for parts **f** and **g**?

3-3 Combining Events

A—DEFINITIONS

In planning their 3 children, suppose that the couple would be disappointed if there were fewer than two girls, or if all were the same sex. Referring to Table 3-2, you can see that this is the event "G or H," also denoted by $G \cup H$, and read "G union H." From the lists of Table 3-2, we can pick out the points that are in G or in H, and so obtain:

$$G \cup H = \{e_1, e_2, e_3, e_5, e_8\}$$

And in general, we define:

$$\boxed{G \cup H \equiv \text{set of points that are in } G, \text{ or in } H, \text{ or in both.}} \quad (3\text{-}8)$$

Figure 3-6a illustrates this definition. Since five outcomes are included in $G \cup H$, its probability is 5/8.

The couple would be doubly disappointed if there were fewer than two girls *and* if all children were the same sex. This clearly is a much more restricted combined event, consisting only of those outcomes that satisfy both G and H. This is denoted[1] by $G \cap H$, and is read "G intersect H" as well as "G and H." From the lists of Table 3-2, we see there is only one point in both G and H:

$$G \cap H = \{e_1\}$$

---

[1]To remember when $\cup$ or $\cap$ is used, it may help to recall that $\cup$ stands for "Union," and that $\cap$ resembles the letter A in the word "And." These technical symbols are used to avoid the ambiguity that might occur if we used ordinary English. For example, the sentence "E $\cup$ F has 5 points" has a precise meaning, but the informal "E or F has 5 points" is ambiguous.

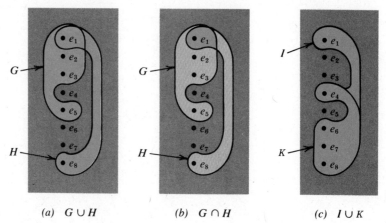

Figure 3-6
Venn diagrams, illustrating combined events. (a) G ∪ H shaded, "G or H." (b)
G ∩ H shaded, "G and H." (c) I ∪ K shaded.

And in general, we define:

$$G \cap H \equiv \text{set of points that are in both } G \text{ and } H$$ (3-9)

Figure 3-6b illustrates this definition. Since only one outcome is included in $G \cap H$, its probability is 1/8.

EXAMPLE **3-4**

From the lists in Table 3-2, construct the lists for the following events and hence find their probability.

$$G \cup F, \qquad G \cap F, \qquad I \cup K, \qquad I \cap K$$

Solution $G \cup F = \{e_1, e_2, e_3, e_5, e_7\}$
$G \cap F = \{e_3\}$
$I \cup K = \{e_1, e_4, e_6, e_7, e_8\}$
$I \cap K = \{ \}$

Hence $\Pr(F \cup G) = 5/8$
    $\Pr(F \cap g) = 1/8$
    $\Pr(I \cup K) = 5/8$
    $\Pr(I \cap K) = 0$

B—PROBABILITY OF G ∪ H

We already have shown how $\Pr(G \cup H)$ may be found from the Venn diagram in Figure 3-6. Now we will develop a formula. First, consider a pair of events that do not have any points in common, such as I and K from Table 3-2. Because these events do not overlap, they are called *mutually exclusive*. From Figure 3-6c, it is obvious that:

$$Pr(I \cup K) = Pr(I) + Pr(K) \qquad (3\text{-}10)$$
$$\frac{5}{8} = \frac{1}{8} + \frac{4}{8}$$

But this simple addition does not always work. For example:

$$Pr(G \cup H) \neq Pr(G) + Pr(H) \qquad (3\text{-}11)$$
$$\frac{5}{8} \neq \frac{4}{8} + \frac{2}{8}$$

What has gone wrong in this case? Since G and H overlap, in summing $Pr(G)$ and $Pr(H)$ we double counted e_1, the intersection $G \cap H$. This is easily corrected. Subtracting $Pr(G \cap H)$ eliminates this double counting. Accordingly, it is generally true that:

$$\boxed{Pr(G \cup H) = Pr(G) + Pr(H) - Pr(G \cap H)} \qquad (3\text{-}12)$$

In our example:

$$\frac{5}{8} = \frac{4}{8} + \frac{2}{8} - \frac{1}{8}$$

Formula (3-12) applies not only to those cases where G and H overlap. It also applies in cases like (3-10), where I and K do not overlap, where $Pr(I \cap K) = 0$, and this last term in (3-12) disappears. Then we obtain the special case:

$$\boxed{\begin{array}{c} \text{If } I \text{ and } K \text{ are mutually exclusive,} \\ Pr(I \cup K) = Pr(I) + Pr(K) \end{array}} \qquad (3\text{-}13)$$

C—COMPLEMENTS

In Table 3-2, note that G consists of exactly those points that are not in E. We therefore call G the "*complement of E*," or "*not E*," and denote it by $\overline{E}$. In general, for any event E:

$$\boxed{\overline{E} \equiv \text{set of points that are } not \text{ in } E} \qquad (3\text{-}14)$$

Since the points in E plus the points in $\overline{E}$ make up 100% of the outcomes,

$$Pr(E) + Pr(\overline{E}) = 1.00$$

This gives us a very useful formula:

$$\boxed{Pr(E) = 1 - Pr(\overline{E})} \qquad (3\text{-}15)$$

EXAMPLE **3-5**

To make sure that they get at least one boy, a couple plans on having 5 children. What are their chances of success?

Solution

It would be tedious to list the sample space and pick out the event of at least one boy. Let us see if we can avoid this mess by working with the complement.

$$\text{Let } E = \text{at least one boy}$$
$$\text{then } \overline{E} = \text{no boys; that is, all girls.}$$

We see that $\text{Pr}(\overline{E})$ is much easier to calculate: the probability of getting a girl every time is:

$$\text{Pr}(\overline{E}) = \frac{1}{2} \times \frac{1}{2} \times \frac{1}{2} \times \frac{1}{2} \times \frac{1}{2} = \frac{1}{32}$$

Finally, we can obtain the required probability of E (at least one boy) as the complement:

$$\text{Pr}(E) = 1 - \text{Pr}(\overline{E}) \qquad\qquad (3\text{-}15)\,\text{repeated}$$
$$= 1 - \frac{1}{32} = \frac{31}{32} = .97$$

In other words, it is 97% certain that they will get at least one boy in five births.

In Example 3-5, we found that complements sometimes provide the easiest way to calculate the required probability. You should be on the alert for similar problems: The key words to watch for are "at least," "more than," "less than," "no more than," and so on.

EXAMPLE **3-6**

In a certain college, men engage in various sports in the following proportions:

Football (F), 60%
Basketball (B), 50%
Both football and basketball, 30%

If a man is selected at random for an interview, what is the chance that he will:

a. Play football or basketball?

b. Play neither sport?

Solution

> **a.** $\Pr(F \cup B) = \Pr(F) + \Pr(B) - \Pr(F \cap B)$ like (3-12)
> $= .60 + .50 - .30 = .80$
>
> **b.** $1 - P(F \cup B) = 1 - .80 = .20$

PROBLEMS

3-9 Suppose that a coin was unfairly tossed 3 times in such a way that over the long run the following relative frequencies were observed:

e	Pr(e)
·(H H H)	.15
·(H H T)	.10
·(H T H)	.10
·(H T T)	.15
·(T H H)	.15
·(T H T)	.10
·(T T H)	.10
·(T T T)	.15

Suppose that we are interested in the following events:

E: fewer than 2 heads
F: all coins the same
G: fewer than 2 tails
H: some coins different

Find the following probabilities.

a. $\Pr(E)$, $\Pr(F)$, $\Pr(E \cup F)$, $\Pr(E \cap F)$.

b. Verify that (3-12) holds true.

c. $\Pr(G)$, $\Pr(H)$, $\Pr(G \cup H)$, $\Pr(G \cap H)$.

d. Verify that (3-12) holds true.

3-10 Of the U.S. population in 1980,

10% were from California,
 6% were of Spanish origin,
 2% were from California and of Spanish origin.

If an American was drawn at random, what is the chance she would be:

a. From California or of Spanish origin?

b. Neither from California nor of Spanish origin?

c. Of Spanish origin, but not from California?

3-11 In a family of 10 children (assuming boys and girls are equally likely), what is the chance there will be:

a. No boys?

b. Both boys and girls occurring?

3-12   Suppose that a class of 100 students consists of four groups, in the following proportions:

	MEN	WOMEN
Taking math	17%	38%
Not taking math	23%	22%
	40%	60%

If a student is chosen by lot to be class president, what is the chance that the student will be:

a. A man?   .4

b. A woman?   .6

c. Taking math?   .17 + .38 = .55

d. A man, or taking math?   .4 + .38 = .78

e. A man, and taking math?   .17

f. If the class president in fact turns out to be a man, what is the chance that he is taking math? not taking math?

3-13   The men in a certain college engage in various sports in the following proportions:

Football, 30% of all men   Both football and basketball, 5%
Basketball, 20%   Both football and baseball, 10%
Baseball, 20%   Both basketball and baseball, 5%
                    All three sports, 2%

If a man is chosen by lot for an interview, use a Venn diagram to calculate the chance that he will be:

a. An athlete (someone who plays at least one sport).

b. A football player only.

c. A football player or a baseball player.

   If an *athlete* is chosen by lot, what is the chance that he will be:

d. A football player only?

e. A football player or a baseball player?

3-14   A salesman makes 12 calls per day, and on each call has a 20% chance of making a sale.

a. What is the chance he will make no sales at all on a given day? $(.8)^{12} = .07$

b. What is the chance he will make at least one sale?
1 - no sales = 1 - $(.8)^{12}$ = 1 - .07 = .93 $\frac{a}{a+b}$ =

c. If he sells for 200 days of the year, about how many of these days will he make no sales at all?   .07 on one day
                            .07 × 200 ≐ 14 days

3-4 Conditional Probability

A—DEFINITION

Conditional probability is just the familiar concept of limiting relative frequency, but with a slight twist—the set of relevant outcomes is restricted by a condition. An example will illustrate.

EXAMPLE 3-7

In a family of 3 children, suppose it is known that G (fewer than two girls) has occurred. What is the probability that H (all the same sex) has occurred? That is, if we imagine many repetitions of this experiment and consider just those cases in which G has occurred, how often will H occur? This is called the *conditional probability* of H, given G, and is denoted $Pr(H/G)$. Answer in two different cases, where the probability of a boy is:

a. 50% (as in Figure 3-2).
b. 52% (as in Figure 3-3).

Solution

a. As shown in Figure 3-7, there are four outcomes in G, and only one of them is in H. Thus, when all outcomes are equally likely:

$$Pr(H/G) = \frac{1}{4} = .25$$

b. When the outcomes are not equally likely, we must be more subtle. Suppose, for example, that the experiment is carried out 100

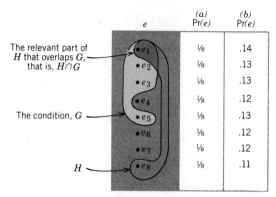

		e	(a) $Pr(e)$	(b) $Pr(e)$
The relevant part of H that overlaps G, that is, $H \cap G$		e_1	$\frac{1}{8}$	.14
		e_2	$\frac{1}{8}$	.13
		e_3	$\frac{1}{8}$	.13
		e_4	$\frac{1}{8}$	.12
The condition, G		e_5	$\frac{1}{8}$	.13
		e_6	$\frac{1}{8}$	.12
		e_7	$\frac{1}{8}$	.12
H		e_8	$\frac{1}{8}$	.11

Figure 3-7
Venn diagram to illustrate the conditional probability $Pr(H/G)$.
(a) If $Pr(Boy) = .50$ (b) If $Pr(Boy) = .52$

million times. Then how often will G occur? The answer is about 53 million times (from the last column of Figure 3-7, $Pr(G)$ = .14 + .13 + .13 + .13 = .53). Of these times, how often will H occur? The answer is about 14 million times. Thus, from our fundamental notion of probability as relative frequency:

$$Pr(H/G) = \frac{14 \text{ million}}{53 \text{ million}} = .27 \qquad (3\text{-}16)$$

Now let us express our answer in general terms. The ratio in (3-16) is $Pr(H \cap G)$ divided by $Pr(G)$. Thus:

$$Pr(H/G) = \frac{Pr(H \cap G)}{Pr(G)} \qquad (3\text{-}17)$$

B—AN APPLICATION OF CONDITIONAL PROBABILITY

Formula (3-17) can be reexpressed. Let us cross-multiply by $Pr(G)$, and note of course that $Pr(H \cap G) = Pr(G \cap H)$. Then:

$$Pr(G \cap H) = Pr(G)\ Pr(H/G) \qquad (3\text{-}18)$$

This formula breaks $Pr(G \cap H)$ into two easy steps: $Pr(G)$ and then $Pr(H/G)$. An example will show how useful it is.

EXAMPLE **3-8**   Three defective light bulbs inadvertently got mixed with 6 good ones. If 2 bulbs are chosen at random for a ceiling lamp, what is the probability that they both are good?

Solution   We can break down this problem very naturally if we imagine the bulbs being picked up one after the other. Then let us denote:

$$G_1 = \text{first bulb is good}$$
$$G_2 = \text{second bulb is good}$$

Thus:

$$Pr(\text{both good}) = Pr(G_1 \cap G_2)$$
$$= Pr(G_1)\ Pr(G_2/G_1) \qquad \text{like (3-18)}$$

Now on the first draw, there are 6 good bulbs among 9 altogether, so that the probability of drawing a good bulb is $Pr(G_1)$ = 6/9. After that, however, there are only 5 good bulbs among the 8 left, so that $\Pr(G_2/G_1) = 5/8$. Thus:

$$\Pr(G_1 \cap G_2) = \frac{6}{9} \times \frac{5}{8} = \frac{5}{12} = .42 \qquad (3\text{-}19)$$

Remarks This problem could have been solved just as well using a tree. If this experiment were repeated many times, 6/9 of the time there would be a good bulb drawn first; of these times, 5/8 of the time there would be another good bulb drawn second. Thus, the probability of drawing two good bulbs altogether would be:

$$\frac{6}{9} \text{ of } \frac{5}{8} = \frac{5}{12} \qquad (3\text{-}19) \text{ confirmed}$$

Thus, the product formula (3-18) has a strong intuitive basis.

PROBLEMS

3-15 In this problem, we will simulate the family planning experiment to show once more how probability is just long-run relative frequency. Flip three coins (or use the random digits in Appendix Table I three at a time, letting an even digit represent a boy, an odd digit a girl).

Repeat this simulation $n = 50$ times, and record the frequency f of the following events (same as in Example 3-7):

G: fewer than 2 girls
H: all the same sex.

Then calculate the following relative frequencies:

a. $\dfrac{f(G)}{n}$

b. $\dfrac{f(H \cap G)}{n}$

c. $\dfrac{f(H \cap G)}{f(G)}$

If you like, pool your data with some friends to get closer to the long-run relative frequencies. What would these frequencies be called? Now calculate them from the sample space.

3-16 In the U.S. in 1974, the population was classified as male or female, and as being in favor of or opposed to abortion.[2] The proportions in each

---

[2] The *Gallup Opinion Index*, April 1974, p. 24. The exact question was, "The U.S. Supreme Court has ruled that a woman may go to a doctor to end pregnancy at any time during the first 3 months of pregnancy. Do you favor or oppose this ruling?" The 10% who had no opinion are not included.

category were approximately as follows (note that all the proportions add up to 1.00 = 100% of the population):

	FAVOR	OPPOSED
Male	.27	.21
Female	.24	.28

.51

What is the probability that an individual drawn at random will be:

a. In favor of abortion? .51

b. In favor of abortion, if male?

c. In favor of abortion, if female?

3-17 In a family of 3 children, what is the chance of:

a. At least one girl?

b. At least 2 girls?

c. At least 2 girls, given at least one girl?

d. At least 2 girls, given that the eldest child is a girl?

3-18 Suppose that 4 defective light bulbs inadvertently have been mixed up with 6 good ones.

a. If 2 bulbs are chosen at random, what is the chance that they both are good?

b. If the first 2 are good, what is the chance that the next 3 are good?

c. If we started all over again and choose 5 bulbs, what is the chance they all would be good?

3-19 Critically discuss the following arguments:

a. In the 3-year period that followed the murder of President Kennedy in 1962, fifteen material witnesses died—six by gunfire, three in motor accidents, two by suicide, one from a cut throat, one from a karate chop to the neck, and two from natural causes. An actuary concluded that on the day of the assassination, the odds against this particular set of witnesses being dead within 3 years were one hundred thousand trillion to one. Since all these things couldn't have just happened, they must reflect an organized coverup of the assassination.

b. A student of statistics, so we're told, once heard that there was one chance in a million of a bomb being on an aircraft. He calculated there would be only one chance in a million million (trillion) of there being two bombs on an aircraft. In order to enjoy these longer odds, therefore, he always carried a bomb on with him (carefully defused, of course—he was no fool.)

3-20 Two dice are thrown and we are interested in the following events:

E: first die is 5

F: total is 7

G: total is 10

By calculating the probabilities using Venn diagrams, show that:

a. $\Pr(F/E) = \Pr(F)$.

b. $\Pr(G/E) \neq \Pr(G)$.

c. Is the following a correct verbal conclusion? If not, correct it.

If I'm going to bet on whether the dice show 10, it will help (change the odds) to peek at the first die to see whether it is a 5. But if I'm going to bet on whether the dice show 7, a peek won't help.

3-21 Two cards are drawn from an ordinary deck. What is the probability of drawing:

$\Pr(A_1) = \frac{4}{52}$ $\Pr(A_2 | A_1) = \frac{\Pr(A_2 \cap A_1)}{\Pr(A_1)}$

a. 2 aces? $\Pr(A_1 \cap A_2) = \Pr(A_1)\Pr(A_2 | A_1)$

b. The 2 black aces?

c. 2 honor cards (ace, king, queen, jack or ten)?

*3-22 A poker hand (5 cards) is drawn from an ordinary deck of cards. What is the chance that:

a. The first 4 cards are the aces?

b. The first 2 cards and the last 2 cards are the aces?

c. The 4 aces are somewhere among the 5 cards ("4 aces")?

d. "4 of a kind" (4 aces, or 4 kings, or 4 queens, . . . , or 4 deuces)?

3-5 Independence

A—DEFINITION

Independence is a very precise concept that we define in terms of certain probabilities. An example will illustrate.

EXAMPLE 3-9

In the U.S. in 1974, a Gallup poll classified the population as white or nonwhite and in favor of or opposed to abortion. The proportions in each category were as follows (note that all the proportions add up to 1.00 = 100% of the population).

	FAVOR (F)	OPPOSED (O)
White (W)	.468	.432
Nonwhite (B)	.052	.048

If a person is drawn at random, what is:

a. $\Pr(F)$?

b. $\Pr(F/W)$?

Solution   **a.**

$$\Pr(F) = .468 + .052 = .52 \qquad \text{like (3-13)}$$

b.

$$\Pr(F/W) = \frac{\Pr(F \cap W)}{\Pr(W)} \qquad \text{like (3-17)}$$

$$= \frac{.468}{.468 + .432} = .52$$

Since both these probabilities are the same, the probability of being F is not in any way affected by being W. This kind of independence, defined in terms of probability, is called *statistical independence*. We can state the exact definition.

F is called *statistically independent* of E if

$$\Pr(F/E) = \Pr(F) \qquad\qquad (3\text{-}20)$$

Of course, if $\Pr(F/E)$ is different from $\Pr(F)$, we call F statistically *dependent* on E. Statistical dependence is the usual case, since it is much easier for two probabilities to be somewhat unequal than to be exactly equal. For example, in Problem 3-16 we found that being in favor of abortion was statistically dependent on being male.[3]

So far we have insisted on the phrase "*statistical* independence," in order to distinguish it from other forms of independence—philosophical, logical, or whatever. For example, we might be tempted to say that for the dice of Problem 3-20, F was "somehow" dependent on E because the total of the two tosses depends on the first die. Although this vague notion of dependence is not used in statistics and will be considered no further, we mention it as a warning that *statistical* independence is a very precise concept, defined by probabilities [as in (3-20)].

Now that we clearly understand statistical independence and agree that it is the only kind of independence we will consider, we can be informal and drop the word "statistical."

B—IMPLICATIONS

If an event F is independent of another event E, we can develop some interesting logical consequences. According to (3-18), it is always true that

---

[3] And if, in Example 3-9, the exact probabilities for the U.S. were quoted instead of the approximate probabilities, $\Pr(F/W)$ would turn out to be slightly different from $\Pr(F)$; that is, F would be slightly dependent on W.

$$\Pr(E \cap F) = \Pr(E) \Pr(F/E)$$

When we substitute (3-20), this becomes the simple multiplication rule:

> **For independent events,**
> $$\mathbf{Pr}(E \cap F) = \mathbf{Pr}(E)\, \mathbf{Pr}(F)$$

(3-21)

Furthermore, by dividing (3-21) by $\Pr(F)$ we obtain

$$\frac{\Pr(E \cap F)}{\Pr(F)} = \Pr(E)$$

that is,

$$\Pr(E/F) = \Pr(E) \tag{3-22}$$

That is, E is independent of F. In other words:

> **Whenever F is independent of E,**
> **then E must be independent of F**

(3-23)

In view of this symmetry, we can simply state that E and F are *independent of each other* whenever any of the three logically equivalent statements (3-20), (3-21), or (3-22) is true. Often the multiplicative form (3-21) is the preferred form, in view of its symmetry—in fact, it could have been taken as the very definition of statistical independence.

C—CONCLUSION

Now we have completed our development of the most important formulas of probability. To review them, Table 3-3 sets out our basic conclusions for $\Pr(E \cup F)$ and $\Pr(E \cap F)$.

TABLE **3-3** **Review of Probability Formulas**

	$\Pr(E \cup F)$	$\Pr(E \cap F)$
General Theorem	$= \Pr(E) + \Pr(F) - \Pr(E \cap F)$	$= \Pr(E) \Pr(F/E)$
Special Case	$= \Pr(E) + \Pr(F)$ if E and F are mutually exclusive; i.e., if $\Pr(E \cap F) = 0$	$= \Pr(E) \Pr(F)$ if E and F are independent; i.e. if $\Pr(F/E) = \Pr(F)$

PROBLEMS

3-23 The 1980 U.S. population, broken down by region and attitude to legalization of marijuana, roughly turned out as follows (note that all proportions add up to 100%):

	IN FAVOR (F)	OPPOSED ($\overline{F}$)
East (E)	7.8%	22.2%
All except East ($\overline{E}$)	18.2%	51.8%

    a. What is Pr(F) (the probability that an individual drawn at random will be in favor of legalization)?

    b. What is Pr(F/E)?

    c. Is F independent of E?

3-24 In Problem 3-23, we found that F was independent of E.

    a. Can you guess, or better still, state for certain on the basis of theoretical reasoning:

        i. Whether E will be independent of F?

       ii. Whether E will be independent of $\overline{F}$?

    b. Calculate the appropriate probabilities to verify your answers in part (a).

3-25 Three coins are fairly tossed, and we define:

E_1: first two coins are heads.
E_2: last coin is a head.
E_3: all three coins are heads.

Try to answer the following questions intuitively. (Does knowledge of the condition affect your betting odds?) Then verify by drawing the sample space and calculating the relevant probabilities.

    a. Are E_1 and E_2 independent?

    b. Are E_1 and E_3 independent?

3-26 Repeat Problem 3-25, using the unfairly tossed coins whose sample space is as follows (as in Problem 3-9):

e	Pr(e)
•(H H H)	.15
•(H H T)	.10
•(H T H)	.10
•(H T T)	.15
•(T H H)	.15
•(T H T)	.10
•(T T H)	.10
•(T T T)	.15

3-27   A single card is drawn from a standard deck, and we define:

E: it is an ace.
F: it is a heart.

Are *E* and *F* independent, when we use:

a.   An ordinary 52-card deck?

b.   An ordinary deck, with all the spades deleted?

c.   An ordinary deck, with all the spades from 2 to 9 deleted?

3-28   If *E* and *F* are two mutually exclusive events, what can be said about their independence? [*Hint:* What is Pr(*E* ∩ *F*)? Then, using (3-17), what is Pr(*E/F*)? Does it equal Pr(*E*)?]

3-6   Bayes Theorem: Tree Reversal

An important branch of applied statistics called *Bayes Analysis* can be developed out of conditional probability and trees. An example will illustrate.

EXAMPLE **3-10**   I am thinking of buying a used Q-car at Honest Ed's. In order to make an informed decision, I look up the records of Q-cars in an auto magazine, and find that, unfortunately, 30% have faulty transmissions.

   To get more information on the particular Q-car at Honest Ed's, I hire a mechanic who can make a shrewd guess on the basis of a quick drive around the block. Of course, he isn't always right; but he does have an excellent record: Of all the faulty cars he has examined in the past he correctly pronounced 90% "faulty"; in other words, he wrongly pronounced only 10% "OK." He has almost as good a record in judging good cars: He has correctly pronounced 80% "OK," while he wrongly pronounced only 20% "faulty." (We emphasize that *"faulty"* in quotation marks describes the mechanic's opinion, while *faulty* with no quotation marks describes the actual state of the car.)

   What is the chance that the Q-car I'm thinking of buying has a faulty transmission:

a.   Before I hire the mechanic?

b.   If the mechanic pronounces it "faulty"?

c.   If the mechanic pronounces it "OK"?

Solution   **a.**   Before any mechanical examination, the chance this car is faulty is 30% (the proportion of all Q-cars that are faulty—the only information I have).

b. Imagine running hundreds of Q-cars past the mechanic to see how he judges them. The first branching of the tree in Figure 3-8 shows the 30% of the cars that *actually* are faulty, and the 70% that are OK. As we move to the right, the second branching shows how well the mechanic is able to *judge* them: For example, in the top branch we see that if a car is faulty, he's 90% sure of correctly judging it "faulty." So 90% of 30% = 27% of all cars are actually faulty and then correctly identified as such, and we mark this 27% in the right-hand column.

Looking down to the third number in the right-hand column, we similarly find that 20% of 70% = 14% of all cars are *good* cars that are judged "faulty".

Thus altogether 27% + 14% = 41% of the cars are judged "faulty", and are encircled in blue in Figure 3-8. Of these cars, about two-thirds (27/41) are actually faulty. Once the mechanic says "faulty," therefore, the probability is about ⅔ that the car actually is faulty. This conditional probability can be calculated more formally from (3-17):

$$Pr(\text{faulty/"faulty"}) \doteq \frac{.27}{.41} = .66 \qquad (3\text{-}24)$$

To sum up: Once the car has been pronounced "faulty" by the mechanic, the chance that it is actually faulty rises from the original 30% (calculated in part **a**) up to 66%.

c. Once the mechanic says "OK," now we know it must be one of the cars in the complementary event—also encircled at the right of Figure 3-8. And only a very small proportion of these cars actually are faulty:

$$Pr(\text{faulty/"OK"}) = \frac{.03}{.59} = .05 \qquad \begin{matrix}(3\text{-}25) \\ \text{like}\,(3\text{-}17)\end{matrix}$$

This also makes good intuitive sense: Once the car has been pronounced "OK," the chance that it actually is faulty drops from 30% down to 5%.

The calculations in Figure 3-8 can be clearly summarized in another tree, the *reverse tree* in Figure 3-9. (We emphasize that this involves no new calculations; it is only a way of conveniently *displaying* the calculations already done.) Notice the reverse order in Figure 3-9: The first branching now shows the test opinion of the mechanic. (The 41% "faulty" and the 59% "OK" appeared as the two encircled events on the right of Figure 3-8.)

Moving to the right in Figure 3-9, the second branching shows the actual condition of the cars—and the answers to our questions. For example, the

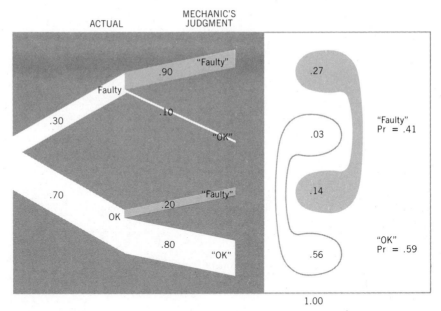

Figure 3-8
Calculation of the Tree and Conditional Probabilities

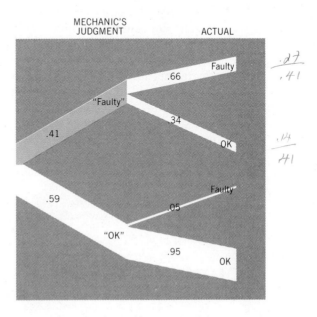

Figure 3-9
Bayes Theorem (Tree Reversal). The results of Figure 3-8 displayed as a reverse tree. The last branching gives the posterior probabilities.

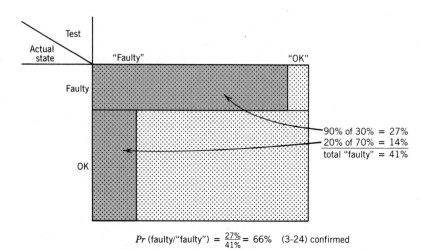

$$Pr\ (\text{faulty/"faulty"}) = \frac{27\%}{41\%} = 66\%\quad (3\text{-}24)\ \text{confirmed}$$

Figure 3-10
Alternative to tree reversal: Bayes Theorem represented as a rectangular sample space. Each of the hundreds of possible cars is shown as a dot, and those that test out "faulty" are shaded.

top branch displays the answer calculated earlier in (3-24): Once the car is judged "faulty" the chance that it actually turns out faulty is .66. And the third branch from the top displays the answer calculated in (3-25): Once the car is judged "OK," the chance that it actually is faulty is just .05. (To complete the tree, the complementary probability .95 is shown on the final branch.)

Alternatively, this whole problem could have been illustrated with a rectangular sample space instead of a pair of trees, as Figure 3-10 shows.

No matter how we represent it—whether Figure 3-10 or Figure 3-9—this technique is called *Bayes Theorem*. The initial probabilities *before* any testing are called the *prior probabilities* (and appear at the first branching in Figure 3-8). The probabilities *after* testing are called the *posterior probabilities* (and appear as the last branching in Figure 3-9); these are the relevant "betting odds" for making the decision about the car *after* we have the mechanic's opinion.

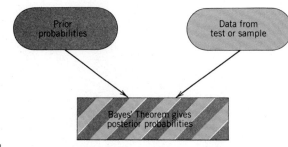

Figure 3-11
The Logic of Bayes Theorem

(highest bar) in Figure 2-2. That is, the mode is where the distribution peaks.

Although the mode is very easy to obtain—at a glance—it is not a very good measure of central tendency, since it often depends on the arbitrary grouping of the data. It is also possible to draw a sample where the largest frequency occurs twice (or more). Then there are two peaks, and the distribution is called *bimodal*.

B—THE MEDIAN

The median we have met already—the 50th percentile. To denote that it is the middle value that splits the distribution into two halves, we use the symbol $\overset{\downarrow}{X}$:

$$\boxed{\text{Median } \overset{\downarrow}{X} \equiv \text{ middle observation or 50th percentile}} \qquad (2\text{-}2)$$

where $\equiv$ means *equals, by definition*. (See the glossary just before the index for the reference list of such symbols.)

For example, suppose 5 rail shipments suffered the following damages (ordered by size of loss): \$10, \$20, \$30, \$50, \$90. Then the median loss would be the middle value, \$30. (When the sample size n is even, there are *two* middle values, and the median is defined as their midpoint. For example, suppose the damages to 4 shipments were: \$10, \$20, \$30, \$50. Then the median loss would be \$25, the midpoint of \$20 and \$30.)

If the array consists of a large number of observations grouped into cells, then the median can be approximated by covering the appropriate distance across the median cell. For example, in the distribution of men's heights in Table 2-2, we can first find the median cell by accumulating the relative frequency as we pass down the cells: 2%, 2% + 6% = 8%, 30%, 62%. That is, by the end of the fourth cell, we have accumulated more than 50% of the observations, and therefore have passed the median. Somewhere back within the fourth cell we will find the median itself. As a rough approximation, we can take the median as the midpoint of the cell, 69 inches. (For a more careful approximation, see Appendix 2-2.)

C—THE AVERAGE OR MEAN

The word *average* is derived from the Arabic root *awar* meaning damaged goods. Even today, in marine law, average means the equitable division of loss among the interested parties. For example, consider again the 5 shippers who suffered losses of \$10, \$20, \$30, \$50, and \$90. Their average loss is found by dividing the total loss among them equally:

$$\text{Average} = \frac{10 + 20 + 30 + 50 + 90}{5} = \frac{200}{5} = \$40$$

The point of Bayes Theorem may be stated more generally: Prior probabilities, combined with some sort of information such as a test or sample, yield posterior probabilities (the relevant betting odds). Figure 3-11 shows this schematically.

PROBLEMS

3-29 A company employs 75 men and 25 women. The accounting department provides jobs for 12% of the men and 20% of the women. If a name is chosen at random from the accounting department, what is the probability that it is a man? That it is a woman?

3-30 In a certain jurisdiction in Alabama (Kaye, 1982), 25% of the men were black, 75% were white. The literacy rates for blacks was 48%, and for whites was 83%. What proportion of the literate men were black?

3-31 The long experience of a clinic is that $1/10$ of its patients have disease A, $2/10$ have disease B, and $7/10$ have neither. Of those with disease A, $9/10$ have headaches; of those with disease B, $1/2$ have headaches; and of those with neither disease, $1/20$ have headaches.

If a patient in this clinic has headaches, what is the probability that he has Disease A? Disease B? Neither?

3-32 The chain saws produced by a large manufacturer for the first 3 months of 1984 were of poor quality. In each of the 300 shipments, 40% of the saws were defective. After tightening up quality control, however, this figure was reduced to 10% defective in each of the 900 shipments produced in the last 9 months.

The manager of a large hardware store was sent a shipment of the 1984 model, and wanted to know whether it was one of the poor shipments produced early in 1984. (If so, he wanted to return it, or at least check it out, saw by saw.)

a. Before opening the shipment, what was the chance it was one of the poor ones?

b. He opened it, randomly drew out a saw, and tested it, and found it was defective. Now what is the chance it was one of the poor shipments?

3-33 Continuing Problem 3-32, calculate the chance it is one of the poor shipments if he:

a. Draws a second defective saw?

b. Draws a third defective saw?

3-7 Other Views of Probability

In Section 3-1, we regarded probability as the limit of relative frequency. There are several other possible approaches, including *symmetric* probability, *axiomatic* probability, and *subjective* probability, which we shall treat in historical order.

A—SYMMETRIC PROBABILITY

Symmetric probability was first developed for fair games of chance such as dice, where the outcomes were equally likely. This permitted probabilities to be calculated even before the dice were thrown; the empirical determination (3-1) was not necessary, although it did provide a reassuring confirmation.

EXAMPLE **3-11**

In throwing a single die, what is the probability of getting an even number?

Solution

We already showed in Example 3-1 that since the die is symmetric, each outcome must have the same probability, ⅙. Thus:

$$\Pr(\text{even number}) = \Pr(2 \text{ or } 4 \text{ or } 6)$$
$$= \frac{1}{6} + \frac{1}{6} + \frac{1}{6} = \frac{3}{6} = .50$$

It is easy to generalize. Suppose an experiment has N equally likely outcomes altogether, and M of them constitute event E. Then

$$\Pr(E) = \frac{M}{N} \tag{3-26}$$

Symmetric probability theory begins with (3-26) as the very definition of probability, and it is a little simpler than the relative frequency approach. However, it is severely limited because it lacks generality—it cannot even handle crooked dice.

Symmetric probability theory also has a major philosophical weakness. Note how the preamble to the definition (3-26) involved the phrase "equally probable." In using the word "probable" in defining probability, we are guilty of circular reasoning.

Our own relative frequency approach to probability suffers from the same philosophical weakness, incidentally. What sort of limit is meant in equation (3-2)? It is *logically* possible that the relative frequency f/n behaves badly, even in the limit; for example, no matter how often we toss a die, it is just conceivable that the ace will keep turning up every time, making $\lim f/n = 1$. Therefore, we should qualify equation (3-2) by stating that the limit occurs with high *probability*, not logical certainty. But then we would be using the concept of probability to define probability in (3-2)—circular reasoning again. To break such a circle, we shall turn next to an axiomatic approach.

B—AXIOMATIC PROBABILITY

All attempts so far to define probability have been weak because they require using probability itself within the definition of probability. The only philosophically satisfactory way to break this circular reasoning is to let probability be a basic undefined term. In such a mathematical model, we make no attempt to say what probability *really* is. We simply state the rules (*axioms*) that it follows:

Perhaps an analogy from chess will help. In chess, no attempt is made to define what a queen, for example, *really* is. Instead a queen is merely characterized by the set of rules (axioms) she must obey: She can move any number of spaces, in any straight direction. She is no more and no less than this. Similar rules (axioms) are made for the other pieces, to complete the definition of the game. It is then possible to draw some conclusions (prove theorems) such as: A king and a rook can win against a king alone; a king and a knight cannot.

Axiomatic probability theory is like the game of chess. It starts with basic undefined terms and axioms, and from them draws conclusions (proves theorems), such as a carefully stated version of equation (3-2)— which is known as the theorem or law of large numbers. An example of how this "game of probability" works is illustrated in Appendix 3-7.

In its intellectual content, of course, axiomatic probability theory is much richer than chess. Moreover, it leads to theorems that are useful in dealing with practical problems. These are, in fact, the same theorems that we have derived earlier in this chapter using our less abstract approach—and that we will find so useful in the rest of this book.

C—SUBJECTIVE PROBABILITY

Subjective or personal probability is an attempt to deal with unique historical events that cannot be repeated, and therefor cannot be given any frequency interpretation. For example, consider events such as a doubling of the stock market average within the next decade, or the overthrow of a certain government within the next month. These events are described as "likely" or "unlikely," even though there is no hope of estimating this by observing their relative frequency. Nevertheless, their likelihood vitally influences policy decisions, and as a consequence must be estimated in some way. Only then can wise decisions be made on what risks are worth taking.

The key question, of course, is: How might we estimate, for example, Mary Smith's subjective probability that the stockmarket (Dow-Jones average) will triple in the next decade? To answer this question, we could get her to compare her subjective probability of this event with some other well-defined *objective* probability. For example, which bet would she prefer: the bet that the stockmarket will triple, or that a penny will turn up heads? If she prefers to bet on the stockmarket tripling, then her personal probability of this happening is clearly more than .50. This general approach can be used to "pin her down" even further, as we illustrate in the Problems.

PROBLEMS

3-34 **a.** If you had your choice today between the following two bets, which would you take?

Bet 1 (Election Bet)
If the Democratic candidate wins the next presidential election, you will then win a $100 prize (and win nothing otherwise).

Bet 2 (Jar Bet)
A chip will be drawn at random from a jar containing 1 black chip and 999 white chips. If the chip turns out black, you will then win the $100 prize (and win nothing otherwise).

    b. Repeat choice **a**, with a change—make the composition of the jar the opposite extreme: 999 black chips and 1 white chip.

    c. Obviously, your answers in parts **a** and **b** depended on your subjective estimate of American politics; there were no objective "right answers." However, the odds in the urn were so lopsided that there undoubtedly is widespread agreement in part **a** to prefer the election bet, and in part **b** to prefer the jar bet. The question is: as you gradually increase the black chips from 1 to 999, at what point do you become indifferent between the two bets? Is it reasonable to call this your personal probability of a Democratic win?

3-35 Using the "calibrating jar" of Problem 3-34, roughly evaluate your personal probability that:

    a. The Dow Jones average (of certain stock-market prices) will advance at least 10% in the next twelve months.

    b. U.S. population will increase by at least 1.6 million in the next 12 months. (In percentage terms, this is an increase of at least 0.7%.)

    c. U.S. population will be double or more at the end of 100 years. (That is, grow at an average rate of at least 0.7% per year.)

    d. The next vice president of the U.S. will be a female.

    e. The next president of your student council will be a female.

    f. The next president of your student council will be a female, if the president is a senior student chosen at random.

    g. At the next Superbowl game, the coin that is flipped (to determine the kickoff) will turn up heads.

3-36 In Problem 3-35, for which answers do you think there will be least agreement among the students in your class? Most agreement? Which questions are amenable to a brief investigation that would result in everyone agreeing? (If you have time in class, check out your answers.)

3-37 Do you think the following conclusions are valid? If not, correct them:

    i. Certain probabilities (as in part **g** of Problem 3-35) are agreed upon by practically everybody; we may call them "objective probabilities."

    ii. Other probabilities (as in parts **a** or **c**) are disagreed upon, even by experts; we may call them "subjective probabilities."

    iii. But in between, there is a continuous range of probabilities that are subjective to a greater or lesser degree.

CHAPTER 3 SUMMARY

3-1 Probability is just the proportion that emerges in the long run, as the experiment is endlessly repeated (the sample size n grows larger and larger).

3-2 Probability trees break a complex experiment into small manageable stages. At each stage, the various outcomes are represented by branches, and so the tree grows into a complete representation of the experiment.

3-3 An event is just a collection of individual outcomes, and its probability is just the sum of the individual probabilities. From this first principle, we can easily find the probability of G *or* H by addition:

$$\Pr(G \cup H) = \Pr(G) + \Pr(H) - \Pr(G \cap H)$$

3-4 The probability of G *and* H is obtained by multiplication:

$$\Pr(G \cap H) = \Pr(G) \cdot \Pr(H/G)$$

This multiplication principle is so natural, in fact, that it was used in probability trees without formal introduction.

3-5 An event F is called independent of another event E, if the probability of F remains the same after E has occurred:

$$\Pr(F/E) = \Pr(F)$$

3-6 Bayes analysis is simply a clever use of conditional probability to reverse probability trees. It combines prior probabilities with sample information (in the original tree) to obtain the posterior probabilities (in the reversed tree).

3-7 As well as the relative-frequency view, other views of probability are of interest: symmetric probability in fair games; axiomatic probability in mathematics; and subjective probability, which is increasingly important whenever human judgement is required—as in business or social science.

REVIEW PROBLEMS

3-38   To reduce theft, suppose that a company screens its workers with a lie-detector test that has been proven correct 90% of the time (for guilty subjects, and also for innocent subjects). The company fires all the workers who fail the test. And suppose 5% of the workers are guilty of theft.

     a.   Of the workers who are fired, what proportion actually will be guilty?

     b.   Of the workers who remain (who are not fired), what proportion will be guilty?

     c.   Write a brief essay on the ethical problems of such a test, outlining its costs and benefits. Would it make a difference if the company was a dairy producing ice cream bars, or the defense department producing top secret documents?

3-39   True or False? If false, correct it:

     a.   When two events are independent, the occurrence of one event will not change the probability of the second event.

     b.   Two events are mutually exclusive if they have no outcomes in common.

     c. A and B are mutually exclusive if $\Pr(A \cap B) = \Pr(A)\,\Pr(B)$.

     d.   If a fair coin has been fairly tossed 5 times and has come up tails each time, on the 6th toss the conditional probability of tails will be 1/64.

3-40   Suppose that A and B are independent events, with $\Pr(A) = .6$ and $\Pr(B) = .2$. What is:

     a. $\Pr(A/B)$?

     b. $\Pr(A \cap B)$?

     c. $\Pr(A \cup B)$?

3-41   Repeat Problem 3-40 if A and B are mutually exclusive instead of independent.

3-42   A national survey of couples showed that 30% of the wives watched a certain TV program, and 50% of the husbands. Also, if the wife watched, the probability that the husband watched increased to 60%. For a couple drawn at random, what is the probability that:

     a.   The couple both watch?

     b.   At least one watches?

     c.   Neither watches?

     d.   If the husband watches, the wife watches?

     e.   If the husband does not watch, the wife watches?

     f.   Answer True or False; if False correct it: The unconditional probability that the wife watches is somewhere between the two conditional probabilities given in parts **d** and **e**.

3-43 Suppose that the last 3 men out of a restaurant all lose their hatchecks, so that the hostess hands back their 3 hats in random order. What is the probability:

    a. That no man will get the right hat? $\frac{2}{3} \times \frac{1}{2} = \frac{1}{3}$

    b. That exactly 1 man will? $\frac{1}{3} \times \frac{1}{2} + \frac{2}{3} \times \frac{1}{2} = .5$

    c. That exactly 2 men will? 0

    d. That all 3 men will?

3-44 Find (without bothering to multiply out the final answer) the probability that:

    a. A group of 3 people (picked at random) all have different birthdays?

    b. A group of 30 people all have different birthdays?

    c. In a group of 30 people there are at least two people with the same birthday?

    d. What assumptions did you make above?

*3-45 On November 24, 1968, two hijackings occurred on the same day, and made the front page of the *New York Times*. How unusual and newsworthy is such a coincidence? (Glick, 1970) To answer this question, the following data are relevant: During the 4 winter months November–February, there are 120 days, and it turns out that there were 22 hijackings. If we assume that the hijackings are independent and apt to occur equally likely on any day, what is the probability that in a 120-day period with 22 hijackings, there will be some day when two or more hijackings occur?

3-46 An electronics firm bought its TV tubes from two suppliers—75% from A and 25% from B. These two suppliers produced different rates of faulty tubes: A's production was 10% faulty, while B's production was only 2% faulty.

    If a tube was selected at random, what is the probability it was produced by supplier A:

    a. If nothing else is known?

    b. If testing shows the tube is faulty?

    c. If testing shows the tube is faultless?

3-47 Use Venn diagrams to determine which of the following statements are true:

    a. $\overline{E \cup F} = \overline{E} \cup \overline{F}$

    b. $\overline{E \cup F} = \overline{E} \cap \overline{F}$

    c. $\overline{E \cap F} = \overline{E} \cap \overline{F}$

    d. $\overline{E \cap F} = \overline{E} \cup \overline{F}$

    Incidentally, the true statements are known as *De Morgan's Laws*.

3-48 To guarantee confidentiality in a sensitive area of survey sampling, *randomized response* is sometimes used. For example, to investigate what proportion of doctors have ever performed an illegal abortion, each doctor interviewed could be asked to retire to his office, flip a coin, and then:

i.   If it turns up tails, answer the question "Have you ever performed an illegal abortion?"

ii. If it turns up heads, flip it again and answer "Did it come up heads the second time?"

Suppose the doctor comes out of his office with the answer "Yes." There is no possible way for the interviewer to know whether the doctor had performed an illegal abortion, or whether he just flipped a second head. Thus confidentiality is guaranteed. This is not only good ethics, but good science too—nonresponse and false response bias can be kept to a minimum.

Although it is impossible for us to ascertain anything about the individual, can we nevertheless find out what we need to know about the population? Let us see.

a.   Suppose the proportion of doctors who had performed illegal abortions is $A = 20\%$, for example. What porportion Y would answer "Yes" to the interviewer?

b.   To generalize part (a), find the equation relating Y to A. Then solve for A in terms of Y.

c.   In the actual survey, suppose 41% answered "Yes" to the interviewer. What proportion had actually performed an illegal abortion?

d.   What assumptions are you making to get the answers above?

3-49 A certain millionaire devised the following "sure-fire" sequential scheme for making $1000 by selecting the right color (black or red) at roulette.

The first time, bet $1000. If you win, stop. If you lose, double your bet.

If you win this second time, stop. If you lose, double your bet, and continue in this way until you win—at which point your net winning will be $1000. What do you think of this idea?

Chapter

Probability Distributions

*The
normal
law of error
stands out in the
experience of mankind
as one of the broadest
generalizations of natural
philosophy. It serves as the
guiding instrument in researches
in the physical and social sciences and
in medicine, agriculture and engineering.
It is an indispensable tool for the analysis and the
interpretation of the basic data obtained by observation and experiment.*

W. J. YOUDEN

4-1 Discrete Random Variables

A—PROBABILITY DISTRIBUTIONS

Suppose a couple plan to have 3 children, and are interested in the number of girls they might have. This is an example of a *random variable* and is customarily denoted by a capital letter:

$$X = \text{the number of girls}$$

The possible values of X are 0, 1, 2, 3; however, they are not equally likely. To find what the probabilities are, we must examine the original sample space. Already calculated in Figure 3-3, it is repeated in Figure 4-1 (taking the more accurate model where the probability of a boy on each birth is $\pi = .52$). Thus, for example, the event "exactly one girl" ($X = 1$) consists of three outcomes, each having probability .13; hence its probability is

$$.13 + .13 + .13 = 3(.13) = .39 \qquad (4\text{-}1)$$

Similarly, the probability of each of the other events is computed. Thus, in Figure 4-1 we obtain the *probability distribution* of X shown on the right. This illustrates the key idea of a random variable:

> **A discrete random variable takes on various values x with probabilities specified by its probability distribution $p(x)$.** $(4\text{-}2)$

In Figure 4-1, the original sample space (outcome set) has been reduced to a much smaller and more convenient numerical sample space. The original sample space was introduced to enable us to calculate the probability distribution $p(x)$ for the new space; having served its purpose, the old unwieldy space is then forgotten. The interesting questions can be answered very easily in the new space. For example, referring to Figure 4-1, what is the probability of fewer than 2 girls? We simply add up the relevant probabilities in the new sample space:

$$\Pr(X < 2) = p(0) + p(1) \qquad (4\text{-}3)$$
$$= .14 + .39 = .53$$

Figure 4-2 shows how a graph makes this calculation easy to see.

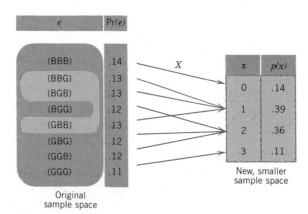

Figure 4-1

The random variable $X =$ number of girls.

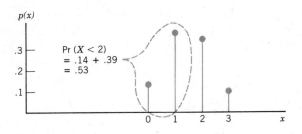

Figure 4-2
The graph of the probability distribution in Figure 4-1.

*B—A RIGOROUS TREATMENT OF RANDOM VARIABLES

Although the intuitive definition of a random variable in (4-2) is usually good enough, in this brief optional section we give a more careful definition. As Figure 4-1 shows, the random variable X associates each outcome e with a number x. This relation is shown by the blue arrows, and is a kind of function. In fact, a random variable is mathematically defined just this way—as a *numerical-valued function defined over a sample space.*

This definition stresses the random variable's relation to the original sample space. Thus, for example, the random variable Y = the number of boys, is clearly a different function (random variable) than X = the number of girls. Yet, if boys and girls were equally likely on each birth, X and Y would have the *same probability distribution*, and anyone who used the loose definition (4-2) might be deceived into thinking that they were the *same random variable*. In conclusion, there is more to a random variable than its probability distribution.

These ideas will be easier to keep in mind if we use appropriate notation. A *capital* letter such as X represents a random variable (function), while small x represents a typical value that it may take. If these values are 0, 1, 2, . . . , then they form the new small sample space. Their probabilities are denoted $p(0)$, $p(1)$, $p(2)$, . . . , or $p(x)$ in general. This notation, like any other, may be regarded simply as an abbreviation for convenience. Thus, for example, to denote "the probability of one girl" we can write:

$$Pr(X = 1) \text{ or just } p(1).$$

Finally, we emphasize the difference between *discrete* and *continuous* variables. A random variable is called *discrete* if it has just a finite (or "countably infinite") set of values. For example, when X = the number of heads in 3 tosses of a coin, then its values are 0, 1, 2, 3—a finite set. Or if X = the number of tosses required to get the first head, then its values are 1, 2, 3, 4, 5, . . .—a "countably infinite" set.

By contrast, a *continuous* random variable takes on a continuum of values. For example, if X = the number of gallons that pass hourly through a meter with a capacity of 50 gallons per hour, then its value could be any number between 0 and 50, not necessarily an integer—for example, 17.2 or 39.826. The mathematics required for continuous random variables is more advanced, requiring integration (the analogue of summation), which is studied in calculus.

PROBLEMS

4-1 In planning a family of 3 children, assume boys and girls are equally likely on each birth. Find the probability distribution of:

a. X = the number of girls

b. Y = the number of runs (where a "run" is a run or string of children of the same sex. For example Y = 2 for the outcome BGG).

4-2 Simulate the experiment in Problem 4-1 by using the random numbers in Appendix Table I (even number = heads, odd = tails). Repeat this experiment 50 times, tabulating the frequency of X. Then calculate:

a. The relative frequency distribution.

b. The mean $\overline{X}$.

c. The variance s^2.

4-3 If the experiment in Problem 4-2 were repeated millions of times (rather than 50 times), to what value would the calculated quantities tend?

4-4 A salesperson for a large pharmaceutical company makes 3 calls per year on a drugstore, with the chance of a sale each time being 80%. Let X denote the total number of sales in a year (0, 1, 2, or 3).

a. Tabulate and graph the probability distribution $p(x)$.

b. What is the chance of at least two sales?

4-5 Repeat Problem 4-4 under the different assumption that "nothing succeeds like success." Specifically, while the chance of a sale on the first call is still 80%, the chance of a sale on later calls depends on what happened on the previous call, being 90% if the previous call was a sale, or 40% if the previous call was no sale.

4-2   Mean and Variance

In Chapter 2 we calculated the mean $\overline{X}$ and variance s^2 of a sample of observations from its relative frequency distribution (f/n). In the same way, it is natural to calculate the mean and variance of a random variable from its probability distribution $p(x)$:

$$\boxed{\begin{array}{l} \textbf{Mean } \mu \equiv \Sigma x p(x) \\[2mm] \textbf{Variance } \sigma^2 \equiv \Sigma (x - \mu)^2\, p(x) \end{array}}$$

(4-4)
like (2-20)
(4-5)
like (2-21) and (2-23)

TABLE **4-1** A Comparison of Sample and Population Moments

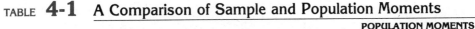

SAMPLE MOMENTS USE RELATIVE FREQUENCIES f/n	POPULATION MOMENTS (GREEK LETTERS) USE PROBABILITIES $p(x)$ (LONG-RUN f/n)
Sample Mean $$\overline{X} = \Sigma x\left(\frac{f}{n}\right)$$	Population Mean $$\mu \equiv \Sigma x p(x)$$
Sample Variance $$s^2 \simeq MSD = \Sigma(x - \overline{X})^2 \left(\frac{f}{n}\right)$$	Population Variance $$\sigma^2 \equiv \Sigma(x - \mu)^2\, p(x)$$

Here we are following the usual custom of reserving Greek letters for theoretical values (μ is the Greek letter mu, equivalent to m for mean, and σ is the Greek letter sigma, equivalent to s for standard deviation). Of course, probabilities can be viewed as just the long-run relative frequencies from the population of all possible repetitions of the experiment. Thus we often call μ and σ^2 the *population* moments, to distinguish them from the *sample* moments $\overline{X}$ and s^2 that are calculated for a mere sample. Table 4-1 makes this clear.

The calculation of σ^2 can often be simplified by using an alternative formula developed in Appendix 4-2:

$$\boxed{\sigma^2 = \Sigma x^2\, p(x) - \mu^2}$$ (4-6)

EXAMPLE **4-1** Again consider X = the number of girls in a family of 3 children, whose probability distribution was derived in Figure 4-1. Calculate the mean, variance, and standard deviation of X.

Solution The computations are similar to Table 2-8, and are set out in Table 4-2.

TABLE **4-2** Calculation of the Mean and Variance of X = number of girls

GIVEN PROBABILITY DISTRIBUTION		CALCULATION OF μ USING (4-4)	CALCULATION OF σ^2 USING (4-5)			EASIER CALCULATION OF σ^2 USING (4-6), MULTIPLY 1ST AND 3RD COLUMNS
x	p(x)	xp(x)	$(x - \mu)$	$(x - \mu)^2$	$(x - \mu)^2 p(x)$	$x^2 p(x)$
0	.14	0	-1.44	2.07	.29	0
1	.39	.39	$-.44$	.19	.08	.39
2	.36	.72	.56	.31	.11	1.44
3	.11	.33	1.56	2.43	.27	.99
		$\mu = 1.44$			$\sigma^2 = .75$	$\Sigma x^2 p(x) = 2.82$ from 3rd column, $\mu^2 = 2.07$ Difference $\sigma^2 = .75\,\checkmark$

Remarks   To calculate σ^2, the alternative formula (4-6) was easier. It required just the last column of Table 4-2, instead of the previous three columns required by the definition (4-5).

  Also note that $\sigma = \sqrt{.75} = .87$ is indeed the typical deviation, lying between the largest deviation 1.56 and smallest deviation .44 (found in the fourth column).

  Since the definitions of μ and σ are similar to those of $\overline{X}$ and s, we find similar interpretations. We continue to think of the mean μ as the balancing point—a weighted average using probability weights rather than relative frequency weights. And the standard deviation σ is the typical deviation.

  We emphasize that the distinction between sample and population moments must not be forgotten: μ is called the population mean since it is based on the population of all possible repetitions of the experiment; on the other hand, we call $\overline{X}$ the sample mean since it is based on a mere sample drawn from the parent population.

PROBLEMS

4-6 Compute μ and σ for each of the following distributions (from Problems 4-4 and 4-5). Graph each distribution, showing the mean as the balancing point, and the standard deviation as a typical deviation (like Figure 2-6).

a.		b.	
X	p(x)	X	p(x)
0	.01	0	.07
1	.09	1	.10
2	.39	2	.18
3	.51	3	.65

4-7 On the basis of past experience, the buyer for a large sports store estimates that the number of 10-speed bicycles sold next year (demand) will be somewhere between 40 and 90—with the following distribution:

DEMAND	PROBABILITY
X	p(x)
40	.05
50	.15
60	.41
70	.34
80	.04
90	.01

a. What is the expected (mean) demand? What is the standard deviation?

b. If 60 are ordered, what is the chance they will all be sold? What is the chance some will be left over (undesired inventory)?

c. To be almost sure (95%) of having enough bicycles, how many should be ordered?

*4-8 In planning a huge outdoor concert for June 16, the producer estimates the attendance will depend on the weather according to the following table. He also finds out from the local weather office what the weather has been like, for June days in the past 10 years.

WEATHER	ATTENDANCE	RELATIVE FREQUENCY
wet, cold	5,000	.20
wet, warm	20,000	.20
dry, cold	30,000	.10
dry, warm	50,000	.50

a. What is the expected (mean) attendance?

b. The tickets will sell for $9 each. The costs will be $2 per person for cleaning and crowd-control, plus $150,000 for the band, plus $60,000 for administration (including the facilities). Would you advise the producer to go ahead with the concert, or not? Why?

*4-9 In Problem 4-8, suppose the producer has gone ahead with his plans, and on June 10 has obtained some rather gloomy long-run weather forecasts: The 4 weather conditions now have probabilities .30, .20, .20, and .30, respectively.

If he cancels the concert, he will still have to pay half of the $60,000 administration cost, plus a $15,000 cancellation penalty to the band.

Would you advise him to cancel, or not?

4-3 The Binomial Distribution

There are many types of discrete random variables, and the commonest is called the *binomial*. The classical example of a binomial variable is:

$$S = \text{number of heads in several tosses of a coin.}$$

There are many random variables of this binomial type, a few of which are listed in Table 4-3. (We have already encountered not only the coin-tossing example, but also another: the number of girls in a family of three children.) To handle all such cases, it will be helpful to state the basic assumptions in general notation:

1. We suppose there are n *trials* (tosses of the coin).
2. In each trial, a certain event of interest can occur, or fail to occur;

TABLE **4-3** **Examples of Binomial Variables**

TRIAL	"SUCCESS"	"FAILURE"	π	n	S
Tossing a fair coin	Head	Tail	1/2	Number of tosses	Number of heads altogether
Birth of a child in a family	Girl	Boy	$.48 \simeq 1/2$	Number of children	Number of girls in the family
Pure guessing on a multiple choice question (with 5 choices, say)	Correct	Wrong	1/5	Number of questions on exam	Number of correct answers
Drawing a voter at random in a poll	Republican	Democrat or other non-Republican	Proportion of Republicans in the population	Sample size of voters	Number of Republicans in the sample
Drawing a woman at random in a fertility survey	Pregnant	Not pregnant	Pregnancy rate in the population of women	Sample size	Number of pregnant women in the sample

then we say a *success* (head) or *failure* (tail) has occurred. Their respective probabilities are denoted by π and $(1 - \pi)$, and these do not change from trial to trial.

3. Finally, we assume the trials are *statistically independent*.

Then S, the total number of successes in n trials, is called a binomial variable. A formula for its probability distribution p(s) can be easily derived—as we already found in Figure 4-1: When each child has a .48 chance of being a girl, the probability of exactly 1 girl in a family of 3 children was:

$$p(1) = 3(.48)^1(.52)^2 = .39$$

(4-7)

like (4-1) and (3-4)

The general binomial formula is similar (a formal derivation is given in Appendix 4-3): When each trial has probability π of success, the probability of exactly s successes in n trials is

$$\boxed{\begin{array}{c} \textbf{Binomial distribution} \\ p(s) = \dbinom{n}{s} \pi^s (1 - \pi)^{n-s} \end{array}}$$

(4-8)

The *binomial coefficient* $\binom{n}{s}$ is defined by[1]

$$\binom{n}{s} \equiv \frac{n!}{s!(n-s)!}$$ (4-9)

where, in turn, the *factorial* n! is given by

$$n! \equiv n(n-1)(n-2)\cdots 1$$ (4-10)

For example, the binomial coefficient $\binom{3}{1}$ is:

$$\binom{3}{1} = \frac{3!}{1!2!}$$
$$= \frac{3 \cdot 2 \cdot 1}{(1)(2 \cdot 1)} = 3$$

We can confirm that the general binomial formula does indeed cover the probability of exactly 1 girl in (4-7) as a special case. We simply substitute $n = 3$, $\pi = .48$, and $s = 1$ into (4-8):

$$p(1) = \binom{3}{1}(.48)^1(.52)^2$$
$$= 3(.13) = .39 \qquad \text{(4-7) verified}$$

In using (4-8), we must emphasize the most important assumption: The binomial distribution is appropriate only if the trials are *independent*. As an example of independence, knowing that the first toss of a coin has come up heads will not affect the probability on the second toss, if the coin is properly tossed.

However, if the coin is poorly tossed (e.g., tossed so that it is likely to turn over just once), then knowing that the first toss is heads will reduce the probability of heads on the second toss; this is an example of dependence.

As another example of independence, suppose you are drawing two cards from a deck and you *replace* the first before drawing the second. Then the chance of, say, an ace on the second draw is 4/52, independent of whatever the card on the first draw happened to be.

On the other hand, if you *keep out* the first card, your chances on the second draw are altered, since you can't get that first card again. Thus if

---

[1] The special cases $\binom{n}{n}$ and $\binom{n}{0}$ are not covered by (4-9); but are independently defined to equal 1. This enables (4-8) to give the correct answer even when $s = n$ or $s = 0$.

the first card is an ace, the chance of the second card being an ace is reduced to 3/51; if the first card is a non-ace, the chance of the second card being an ace is increased to 4/51. This is another example of dependence, where the bionomial (4-8) cannot be applied.

EXAMPLE **4-2**   In a family of 8 children, what is the probability of getting exactly 3 girls? First list the assumptions you find it necessary to make.

Solution   1. We assume parents do not have any control over the sex of their children. (For example, they have no means of increasing the probability of a girl, even after a run of 7 boys). That is, each birth is a repeated random event like the toss of a coin. This assumption assures statistical independence, so that the binomial formula may be applied.

2. We also assume that girls and boys are equally likely, so that the probability of a "success" (girl) is $\pi = .50$. (And unless stated otherwise, it is the assumption we will continue to make throughout the book.) This assumption keeps computations simple, and is accurate enough for most purposes.

The probability of exactly 3 girls, then, is found by substituting $n = 8$, $s = 3$, and $\pi = .5$ into (4-8):

$$p(3) = \binom{8}{3} .5^3 \, .5^5$$

$$= \frac{8 \cdot 7 \cdot 6 \cdot 5 \cdot 4 \cdot 3 \cdot 2 \cdot 1}{(3 \cdot 2 \cdot 1)(5 \cdot 4 \cdot 3 \cdot 2 \cdot 1)} .5^3 \, .5^5$$

$$= 56(.5)^8 = .219 \qquad\qquad (4\text{-}11)$$

In practice, it is a nuisance calculating binomial coefficients and binomial probabilities like this by hand. Why not let the computer do them once and for all? In Appendix Table IIIa we give the printout for some binomial coefficients $\binom{n}{s}$. And in Appendix Table IIIb we give the printout for some binomial probabilities $p(s)$ computed from (4-8). With these, we could find the answer in Example 4-2 very easily. In Table IIIb we simply look up $n = 8$, $\pi = .50$, and $s = 3$. This immediately gives the answer .219, which confirms (4-11).

Another example will illustrate Table IIIb clearly.

EXAMPLE **4-3**   A sample of 5 voters is to be randomly drawn from a large population, 60% of whom are Democrats (as in the 1964 U.S. presidential election, for example, when Johnson defeated Goldwater).

a. The number of Democrats in the sample can vary anywhere from 0 to 5. Tabulate its probability distribution.
b. Calculate the mean and standard deviation.
c. What is the probability of exactly 3 Democrats in the sample?
d. Calculate the probability that the sample will correctly reflect the population, in having a majority of Democrats. That is, calculate the probability of at least 3 Democrats in the sample.
e. Graph your answers above.

Solution

a. Each voter that is drawn constitutes a trial. On each such trial, the probability of a Democratic voter (success) is $\pi = .60$. In a total of $n = 5$ trials, we want the probability of S successes, where $S = 0, 1, \ldots , 5$. So we simply look up Appendix Table IIIb and copy down the distribution for $n = 5$ and $\pi = .60$.

(a) DISTRIBUTION		(b) MEAN	VARIANCE		
s	$p(s)$	$sp(s)$	$s - \mu$	$(s - \mu)^2$	$(s - \mu)^2 p(s)$
0	.010	0	-3	9	.090
1	.077	.077	-2	4	.308
2	.230	.460	-1	1	.230
3	.346	1.038	0	0	0
4	.259	1.036	1	1	.259
5	.078	.390	2	4	.312
	1.000	$\mu = 3.00$			$\sigma^2 = 1.20$
					$\sigma = 1.10$

c. From the table, we find:

$$p(3) = .346 \simeq 35\%$$

d. From the table, we add up:

$$p(3) + p(4) + p(5) = .346 + .259 + .078$$
$$= .683 \simeq 68\% \qquad (4\text{-}12)$$

e.

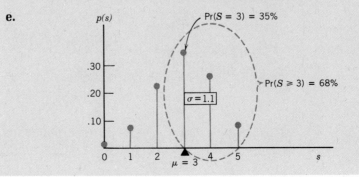

In practice, we often want to add up a string of probabilities as in Example 4-3**d.** Again, why not let the computer do it once and for all? In Appendix Table IIIc, we give the computer output for the sum of the probabilities in the right-hand tail—called the *cumulative* binomial probability. For instance, we could find the answer in Example 4-3**d** very easily. In Table IIIc we simply look up $n = 5$, $\pi = .60$, and $s_0 = 3$. This immediately gives the answer .683, which confirms (4-12).

A final example will illustrate some of the limitations of the bionomial.

EXAMPLE 4-4

a. Suppose a warship takes 10 shots at a target, and it takes at least 4 hits to sink it. If the warship has a record of hitting on 20% of its shots in the long run, what is the chance of sinking the target?

b. What crucial assumptions did you make in part (a)? Why might they be questionable?

c. To appreciate this crucial assumption, put yourself in the position of the captain of the British battleship *Prince of Wales* in 1941. The gunners on the German *Bismark* have just homed in on the British *Hood*, and sunk it after several shots. They now turn their fire on you, and after an initial miss they make a direct hit. Do you leave the probability they will hit you on the next shot unchanged, or do you revise it?

Solution

a. For $n = 10$ and $\pi = .20$, we look up $s_0 = 4$ in the cumulative binomial, Table IIIc. It immediately gives the answer, $.121 \simeq 12\%$.

b. Since naval gunners adjust each shot to allow for previous mistakes, their probability of hitting the target improves over time (rather than remaining a constant $\pi = .20$, say, as the binomial assumes). Also, once they succeed in hitting the target, the gunners will be more likely to stay on target thereafter; that is, the shots may be dependent (rather than independent, as the binomial assumes).

c. We agree with the captain of the *Prince of Wales*—he revised upward the probability that he would be hit on the next shot, and prudently broke off the action. (Other British ships picked up the trail of the *Bismark*, however, and sank it a few days later.)

This example leaves us with an important word of caution: The binomial is only valid for *independent* trials with a *constant* probability of success.

PROBLEMS

4-10 In families with 6 children, let X = the number of boys. For simplicity, assume that births are independent and boys and girls are equally likely.

    a. Graph the probability distribution of X.

    b. Calculate the mean and standard deviation, and show them on the graph.

    c. Of all families with 6 children, what proportion have:

        i. 3 boys and 3 girls?

        ii. 3 or more boys?

        iii. A boy as both the oldest and youngest child?

4-11 A salesperson has an 80% chance of making a sale when she calls on a certain drugstore (as in Problem 4-4). She makes 3 calls per year, and is interested in S = the total number of times she makes a sale in the next 3 years.

    a. Graph the distribution of S.

    b. What is the chance she will make a sale at least 5 times?

4-12 **a.** One hundred coins are spilled at random on the table, and the total number of heads S is counted. The distribution of S is binomial, with n = _____ and π = _____. Although this distribution would be tedious to tabulate, the *average* (mean) of S is easily guessed to be _____.

    b. Repeat part (a) for S = the number of aces when 30 dice are spilled at random.

    c. Repeat part (a) for S = the number of correct answers when 50 true–false questions are answered by pure guessing.

    d. Guess what the mean is for *any* binomial variable, in terms of n and π.

4-13 In Problem 4-12, you guessed the mean of a binomial variable. There is also a formula for the standard deviation (both formulas will be proved later, in Problem 6-23):

$$\boxed{\begin{array}{l} \text{Binomial mean: } \mu = n\,\pi \\ \text{Standard deviation: } \quad \sigma = \sqrt{n\pi(1-\pi)} \end{array}} \qquad (4\text{-}13)$$

Now that you understand these formulas, you can use them whenever it is convenient. For example:

    a. Verify μ and σ found in Problem 4-10**b**.

    b. Calculate μ and σ for Problem 4-11.

4-14 A multiple choice exam consists of 10 questions, each having 5 possible answers to choose among. To pass, a mark of 50% (5 out of 10 questions correct) is required. What is the chance of passing if:

a. You go into the exam without knowing a thing, and have to resort to pure guessing?

b. You have studied enough so that on each question, 3 choices can be eliminated. Then you have to make a pure guess between the remaining 2 choices.

4-15 **a.** In World War II, roughly 2% of the bombers were shot down in each raid. If a pilot flew on 50 raids, is he sure to be shot down? If so, why? If not, what is the correct probability that he would survive (assuming statistical independence of the 50 successive raids).

b. Do you think the assumption of independence in part **a** is roughly valid? If so, why? If not, why not—and what difference would it make to your answer?

4-4 Continuous Distributions

In Figure 2-2 we saw how a continuous variable such as men's height could be represented by a bar graph showing relative frequencies. This graph is reproduced in Figure 4-3a (with men's height now measured in feet rather than inches; furthermore, the Y-axis has been shrunk to the same scale as the X-axis). The sum of all the relative frequencies (i.e., the

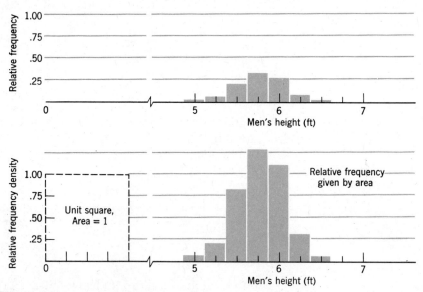

Figure 4-3
(a) Relative frequency histogram with the sum of the lengths of the bars = 1.
(b) Rescaled into relative frequency density, making the sum of the areas of the bars = 1.

sum of all the *lengths* of the bars) in Figure 4-3a is of course 1, as first noted in Table 2-2.

We now find it convenient to change the vertical scale to relative frequency *density* in panel (b), a rescaling that makes the total area (i.e., the sum of all the *areas* of the bars) equal to 1. We accomplish this by defining:

$$\text{Relative frequency density} \equiv \frac{\text{relative frequency}}{\text{cell width}} \tag{4-14}$$

$$= \frac{\text{relative frequency}}{1/4} \tag{4-15}$$
$$= 4 \text{ (relative frequency)}$$

Thus in Figure 4-3, panel (b) is 4 times as high as panel (a); we also see that panel (b) now has an area equal to 1.

In Figure 4-4 we show what happens to the relative frequency density of a continuous random variable as sample size increases. With a small sample, chance fluctuations influence the picture. But as sample size increases, chance is averaged out, and relative frequencies settle down to probabilities. At the same time, the increase in sample size allows a finer definition of cells. While the area remains fixed at 1, the relative frequency density becomes approximately a curve, the *probability density function*, $p(x)$, which we informally call the *density*, or the *probability distribution*. (To calculate probabilities and moments of continuous distributions, see Appendix 4-4.)

PROBLEMS

4-16 At a busy switchboard, the time t (in seconds) that elapsed between one incoming call and the next was recorded thousands of times. When graphed, the relative frequency density of t turned out as shown.

a. What is the approximate probability that t exceeds half a minute (30 seconds)?

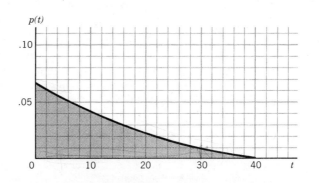

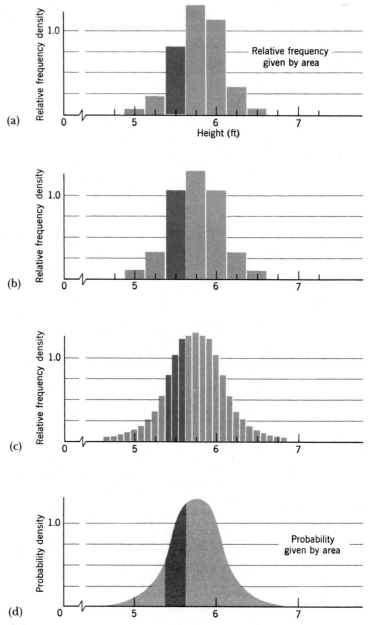

Figure 4-4
How relative frequency density may be approximated by a probability density as sample size increases, and cell size decreases. (a) Small n, as in Figure 4-3b. (b) Large enough n to stabilize relative frequencies. (c) Even larger n, to permit finer cells while keeping relative frequencies stable. (d) For very large n, this becomes (approximately) a smooth probability density curve.

b. Find the modal time and the median time. In relation to these two, where is the mean?

4-5 The Normal Distribution

For many random variables, the probability distribution is a specific bell-shaped curve, called the *normal* curve, or *Gaussian* curve (in honor of the great German mathematician Karl Friedrick Gauss, 1777–1855). This is the most common and useful distribution in statistics. For example, errors made in measuring physical and economic phenomena often are normally distributed. In addition, many other probability distributions (such as the binomial) often can be approximated by the normal curve.

A—STANDARD NORMAL DISTRIBUTION

The simplest of the normal distributions is the *standard normal* distribution shown in Figure 4-5, and discussed in detail in part c. Called the Z distribution, it is distributed around a mean $\mu = 0$ with a standard deviation $\sigma = 1$. Thus, for example, the value $Z = 1.5$ is one-and-a-half standard deviations above the mean, and in general:

> Each Z value is *the number of standard deviations away from the mean.* $\qquad$ (4-16)

We often want to calculate the probability (i.e., the area under the curve) beyond a given value of Z, like the value $Z = 1.5$ in Figure 4-5. This, and all other such tail probabilities, have already been calculated by statisticians and are set out in Appendix Table IV (which is very similar to the binomial tail probabilities in Table III).[2]

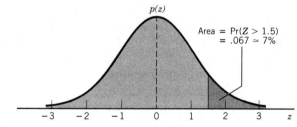

Figure 4-5
Standard normal distribution, illustrating the tabulated probability beyond a given point.

---

[2]There is one new feature, however, in looking up probabilities for a continuous variable: The probability of a single point is zero (since the "area" above a single point reduces to a line of zero width). It therefore makes no difference whether we include or exclude a single point in calculating a probability:

$$\text{If } X \text{ is continuous, } \Pr(X \geq c) = \Pr(X > c) \qquad (4\text{-}17)$$

In other words, $\geq$ and $>$ can be used interchangeably for any continuous random variable, including the standard normal Z.

EXAMPLE **4-5**   If Z has a standard normal distribution, find:

    a. $\Pr(Z > 1.64)$
    b. $\Pr(Z < -1.64)$
    c. $\Pr(1.0 < Z < 1.5)$
    d. $\Pr(-1 < Z < 2)$
    e. $\Pr(-2 < Z < 2)$

Solution   Since the normal probabilities in Appendix Table IV are so useful, we record them also inside the front cover where they are easy to find. Using them, we calculate each probability and illustrate it in the corresponding panel of Figure 4-6.

    a. $\Pr(Z > 1.64) = .051 \simeq 5\%$

    b.   By symmetry,

$$\Pr(Z < -1.64) = \Pr(Z > 1.64) \tag{4-18}$$
$$= .051 \simeq 5\%$$

    c.   Take the probability above 1.0, and subtract from it the probability above 1.5:

$$\Pr(1.0 < Z < 1.5) = \Pr(Z > 1.0) - \Pr(Z > 1.5)$$
$$= .159 - .067$$
$$= .092 \simeq 9\% \tag{4-19}$$

    d.   Subtract the two tail areas from the total area of 1:

$$\Pr(-1 < Z < 2) = 1 - \Pr(Z < -1) - \Pr(Z > 2)$$
$$= 1 - .159 - .023 \tag{4-20}$$
$$= .818 \simeq 82\%$$

    e.

$$\Pr(-2 < Z < 2) = 1 - \Pr(Z < -2) - \Pr(Z > 2)$$
$$= 1 - 2(.023)$$
$$= .954 \simeq 95\%$$

B—GENERAL NORMAL DISTRIBUTION

So far we have considered only a very special normal distribution—the standard normal Z with mean zero and standard deviation 1. But in general, a normal distribution may have any mean μ, and any standard deviation σ.

    For example, when the population of American men have their height X arrayed into a frequency distribution, it looks about like Figure 4-7—a normal distribution with mean $\mu = 69$ inches, and standard deviation $\sigma = 3$ inches. At the bottom of the figure, we lay out the standard deviation

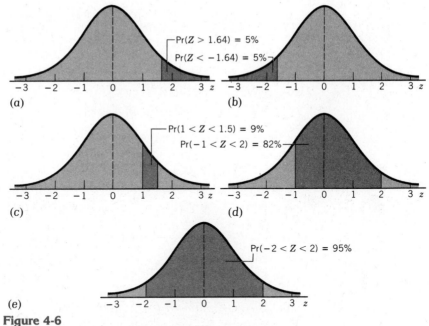

Figure 4-6
Standard normal probabilities illustrated

as a "yardstick," which shows how many standard deviations from the mean any given height may be—the standardized score Z.

How could we calculate the proportion of men above 6'2" (74 inches), for example? Figure 4-7 shows us at a glance that a height of 74 inches is nearly 2 standard deviations above the mean. To calculate exactly how far—the Z score—we proceed in two easy steps:

1. The critical value X = 74 differs from its mean μ = 69 by:

$$\text{Deviation} = 74 - 69 = 5 \text{ inches} \tag{4-21}$$

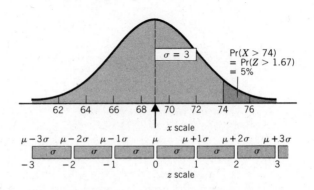

Figure 4-7
Standardization: A general normal variable (men's heights) rescaled to a standard normal.

2. How many standard deviations is this? Since there are 3 inches in each standard deviation, the deviation of 5 inches represents:

$$Z = \frac{5}{3} = 1.67 \text{ standard deviations} \tag{4-22}$$

Finally we can refer this to Table IV to get the required probability:

$$\Pr(Z > 1.67) = .047 \simeq 5\%$$

That is, the proportion is about 5%, which is graphed in Figure 4-7.

To formalize how we derived the Z value of 1.67 in (4-22), we first calculated the deviation $(X - \mu)$, and then compared it to the standard deviation σ:

$$\boxed{Z = \frac{X - \mu}{\sigma}} \tag{4-23}$$

That is, the Z value gives the number of standard deviations away from the mean—as we first saw in (4-16).

EXAMPLE **4-6**   Suppose that scores on an aptitude test are normally distributed about a mean $\mu = 60$ with a standard deviation $\sigma = 20$. What proportion of the scores:

Solution   a. Exceed 85?

b. Fall below 50?

a. As shown in Figure 4-8, we first must *standardize* the score $X = 85$ (that is, express it as a standard Z value):

$$Z = \frac{X - \mu}{\sigma} \qquad \text{(4-23) repeated}$$

$$= \frac{85 - 60}{20} = \frac{25}{20} = 1.25$$

Thus

$$\Pr(X > 85) = \Pr(Z > 1.25)$$
$$= .106 \simeq 11\%$$

b.

$$Z = \frac{X - \mu}{\sigma} \qquad \text{(4-23) repeated}$$

$$= \frac{50 - 60}{20} = \frac{-10}{20} = -.50$$

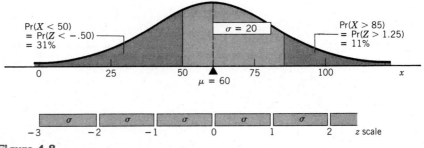

Figure 4-8
General normal rescaled to a standard normal.

Thus $\qquad Pr(X < 50) = Pr(Z < -.50)$
By symmetry, $\qquad\qquad\qquad = Pr(Z > +.50)$
$\qquad\qquad\qquad\qquad\qquad = .309 \simeq 31\%$ $\qquad$ (4-24)

As these examples have shown, we can always calculate probabilities for a normal distribution, by first finding the standard Z value.

*C—FORMULAS AND GRAPHS

In this optional section, we will carefully lay out some of the details of the normal distribution. Let us start with a precise definition. The standard normal, for example, is the distribution that has the probability function (density):

$$p(z) = \frac{1}{\sqrt{2\pi}} e^{-\frac{1}{2}z^2} \qquad (4-25)$$

The constant $1/\sqrt{2\pi}$ is a scale factor required to make the total area 1. The symbols π and e denote important mathematical constants, approximately 3.14 and 2.72, respectively.

The graph of (4-25) is shown in Figure 4-9, and has the following features:

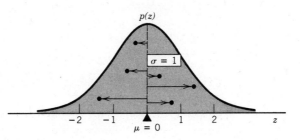

Figure 4-9
The graph of the standard normal Z. The mean (balancing point) is 0, and the standard deviation (typical deviation) is 1.

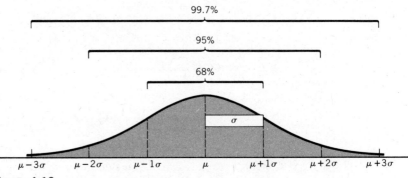

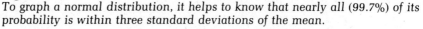

Figure 4-10
To graph a normal distribution, it helps to know that nearly all (99.7%) of its probability is within three standard deviations of the mean.

1. As we move to the left or right of 0, z^2 increases in the negative exponent. Therefore, p(z) decreases, approaching 0 symmetrically in both tails.
2. The mean is 0—the balancing point is the center of symmetry.
3. The standard deviation is 1. This can be rigorously proved by advanced calculus. Or, the intuitive reason is shown in Figure 4-9. Six typical values of Z are shown as dots. Their deviations from the mean are sometimes less than 1, sometimes more than 1. Thus 1 seems to be the typical deviation, or standard deviation.

For occasional theoretical purposes, we also need the formula for the *general* normal distribution:

$$p(x) = \frac{1}{\sqrt{2\pi}\sigma} e^{-\frac{1}{2}\left(\frac{x-\mu}{\sigma}\right)^2} \tag{4-26}$$

In the exponent we find the familiar $z = (x - \mu)/\sigma$ as in (4-23). The shape of this distribution is therefore the same bell shape as the standard normal (except for being spread out by the factor σ).

To graph the normal distribution, it helps to know that nearly all of the area or probability lies within 3 standard deviations of the mean. (By looking up z = 3 in the standard normal table, we find the probability is 99.7%.) This is shown, along with several other areas, in Figure 4-10.

PROBLEMS

Recall that Table IV is repeated inside the front cover where it is easy to find.

4-17 If Z is a standard normal variable, calculate:

(a) $\Pr(Z > 1.60)$ (e) $\Pr(0 < Z < 1.96)$

(b) $\Pr(1.6 < Z < 2.3)$ (f) $\Pr(-1.96 < Z < 1.96)$

(c) $\Pr(Z < 1.64)$ (g) $\Pr(-1.50 < Z < .67)$

(d) $\Pr(-1.64 < Z < -1.02)$ (h) $\Pr(Z < -2.50)$

4-18   **a.** How far above the mean of the Z distribution must we go so that only 1% of the probability remains in the right-hand tail? That is:

$$\text{If } \Pr(Z \geq z_0) = .01, \text{ what is } z_0?$$

Then what percentile is z_0 called?

b. How far on either side of the mean of the Z distribution must we go to include 95% of the probability? That is:

$$\text{If } \Pr(-z_0 < Z < z_0) = .95, \text{ what is } z_0?$$

c. What is the IQR (Interquartile Range) of the Z distribution?

4-19 The Swedish population of men's heights is approximately normally distributed with a mean of 69 inches and standard deviation of 3 inches. Find the proportion of the men who are:

a. Under 5 feet (60 inches).

b. Over 6 feet (72 inches).

c. Between 5 feet and 6 feet.

4-20 The mathematical Scholastic Aptitude Test (SAT, math) gives scores that range from 200 to 800. They are approximately normally distributed about a mean of 470, with a standard deviation of 120 (for the population of U.S. college-bound seniors in 1978).

Graph this distribution, and then calculate and illustrate the following:

a. What percentage of students scored between 500 and 600?

b. What score is the 75th percentile (upper quartile)?

c. What score is the 25th percentile (lower quartile)?

d. What is the interquartile range?

4-21 A builder requires his electrical subcontractor to finish the electrical wiring in a renovated office building within 28 days. The subcontractor feels he is most likely to finish in about 20 days, but because of unforeseen events, he guesses it might take him about 5 days more or less than this. More precisely, suppose the required time has a normal distribution with $\mu = 20$ and $\sigma = 5$.

a. To provide an incentive to finish on time, the builder specifies a $1000 penalty if the electrical work takes longer than 28 days. What is the chance of this?

b. To provide an even stronger incentive, the contractor specifies an additional $1000 penalty for each additional delay of 3 days. (For example, if the contractor finished on day 35, he would pay a penalty of $1000 three times: on days 28, 31, and 34.) What is the expected (mean) penalty?

4-22　**a.**　If X has mean 5 and standard deviation 10, what are the moments of:

　　　i.　$U \equiv 2X + 3$

　　　ii.　$V \equiv .4X - .7$

　　　iii.　$Z \equiv \dfrac{X - 5}{10}$

b.　If X has mean μ and standard deviation σ, what are the moments of

$$Z = \frac{X - \mu}{\sigma}$$

c.　In part **b,** if X has a distribution that is normally shaped, what will be the shape of the distribution of Z?

4-6　A Function of a Random Variable

A—FUNCTIONS IN GENERAL

Consider again the planning of a family of three children; suppose that the annual cost of clothing (R) is a function of the number of girls (X) in the family, that is:

$$R = g(X)$$

Specifically, suppose that:

$$R = -X^2 + 3X + 5. \tag{4-27}$$

which is equally well given[3] by Table 4-4.

The values of R customarily are rearranged in order, as shown in the middle column of Table 4-5. Furthermore, the values of R have certain probabilities that may be deduced from the probabilities of X (just as the probabilities of X were deduced from the original sample space in Figure 4-1). For example, the probability that $R = 7$ is just the sum of the probabilities that $X = 1$ and $X = 2$ (.39 and .36 give a total of .75).

---

*[3]Why might annual clothing cost R (in $100) have a *parabolic* relation to X such as shown in Table 4-4? Perhaps certain savings are possible if all children are the same sex, so that R drops when X = 0 or 3.

TABLE **4-4** Tabled Form of
$$R = g(X) = -X^2 + 3X + 5$$

VALUE OF X	VALUE OF $R = g(X)$
0	5
1	7
2	7
3	5

TABLE **4-5** Mean of R Calculated by First Deriving the Probability Distribution of R

x	$p(x)$	$r = g(x)$	$p(r)$	$rp(r)$
0	.14	5	.25	1.25
1	.39	7	.75	5.25
2	.36			
3	.11			$\mu_R = 6.50$

What will be the average cost μ_R? From the distribution of R in Table 4-5, this is calculated in the last column, and is $\mu_R = 6.5$. However, often it is more convenient to bypass the distribution of R and calculate the mean of R directly from the distribution of X, as shown in Table 4-6.

It is easy to see why this works: In a disguised way, we are calculating μ_R in the same way as in Table 4-5. For example, the two rows for $X = 1$ and $X = 2$ in Table 4-6 appear condensed together as the single row for $R = 7$ in Table 4-5. Similarly, the other two rows in Table 4-6 correspond to the row for $R = 5$ in Table 4-5, so that both tables yield the same value for μ_R. The only difference, really, is that Table 4-6 is ordered according to X values, while Table 4-5 is ordered (and condensed) according to R values.

This example can easily be generalized. If X is a random variable and g is any function, then $R = g(X)$ is a random variable. μ_R may be calculated either from the probability function of R, or alternatively from the probability function of X according to:

TABLE **4-6** Mean of R Calculated Directly from $p(x)$, as an Easier Alternative to Table 4-5

x	$p(x)$	$r = g(x)$	$g(x)p(x)$
0	.14	5	.70
1	.39	7	2.73
2	.36	7	2.52
3	.11	5	.55
			$\mu_R = 6.50$

$$\mu_R = \Sigma\, g(x)p(x) \qquad\qquad (4\text{-}28)$$

An example will illustrate how useful this formula is.

EXAMPLE **4-7**   The wear on a railroad axle increases as the weight on it increases. Roughly speaking, a doubling of the weight quadruples the wear; and in general, if we denote wear by R and weight by X,

$$R = X^2$$

Now suppose a passenger car passes over a very irregular section of track, so that the weight on each axle fluctuates widely according to the following distribution, approximately:

WEIGHT X	PROPORTION OF TIME $p(x)$
8	.25
20	.70
80	.05

a. Calculate the average weight, μ_X.
b. Calculate the average wear, μ_R.
c. If the railroad made its rails perfectly smooth (so the weight on each axle was always exactly 20 tons), it would not only reduce the wear and tear on its passengers, it would reduce the wear and tear on its axles. By how much?

Solution

GIVEN		a. μ_X	b. μ_R	
x	$p(x)$	$xp(x)$	$g(x) = x^2$	$g(x)p(x)$
8	.25	2	64	16
20	.70	14	400	280
80	.05	4	6400	320
		$\mu_X = 20$		$\mu_R = 616$

c. With the weight a constant 20 tons on each axle, the wear would be $20^2 = 400$. This is 216 less than before—a substantial reduction, of 35%.

B—LINEAR FUNCTIONS

The simplest function is the linear transformation, $Y = a + bX$. It produces some very convenient formulas for the mean and variance of the new random variable Y (just as it did for the linear transformation of a sample in Chapter 2):

$$\boxed{\begin{aligned} \text{If } Y &= a + bx \\ \text{then } \mu_Y &= a + b\mu_X \\ \text{and } \sigma_Y &= |b|\sigma_X \end{aligned}}$$

(4-29)
like (2-19)

An example will illustrate.

EXAMPLE 4-8

In a family of 3 children, assume the probability of a girl on each birth is 48%. Then suppose the cost of sports equipment Y (in \$100, annually) depends on the number of girls X according to a linear relation:

$$Y = 2 + .4X \qquad (4\text{-}30)$$

Calculate the mean cost μ_Y in two ways:

a. The easy way, using (4-29).
b. To confirm, calculate μ_Y from the general formula (4-28).

Solution

a. First, the mean of X can be found from the binomial formula:

$$\mu_X = n\pi \qquad \text{(4-13) repeated}$$
$$= 3(.48) = 1.44$$

Then
$$\mu_Y = a + b\mu_X \qquad \text{(4-29) repeated}$$
$$= 2 + .4(1.44)$$
$$= 2.58 \text{ hundred dollars annually}$$

b. X has a binomial distribution, and from this we can calculate μ_Y:

x	$p(x)$	$y = g(x)$	$g(x)p(x)$
0	.14	2.0	.280
1	.39	2.4	.936
2	.36	2.8	1.008
3	.11	3.2	.352
			$\mu_Y = 2.58$

This example has shown how simple it is to find the mean μ_Y of a linear transformation: We merely transform μ_X with the same formula.

To really appreciate this point, we might ask whether a nonlinear transformation works this easily. An example will give us the answer.

EXAMPLE 4-9

Consider the nonlinear cost function:

$$R = g(X) = -X^2 + 3X + 5 \qquad \text{(4-27) repeated}$$

Using Σ g(x)p(x) in Table 4-6, we found $\mu_R = 6.50$. Could we have obtained it more easily by transforming μ_X?

Solution   The mean of X was $\mu_X = 1.44$. When we transform this, we obtain:

$$g(1.44) = -(1.44)^2 + 3(1.44) + 5$$
$$= 7.25$$

But this does not give the correct answer, $\mu_R = 6.50$. Thus the short-cut does not work.

In conclusion, then, we see why statisticians prefer, if possible, to work with linear transformations: They are the only ones that permit the easy calculation of μ and σ, with a formula like (4-29).

C—NOTATION FOR THE MEAN

Since means play a key role in statistics, they have been calculated by all sorts of people, who sometimes use different names for the same concept. For example, geographers use the term "*mean annual rainfall*," teachers use the term "*average grade*," and gamblers and economists use the term "*expected gain*." For any random variable X, therefore, we must recognize that all the following terms have exactly the same mathematical meaning:

$$\mu_X = \text{mean of } X$$
$$= \text{average of } X$$
$$= \text{expected value of } X, \text{ or } E(X)$$

Thus we can rewrite (4-28) as:

$$E(R) = \sum g(x)p(x) \tag{4-31}$$

This expected value symbol E is often used, as a reminder that it represents a weighted sum (E looks like Σ).

Finally, since we recall that R was just an abbreviation for g(X), we can equally well write (4-31) as:

$$\boxed{E[g(X)] = \sum g(x)p(x)} \tag{4-32}$$

As an example of this notation, we may write:

$$E(X - \mu)^2 = \sum (x - \mu)^2 p(x) \tag{4-33}$$

That is, noting (4-5):

$$E(X - \mu)^2 = \sigma^2 \tag{4-34}$$

This emphasizes that σ^2 may be regarded as just a kind of expected value—the expected squared deviation [like the mean squared deviation in (2-10)]. In this new expected value notation, we also may rewrite (4-6) as:

$$\sigma^2 = E(X^2) - \mu^2 \tag{4-35}$$

Sometimes we find it useful to solve this for $E(X^2)$:

$$E(X^2) = \mu^2 + \sigma^2 \tag{4-36}$$

The E notation is so useful that we shall continue to use it throughout the book.

PROBLEMS

4-23 **a.** A certain game depends on the number of dots X when a die is rolled. Tabulate the distribution of X, and calculate the average $E(X)$.

b. The payoff in this game is a linear function, $Y = 2X + 8$. Tabulate the distribution of Y, and from it calculate the expected payoff $E(Y)$.

c. Is it true that the means satisfy the same relation as the individual values; that is:

$$\text{If } Y = 2X + 8$$
$$\text{then does } E(Y) = 2E(X) + 8?$$

d. Graph the distributions of X and of Y, illustrating what you found in part **c.**

4-24 Repeat Problem 4-23 for the nonlinear function $Y = X^2$.

4-25 The time T, in seconds, required for a rat to run a maze, is a random variable with the following probability distribution

t	2	3	4	5	6	7
$p(t)$	.1	.1	.3	.2	.2	.1

a. Find the average time, and variance.

b. Suppose that the rat is rewarded with 1 biscuit for each second faster than 6. (For example, if he takes just 4 seconds, he gets a reward of 2 biscuits. Of course, if he takes 6 seconds or longer, he gets no reward.) What is the rat's average reward?

c. Suppose that the rat is punished by getting a shock that increases sharply as his time increases—specifically, a shock of T^2 volts for a time of T seconds. What is the rat's average punishment? Calculate two ways:

   i. using (4-32)

   ii. using (4-36).

4-26 The owner of a small motel with 10 units has 3 TVs that he will install for the night on request for a small fee. Long-run past experience shows that only 20% of his guests request a TV, so he hopes his supply of 3 TVs will ordinarily be adequate.

Considering just those nights when all 10 units of the motel are rented:

a. On what proportion of the nights will the requests exceed the available supply of 3 TVs?

b. What is the average number of TVs requested?

*c. What is the average number of TVs actually rented?

*d. What is the average revenue, if the owner charges $3 per night for each TV rented?

e. What assuptions did you make?

CHAPTER 4 SUMMARY

4-1 A random variable has a probability distribution that can often be derived from the original sample space of elementary outcomes.

4-2 The mean μ and standard deviation σ are defined for a probability distribution in the same way as $\overline{X}$ and s were defined for a relative frequency distribution in Chapter 2. The Greek letters emphasize their theoretical (ideal) nature.

4-3 When an experiment consists of *independent* trials with the *same* chance π of success on each trial, the total number of successes S is called a binomial variable. Its probability distribution can be easily found from a formula, or Table IIIb.

4-4 For random variables that vary continuously, probabilities are given by areas under a continuous distribution (probability density curve).

4-5 The commonest continuous distribution is the bell-shaped normal (Gaussian) distribution. To measure the number of standard deviations from the mean, we calculate:

$$Z = \frac{X - \mu}{\sigma}$$

We can then use this calculated value of Z to read off probabilities from the standard normal table (Table IV, inside the front cover).

4-6   The mean has several common aliases—average, expected value, or $E(X)$. Whatever the name, they all have exactly the same meaning. A new random variable obtained from X, let us say $R = g(X)$, can have its expected value easily calculated from the distribution $p(x)$:

$$E[g(X)] = \Sigma\, g(x)p(x)$$

REVIEW PROBLEMS

4-27   IQ scores are distributed approximately normally, with a mean of 100 and standard deviation of 15.

   a.   What proportion of IQs are over 120?

   b.   Suppose Fred Miller has an IQ of 120. What percentile is this?

   c.   Suppose Mary Tibbs has an IQ of 135. What percentile is this?

4-28   In the 1964 U.S. presidential election, 60% voted Democratic and 40% voted Republican. Calculate the probability that a random sample would correctly forecast the election winner—that is, that a majority of the sample would be Democrats, if the sample size were:

   a. $n = 1$

   b. $n = 3$

   c. $n = 9$

   Note how the larger sample increases the probability of a correct forecast.

4-29   The following table classifies the 31 million American women ever married, according to family size, in the 1970 census. (At the end of the table, 17% of the women had more than 4 children, and they are counted as having 6 children—a roughly representative figure.)

Children Born to AmericanWomen
(ever married, aged 45 or over, 1970 Census)

FAMILY SIZE (CHILDREN)	RELATIVE FREQUENCY (%)
0	16
1	17
2	24
3	16
4	10
6	17
Total	100%

   a.   Find:

   i.   Mean family size.

   ii.   Median family size.

   iii.   Modal family size.

    iv. Expected family size.

    v. Standard deviation of family size.

    To determine population growth, which of these measures is most appropriate?

b. Graph the relative frequency distribution. Mark the mean as the balancing point. Show the standard deviation as the unit of scale as in Figure 2-6.

c. What family size is commonest? What proportion of the families are this size? How many million families is this?

d. What proportion of the families have 4 or more children? Less than 4?

4-30 To see how useful means can be, let us consider some questions based on Problem 4-29.

  a. How many Americans would there be a generation later if American women continued to reproduce at the rate shown in the given table? The answer depends not only on the mean figure of 2.55; it also depends on many other factors such as whether unmarried women reproduce at the same rate, what proportion of children live to maturity, and what proportion of babies are female.

    A more meaningful figure therefore is the mean number of surviving female babies born to all women. This average (with a few minor changes, described in Haupt and Kane, 1978) is called the net reproduction rate (NRR) and, to the extent that the NRR stays above 1.00, there will be long-run population growth.

    In America in 1970, the NRR rate was 1.17 and there were about 200 million people. If we ignore immigration, changes in life expectancy, and other changes in age distribution, and if we also assume the NRR stays at 1.17, we can get a *projection* of how the population would grow.

    What would be the projected population size one generation later? Two generations later? Ten generations later (about 250 years)? This is called *exponential growth*.

  b. In fact, the NRR does not stay constant at all. By 1975 it had dropped to an astonishing .83, less than half its postwar peak of 1.76 in 1957.

    Answer part **a** assuming the NRR stays at .83. This is called *exponential decline*.

  c. How useful do you think the projections are in parts **a** and **b**?

4-31 Hawaii contains 770,000 people, 60% of whom are Asian, 39% white, and 1% black. If a random sample of 7 persons is drawn:

  a. What is the chance that a majority will be Asians?

  b. What is the chance that none will be black?

  c. For the number of Asians in the sample, what is the expected value? And the standard deviation?

4-32 The time required to complete a college achievement test was found to be normally distributed, with a mean of 90 minutes and a standard deviation of 15 minutes.

   a. What proportion of the students will finish in 2 hours (120 minutes)?

   b. When should the test be terminated to allow just enough time for 90% of the students to complete the test?

4-33 A family has four smoke alarms in their home, all battery operated and working independently of each other. Each has a reliability of 90%—that is, has a 90% chance of working. If fire breaks out that engulfs all four in smoke, what is the chance that at least one of them will sound the alarm?

4-34 Suppose that, of all patients diagnosed with a rare tropical disease, 40% survive for a year. In a random sample of 8 such patients, what is the probability that 6 or more would survive?

4-35 A thermostat requires a small metal part, whose length is crucial: it must be between 11.8 and 12.2 mm long or the thermostat won't work. The supplier of this part is having difficulty staying within this tolerance. In fact, the last thousand he produced had the following distribution:

LENGTH X (CELL MIDPOINT)	FREQUENCY	xf	$(x-\bar{x})^2$	$(x-\bar{x})^2 f$
11.8	30	354		1.728
11.9	110	1309		2.156
12.0	400	4800		.64
12.1	360	4356		1.296
12.2	90	1098		2.304
12.3	10	123		.676
Total	1000	$\frac{12040}{1000} = 12.04$		

   a. Calculate the mean length and standard deviation. Show those on a graph of the distribution.

   b. Suppose the underlying distribution (without crude grouping) is normal, with the mean and standard deviation found in part **a.** What proportion of the output would be unacceptable? Show this on a graph of the normal distribution.

4-36 Mercury Mufflers guarantees its mufflers for 5 years. On each muffler they sell they make $8, out of which they must pay for possible replacements under the guarantee. Each time a muffler is replaced, it costs $40.

   Their muffler life is approximately normally distributed around a mean of 7 years, with a standard deviation of 3 years.

   a. What is the average profit per sale (net, after paying for replacements)?

   b. If they want their average profit per sale to be $3, to what time period should they reduce the guarantee?

4-37   (The *Sign Test*) Eight volunteers are to have their breathing capacity measured before and after a certain treatment, and recorded in a layout like
the following:

| PERSON | BREATHING CAPACITY | | |
	BEFORE	AFTER	IMPROVEMENT
H.J.	2250	2350	+ 100
K.L.	2300	2380	+ 80
M.M.	2830	2800	− 30
.	.	.	.
.	.	.	.
.	.	.	.

a.   Suppose that the treatment has no effect whatever, so that the "improvements" represent random fluctuations (resulting from measurement error or minor variation in a person's performance). Also assume
that measurement is so precise that an improvement of exactly zero
is never observed.

What is the probability that seven or more signs will be +?

b.   If it actually turned out that seven of the eight signs were +, would
you question the hypothesis in part **a** that the treatment has no effect
whatever?

4-38   At a small New England college, there are only 10 classes of statistics—
with the following distribution of class size:

b. $$\frac{10^2 \times .5 + 20^2 \times .3 + 90^2 \times .2}{90 \times .2 + 20 \times .3 + 10 .5}$$

CLASS SIZE	RELATIVE FREQUENCY
10	.50
20	.30
90	.20
Total	1.00

The student newspaper reported that the average statistics student faced
a class size "over 50." Alarmed, the Dean asked the statistics professors
to calculate their average class size, and they reported, "under 30." Who's
telling the truth? Or are there two truths? Specifically, calculate:

a.   What class size does the average professor have?

b.   What class size does the average student have? (Hint: Problem 2-33)

4-39   In the 1972 presidential election, 60% voted for Nixon (Republican). A
small random sample of 4 voters was selected, and the number who voted
Republican (R) turned out to be 3. This sampling experiment was repeated
fifty times; R turned out to be usually 2 or 3, but occasionally as extreme
as 4 or 1 or even 0. The results were arrayed in the following table:

R	FREQUENCY
0	1
1	1
2	9
3	25
4	14
	50

a.   Graph the relative frequency distribution of R, and calculate the mean and standard deviation.

b.   If the sampling experiment were repeated millions of times (not merely 50), what would be your answers to part **a**?

4-40 a.   The governing board of a small corporation consists of the chairman and 6 other members, with a majority vote among these seven deciding any given issue. Suppose that the chairman wants to pass a certain motion, but is not sure of its support. Suppose the other six members vote independently, each with probability 40% of voting for the motion. What is the chance that it will pass?

b.   If the chairman had two firm allies who were certain to vote for the motion, how would that improve the chances of its passing? [Assume that the other four members are the same as in part **a**.]

c.   Go back to the same model as in part **a**, except now assume that two board members are political allies of the chairman, and gather privately with him beforehand. All three agree to vote within this caucus to determine their majority position, and then go into the general meeting with a solid block of three votes to support that position. Does this help or hinder the chances of the chairman's motion passing? How much?

d.   Does this illustrate the motto "united we stand, divided we fall"? Why?

e.   Do you think this furthers the democratic process? Why?

Chapter

Two Random Variables

If you bet on a horse, that's gambling. If you bet you can make three spades, that's entertainment. If you 'bet cotton will go up three points, that's business. See the difference?

5-1 Distributions

This first section is a simple extension of the last two chapters. The main problem will be to recognize the old ideas behind the new names. We therefore give an outline in Table 5-1, both as an introduction and a review.

A—JOINT DISTRIBUTIONS

In the planning of three children, let us define two random variables:

$$X = \text{number of girls}$$
$$Y = \text{number of runs}$$

where a run is an unbroken string of children of the same sex. For example, $Y = 1$ for the outcome BBB, while $Y = 2$ for the outcome BBG.

Suppose that we are interested in the probability that a family would have 1 girl and 2 runs. As usual, we refer to the sample space of the experiment copied down from Figure 4-1 into Figure 5-1. These two events—$X = 1$ and $Y = 2$—are colored, and their intersection has probability:

$$\Pr(X = 1 \cap Y = 2) = .13 + .13 = .26 \tag{5-1}$$

which we denote $p(1, 2)$ for simplicity.

TABLE **5-1** **Outline of Section 5-1**

OLD IDEA		NEW TERMINOLOGY	
Pr(G ∩ H)	(3-9)	Joint distribution:	
applied to:			
Pr(X = 1 ∩ Y = 2)		p(1, 2)	
Pr(X = x ∩ Y = y) in general		p(x,y) in general	(5-2)
Event E is independent of F if:		Variable X is independent of Y if, for all x and y:	
Pr(E ∩ F) = Pr(E)Pr(F)	(3-21)	p(x,y) = p(x)p(y)	(5-6)

Similarly, we could compute p(0,1), p(0,2), p(0,3), p(1,2), . . . , obtaining the *joint (or bivariate) probability distribution* of X and Y. Actually, the easiest way to derive the joint distribution is to run down the columns of X and Y values, tabulating all this information line by line into the appropriate cell on the right of Figure 5-1—as shown by the two typical arrows.

The distribution p(x,y) may be graphed, as shown in Figure 5-2. Each probability can be represented by an appropriately sized dot as shown in panel (a), or by a height as shown in panel (b). [Note that the graph in panel (a) is very similar to the right-hand table in Figure 5-1 rotated 90° counterclockwise. This rotation is a nuisance, but cannot be avoided because the conventions for laying out tables and graphs are different.]

The formal definition of the joint probability distribution is:

$$p(x,y) \equiv Pr(X = x \cap Y = y) \tag{5-2}$$

The general case is illustrated in Figure 5-3. The events $X = 0$, $X = 1$, $X = 2, \ldots$, are shown schematically as a horizontal slicing. Similarly, the events $Y = 0$, $Y = 1, \ldots$, are shown as a vertical slicing. The inter-

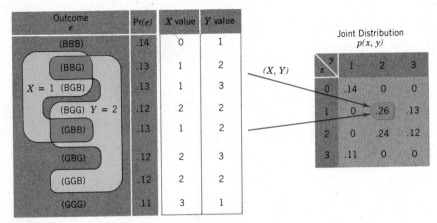

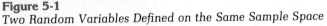

Figure 5-1
Two Random Variables Defined on the Same Sample Space

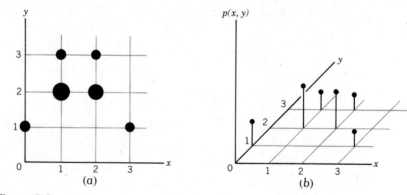

Figure 5-2

Two graphic presentations of the joint distribution in Table 5-2. (a) p(x,y) is represented by the size of the dot. (b) p(x,y) is represented by height.

section of the slice $X = x$ and the slice $Y = y$ is the event $(X = x \cap Y = y)$. Its probability is collected into the table, and denoted $p(x,y)$.

B—MARGINAL DISTRIBUTIONS

Suppose that we are interested only in X, yet have to work with the joint distribution of X and Y. Specifically, suppose we are interested in the event $X = 2$, which is a horizontal slice in the schematic sample space of Figure 5-3. Of course, its probability $p(2)$ is the sum of the probabilities of all those chunks comprising it:

$$p(2) = p(2, 0) + p(2, 1) + p(2, 2) + p(2, 3)$$
$$+ \cdots + p(2, y) + \cdots \qquad (5\text{-}3)$$
$$= \Sigma \, p(2, y)$$

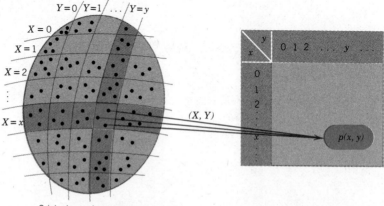

Figure 5-3

Two random variables (X, Y), showing their joint probability distribution derived from the original sample space (compare with Figure 5-1)

TABLE **5-2** **The Marginal Distributions Obtained by Summing Rows and Columns of the Joint Distribution**

x	y			p(x)
	1	2	3	
0	.14	0	0	.14
1	0	.26	.13	.39
2	0	.24	.12	.36
3	.11	0	0	.11
p(y)	.25	.50	.25	1.00√

and, in general, for any given x:

$$p(x) = \sum_y p(x,y) \qquad (5\text{-}4)$$

This idea may be applied to the distribution in Figure 5-1, which we repeat in Table 5-2. For example,

$$p(2) = 0 + .24 + .12 = .36 \qquad (5\text{-}5)$$

We place this row sum in the right-hand margin. When all the other row sums are likewise recorded in the margin, they make up the complete distribution p(x). This is sometimes called the *marginal* distribution of X, to describe how it was obtained. But, of course, it is just the ordinary distribution of X (which could have been found without any reference to Y, as indeed it was in Figure 4-1).

In conclusion, the word "marginal" merely describes how the distribution of X can be calculated from the joint distribution of X and another variable Y; row sums are calculated and placed "in the margin."

Of course, the distribution of Y can be calculated in a similar way. The sum of each column is placed in the bottom margin, giving us the marginal distribution p(y).[1]

C—INDEPENDENCE

Two random variables X and Y are called independent if the events (X = x) and (Y = y) are independent. That is,

---

[1]Strictly speaking, p(y) is not an adequate notation, and may cause ambiguity on occasion. For example, from Table 5-2 we can find any particular value for the marginal distribution of Y; when y = 2, for example, we find p(2) = .50. This seems to contradict (5-5), where we wrote p(2) = .36. (There, of course, we were referring to X, not Y, but the notation did not distinguish.) We could resolve this contradiction by more careful notation:

$$p_Y(2) = .50 \text{ whereas } p_X(2) = .36$$

Such subscripts are so awkward, however, that we will avoid them unless they are absolutely necessary.

$$\boxed{\begin{array}{c} X \text{ and } Y \text{ are independent if} \\ p(x,y) = p(x)p(y) \\ \text{for all } x \text{ and } y \end{array}} \qquad \begin{array}{l} (5\text{-}6) \\ \text{like } (3\text{-}21) \end{array}$$

For example, for the distribution in Table 5-2, are X and Y independent? For independence, (5-6) must hold for every (x,y) combination. We ask whether it holds, for example, when $x = 0$ and $y = 1$. In other words, is it true that:

$$p(0,1) \overset{?}{=} p(0)p(1)$$

The answer is no, because

$$.14 \neq (.14)(.25)$$

Thus X and Y fail to be independent; that is, they are *dependent*.
Another example will illustrate further.

EXAMPLE **5-1**   Suppose that X and Y have the following joint distribution $p(x,y)$:

x	10	20	30	40
20	.04	.08	.08	.05
40	.12	.24	.24	.15

(y across top)

Are X and Y independent?

Solution   To check the condition for independence (5-6), we first calculate the marginal distributions $p(x)$ and $p(y)$:

x	10	20	40	80	p(x)
20	.04	.08	.08	.05	.25
40	.12	.24	.24	.15	.75
p(y)	.16	.32	.32	.20	1.00✓

(y across top)

Then we systematically multiply $p(x)p(y)$, which is easiest to do in a table again:

Table of $p(x)p(y)$

x	10	20	40	80
20	(.25)(.16) = .04	(.25)(.32) = .08	(.25)(.32) = .08	(.25)(.20) = .05
40	(.75)(.16) = .12	(.75)(.32) = .24	(.75)(.32) = .24	(.75)(.20) = .15

(y across top)

Since this table of $p(x)p(y)$ agrees everywhere with the original table of $p(x,y)$, we have proved that (5-6) is true, and so X and Y are indeed independent.

In Example 5-1, an interesting pattern is evident: The rows are proportional (the second row is 3 times the first). The columns too are proportional (the second column is 2 times the first, etc.). And this pattern is generally true.

> Whenever X and Y are independent, then the rows of the table $p(x,y)$ will be proportional, and so will the columns.   (5-7)

PROBLEMS

5-1 Suppose a student's grade point average for grade 11 is denoted by X, and for grade 12 by Y. A population of 10,000 high school graduates applying to a certain college yielded the following bivariate distribution:

		y	
x	70	80	90
70	.10	.10	0
80	.10	.40	.10
90	0	.10	.10

  a. Calculate $p(x)$ and $p(y)$. Are they the same distribution?

  b. Are X and Y independent?

  c. Calculate $E(X)$ and $E(Y)$.

  d. Calculate the variance of X and of Y.

  e. Graph the joint distribution $p(x,y)$ showing each probability as an appropriately sized dot on the (x,y) plane (as in Figure 5-2a).

5-2 **a.** Continuing Problem 5-1, look at just those students who had $X = 70$. What is their distribution of Y marks? What is their average Y mark? [Call it $E(Y/X = 70)$.]

  b. Calculate also $E(Y/X = 80)$ and $E(Y/X = 90)$ and mark them all on the graph of $p(x,y)$. Describe in words what this shows.

  c. In view of part **b**, would you say Y is independent of X? Is this consistent with your answer in Problem 5-1**b**?

5-3 What we calculated in Problem 5-2 can alternatively be done with formulas. The conditional distribution of Y is defined as:

$$p(y/x) \equiv \frac{p(x,y)}{p(x)} \qquad \text{like (3-17)}$$

Use this to calculate $p(y/x)$ for $X = 70$. Does it agree with the previous answer in Problem 5-2a?

5-4 For which of the following joint distributions are X and Y independent?

a.

	y	
x	1	2
1	.10	.20
2	.30	.40

b.

	y		
x	0	1	2
0	.10	.20	.10
1	.15	.20	.25

c.

	y		
x	0	1	2
1	0	.1	.1
2	.1	.4	.1
3	.1	.1	0

d.

	y			
x	1	2	3	4
1	.12	.03	.06	.09
2	.20	.05	.10	.15
3	.08	.02	.04	.06

5-5 Using the table in Problem 5-4c, calculate and tabulate $p(x/y)$ for $Y = 1$. Is it the same as the unconditional distribution $p(x)$? What does this indicate about the independence of X and Y? Is this consistent with your earlier answer?

5-6 Repeat Problem 5-5, using the table in Problem 5-4d.

5-7 A salesman has an 80% chance of making a sale on each call. If three calls are made, let

$$X = \text{total number of sales}$$
$$Y = \text{total profit from the sales}$$

where the profit Y is calculated as follows: Any sales on the first two calls yield a profit of $100 each. By the time the third call is made, the original product has been replaced by a new product whose sale yields a profit of $200. Thus, for example, the sequence (sale, no sale, sale) would give $Y = \$300$.

List the sample space, and then:

a. Tabulate and graph the bivariate distribution.

b. Calculate the marginal distribution of X and of Y. Does the distribution of X agree with Problem 4-4?

c. What is the mean of X and of Y?

d. Are X and Y independent?

5-2 A Function of Two Random Variables

Once more let us consider the planning of a 3 child family. Suppose the cost of clothing R depends upon both the number of girls X and the number of runs Y; that is:

$$R = g(X,Y)$$

What will be the average cost, μ_R? The answer is very similar to what we worked out in Section 4-6; there are two alternatives:

a. We could tabulate the distribution of R, and then calculate

$$E(R) = \Sigma r p(r) \tag{5-8}$$

b. An alternative one-step method is to use the distribution $p(x,y)$ directly:

$$\boxed{E[g(X,Y)] = \sum_x \sum_y g(x,y)p(x,y)} \tag{5-9}$$

Note how similar this is to the earlier formula:

$$E[g(X)] = \sum_x g(x)p(x) \tag{4-32 repeated}$$

An example will illustrate that these two methods are equivalent.

EXAMPLE 5-2 In planning the 3 child family, suppose specifically that annual clothing costs are:

$$R = g(X,Y) = 10 + X + Y$$

Calculate the expected cost $E(R)$ in the two alternative ways:

a. Using (5-8).
b. Using (5-9).

Solution **a.** In Table 5-3 below, we first derive the distribution of R. (Compare to Table 4-5.) For example, the three arrows represent the three ways that $R = 10 + X + Y$ can equal 14: When X and Y are 3 and 1; or 2 and 2; or 1 and 3. Finally, on the extreme right,

TABLE 5-3 **Mean of R calculated by first deriving the distribution of R**

		y			DISTRIBUTION OF R		E(R)
x	1	2	3	$r = g(x,y)$ $= 10 + x + y$	$p(r)$	$rp(r)$
0	.14	0	0	11	.14	1.54
1	0	.26	.13	12	0	0
2	0	.24	.12	13	.26	3.38
3	.11	0	0	14	.48	6.72
				15	.12	1.80
					$E(R) =$	13.44

Mean of R calculated directly from $p(x,y)$

x	y		
	1	2	3
0	$(10 + 0 + 1).14 = 1.54$	0	0
1	0	$(10 + 1 + 2).26 = 3.38$	$(10 + 1 + 3).13 = 1.82$
2	0	$(10 + 2 + 2).24 = 3.36$	$(10 + 2 + 3).12 = 1.80$
3	$(10 + 3 + 1).11 = 1.54$	0	0

Overall Sum $= E(R) = 13.44$

we calculate the expected value as usual, weighting each value of r with its probability.

b. In Table 5-4 above, we calculate $E(R)$ directly by calculating $g(x,y)p(x,y)$ for every combination of x and y. (Compare to Table 4-6.) For example, in the first cell, when $x = 0$ and $y = 1$, we calculate $g(x,y) = 10 + x + y = 10 + 0 + 1$, and multiply by its probability $p(x,y) = .14$, to get 1.54. We sum these values in all the rows and columns (over all x and y) to finally obtain $E(R) = 13.44$. This agrees with the answer to part (a) as promised.

PROBLEMS

5-8 Suppose that:

$$R = XY$$
$$T = (X - 2)(Y - 2)$$
$$U = (2X - Y)^2$$
$$V = 4X + 2Y$$

and that X and Y have the following joint distribution:

x	y		
	0	2	4
0	.1	.1	0
2	.1	.4	.1
4	0	.1	.1

a. Find the mean of R, by first finding its distribution $p(r)$ and then calculating $\Sigma rp(r)$, as in (5-8).

b. Find the mean of R using $\Sigma\Sigma \, g(x,y) \, p(x,y)$, as in (5-9).

c. Find $E(T)$, $E(U)$, and $E(V)$ any way you like.

5-9 In a stand of 380 hardwood trees of marketable size, the diameter at waist height (D) and usable height (H) of each tree was measured (in feet). The

380 pairs of measurements were tallied in the following table, and then converted to relative frequency:

d	h 20	25
1.0	.16	.09
1.25	.15	.30
1.50	.03	.17
1.75	.00	.10

Since a tree trunk is roughly a cylinder, the volume V of wood that it contains is given approximately by:

$$V = .4D^2H$$

a. Calculate the average volume per tree $E(V)$.

b. Calculate the standard deviation of V.

c. What is the total volume in the whole stand of 380 trees?

d. Calculate the average diameter $E(D)$ and the average height $E(H)$. Is it true that $E(V) = .4[E(D)]^2[E(H)]$?

5-10 In a certain gambling game, a pair of dice have their faces renumbered: The sides marked 4, 5, and 6 are renumbered with 1, 2, and 3, so that each die can have only three outcomes—each with the same probability $2/6 = 1/3$.

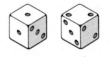

If the two dice are thrown independently, tabulate the joint distribution of:

$$X = \text{number on the first die}$$
$$Y = \text{number on the second die}$$

Then find the mean, variance, and standard deviation for each of the following:

a. X

b. Y

c. $S = X + Y$

d. Note that S is just the sum of X and Y. Is the mean of S similarly just the sum of the means? Is the variance just the sum of the variances? And is the standard deviation just the sum of the standard deviations?

5-11 Suppose the two dice in Problem 5-10 are differently renumbered, so that the probabilities for X and Y are as follows:

x	$p(x)$	y	$p(y)$
1	2/6	1	1/6
2	1/6	2	3/6
3	3/6	3	2/6

Answer the same questions as before.

5-3 Covariance

A—COVARIANCE IS LIKE VARIANCE

In this section we will develop the *covariance* to measure how two variables X and Y vary together, using the familiar concept of $E[g(X,Y)]$. (In fact, we have already calculated a covariance in Problem 5-8c. Now we will explain it.)

First recall our measure of how X alone varies: We started with the deviations $X - \mu$, squared them, and then took the expectation (average):

$$\text{Variance } \sigma_X^2 \equiv E(X - \mu)^2 \qquad \text{(4-34) repeated}$$

To measure how two variables X and Y vary together, we again start with the deviations, $X - \mu_X$ and $Y - \mu_Y$. We multiply them, and then take the expectation:

$$\boxed{\text{Covariance } \sigma_{XY} \equiv E(X - \mu_X)(Y - \mu_Y)} \qquad \text{(5-10)}$$

An example will illustrate the similarity of the variance and covariance concepts.

EXAMPLE **5-3**   Suppose a population of working couples has the bivariate distribution of income shown in Table 5-5.

TABLE **5-5**   **Joint Distribution of Husband's Income X and Wife's Income Y.**

x	y 10	20	30	40	$p(x)$
10	.20	.04	.01		.25
20	.10	.36	.09		.55
30		.05	.10		.15
40				.05	.05
$p(y)$	.30	.45	.20	.05	

a. Calculate the variance of X, and of Y.

b. Calculate the covariance of X and Y.

Solution a. The calculation of σ_X^2 is based on p(x)—the marginal distribution of X in the last column of Table 5-5 above. We first find $\mu_X = 20$, and then:

GIVEN		VARIANCE
x	p(x)	$(x - \mu)^2\ p(x)$
10	.25	$(-10)^2 .25 = 25$
20	.55	$0^2 .55 = 0$
30	.15	$(10)^2 .15 = 15$
40	.05	$(20)^2 .05 = 20$
		$\sigma_X^2 = 60$

Similarly, using the marginal distribution of Y along the bottom row of Table 5-5, we would find $\mu_Y = 20$, and then $\sigma_Y^2 = 70$. We have gone through the familiar calculations for σ^2 in this explicit form in order to clarify the calculation in part **b**.

b. The covariance calculation is much the same, except that we now use the whole bivariate distribution (rather than just the marginal distribution that we used in part **a**.) The calculations are shown in Table 5-6. For example, in the northeast cell, where x = 10 and y = 30, we calculate $(x - \mu_X)(y - \mu_X)p(x,y) = (10 - 20)(30 - 20)(.01) = -1$. We sum all such values, to finally obtain $\sigma_{XY} = 49$.

TABLE **5-6** **Sum of $(x - \mu_X)(y - \mu_Y)p(x,y)$ yields covariance σ_{XY}**

x	y 10	20	30	40
10	$(-10)(-10).20$ $= +20$	$(-10)(0).04$ $= 0$	$(-10)(10).01$ $= -1$	
20	$(0)(-10).10$ $= 0$	$(0)(0).35$ $= 0$	$(0)(10).09$ $= 0$	
30		$(10)(0).05$ $= 0$	$(10)(10).10$ $= +10$	
40				$(20)(20).05$ $= +20$

Overall sum = $\sigma_{XY} = 49$

The calculation of the covariance can often be simplified by using an alternative formula:

$$\sigma_{XY} = E(XY) - \mu_X\mu_Y$$

(5-11)
like (4-35)

B—THE COVARIANCE OF X AND Y INDICATES HOW THEY ARE RELATED

The mean levels of X and Y in Table 5-6 were shown by dotted lines, which break the table into four quadrants. When large values of X and of Y occur together (in the southeast quadrant), the deviations $(X - \mu_X)$ and $(Y - \mu_Y)$ are both positive, and so their product is positive. Similarly, when small values of X and of Y occur together (in the northwest quadrant), both deviations are negative, so that their product is positive in this case as well.

In Table 5-6, most of the probability is in these two quadrants. Therefore, the sum of all the calculated values—the covariance—is positive, and nicely summarizes the positive relation between the two variables (when one is large, the other tends to be large also).

But now consider the other two quadrants, where one deviation is negative and the other is positive—and hence their calculated product is negative. If most of the probability lies in these two quadrants, then the sum of all the calculated values—the covariance—would be negative, and would nicely summarize the negative relation between the two variables (when one is large, the other tends to be small).

Finally, what if all four corners of the table had substantial probabilities so that the positive and negative calculations cancelled each other out? The covariance σ_{XY} would be zero, and X and Y would be called *uncorrelated*. One way this can happen is for X and Y to be independent. Stated more formally (and proved in Appendix 5-3):

> If X and Y are independent, then they are uncorrelated ($\sigma_{XY} = 0$).   (5-12)

C—CORRELATION: THE COVARIANCE RESCALED

We have seen how the covariance nicely indicates whether X and Y have a positive, negative, or zero relation. Yet σ_{XY} can still be improved, because it depends upon the units in which X and Y are measured. If X, for example, were measured in inches instead of feet, each X-deviation in (5-10), and hence σ_{XY} itself, would unfortunately increase by 12 times. To eliminate this difficulty, we define the correlation ρ (rho, the Greek equivalent of r for relation):

$$\text{Correlation, } \rho \equiv \frac{\sigma_{XY}}{\sigma_X \sigma_Y} \qquad (5\text{-}13)$$

This does, in fact, work: Measuring X in terms of inches rather than feet still increases σ_{XY} in the numerator by 12 times, but this is exactly cancelled by the same 12 times increase in σ_X in the denominator. Similarly, (5-13) neutralizes any change in the scale of Y; thus correlation is completely independent of the scale in which either X or Y is measured.

Another reason that ρ is a very useful measure of the relation between X and Y is that it is always bounded:

$$-1 \le \rho \le +1 \qquad\qquad (5\text{-}14)$$

Now suppose X and Y have a perfect positive linear relation. For example, suppose they always take on the same values, so that $X = Y$. In this case, all the probabilities in Table 5-6 would lie along the northwest–southeast diagonal of the table (e.g., some probability that $X = 30$ and $Y = 30$, but no probability that $X = 30$ and $Y = 20$). Then it is easy to show that ρ takes on the limiting value of $+1$.[2] Similarly, if there is a perfect negative linear relation, then ρ would be -1. To illustrate these bounds, we can calculate ρ for the data in Example 5-3:

$$\rho = \frac{49}{\sqrt{60}\ \sqrt{70}} = .76$$

which is indeed less than 1.

PROBLEMS

5-12 For each of the following joint distributions, calculate σ_{XY} from the definition (5-10), and from the easier formula (5-11). Then calculate ρ_{XY}.

a.

x	y = 0	1	
0	.2	0	.2
1	.4	.2	.6
2	0	.2	.2
	.6	.4	1.00

b.

x	y = 0	5	10
2	0	.1	.1
4	.1	.4	.1
6	.1	.1	0

c.

x	y = 1	2	
0	.06	.04	.1
1	.30	.20	.5
2	.24	.16	.4
	.6	.4	1.00

d.

x	y = 1	2	3
0	1/8	0	0
1	0	2/8	1/8
2	0	2/8	1/8
3	1/8	0	0

5-13 **a.** In Problem 5-12, parts **c** and **d**, what is ρ_{XY}? Are X and Y independent?

b. Looking beyond these particular examples, which of the following statements are true for any X and Y?

---

[2]Since $X = Y$, (5-10) becomes
$$\sigma_{XY} = E(X - \mu_X)(X - \mu_X)$$
$$= E(X - \mu_X)^2 = \sigma_X^2$$
Also note that, since $X = Y$, $\sigma_X = \sigma_Y$. When we substitute all this into (5-13), ρ becomes $\sigma_X^2/\sigma_X\sigma_X = 1$.

(1)   If X and Y are independent, then they must be uncorrelated.

(2)   If X and Y are uncorrelated, then they must be independent.

5-14   To try to get some idea of what happiness is related to, Americans were asked in 1971 to rank themselves on a "happiness index" as follows: $H = 0$ (not happy), $H = 1$ (fairly happy), or $H = 2$ (very happy).

Annual household income X (in thousands of dollars) was also recorded for each individual. Then the relative frequencies of various combinations of H and X were roughly as follows (Gallup, 1971):

Joint distribution of happiness H and income X

x	h		
	0	1	2
2.5	.03	.12	.07
7.5	.02	.13	.11
12.5	.01	.13	.14
17.5	.01	.09	.14

a.   Calculate $E(H/X)$ for the various levels of X, and mark them on the graph of the bivariate distribution (as in Problem 5-2).

b.   Calculate the covariance σ_{XH} and correlation ρ_{XH}.

c.   Answer True or False; if False, correct it:

   i.   As X increases, the average level of H increases. This positive relation is reflected in a positive correlation ρ.

   ii.   Yet the relation is just a *tendency* (H fluctuates around its average level), so that ρ is less than 1.

   iii.   This shows that richer people tend to be happier than the poor. That is, money makes people happier. (If you think this doesn't necessarily follow, explain why.)

5-15   Continuing Problem 5-14, the amount of education was also measured for each individual: $X = 1$ (elementary school completed), $X = 2$ (high school completed), or $X = 3$ (college completed, or more). Thus $X = $ number of schools completed. Then the relative frequencies of various combinations were roughly as follows (Gallup, 1971):

Joint distribution of happiness H and education X

x	h		
	0	1	2
1	.02	.08	.05
2	.02	.28	.25
3	.01	.13	.16

Repeat the same questions as in Problem 5-14 (with education substituted for happiness, of course).

5-4 Linear Combination of Two Random Variables

Suppose that a couple is drawn at random from a certain population of working couples. In thousand dollar units, let:

$$X = \text{man's income}$$
$$Y = \text{woman's income}$$

Then the couple's total income is the *sum*:

$$S = X + Y$$

Or, suppose that the tax rates on X and Y are fixed at 20% and 30%, respectively; then the couple's income after taxes is a *weighted sum* (also called a *linear combination*):

$$W = .8X + .7Y \tag{5-15}$$

Of course, since X and Y are random variables, S and W are also. How can we find their moments easily?

A—MEAN

Continuing our example, suppose we know that:

$$E(X) = 12 \quad \text{and} \quad E(Y) = 9$$

It is natural to guess that:

$$E(S) = 12 + 9 = 21$$

This guess is correct, because it is generally true that[3]:

---

[3]For a proof of (5-16), we start with the formula (5-9), noting that $g(X,Y) = (X + Y)$ in this case. Thus,

$$E(X + Y) = \sum_x \sum_y (x + y)p(x,y)$$
$$= \sum_x \sum_y xp(x,y) + \sum_x \sum_y yp(x,y)$$

Considering the first term, we may write it as:

$$\sum_x \sum_y xp(x,y) = \sum_x x \left[\sum_y p(x,y) \right]$$
$$= \sum_x x[p(x)] \qquad \text{noting (5-4)}$$
$$= E(X)$$

Similarly, the second term reduces to $E(Y)$, so that:

$$E(X + Y) = E(X) + E(Y) \qquad \text{(5-16) proved}$$

The proof for the weighted sum (5-17) is similar.

$$\boxed{E(X + Y) = E(X) + E(Y)} \tag{5-16}$$

(This is the same result that we showed in Problems 5-10 and 5-11 already.) Similarly, it is natural to guess that:

$$E(W) = .8(12) + .7(9) = 15.9$$

This guess also is correct, because it is generally true that:

$$\boxed{E(aX + bY) = aE(X) + bE(Y)} \tag{5-17}$$
$$\text{like (4-29)}$$

Statisticians often refer to this important property as the *additivity* or *linearity* of the expectation operator.

To appreciate the simplicity of (5-17), we should compare it to (5-9). Both formulas provide a way to calculate the expected value of a function of X and Y. However, (5-9) applies to *any* function of X and Y, whereas (5-17) is restricted to *linear* functions only. But when we are dealing with this restricted class of functions, (5-17) is much simpler. [Whereas evaluation of (5-9) involves working through the whole joint probability distribution of X and Y, (5-17) requires only the marginal distributions of X and Y.]

B—VARIANCE

The variance of a sum is a little more complicated than its mean[4]:

$$\boxed{\text{var}(X + Y) = \text{var } X + \text{var } Y + 2 \text{ cov } (X, Y)} \tag{5-18}$$

where var and cov are abbreviations for variance and covariance, of course. Similarly, for a weighted sum:

---

[4]For a proof of (5-18), it is time to simplify our proofs by using brief notation—that is, by using $E(W)$ rather than the awkward $\Sigma w p(w)$, or the even more awkward $\underset{x\,y}{\Sigma\Sigma}\, w(x,y)p(x,y)$. In this new notation, the variance of $S = X + Y$ is defined as:

$$\text{var } S \equiv E(S - \mu_S)^2 \qquad\qquad \text{like (4-34)}$$

Substituting S and μ_S:

$$\begin{aligned}
\text{var } S &= E[(X + Y) - (\mu_X + \mu_Y)]^2 \\
&= E[(X - \mu_X) + (Y - \mu_Y)]^2 \\
&= E[(X - \mu_X)^2 + 2(X - \mu_X)(Y - \mu_Y) + (Y - \mu_Y)^2]
\end{aligned}$$

According to (5-17), this linear combination of several terms can be averaged term by term. Thus,

$$\begin{aligned}
\text{var } S &= E(X - \mu_X)^2 + 2E(X - \mu_X)(Y - \mu_Y) + E(Y - \mu_Y)^2 \\
&= \text{var } X + 2 \text{ cov } (X, Y) + \text{var } Y \qquad\qquad \text{(5-18) proved}
\end{aligned}$$

The proof of the weighted sum (5-19) is similar.

$$\boxed{\text{var}(aX + bY) = a^2 \text{ var } X + b^2 \text{ var } Y + 2ab \text{ cov }(X, Y)} \quad (5\text{-}19)$$

It is helpful to briefly sketch just why the covariance term appears in these two formulas. It is because variance is defined by *squared* deviations. And, as everyone knows:

$$(a + b)^2 = a^2 + b^2 + 2ab \quad (5\text{-}20)$$

A more intuitive reason for the covariance in (5-18) can be developed. Consider, for example, what happens if X and Y are positively related. When X is high, Y tends to be high as well, making the sum $(X + Y)$ very high. Similarly, when X is low, Y tends to be low, making the sum $(X + Y)$ very low. These extreme values of $(X + Y)$ make its variance large, and the formula for var $(X + Y)$ reflects this—by adding on the final term (cov X, Y).

An example will illustrate how the variance formulas work.

EXAMPLE **5-4**   Recall the example of the husband's and wife's income, X and Y. Suppose

$$\text{var } X = 16, \quad \text{var } Y = 10, \quad \text{and} \quad \text{cov}(X,Y) = 8$$

Calculate the variance of:

a. Total income $S = X + Y$

b. After-tax income $W = .8X + .7Y$

Solution

a. $\text{var}(X + Y) = \text{var } X + \text{var } Y + 2 \text{ cov}(X,Y)$
$\phantom{\text{var}(X + Y)} = 16 + 10 + 2(8) = 42$

b. $\text{var}(.8X + .7Y) = .8^2 \text{ var } X + .7^2 \text{ var } Y + 2(.8)(.7) \text{ cov}(X,Y)$
$\phantom{\text{var}(.8X + .7Y)} = .64(16) + .49(10) + 1.12(8) = 24.1$

Finally, consider the very simple and common case that occurs when X and Y are uncorrelated—that is, $\text{cov}(X,Y) = 0$. Then (5-18) reduces to:

$$\text{var}(X + Y) = \text{var } X + \text{var } Y$$

Since independence assures us that X and Y are uncorrelated according to (5-12), we may finally conclude that:

If X and Y are independent:
$$\text{var }(X + Y) = \text{var } X + \text{var } Y \quad (5\text{-}21)$$
Similarly:
$$\text{var }(aX + bY) = a^2 \text{ var } X + b^2 \text{ var } Y \quad (5\text{-}22)$$

An example will illustrate how these formulas simplify an otherwise tedious task.

EXAMPLE **5-5**

When a pair of dice are thrown, we are customarily interested in the total number of dots T. Calculate:

a. Its mean.

b. Its standard deviation (SD).

Solution

The hard way to solve this problem is with a tree, or, equivalently, a bivariate distribution. But we now have an easier way: Break down the problem into simple components by writing:

$$T = X_1 + X_2$$

where X_1 is the number of dots on the first die, and X_2 is the number of dots on the second die. For *each* die, the distribution (and hence the moments) are easy to calculate:

FAIR DIE		MEAN	VARIANCE
x	$p(x)$	$xp(x)$	$(x - \mu)^2\, p(x)$
1	1/6	1/6	6.25/6
2	1/6	2/6	2.25/6
3	1/6	3/6	.25/6
4	1/6	4/6	.25/6
5	1/6	5/6	2.25/6
6	1/6	6/6	6.25/6
		21/6	17.50/6
		$E(X) = 3.5$	$\sigma^2 = 2.92$
			$\sigma = 1.71$

Now we can put these components together:

a. $E(X_1 + X_2) = E(X_1) + E(X_2)$
$$= 3.5 + 3.5 = 7.0 \qquad\qquad \text{like (5-16)}$$

b. Since the two dice are independent, (5-21) gives:

$$\text{var } (X_1 + X_2) = \text{var } X_1 + \text{var } X_2 \qquad\qquad (5\text{-}23)$$
$$= 2.92 + 2.92 = 5.84$$
$$\text{Hence SD}(X_1 + X_2) = \sqrt{5.84} = 2.42$$

Remarks

We might be tempted to calculate the standard deviation by a formula similar to (5-23), namely:

$$SD(X_1 + X_2) \overset{?}{=} SD(X_1) + SD(X_2)$$

We can easily confirm that this will not work:

$$2.42 \neq 1.71 + 1.71$$

In conclusion, then, it is *variances* that we add, not standard deviations.

Another example will confirm the simplicity of these formulas for analyzing sums.

EXAMPLE **5-6** Suppose a die is thrown 10 times, instead of just twice. For the total number of dots, calculate:

    a. The mean.

    b. The standard deviation.

Solution The hard way to do this problem is with a tenfold multivariate distribution—a gargantuan task, even for a computer.

    The easy way, once more, is to break down the problem into simple components. Again let X_1 be the number of dots on the first toss, X_2 the number on the second toss, and so on. Also note the mean and variance for each toss calculated in the previous example.

    a. $E(X_1 + X_2 + \cdots + X_{10}) = 3.5 + 3.5 + \cdots$
$$= 10(3.5) = 35$$

    b. $\text{var}(X_1 + X_2 + \cdots + X_{10}) = 2.92 + 2.92 + \cdots$
$$= 10(2.92) = 29.2$$

$$SD = \sqrt{10}\,\sqrt{2.92} = 5.40 \qquad (5\text{-}24)$$

TABLE **5-7** **The Mean and Variance of Various Functions of Two Random Variables**

FUNCTION OF X AND Y	MEAN AND VARIANCE DERIVED BY:	MEAN	VARIANCE
1. Any function $g(X, Y)$		$E[g(X, Y)]$ $= \sum_x \sum_y g(x,y)p(x,y)$	
2. Linear combination $aX + bY$		$E(aX + bY)$ $= aE(X) + bE(Y)$	$\text{var}(aX + bY)$ $= a^2\text{var}\,X + b^2\text{var}\,Y$ $+ 2ab\,\text{cov}(X, Y)$
3. Simple sum $X + Y$	Setting $a = b = 1$ in row 2	$E(X + Y)$ $= E(X) + E(Y)$	$\text{var}(X + Y)$ $= \text{var}\,X + \text{var}\,Y$ $+ 2\,\text{cov}(X, Y)$
4. Function of one variable, aX	Setting $b = 0$ in row 2	$E(aX) = aE(X)$	$\text{var}(aX) = a^2\text{var}\,X$

The theorems of this chapter are summarized in Table 5-7 for future reference. The general function g(X, Y) is dealt with in the first row, while the succeeding rows represent increasingly simpler special cases.

PROBLEMS

5-16 Following (5-20), it was explained why the formula for var(X + Y) includes the covariance term—when the covariance was positive. Give a similar explanation when the covariance is negative.

5-17 A small college had a faculty of 10 couples. The annual incomes (in thousands of dollars) of the men and women were as follows:

COUPLE	MAN	WOMAN
MacIntyre	20	15
Sproule	30	35
Carney	30	25
Devita	20	25
Peat	20	25
Matias	30	15
Steinberg	40	25
Aldis	30	25
Yablonsky	40	35
Singh	40	25

A couple is drawn by lot to represent the college at a workshop on personal finances. Let X and Y denote the randomly drawn income of the man and woman. Then find:

a. The bivariate probability distribution, and its graph.

b. For X the distribution, mean, and variance. Also for Y.

c. The covariance σ_{XY}.

d. If S is the combined income of the couple, what is its mean and variance? Calculate two ways: from the distribution of S, as in (5-8), and then using the easy formulas for sums.

e. Suppose that $W = .6X + .8Y$ is the couple's income after taxes. What is its mean and variance?

f. To measure the degree of sex discrimination against wives, a sociologist measured the difference $D = X - Y$. What is its mean and variance?

g. How good a measure of sexual discrimination is $E(D)$?

5-18 Continuing Problem 5-17, we shall consider some alternative schemes for collecting the tax T on the couple's income S. Find the mean and standard deviation of T,

a. If S is taxed at a straight 20%—that is,

$$T = .20S$$

b. If S is taxed at 50%, with the first 15 thousand exempt—that is,

$$T = .5(S - 15)$$

c. If S is taxed according to the following progressive tax table:

S	T	S	T
30	4	55	11
35	5	60	13
40	6	65	16
45	7	70	19
50	9	75	22

5-19 Compare the three tax schemes in Problem 5-18 in terms of the following criteria:

    i. Which scheme yields the most revenue to the government?

    ii. Which scheme is most egalitarian; that is, which scheme results in the smallest standard deviation in net income left after taxes?

CHAPTER 5 SUMMARY

5-1 A pair of random variables X and Y has a joint probability distribution $p(x,y)$, from which the distributions $p(x)$ and $p(y)$ can be found in the margin. X and Y are then called independent if the simple multiplication rule holds: $p(x,y) = p(x)p(y)$ for all x and y.

5-2 A new random variable, $R = g(X,Y)$ let us say, has an expected value that can be calculated from the joint distribution $p(x,y)$:
$E[g(X,Y)] = \Sigma\Sigma\ g(x,y)p(x,y)$.

5-3 Just as variance measures how one variable varies, so covariance measures how two variables vary together. Its standardized version is the correlation $\rho = \sigma_{XY}/\sigma_X\sigma_Y$, which measures the degree of relation between X and Y.

5-4 The sum $X + Y$ is a particularly convenient function. Its expected value is simply $E(X) + E(Y)$, and its variance is $\text{var}(X) + \text{var}(Y) + 2\ \text{cov}(X,Y)$.

REVIEW PROBLEMS

5-20 In planning a new shopping mall, the developer has to allow time for two consecutive stages: time for buying the land (B), and then for construction (C). Based on past experience, the developer feels that these times will be uncertain, with the distributions given below (in years). He also feels that B and C are statistically independent (since a delay in buying the land does not change the betting odds on the construction time).

b	p(b)	c	p(c)
1	.10	1	.50
2	.30	2	.40
3	.30	3	.10
4	.20		
5	.10		

The developer will enjoy a big tax advantage if he can finish in 4 years or less. What is the chance of this?

5-21 Answer Problem 5-20 under the following alternative assumptions about the distributions of B and C:

a. Assume B and C are independent, with distributions given by the following formulas (geometric distributions):

$$p(b) = .4(.6)^{b-1}, \quad b = 1,2,3, \ldots$$
$$p(c) = .7(.3)^{c-2}, \quad c = 2,3,4, \ldots$$

b. Assume B and C are independent and normally distributed, with means of 2 and 1.5 years, and standard deviations of .3 and .4 years, respectively. (*Hint:* As we will see in Chapter 6, the sum B + C will be normally distributed too.)

c. Assume B and C are as in part **b,** except that now they are correlated: If B is especially long, the developer will work hard to shorten C, so that B and C have a negative correlation, $\rho = -.40$.

5-22 The students in a large class wrote two exams, obtaining a distribution of grades with the characteristics shown in the table below.

	CLASS MEAN μ	STANDARD DEVIATION σ
1st exam, X_1	50	20
2nd exam, X_2	80	20
a. Average, $\overline{X}$	?	?
b. Weighted Average, W	?	?
c. Improvement, I	?	?

If the covariance of the two grades was $\sigma_{12} = 50$, fill in the table for the following three cases:

a. The instructor calculated a simple average of the two grades:

$$\overline{X} = (X_1 + X_2)/2 \qquad E(x_1 + x_2)$$

b. The instructor thought the second exam was twice as important, so she took a weighted average:

$$E\left(\frac{1}{3}x_1 + \frac{2}{3}x_2\right)$$

$$W = \frac{1}{3}X_1 + \frac{2}{3}X_2$$

c.   The instructor wanted to find how much X_2 was an improvement over X_1:

$$I = X_2 - X_1$$

5-23   Repeat Problem 5-22:

   i.   If the covariance $\sigma_{12} = -200$. How might you interpret such a negative covariance? What has it done to the variance of the average grade? And the variance of the difference I?

   ii.   If the covariance $\sigma_{12} = 0$.

REVIEW PROBLEMS, CHAPTERS 1 TO 5

5-24   Circle the correct choice in each of the five following brackets. Note the following abbreviations:

$$RCE = \text{randomized controlled experiment}$$
$$OS = \text{observational study}$$

To really find out how effective a treatment is, a [*RCE, OS*] is better than a [*RCE, OS*], where feasible. This is because a [*RCE, OS*] actually makes the treatment and control groups equal on average in every respect—except for the treatment itself, of course; whereas a [*RCE, OS*] is usually cluttered up with extraneous factors that bias the answer. To the extent these extraneous factors can be measured and analyzed with a [*multiple regression, confidence interval*], however, bias can be reduced.

5-25   Suggest some plausible reasons for each of the following relations:

   a.   Over a period of several years, the relation between the number of colds reported weekly in a city and the amount of beer sold was found to be negative (low number of colds associated with high beer sales). Does beer prevent colds?

   b.   Over a period of 50 years, the relation between clergymen's annual salaries and annual alcohol sales in the U.S. was found to be positive. Does paying clergymen more improve alcohol sales?

   c.   The Irish humorist George Bernard Shaw once observed that there was a strong positive relation between a young man's income and clothes. His advice to a young man aspiring to be rich was therefore simple: Buy a black umbrella and top hat.

5-26   When asked to predict next year's inflation rate (in %), five economists gave the following answers:

$$X = 5, 3, 6, 5, 2$$

a. What is the average, $\overline{X}$?

b. The first economist was known for his hard-headed realism and superior track record; the last was known for occasional ventures into unrealistic theorizing. It was therefore decided to give relative weights to the five answers as follows:

$$w = .3, .2, .2, .2, .1$$

Calculate the weighted mean ΣwX.

c. Which is closer to the first, most reliable observation ($X = 5$)—the simple average or the weighted average?

5-27 The income distribution of the population of American men in 1975 was approximately as follows:

ANNUAL INCOME ($000)	PROPORTION OF MEN AT THAT INCOME
0–5	36%
5–10	23%
10–15	20%
15–20	11%
20–25	5%
25–30	3%
30–35	2%

a. Graph this distribution.

b. Calculate the mean income, and the standard deviation, and show them on the graph.

c. What is the mode? Where will the median be in relation to the mean and mode?

5-28 An exam consists of 8 multiple-choice questions, each with a choice of 5 answers. Let X be the number of correct answers for a student who just guesses each answer.

a. Graph the distribution of X.

b. Calculate μ_X and σ_X, and show them on the graph.

c. If the instructor calculates a rescaled mark $Y = 10X + 20$, what are μ_Y and σ_Y?

d. If the passing mark is $Y = 50$, what is the chance the student who resorts to pure guessing will pass?

5-29 To find out who won the Reagan-Carter TV debate before the 1980 Presidential election, ABC News conducted a phone-in survey. Nearly a million people volunteered to call in: 723,000 believed that Reagan had won the debate, and 244,000 believed Carter had (London Free Press, Oct 29, 1980).

CBS on the other hand conducted a very small sample survey. Of all the registered voters who had watched the debate, 1,019 were polled at random by telephone: 44% believed that Reagan had won the debate, 36% believed Carter had, 14% called it a tie, and 6% had no opinion.

Do these two polls agree, allowing for sampling fluctuation? Explain why, or why not. Include a critique of each poll.

5-30   A politician took a random sample of 5 voters from a suburb that is 60% Republican, 20% Democratic, and 20% Independent. What is the chance that a majority in the sample are Republican?

5-31   The following table gives, for a thousand newborn American baby boys (white U.S. males, 1970), the approximate number dying, in successive decades:

AGE, x	$n(x)$ = NUMBER DYING WITHIN THE DECADE	$L(x)$ = NUMBER LIVING TO THE BEGINNING OF THE DECADE	MORTALITY RATE, PER DECADE $m(x) = \dfrac{n(x)}{L(x)}$
0 to 10	26	1,000	.026
10 to 20	9	974	.009
20 to 30	18	965	.019
30 to 40	21	947	.022
40 to 50	49	.	.
50 to 60	117	.	.
60 to 70	219	.	.
70 to 80	282		
80 to 90	208		
90 to 100	51		

Total = 1,000

a.   The third column shows the number surviving at the beginning of each decade. Complete this tabulation of $L(x)$.

b.   The mortality rate in the last column is just the number dying during the decade, relative to the number living at the beginning of the decade. Complete the tabulation of the mortality rate $m(x)$. Then answer True or False. If False, correct it.

   i.   The mortality rate is lowest during the first decade of life.

   ii.   Roughly speaking, the mortality rate nearly doubles every decade.

c.   Ten-year term insurance is an agreement whereby a man pays the insurance company $x (the "premium," which usually is spread throughout the decade in 120 monthly payments) in return for a payment of $1,000 to the man's estate if he dies. In order for this to be a "fair bet" (and ignoring interest), what should the premium x be for a man:

   i.   In his 20s?

   ii.   In his 40s?

    d. The work that we have done so far could be just as well expressed in probability terms. For example, find:

        i. The probability of a man dying in his 40s.

        ii. The probability of a man surviving to age 40.

        iii. The conditional probability of a person dying in his 40s, given that he has survived to 40.

5-32   Referring to the mortality table in Problem 5-31:

    a. Find the mean age at death (i.e., mean length of life, also called life expectancy).

    *b. If the mortality rate were 50% higher after age 40, how much would this reduce life expectancy? (Incidentally, heavy cigarette smoking— two or more packs per day—seems to be associated with roughly this much increase in mortality according to the U.S. Department of Health, Education, and Welfare, 1964.)

5-33   An apartment manager in Cincinnati orders three new refrigerators, which the seller guarantees. Each refrigerator has a 20% probability of being defective.

    a. Tabulate the probability distribution for the total number of defective refrigerators, X.

    b. What are the mean and variance of X?

    c. Suppose that the cost of repair, in order to honor the guarantee, consists of a fixed fee ($10) plus a variable component ($15 per defective refrigerator). That is:

$$c(x) = 0 \qquad\quad \text{if } x = 0$$
$$= 10 + 15x \quad \text{if } x > 0$$

    Find the average cost of repair.

5-34   A company produces TV tubes with a length of life that is normally distributed with a standard deviation of 5 months. How large should the mean μ be in order that 90% of the tubes last for the guarantee period (18 months)?

5-35   A clinic plans to set up a mass testing program for diabetes using an inexpensive test of high reliability: Of all people with diabetes, 95% are correctly diagnosed positive (as having diabetes); of all people free of diabetes, 98% are correctly diagnosed negative (the remaining 2% being erroneously diagnosed positive).

    The community served by the clinic has about 10,000 patients, and the diabetes rate runs at about 1%. The clinic director wants to know three things:

a. About how many patients will have diabetes and be missed (i.e., get a negative diagnosis)?

b. About how many patients will be diagnosed positive (and therefore require followup)?

c. What proportion of the patients in part **b** will actually have diabetes?

d. The director cannot believe your answer to part **c**. Explain why it is so low, as simply as you can.

5-36 When is it true, or approximately true, that:

a. $E(X^2) = [E(X)]^2$?

b. $E(XY) = E(X)E(Y)$?

5-37 Because of unforeseen delays, the last two stages for manufacturing a complex plastic were of very uncertain duration. In fact, the times required for completion (X and Y, in hours) had the following probability distributions:

x	p(x)		y	p(y)
			0	.4
1	.4		1	.2
2	.2		2	0
3	0		3	0
4	.4		4	.4

Two questions were of interest to management: (1) the total time T required; and (2) the cost C of running the first process, which was $200 per hour, plus a fixed cost of $300. In formulas,

$$T = X + Y$$
$$C = 200X + 300$$

Assuming X and Y are independent, tabulate the distribution of T, and of C, and then answer the following questions:

a. What are the medians $M(X)$, $M(Y)$, $M(T)$, and $M(C)$.

b. Is it true that $M(T) = M(X) + M(Y)$? That is,

$$M(X + Y) = M(X) + M(Y)?$$

c. Is it true that $M(C) = 200M(X) + 300$? That is,

$$M(200X + 300) = 200M(X) + 300?$$

5-38 Repeat Problem 5-37 for the means, instead of the medians.

5-39 In 1972, the population of American families yielded the following table for family size:

CHILDREN UNDER 18	0	1	2	3	4
Proportion of families	.44	.19	.18	.10	.09

(This table includes young parents who have not yet had all their children. *Completed* family size would be larger, of course.)

a. Let X be the number of children in a family selected at random. Find μ_X and σ_X.

b. Now let a child be selected at random (rather than a family), and let Y be the number of children in his or her family. What are the possible values of Y? Find the probability distribution, and then μ_Y and σ_Y.

5-40 Continuing Problem 5-39, suppose a new subdivision of 500 homes is being built, and the school board needs to know how big a school to build. Estimate the total number of children.

5-41 Continuing Problem 5-40, suppose the new school has been built, and the junior boys basketball team has made it to the state finals. The coach wants to give each of the 15 players tickets for his family—that is, mother, father, and all brothers and sisters.

a. Estimate how many tickets would be required.

b. List some of the assumptions you are making.

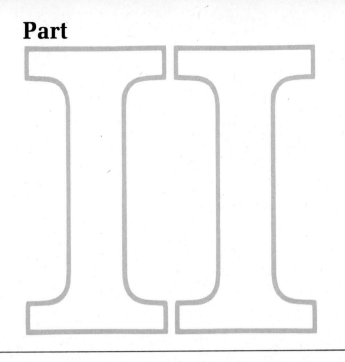

INFERENCE FOR MEANS AND PROPORTIONS

Chapter

6

Sampling

You don't have to eat the whole ox to know that it is tough.

SAMUEL JOHNSON

In the last three chapters, we developed the mathematical tools—probability and random variables—so that we can now answer the basic deductive question in statistics: What can we expect of a random sample drawn from a known population?

Before pursuing this question, however, we must repeat an important warning: How we *collect* data is at least as important as how we analyze it. In particular, a sample should be *representative* of the population, and *random sampling* is the best way to achieve this.

If a sample is not random, it may be worse than useless. In Chapter 1, for example, we already cited the infamous *Literary Digest* poll. As another example, a sociologist surveyed the teachers in a certain state to find out how they spent their time—how much in teaching, administration, and professional reading. Realizing that the response rate would be low, he mailed the questionnaire to a large sample of 1,000 teachers in the state, and was pleased to receive 200 replies.

Even if his mailing list was randomly selected, however, the replies were not—and this is what ultimately counts. Since the teachers who spent a lot of their time on professional reading were too busy to answer the questionnaire, they were not represented in the sample. The sample gave a very misleading picture of the population. The *nonresponse bias* was particularly serious since the reason for nonresponse (being too busy) is closely related to the issue being studied (how teachers spend their time).

A better approach would have been to select a much smaller random sample of teachers, and take great pains to get a complete (or at least a very high) response rate. If he got a truly random sample of 20 replies in this way, it would be much better than the biased sample of 200 replies. Quality is more important than quantity.

In fact, quantity alone can be very deceptive. A large but biased sample may look good because of its size, but in fact, it just consists of the same bias being repeated over and over—the sort of mistake the *Literary Digest* repeated millions of times. By contrast, modern polling organizations like Gallup or Harris essentially use small random samples of about a thousand, and get far better answers: Their samples are unbiased, and can be analyzed by probability theory to ascertain how much sampling uncertainty they have.

It is precisely this issue of sampling uncertainty for random samples that will take up the next few chapters.

6-1   Random Sampling

A—THE POPULATION

The word *population* has a very specific meaning in statistics: It is the total collection of objects or people to be studied, from which a sample is to be drawn. For example, the population of interest might be all the American voters (if we wished to predict an election). Or it might be the population of all the students at a certain university (if we wished to estimate how many would attend a concert).

The population can be any size. For example, it might be only 100 students in a certain men's physical education class (if we wished to estimate their average height). To be specific, suppose this population of 100 heights has the frequency distribution shown in Table 6-1. When we draw a random observation, what is the chance it will be 63 inches, for example? From the second line of Table 6-1, we find the probability is 6 in 100, or .06. Similarly, the probabilities for all the possible outcomes can be found in the relative frequency column. Therefore, we label this column $p(x)$ to show that it can be viewed as the probability distribution not only of the population, but also of a single observation taken at random. This is an important insight:

> **Each individual observation in a random sample has the population probability distribution $p(x)$.**   (6-1)

From $p(x)$, we calculate the population mean μ and standard deviation σ, which we emphasize are the mean and standard deviation of an individual observation. Then in Figure 6-1a we graph this distribution $p(x)$, sketched also as a smooth curve.

TABLE **6-1** **A Population of 100 Students' Heights, and the Calculation of μ and σ**

POPULATION DISTRIBUTION			CALCULATION OF MEAN μ	CALCULATION OF VARIANCE σ^2
HEIGHT x	FREQUENCY	RELATIVE FREQUENCY, ALSO $p(x)$	$xp(x)$	$(x - \mu)^2 p(x)$
60	1	.01	.60	.81
63	6	.06	3.78	2.16
66	24	.24	15.84	2.16
69	38	.38	26.22	0
72	24	.24	17.28	2.16
75	6	.06	4.50	2.16
78	1	.01	.78	.81
	$N = 100\checkmark$	$1.00\checkmark$	$\mu = 69.00$	$\sigma^2 = 10.26$
				$\sigma = 3.20$

B—THE RANDOM SAMPLE

Now let us draw a sample from this population, a sample of $n = 5$ students, for example. As noted in Chapter 1, a sample is called a *simple random sample (SRS)* if each individual in the population is equally likely to be chosen every time we draw an observation. For example, as suggested earlier, two ways we could take a random sample of 5 students in the physical education class of 100 men are:

1. *Draw chips from a bowl.* As illustrated in Figure 6-1b, the most graphic method is to record each student's height on a chip, mix all these 100 chips in a large bowl, and then draw the sample of $n = 5$ chips.

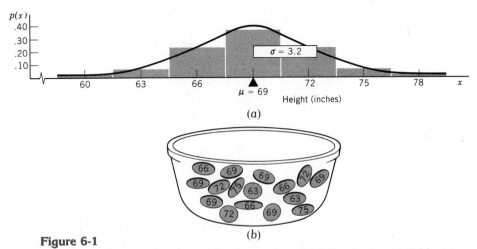

Figure 6-1

Population of 100 students' heights. (a) Bar graph of distribution in Table 6-1. (b) Bowl-of-chips equivalent.

2. *Assign each student a serial number, and select numbers at random.*
    For example, let the class assign itself numbers by counting off from
    00 to 99. Then we could get our sample from the random digits in
    Appendix Table I. For example, if we start at the third line, we would
    select the 5 students with numbers 72, 20, 47, 33, and 84.

These two sampling methods are mathematically equivalent. Since the
second is simpler to employ, it is commonly used in practical sampling.
However, the bowlful of chips is conceptually easier; consequently, in
our theoretical development of random sampling, we will often visualize
drawing chips from a bowl.

Randomness makes it likely that a sample is representative of the underlying population from which it is drawn. Figure 6-2 illustrates this,
with the gray dots, as usual representing all of the actual individuals in
the population, ordered according to their height. The 5 blue dots then
show the sample of individuals that we might happen to draw. The *average* of these sample values (the sample mean $\overline{X}$) is also shown in blue.
Note that it is not as extreme; that is, it is closer to the population mean
μ than most of the individual observations in the sample. This is because
in calculating the sample mean, an extreme individual observation such
as $X = 74$ tends to be diluted by more typical observations like $X = 70$,
or even tends to be offset by an observation at the other end like $X = 68$.
We therefore conclude:

> Because of averaging, the sample mean $\overline{X}$ is not as
> extreme (doesn't vary so widely) as the individuals
> in the population.   (6-2)

C—SAMPLING WITH OR WITHOUT REPLACEMENT

In large populations, such as all American men, the "population bowl"
in Figure 6-1 would contain millions of chips, and it would make practically no difference whether or not we replace each chip before drawing
the next. After all, what is one chip in millions? It cannot substantially
change the relative frequencies, p(x).

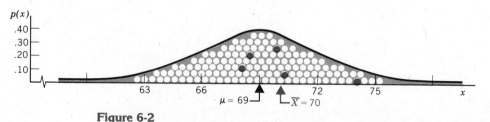

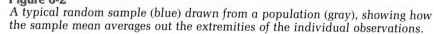

Figure 6-2
*A typical random sample (blue) drawn from a population (gray), showing how
the sample mean averages out the extremities of the individual observations.*

However, in small populations such as the 100 heights in Figure 6-1, replacement of each sampled chip becomes an important issue. If each chip drawn is recorded and then replaced, it restores the population to exactly its original state. Thus, later chips are completely independent of each chip drawn earlier. On the other hand, if each chip is *not* replaced, the probabilities involved in the draw of later chips *will* change. (For example, if the first chip drawn happens to be the only 78″ chip in the bowl, then the probability of getting that chip in succeeding draws becomes zero; it's no longer in the bowl.) In this case, later chips *are* dependent on each chip drawn earlier.

D—CONCLUDING DEFINITION

If we sample with replacement, the *n* observations in a sample are independent. And in large populations, even if we sample without replacement, it is practically the same as with replacement, so that we still essentially have independence. All these cases where the observations are independent are easy to analyze, and lead to very simple formulas. We therefore call them *very simple random samples*:

> A *very simple random sample* (VSRS) is a sample whose *n* observations $X_1, X_2, \ldots, X_n$ are independent. The distribution of each X is the population distribution $p(x)$; that is,
>
> $$p(x_1) = p(x_2) = \cdots = p(x_n) \qquad (6\text{-}3)$$
> $$= \text{population distribution, } p(x)$$
>
> Then each observation has the mean μ and standard deviation σ of the population.

The exception to this independence occurs when the chips are drawn from a small population, *and* are not replaced. This procedure of course is more efficient: Because it ensures that no chip can be repeated, each observation brings fresh information. However, the gains in efficiency from keeping out the chips require a complicated formula, and are substantial only when the population is small. We therefore defer this type of sampling to Section 6-5. Everywhere else, we will assume a VSRS, and often refer to it simply as a *random sample*.

E—HOW RELIABLE IS THE SAMPLE?

The purpose of random sampling, of course, is to make an inference about the underlying population. As a familiar example, we hope the sample mean $\overline{X}$ is a close estimate of the population mean μ. There are two ways we can study just how close $\overline{X}$ comes to μ:

1. Recall how we sampled the blue dots from the gray population in Figure 6-2, and then calculated the sample mean $\overline{X}$. We could repeat this and get a new $\overline{X}$, over and over. By recording how $\overline{X}$ varies from sample to sample, we would build up the *sampling distribution* of $\overline{X}$, denoted $p(\overline{x})$.

    Rather than *actually* sampling a physical population, we can *simulate* this sampling (just as we simulated the roll of a die earlier). This is normally done on a computer, where even complicated sampling can be repeated hundreds of times every second. Since this process—described in detail in Section 6-6—may be viewed as something like "rolling dice," it is called *Monte Carlo* sampling. (The relationship between gambling games and statistics has a long history; in fact, it was gamblers who provided the impetus for probability theory about 300 years ago.)

2. A more precise and useful (but often more difficult) alternative is to derive mathematical formulas for the sampling distribution of $\overline{X}$. Once we have derived such formulas (as we do in the next section) they can be applied broadly to a whole multitude of sampling problems.

PROBLEMS

6-1 Suppose that X, the number of hours per month spent on professional reading by the 2,000 teachers in a certain city, has the distribution shown in the illustration.

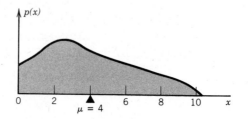

a. Suppose a questionnaire is sent to a random sample of 15 teachers and only the 5 who read the least have time to return it. Sketch on the graph where this sample of 15 might be typically, and then the 5 replies. Where would the average reply be? And how close to the target μ?

b. Repeat part **a** for a random sample of 5 teachers, carefully followed up to obtain a 100% response rate. Would $\overline{X}$ be closer or further from μ?

6-2 Suppose the annual incomes (in $000) of an MBA class of 80 students ten years later had the distribution shown in the illustration.

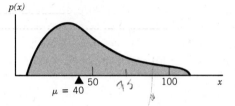

a. Suppose the 12 alumni who returned to their 10th reunion tended to be those who were more prosperous. (The least prosperous couldn't afford it; indeed, a few didn't have a mailing address and therefore didn't hear about the reunion.) Sketch on the graph where this "convenience" sample of a dozen incomes might be. Then where would their average be? How close to the target μ?

b. Repeat part a, if instead a *random* sample of 12 alumni had been chosen, and then carefully followed up to obtain a 100% response rate.

6-3 In a small and hypothetical midwestern city, the 10,000 adults watch football on TV in widely varying amounts. The weekly amount X varies from 0, 1, 2, . . . , 9 hours, and for each of these levels of X, there are 1,000 adults.

a. Graph the population distribution p(x).

b. To simulate drawing one adult (a value of X) at random, take the first digit from Table I. Similarly, to simulate a small sample survey of 5 adults, take the first 5 digits from Table I. Mark these 5 observations in the graph in part a.

c. Calculate the population mean μ. (Or, by symmetry, what must it be?)

d. Calculate the sample mean $\overline{X}$. Is it closer to μ than most of the individual observations? Does this illustrate equation (6-2)?

*6-4 a. Repeat the simulation of Problem 6-3, a dozen times in all (each time starting at a different place in Table I, of course).

b. Pool your data with a few other students, and tabulate the few dozen values of $\overline{X}$. (We suggest grouping the values of $\overline{X}$ in cells of width .6, taking the cell midpoints to be 0.3, 0.9, 1.5, . . .)

c. Graph the distribution of $\overline{X}$ tabulated in part b.

d. If millions of values of $\overline{X}$ had been graphed, instead of just a few dozen, the distribution of $\overline{X}$ would be complete and accurate, with the irregularities smoothed out. Over the graph in part b, roughly sketch what this smooth distribution would be. It is called the *sampling distribution* of $\overline{X}$.

e. About where is this sampling distribution of $\overline{X}$ centered? How does that compare to μ calculated in Problem 6-3?

f. Is the sampling distribution of $\overline{X}$ spread out more than, equally, or less than the population distribution $p(x)$?

6-2 Moments of the Sample Mean

We now are ready to use the theory developed in earlier chapters. The sample mean $\overline{X}$ was defined as:

$$\overline{X} \equiv \frac{1}{n}[X_1 + X_2 + \cdots + X_n] \qquad (6\text{-}4)$$
$$(2\text{-}3) \text{ repeated}$$

Being a linear combination of random variables, $\overline{X}$ itself will also be a random variable. How does it fluctuate? In particular, what is its expectation and variance?

By applying (5-17) we can easily calculate the expectation:

$$E(\overline{X}) = \frac{1}{n}[E(X_1) + E(X_2) + \cdots + E(X_n)]$$

Recall from (6-3) that each observation X has the population distribution $p(x)$ with mean μ. Thus, $E(X_1) = E(X_2) = \cdots = \mu$, and therefore:

$$E(\overline{X}) = \frac{1}{n}[\mu + \mu + \cdots + \mu]$$
$$= \frac{1}{n}[n\mu] = \mu$$

$$\boxed{E(\overline{X}) = \mu} \qquad (6\text{-}5)$$

Similarly, we can obtain the variance of $\overline{X}$. According to (6-3) again, all the observations $X_1, X_2, \ldots$ are independent, and so the simple formula (5-22) can be applied to (6-4):

$$\text{var } \overline{X} = \frac{1}{n^2}[\text{var}(X_1) + \text{var}(X_2) + \cdots + \text{var }(X_n)]$$

Again, we note that each observation X has the population distribution $p(x)$, with variance σ^2, so that:

$$\text{var } \overline{X} = \frac{1}{n^2}[\sigma^2 + \sigma^2 + \cdots + \sigma^2]$$

$$= \frac{1}{n^2}[n\sigma^2] = \frac{\sigma^2}{n} \qquad (6\text{-}6)$$

$$\text{Standard deviation of } \overline{X} = \frac{\sigma}{\sqrt{n}}$$

This typical deviation of $\overline{X}$ from its target μ represents the estimation error, and so it is commonly called the *standard error*, or SE:

$$\boxed{\begin{array}{c} \textbf{Standard error of } \overline{X} \\[4pt] SE = \dfrac{\sigma}{\sqrt{n}} \end{array}} \qquad (6\text{-}7)$$

This formula shows explicitly how the standard error shrinks as the sample size n increases. Compared to a single observation (with spread σ), a sample mean $\overline{X}$ based on $n = 25$ observations, for example, has only 1/5 as much error. Or, if the sample is increased to $n = 100$ observations, there will only be 1/10 as much error. [Notice that the quadrupling of sample size from 25 to 100 results in just a doubling of accuracy—a reflection of the *square root* divisor in (6-7).]

An example will illustrate how useful these two formulas are.

EXAMPLE **6-1** The population of men's heights shown in Figure 6-1 had $\mu = 69$ inches and $\sigma = 3.2$ inches.

a. If many random samples of size $n = 5$ were collected, how would the sample mean $\overline{X}$ fluctuate?
b. In Figure 6-2, the one sample shown in color had $\overline{X} = 70$. Is this a fairly typical sample, or was it a lucky one, particularly close to μ?

Solution a.

$$E(\overline{X}) = \mu = 69 \qquad (6\text{-}5) \text{ repeated}$$

$$SE = \frac{\sigma}{\sqrt{n}} = \frac{3.2}{\sqrt{5}} = 1.4 \qquad (6\text{-}7) \text{ repeated}$$

Thus, the many possible values of the sample mean $\overline{X}$ would fluctuate around the target of 69 inches, with a standard error of 1.4 inches.

b. This particular sample mean $\overline{X} = 70$ deviates from its target $\mu = 69$ by 1 inch. Since 1 inch is not far from the standard error of 1.4 inches, this sample is fairly typical.

PROBLEMS

6-5 Suppose that 10 men were sampled randomly (from the population such as Table 6-1) and their average height $\overline{X}$ was calculated. Then imagine the experiment repeated many times to build up the sampling distribution of $\overline{X}$. Answer True or False; if False, correct it.

  a. The tall and short men in the sample tend to average out, making $\overline{X}$ fluctuate less than a single observation.

  b. To be specific, the sample mean $\overline{X}$ fluctuates around its target μ with a standard error of only σ/n.

6-6 **a.** The population of American men in 1975 had incomes that averaged μ = 10 thousand, with a standard deviation of 8 thousand, approximately. If a random sample of n = 100 men was drawn to estimate μ, what would be the standard error of $\overline{X}$?

  b. The population of men in California is about 1/10 as large, but suppose it had the same mean and standard deviation. If a random sample of n = 100 was drawn, what would be the standard error of $\overline{X}$ now?

6-7 **a.** Continuing Problem 6-6, the population size was 66 million. If a 1% sample was taken (i.e., n = 1% of 66 million = 660 thousand), what would be the standard error of $\overline{X}$?

  b. If a 1% sample of men in California was drawn, what would be the standard error of $\overline{X}$?

*6-8 Suppose the lawyers in a certain state earn the following incomes (thousands of dollars per month):

INCOME x	PROPORTION $p(x)$
2	.50
4	.30
6	.20

  a. Calculate the mean μ and standard deviation σ, and show them on a graph of the population distribution.

  b. Now suppose a random sample of n = 2 incomes is drawn, X_1 and X_2, say. Since each observation has the population distribution, it is easy to tabulate the joint distribution of X_1 and X_2. Then calculate the sampling distribution of $\overline{X}$.

  c. From the distribution of $\overline{X}$, calculate the expected value and standard error, and show them on a graph of the sampling distribution of $\overline{X}$.

  d. From formulas (6-5) and (6-7), calculate the expected value and standard error of the sampling distribution:

   i. When n = 2. Does this agree with part (c)?

   ii. When n = 5.

   iii. When n = 20.

6-3 The Shape of the Sampling Distribution

In Section 6-2 we found the expected value and standard error of $\overline{X}$. The remaining issue is the *shape* of the sampling distribution.

Across the top of Figure 6-3, we show three different populations. In the column below each, successive graphs show how the sampling distribution of $\overline{X}$ changes shape as sample size n increases. Thus, the first column shows how $\overline{X}$ behaves when sampling from a normal population. The sampling distribution of $\overline{X}$ is normal too (fluctuating around μ with less and less error as n increases, as noted already in Section 6-2).

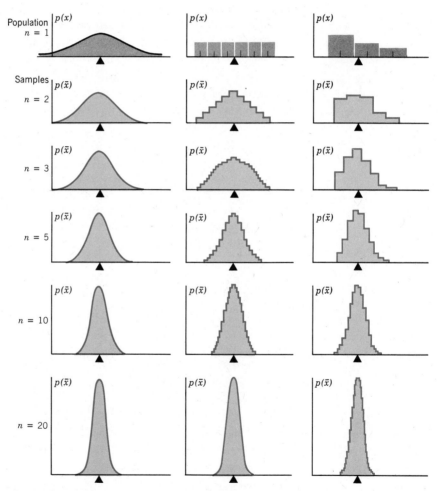

Figure 6-3
The sampling distribution of $\overline{X}$ (in color) contrasted with the parent population distribution (in gray). The left column shows sampling from a normal population. As sample size n increases, the standard error of $\overline{X}$ decreases. The next two columns show that even though the population is not normal, the sampling distribution still becomes approximately normal.

In the second column, the population distribution is rectangular, and, as expected, the sampling distribution of $\overline{X}$ takes a more concentrated shape, tapering off at the ends. What is surprising is that this distribution eventually also becomes normal; indeed it becomes practically normal when n is only 5 or 10.

The third column is astounding: Even though the population distribution is skewed, the sampling distribution of $\overline{X}$ still becomes normal—although it takes a little larger sample size this time. (Notice that it becomes practically normal by the time $n = 10$ or 20.) This remarkable phenomenon can be proved to be generally true:

> If the parent population is normal, *or* the sample size is large (often $n = 10$ or 20 will be large enough), then *in either case* the sampling distribution of $\overline{X}$ has an approximately normal shape. (6-8)

Our conclusions so far on random sampling may be summarized into one statement:

> *The Normal Approximation Rule.* In random samples (VSRS) of size n, the sample mean $\overline{X}$ fluctuates around the population mean μ with a standard error of $\sigma/\sqrt{n}$ (where σ is the population standard deviation).
>
> Therefore, as n increases, the sampling distribution of $\overline{X}$ concentrates more and more around its target μ. It also gets closer and closer to normal (bell-shaped). (6-9)

This rule is customarily called the *Central Limit Theorem*, and is of such fundamental importance that it is illustrated in Figure 6-4 (and discussed further in Appendix 6-3). In Figure 6-4 we continue a convention that distinguishes the sampling distribution from the population distribution:

> Populations are gray.
> Samples and sampling distributions are blue. (6-10)

The Normal Approximation Rule allows us to use the familiar normal tables to determine how closely a sample mean $\overline{X}$ will estimate a population mean μ, as the following example illustrates:

EXAMPLE 6-2

A population of men on a large midwestern campus has a mean height $\mu = 69$ inches, and a standard deviation $\sigma = 3.22$ inches. If a random sample of $n = 10$ men is drawn, what is the chance the sample mean $\overline{X}$ will be within 2 inches of the population mean μ?

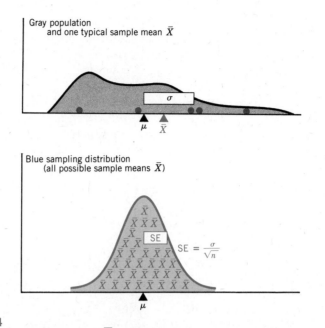

Figure 6-4

The sampling distribution of $\overline{X}$ is approximately normal, and more concentrated around μ than is the population. It is colored blue to distinguish it from the population in gray.

Solution

According to the Normal Approximation Rule (6-9), $\overline{X}$ is normally distributed, with

$$\text{Expected value} = \mu = 69$$

$$\text{Standard error, SE} = \frac{\sigma}{\sqrt{n}} = \frac{3.22}{\sqrt{10}} = 1.02$$

We want to find the probability that $\overline{X}$ is within 2 inches of $\mu = 69$—that is, between 67 and 71. So we first calculate the probability above 71, beginning with its standardization:

$$Z = \frac{\overline{X} - \mu}{SE} \qquad \text{like (4-23)}$$

$$= \frac{71 - 69}{1.02} = 1.96$$

That is, the critical value of 71 for the sample mean is nearly 2 standard errors above its expected value of 69, as Figure 6-5 shows. From the standard normal Table IV, we find the probability that Z will exceed 1.96 is only .025. This is the shaded right-hand tail shown in the figure. Because the normal distribution is symmetric, the left-hand tail has the same probability .025. Thus, we can find the probability we want in the central chunk:

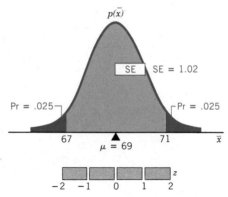

Figure 6-5
Standardization of $\overline{X}$.

$$\text{Probability} = 1.000 - .025 - .025$$
$$= .950$$

We therefore conclude that there is a 95% chance that the sample mean will be within 2 inches of the population mean.

Notice that the key to such questions is to standardize the critical $\overline{X}$ value into a Z value:

$$Z = \frac{\overline{X} - \mu}{\text{SE}}$$

(6-11)
like (4-23)

Then, for this Z value, we look up the probability in the normal Table IV.

We have often remarked that "averaging out" reduces the chance of an extreme value of $\overline{X}$. Another example will make this clear.

EXAMPLE **6-3**

a. Suppose a large class in statistics has marks normally distributed around a mean of 72 with a standard deviation of 9. Find the probability that an individual student drawn at random will have a mark over 80.

b. Find the probability that a random sample of 10 students will have an average mark over 80.

c. If the population were not normal, what would be your answer to part **b**?

Solution

a. The distribution for an individual student (i.e., the population distribution) is shown as the flat gray distribution p(x) in Figure 6-6. The score X = 80 is first standardized with its mean $\mu = 72$ and standard deviation $\sigma = 9$:

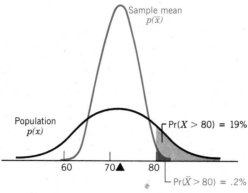

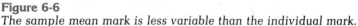

Figure 6-6
The sample mean mark is less variable than the individual mark.

$$Z = \frac{X - \mu}{\sigma} = \frac{80 - 72}{9} = .89 \qquad \text{(4-23) repeated}$$

Thus, $\qquad\qquad \Pr(X > 80) = \Pr(Z > .89)$
$$= .187 \simeq 19\%$$

b. The Normal Approximation Rule (6-9) assures us that $\bar{X}$ has an approximately normal distribution shown by the blue curve, with expected value $= \mu = 72$ and standard error SE $= \sigma/\sqrt{n} = 9/\sqrt{10} = 2.85$. We use these to standardize the critical score $\bar{X} = 80$:

$$Z = \frac{\bar{X} - \mu}{SE} \qquad\qquad \text{(6-11) repeated}$$

$$= \frac{80 - 72}{2.85} = 2.81$$

Thus, $\qquad\qquad \Pr(\bar{X} > 80) = \Pr(Z > 2.81)$
$$= .002$$

These two probability calculations may be easily compared in Figure 6-6. Although there is a reasonable chance (about 19%) that a single student will get over 80, there is very little chance (only about .2%) that a sample average of 10 students will perform this well. Once again we see how "averaging out" tends to reduce the extremes.

c. The Normal Approximation Rule (6-9) tells us that, when n is large, $\bar{X}$ has an approximately normal shape, no matter what the shape of the parent population. We would therefore get approximately the same answer.

The next example will show how a problem involving a *total* can be solved by simply rephrasing it in terms of a mean.

EXAMPLE **6-4**

A ski lift is designed with a total load limit of 10,000 pounds. It claims a capacity of 50 persons. Suppose the weights of all the people using the lift have a mean of 190 pounds and a standard deviation of 25 pounds. What is the probability that a random group of 50 persons will total more than the load limit of 10,000 pounds?

Solution

First, we rephrase the question: "A random sample of 50 persons will total more than 10,000 pounds" is exactly the same as "A random sample of 50 persons will *average* more than 10,000/50 = 200 pounds each."

From the Normal Approximation Rule, the sample average $\overline{X}$ has an approximately normal distribution with expected value = μ = 190, and a standard error = $\sigma/\sqrt{n}$ = 25/$\sqrt{50}$ = 3.54. We use these to standardize the critical average of 200 pounds:

$$Z = \frac{\overline{X} - \mu}{SE} = \frac{200 - 190}{3.54} = 2.83$$

Thus,
$$Pr(\overline{X} > 200) = Pr(Z > 2.83)$$
$$= .002$$

Thus the chance of an overload is only .2%.

The knowledge that the sample mean is normally distributed is very important—not only in making statements about a sample *mean*, but also in making statements about a sample *total* (such as the load limit in the last example).

PROBLEMS

6-9 **a.** A population of incomes of workers between the ages of 18 and 30 is skewed with mean μ = $20,000 and standard deviation σ = $6,000. A sociologist plans to randomly sample 25 incomes from this population. Her sample mean $\overline{X}$ will be a random variable that will only imperfectly reflect the population mean μ. In fact, the possible values of $\overline{X}$ will fluctuate around an expected value of_____with a standard error of_____, and with a distribution shape that is_____.

b. The sociologist is worried that her sample mean will be misleadingly high. A statistician assures her that it is unlikely that $\overline{X}$ will exceed μ by more than 10%. Calculate just how unlikely this is.

6-10 **a.** The millions of SAT math scores of the population of U.S. college-bound seniors in 1978 were approximately normally distributed—around a mean of 470, with a standard deviation of 120.

For a student drawn at random, what is the chance of a score above 500? Show this on a graph of the distribution.

b. Suppose that the registrar of Elora College does not know the population mean, so she draws a random sample of 250 scores, whose average $\overline{X}$ she hopes will crudely estimate the whole population mean. In fact, she hopes $\overline{X}$ will be no more than 10 points off. What are the chances of this?

Show this on a graph of the distribution that shows how $\overline{X}$ fluctuates from sample to sample.

6-11 "One pound" packages filled by a well-worn machine have weights that vary normally around a mean of 16.2 oz, with a standard deviation of .12 oz. An inspector takes a sample of n packages off the production line to see whether their average weight is at least 16 oz. If not, the firm faces a $500 fine. What is the chance of such a fine if the sample size is:

a. n = 1?

b. n = 4?

c. n = 16?

6-12 Suppose that the population of weights of airline passengers has mean 150 pounds and standard deviation 25 pounds. A certain plane has a capacity of 7,800 pounds. What is the chance that a flight of 50 passengers will overload it?

6-13 The managers of Mercury Mufflers find that the time t (in minutes) required for a worker to replace a muffler varies. Over a period of a year, they collected the following data:

t	RELATIVE FREQUENCY
20	10%
30	50%
40	30%
50	10%

a. Calculate the mean and standard deviation of the replacement time t.

b. They plan to do 50 mufflers with 4 men in a day and hope to finish them all between 9 a.m. and 5 p.m. What proportion of the days will they fail to finish on time?

6-14 In each of Problems 6-11 to 6-13, what crucial assumption did you implicitly make? Suggest some circumstances where it would be seriously violated. Then how would the correct answer be different?

6-15 Suppose that in a large club of young couples, the number of children in a family varies over a very small range—0 and 1—as follows:

NUMBER OF CHILDREN, x	RELATIVE FREQUENCY, $p(x)$
0	.70
1	.30
Total	1.00

a. What is the population mean? How is it related to the proportion of familes that have $X = 1$ child?

b. Five couples were chosen as delegates to a conference on family relations. What is the chance that this group of 5 couples would have more than an average of 1/2 child per family:

    i. If they were chosen at random?

    ii. If they were the first 5 couples to volunteer?

6-4 Proportions

A—PROPORTIONS ARE NORMAL, LIKE MEANS

As we first mentioned in Chapter 1, the sample proportion P is an estimate of the population proportion π. Like $\overline{X}$, P also fluctuates from sample to sample in a pattern (sampling distribution) that is easily summarized:

> *The Normal Approximation Rule for Proportions.* In random samples (VSRS) of size n, the sample proportion P fluctuates around the population proportion π with a standard error of $\sqrt{\pi(1 - \pi)/n}$.
>
>   Therefore, as n increases, the sampling distribution of P concentrates more and more around its target π. It also gets closer and closer to normal (bell-shaped)

(6-12)
like (6-9)

(Notice how this Normal Approximation Rule for proportions is very similar to the rule for means. This is no accident: A proportion in fact is just a disguised mean, as we will show in part C later.)

To illustrate the Normal Approximation Rule, in Figure 6-7 we show a population of voters—60% Republican (dark) and 40% Democratic (light). Several random samples (polls) of size $n = 100$ are indicated schematically as subsets, each with its different sample proportion P. All possible values of P form the sampling distribution—normally distributed around the target $\pi = .60$, with a standard error $SE = \sqrt{\pi(1 - \pi)/n} = \sqrt{.60(.40)/100} = .05$. From this, we can calculate any desired probability, as the next example illustrates.

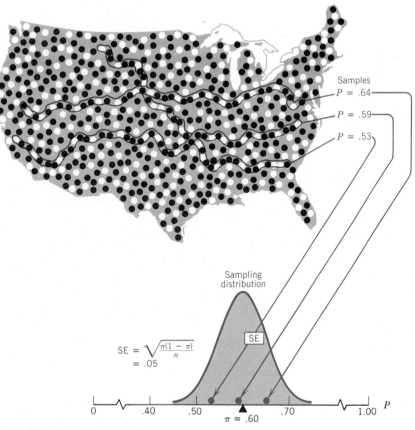

Figure 6-7
*Various possible samples give different proportions of Republicans, P. All possible values of P form the **sampling distribution**.*

EXAMPLE **6-5**   How likely is it that the poll above would contain a minority of Republicans?

Solution   A minority means the proportion of Republicans P is less than 50%. To calculate this probability under the normal curve, we first standardize the critical value $P = .50$:

$$Z = \frac{P - \pi}{SE} \qquad \text{like (6-11)}$$

$$= \frac{.50 - .60}{.05} = -2.00$$

Thus, $\Pr(P < .50) = \Pr(Z < -2.00)$
                                  $= .023 \simeq 2\%$

Thus the chance of an erroneous conclusion from the poll is only 2%.

The next example will illustrate how the sample proportion can be used to solve a wide variety of problems.

EXAMPLE **6-6**   Of your first 15 grandchildren, what is the chance there will be more than 10 boys?

Solution   First we rephrase the question: "More than 10 boys" is exactly the same as "The *proportion* of boys is more than 10/15." In this form, we can use the Normal Approximation theorem (6-11): P fluctuates normally around $\pi = .50$ with a standard error $= \sqrt{\pi(1-\pi)/n} = \sqrt{.50(.50)/15} = .129$. We use these to standardize the critical proportion $P = 10/15$,

$$Z = \frac{P - \pi}{SE}$$

$$= \frac{\dfrac{10}{15} - .50}{.129} = 1.29 \tag{6-13}$$

Thus, $\Pr\left(P > \dfrac{10}{15}\right) = \Pr(Z > 1.29)$

$$= .099 \simeq 10\% \tag{6-14}$$

The chance of more than 10 boys is about 10%.

Example 6-7 is recognized as a binomial problem. It could have been answered more precisely, but with a lot more work, by adding up the probability of getting 11 boys, plus the probability of getting 12 boys, and so on—with each of these probabilities calculated from the binomial formula (4-8). Since the normal approximation is so much easier to evaluate we shall use it henceforth, and refer to it as the *normal approximation to the binomial.*

*B—CONTINUITY CORRECTION

The Normal Approximation Rule (6-12) is only an approximation, and therefore does involve some error. To illustrate this error, note that we

could have answered the question in Example 6-6 *equally well* by rephrasing it differently at the beginning: "More than 10 boys" is exactly
the same as "11 or more boys." If we had begun in this way, in (6-13) we
would have been evaluating a critical proportion of 11/15 rather than
10/15. Then we would, of course, arrive at an answer slightly different
from (6-14), but nevertheless an equally good approximation. Rather than
having to choose between these two approximations, we could get an even
better approximation by striking a compromise right at the beginning:
Instead of 10 boys or 11 boys, use $10\frac{1}{2}$ boys. When this is used in (6-13),
we obtain

$$Z = \frac{\frac{10\frac{1}{2}}{15} - .50}{.129} = 1.55$$

Thus,
$$\Pr\left(P > \frac{10\frac{1}{2}}{15}\right) = \Pr(Z > 1.55)$$
$$= .061 \tag{6-15}$$

This compromise agrees very well with the correct answer of .059, which
we would have obtained if we had gone back and answered the question
the absolutely correct (but tedious) way of fighting our way through the
binomial distribution.

Why is this compromise so good? We arrived at the crude answer of
.098 in (6-14) by approximating a discrete distribution (the binomial) with
a continuous distribution (the normal). This involves a slight mismatch,
and the compromise (6-15) allowed us to correct for this. Thus the compromise is customarily called the *continuity correction* (and is further
explained in Appendix 6-4). It is so helpful that it is worth stating carefully.

> *The continuity correction* (C.C) to the binomial is ob-
> tained by first phrasing the question in the two possible
> ways (for example, "more than 10" or "11 or more.")
> Then the half-way value ($10\frac{1}{2}$) is used in the subsequent
> normal approximation.   (6-16)

*C—PROPORTIONS AS SAMPLE MEANS

We can show that the sample proportion is just a disguised sample mean,
and this has two benefits: (1) It provides a simple introduction to dummy
variables, an indispensable concept in applied statistics. (2) It gives us
"two concepts for the price of one," and so halves the new material we
have to learn. In particular, we will show how the normal approximation

rules are related—that is, how proportions in (6-12) follow from means in (6-9).

Once more we think of the population as a bowl of chips. Sometimes, as in our example of men's heights, the chips have many possible numbers written on them. At other times, as in the family size of Problem 6-15, there are only two kinds of chips. Populations like this, with only two alternatives, lead to the binomial distribution where we call the two alternatives "success" and "failure." As another example, let us next look at a poll where there are only two parties.

EXAMPLE **6-7**   An election may be interpreted as asking every voter in the population, "How many votes do you cast for the Republican candidate?" If this is an honest election, the voter has to reply either 0 or 1. In the 1972 Presidential election (the Republican Nixon versus the Democrat McGovern, ignoring third parties as usual), 60% voted Republican ($X = 1$), while 40% voted Democratic ($X = 0$). The population distribution was therefore as follows:

x = NUMBER OF REPUBLICAN VOTES BY AN INDIVIDUAL	$p(x)$ = RELATIVE FREQUENCY
0	.40
1	.60

a. What is the population mean? The population variance? The population proportion of Republicans?

b. When 10 voters were randomly polled, they gave the following answers:

$$1, 0, 1, 1, 1, 0, 1, 0, 1, 1$$

What is the sample mean? The sample proportion of Republicans?

Solution   **a.** We calculate the population mean and variance in Table 6-2—a short and peculiar table. In fact, it's as short as a table can possibly be. This makes the arithmetic very simple, however, and we easily find the mean is .60—which exactly equals the proportion of Republicans.

We also find the variance is .24—which equals .60 × .40 (the proportion of Republicans and Democrats, respectively). It is remarkable that a table can be so simple to calculate, and at the same time yield such interesting answers.

b. The sample mean is $\overline{X} = \Sigma X/n = 7/10 = .70$. The sample proportion is $P = 7/10$, which exactly equals the sample mean. Once again, just as in part a, we have found that *the proportion coincides with the mean.*

TABLE **6-2** **Population Mean and Variance for a 0-1 Variable (a) When the Proportion of Republicans is 60%, as in Example 6-7 (b) In General, When the Proportion is π**

(a) x = NUMBER OF REPUBLICAN VOTES AN INDIVIDUAL CASTS	p(x) = RELATIVE FREQUENCY = POPULATION PROPORTION	$xp(x)$	$x^2p(x)$
0	.40	0	0
1	.60	.60	.60

$$\mu = \Sigma xp(x) = .60 \qquad E(X^2) = .60$$
$$\text{Therefore, from (4-35),}$$
$$\sigma^2 = E(X^2) - \mu^2$$
$$= .60 - .60^2 = .24$$

(b) x	p(x)	$xp(x)$	$x^2p(x)$
0	$(1 - \pi)$	0	0
1	π	π	π

$$\mu = \Sigma x\, p(x) = \pi \qquad E(X^2) = \pi$$
$$\text{Therefore } \sigma^2 = E(X^2) - \mu^2$$
$$= \pi - \pi^2 = \pi(1 - \pi)$$
$$\sigma = \sqrt{\pi(1 - \pi)}$$

From this example, we can easily generalize. To set things up, for a single voter drawn at random, let

$$X = \text{the number of Republican votes this individual casts} \quad (6\text{-}17)$$

If this seems a strange way to define such a simple random variable, we could explicitly define it with a formula:

$$X = 1 \quad \text{if the individual votes Republican} \qquad (6\text{-}18)$$
$$= 0 \quad \text{otherwise}$$

Thus the population of voters may be thought of as the bowl of chips, marked 0 or 1, shown in Figure 6-8. As usual, we shall let π denote the population proportion that is Republican (proportion of chips marked 1). Then, as shown in Table 6-2, the moments of the population can be expressed in terms of this proportion,

$$\text{Population mean } \mu = \text{population proportion } \pi \qquad (6\text{-}19)$$

$$\text{Population standard deviation } \sigma = \sqrt{\pi(1 - \pi)} \qquad (6\text{-}20)$$

What is true for the population (6-19) is also true for the sample:

$$\boxed{\text{Sample mean } \overline{X} = \text{sample proportion } P} \qquad (6\text{-}21)$$
$$\text{like (6-19)}$$

Figure 6-8
A 0-1 population (population of voters).

To confirm (6-21), note that when we take a sample, its mean is calculated by adding up the 1's (counting up the Republicans), and dividing by n. But, of course, this is just the sample proportion of Republicans.

Thus we have found an ingenious way to handle a proportion: It is simply a disguised mean—the mean of a 0-1 variable. (Since it provided such a simple way of counting the number of Republicans, the 0-1 variable is also called a *counting variable*. Other names are *on-off variable*, *binary variable*, or, most commonly, *dummy variable*.)

Since P is just the sample mean $\overline{X}$ in disguise, we can find its expected value by recalling the general theory of sampling already developed:

$$\text{Expected value of } \overline{X} = \text{population mean } \mu \qquad \text{like (6-5)}$$

Therefore, substituting (6-21) and (6-19):

$$\boxed{\text{Expected value of } P = \text{population proportion } \pi} \qquad \text{(6-22)}$$

The standard error of P is similarly obtained by recalling (6-7), and then substituting into it (6-21) and (6-20):

$$\boxed{\text{Standard error of } P = \sqrt{\frac{\pi(1 - \pi)}{n}}} \qquad \text{(6-23)}$$

With these moments of the sample proportion P in hand, what is the *shape* of its distribution? Recall that the Normal Approximation Rule (6-9) tells us that, if a large sample is taken from essentially *any* population (even the 0-1 population we are now considering), the sample mean will be approximately normal. In this case the sample mean is the sample proportion; thus the normal approximation rule is true for proportions too.

PROBLEMS

If you want high accuracy in your answers (optional), you should do them with continuity correction (wcc) as given in (6-16), especially if the sample size n is small.

6-16 Los Angeles has about four times as many voters as San Diego. To estimate the proportion who will vote for a bond to finance bicycle paths, suppose a random sample is taken in each city. A sample of 1,000 in L.A. will be about _____ as accurate as a sample of 250 in San Diego.

6-17 In a certain city, 55% of the eligible jurors are women. If a jury of 12 is picked fairly (at random), what is the probability that there would be 3 or fewer women?

6-18 In a large production run of millions of electronic integrated circuits, only 2% are defective. What is the chance that of 1,000 circuits pulled off the assembly line, 40 or more would be defective?

6-19 The 1982 model Q-cars had to be recalled because of a slight defect in the suspension. Of those recalled, 20% required repairs. A small dealer with an overworked service department hopes that no more than 5 of his 50 recalled cars will require repairs. What is the chance of his being so lucky?

6-20 In Problems 6-18 and 6-19, what assumption did you make to get your answer? Suggest some circumstances when it would be seriously violated. Then how would the correct answer be different?

6-21 **a.** In the 1976 Presidential election, 51% of the voters were for Carter. If a Gallup poll of 1,000 voters had been randomly sampled from this population, what is the chance it would have erroneously predicted Carter to have a minority?

*b. If the chance of error in part **a** is to be reduced to 1%, how large should the sample be?

6-22 What is the chance that, of the first 10 babies born in the New Year, 7 or more will be boys? Answer in three ways (assuming the chance of a boy on each birth is .50):

  a. Exactly, using the binomial distribution.

  b. Approximately, using the normal distribution.

  *c. Approximately, using the normal approximation with continuity correction.

  *d. Answer True or False; if False, correct it: Part **c** illustrates that the normal approximation with continuity correction can be an excellent approximation, even for n as small as 10.

*6-23 In sampling from a 0-1 population, the *proportion* of successes P is related to the *total number* of successes S. For example, when the total number of Republicans was $S = 7$ in a sample of $n = 10$, we calculated $P = 7/10$, and, in general:

$$P = \frac{S}{n}$$

$$\text{or } S = nP$$

Using the moments of P and the theory of linear transformations, prove that the moments of S are those given in (4-13).

*6-5   Small-Population Sampling

Recall that random sampling [VSRS as defined in (6-3)] does not hold if the population size N is small and the observations (or chips from the bowl) are kept out as they are drawn. Keeping out the chips (sampling without replacement) is more efficient because we do not risk "drawing the same chip" over again and repeating information already known.

For example, suppose we sample the heights of ten men on a small college campus; suppose further that the first student we sample happens to be the 7-foot star of the basketball team. Clearly, we now face the problem of a sample average that is too high. If we replace, then in the next nine men who are chosen, the star *could* turn up again, thus distorting our sample mean for the second time. But if we don't replace, then we don't have to worry about this tall individual again. In summary, sampling without replacement yields a less variable sample mean because extreme values, once sampled, cannot return to haunt us again.

In general, if N and n denote the population and sample size, the gain in efficiency may be expressed as follows.

> If observations are kept out as they are drawn (sampling without replacement), the sampling fluctuation in $\overline{X}$ or P is reduced by the factor:
>
> $$\text{Reduction factor} = \sqrt{\frac{N - n}{N - 1}}$$ (6-24)

In other words, the standard error of $\overline{X}$ changes from $\sigma/\sqrt{n}$ to

$$\text{Standard error of } \overline{X} = \frac{\sigma}{\sqrt{n}} \sqrt{\frac{N - n}{N - 1}}$$ (6-25)

Similarly,

$$\text{Standard error of } P = \frac{\sqrt{\pi(1 - \pi)}}{\sqrt{n}} \sqrt{\frac{N - n}{N - 1}}$$ (6-26)

Certain special sample sizes shed a great deal of light on the reduction factor (6-24):

1. When there is only $n = 1$ chip sampled, it does not matter whether or not it is replaced (because we never take another chip out of the bowl). This is reflected in the reduction factor becoming 1.00. [If you have wondered where the 1 came from in the denominator of (6-24), you can see that it is needed to logically make (6-24) and (6-7) equivalent—as they must be—for a sample size of one.]

2. When $n = N$, the sample coincides with the whole population, every time. Hence every sample mean must be the same—the population mean. The variance of the sample mean, being a measure of its fluctuation, must be zero. This is reflected in the reduction factor becoming 0.

3. On the other hand, in a large population where the population size N is far greater than the sample size n, then the reduction factor is practically 1, and therefore can be ignored. For example, in the typical poll where $N \simeq 100,000,000$ and $n \simeq 1,000$,

$$\text{Reduction factor} = \sqrt{\frac{N - n}{N - 1}}$$

$$= \sqrt{\frac{100,000,000 - 1,000}{100,000,000 - 1}} = .999995 \simeq 1.00$$

We conclude that (6-25) and (6-26) apply to a population of *any* size. But in the large populations we consider hereafter in this book, the reduction factor becomes approximately 1, and can therefore be dropped.

*PROBLEMS

6-24 Rework each of the following problems, assuming the given population size N (and sampling is without replacement):

a. Problem 6-9, assuming $N = 80$.

b. Problem 6-18, assuming $N = 10,000$.

c. Problem 6-19, assuming $N = 50,000$.

6-25 Answer True or False; if False, correct it:

a. When sampling from a population without replacement, the SE of $\bar{X}$ or of P contains the reduction factor $\sqrt{(N - n)/(N - 1)}$.

b. However, if the population is large relative to the sample, this factor may be ignored.

c. To be specific, if $N \geq 100n$, then the reduction factor is between .99 and 1.00, and so changes the SE less than 1%.

6-26　**a.**　In a bridge hand (13 out of 52 cards), what is the probability that there will be at least 7 spades? At least 7 of one suit?

　　　　b.　In a poker hand (5 out of 52 cards), what is the probability that there will be at least 2 aces?

6-27　In the game of bridge, a simple scoring system allots points to each card as follows:

CARD	POINTS
Each card below Jack	0
Jack	1
Queen	2
King	3
Ace	4

　　　　a.　For the population of 52 cards, find the mean number of points, and the standard deviation.

　　　　b.　In a randomly dealt hand of 13 cards, what is the probability that there will be at least 13 points? (Bridge players beware: no points counted for distribution.)

*6-6　Monte Carlo

Although the star on this section indicates it may be skipped, it nonetheless provides some useful insights into sampling—in particular, how "averaging out" makes $\overline{X}$ a reliable estimate of μ. By taking repeated samples, and in each case calculating $\overline{X}$, we will see how $\overline{X}$ fluctuates around its target μ—and so confirm the Normal Approximation Rule.

　　Monte Carlo does far more than confirm, however. In situations that are mathematically intractable, Monte Carlo often provides the *only* practical way to determine sampling distributions. For example, the sampling distribution of the median $\overset{\cdot}{X}$, or of the interquartile range, are both commonly studied by Monte Carlo.

A—SAMPLING OF A SMALL POPULATION USING RANDOM DIGITS

As an example, suppose Table 6-3 gives the known distribution of the desired family size X for the total population of 100 young people in Middletown, USA, aged 18–24. We calculated its mean $\mu = 2.64$ and show it as the customary center of gravity in Figure 6-9. We also give a serial number to everybody in the population, starting at $X = 0$, so that the serial numbers in Table 6-9 correspond to the frequencies in the earlier column.

$$\text{Population mean } \mu = \Sigma\, xp(x) = 2.64 \text{ children}$$
$$\text{and } \sigma = \Sigma(x - \mu)^2 p(x) = 1.2 \text{ children}$$

TABLE **6-3**   **Desired family size X, from the population of 100 Americans in Middletown, aged 18–24 in 1980**

| | POPULATION DISTRIBUTION | | |
| | | RELATIVE FREQUENCY | |
x	FREQUENCY	p(x)	SERIALNUMBERS
0	2	.02	01–02
1	6	.06	03–08
2	49	.49	09–57
3	22	.22	58–79
4	15	.15	80–94
5	3	.03	95–97
6	2	.02	98–99
7	1	.01	00
	N = 100	1.00	

Now we can see what would happen if, as is customary for statisticians, we were to come into this population "blind"—without any knowledge of it—and try to estimate μ by taking a small random sample (VSRS) of n = 5 observations. We draw 5 pairs of random digits in Appendix Table I. We can start anywhere, since all numbers in this table are purely random. If we start at the last row, for example, the first pair of random digits is 77, which is found in Table 6-3 to be the serial number of someone wanting

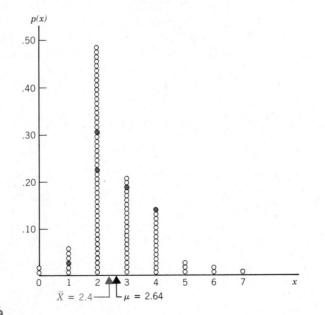

Figure 6-9
Desired family size X. Population of 100 Americans (as in Table 6-3) shown as dots. Random sample of n = 5 shown in color.

TABLE **6-4**   **A sample of _n_ = 5 observations drawn from the population in Table 6-3**

SERIAL NUMBER	DESIRED CHILDREN X
77	3
94	4
30	2
05	1
39	2

$$\overline{X} = 12/5$$
$$\overline{X} = 2.4$$

X = 3 children. Continuing to draw the remaining observations in the same way, we obtain the sample of n = 5 observations in Table 6-4. [Because we are sampling with replacement (VSRS), the luck of the draw might give us the same serial number twice, in which case that individual would be included both times in our sample.]

To summarize this sample, the sample mean $\overline{X}$ = 2.4 is calculated in Table 6-4. Then $\overline{X}$, along with the whole sample of 5 colored dots, is graphed in Figure 6-9. Just as in Figure 6-2, we note that the random sample is quite representative of the population, and so the sample mean $\overline{X}$ = 2.4 is quite a good estimate of the population mean μ = 2.64.

Now we can repeat the sampling over and over. Each time we draw a sample of 5 observations, calculate the mean $\overline{X}$, and mark it in a diagram like Figure 6-10. Thus we build up the sampling distribution, which shows us how $\overline{X}$ fluctuates around its target μ. An example will show the details.

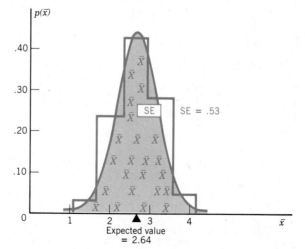

Figure 6-10
Sampling distribution of $\overline{X}$, for n = 5.

EXAMPLE **6-8**

a. Consider again the population in Table 6-3. Let each student in your class draw a sample of $n = 5$ observations (starting at a different place in Appendix Table I, of course), and then calculate the sample mean $\overline{X}$.

   If your class is small, each student can repeat the experiment several times, so that the class obtains at least 50 values of $\overline{X}$. Then let the instructor tabulate and graph the frequency distribution of $\overline{X}$. (We recommend grouping the values of $\overline{X}$ into cells of width .6, taking the cell midpoints to be 2.0, 2.6, 3.2, ...)

b. Calculate the mean and standard error of the distribution of $\overline{X}$. How are they approximately related to $\mu = 2.64$ and $\sigma = 1.2$ for the parent population?

Solution

a. By repeating the sampling experiment over and over like this, we begin to understand the luck of the draw, just as in repeated spinning of a roulette wheel. If this Monte Carlo experiment were carried out for millions of values of $\overline{X}$ (rather than just 50 or so), then the relative frequencies would settle down to the probabilities calculated in the table below and shown as the bars in Figure 6-10. (Your relative frequencies, based on 50 or so values of $\overline{X}$, will roughly resemble these probabilities.) Also shown is the approximate normal distribution that would be obtained if we had used many more finely divided cells.

SAMPLING DISTRIBUTION		CALCULATION OF THE EXPECTED VALUE	CALCULATION OF THE STANDARD ERROR	
$\overline{X}$ (MIDPOINT)	$p(\overline{x})$ (REL FREQ)	$\overline{x}p(\overline{x})$	$(\overline{x} - 2.64)^2$	$(\overline{x} - 2.64)^2 p(\overline{x})$
1.4	.03	.042	1.54	.046
2.0	.23	.460	.41	.094
2.6	.42	1.092	.00	.000
3.2	.28	.896	.31	.088
3.8	.04	.152	1.35	.054
	1.00	Expected value = 2.64	Variance = .28 SE = $\sqrt{.28}$ = .53	

b. The expected value of $\overline{X}$ is calculated from the sampling distribution in the table above, and turns out to be 2.64, the same as the population mean ($\mu = 2.64$). The standard error of $\overline{X}$ is also calculated, and turns out to be .53, which is $1/\sqrt{5}$ times the standard deviation of the population ($\sigma = 1.2$). Your classroom calculations (based on 50 or so values of $\overline{X}$) will be fairly close to this.

Example 6-8 has nicely demonstrated the Normal Approximation Rule: Even though the population is quite skewed, nevertheless the sample mean $\overline{X}$ fluctuates approximately normally around the target μ, with standard error $\sigma/\sqrt{n}$.

B—SAMPLING A POPULATION OF ANY SIZE

Now let us enlarge our horizon beyond Middletown, USA, and consider the population of *all* Americans in 1980, aged 18–24. Note that their relative frequencies given in Table 6-5 are the same as in Table 6-3. (In fact, the hypothetical population of 100 in Middletown was taken to be an exact cross-section of America.)

When we sample our first observation from the population of 30 million, what is the chance it will be $X = 1$, for example? The relative frequency, of course, which is 1,800,000/30,000,000 = .06. This is the same chance as in Middletown. So the same simulation will work: You can visualize the same serial numbers in Table 6-5 as in Table 6-3. Thus the sampling simulation will be exactly the same, and this is a point worth emphasizing in general.

> In sampling with replacement, only the *relative* frequencies matter. The size of the population N is irrelevant.   (6-27)

C—MONTE CARLO BY COMPUTER

The sampling by hand calculation that we have done so far has provided a lot of insight. In practice, however, a computer is much better at such repetitive calculations. A computer can easily draw a sample and calculate its mean $\overline{X}$, then repeat this procedure over and over, hundreds of times a second. It takes only a few seconds, therefore, to build up a sampling distribution of $\overline{X}$.

TABLE **6-5** **Desired family size X, for the population of all 30,000,000 Americans aged 18–24 in 1980 (from Gallup, 1980, approx.)**

X	FREQUENCY	RELATIVE FREQUENCY
0	600,000	.02
1	1,800,000	.06
2	14,700,000	.49
3	6,600,000	.22
4	4,500,000	.15
5	900,000	.03
6	600,000	.02
7	300,000	.01
	30,000,000	1.00$\sqrt{}$

   As an example, suppose the computer is to simulate a random sample
(of n = 10 observations) from a normal population of men's heights with
μ = 69 and σ = 3.2. This involves 4 steps:

1. The "Monte Carlo wheel is spun" in the computer to pick a standard
    normal value at random, such as Z = 1.5 (just as we might select a
    random Z value from Table II).
2. We don't want to stop there, with a *standard* normal value Z; instead
    we want a normally distributed *height* X. We can easily convert Z
    into X by recalling the equation relating the two:

$$Z = \frac{X - \mu}{\sigma}$$

(4-23) repeated

Rearranging this yields

$$X = \mu + Z\sigma$$

(6-28)

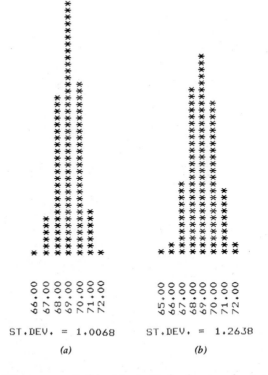

ST.DEV. = 1.0068   ST.DEV. = 1.2638

                (a)   (b)

Figure 6-11
*Sampling distributions computed by Monte Carlo, for sample size n = 10 from
a normal population. (a) the sample mean X̄. (b) The sample median X̆ (more
unreliable, since its distribution is more spread out; compare its standard error
of 1.26 with 1.01).*

Thus the standard value $Z = 1.5$ translates into a height of:

$$X = 69 + 3.2(1.5) = 73.8 \text{ inches}$$

3. This process is repeated 10 times altogether, to get a random sample of $n = 10$ men's heights. The computer then calculates their mean $\overline{X}$.

4. In the same way the computer can then compute a lot more sample means $\overline{X}$, and array them in a graph. This finally gives us the sampling distribution of $\overline{X}$—as shown in Figure 6-11a.

Of course, the computer could just as easily calculate the sampling distribution of some other statistic, such as the median $\overset{\vee}{X}$—as shown in Figure 6-11b. This shows us that for estimating the center of a normal population, $\overset{\vee}{X}$ is less reliable than $\overline{X}$.

The computer could simulate sampling from non-normal populations as well. Such Monte Carlo studies are useful not only as a check on theoretical results such as the normal approximation rule (6-9). They are particularly valuable in providing sampling distributions that cannot be derived theoretically.

*PROBLEMS

6-28   The 1978 populations of the 50 states were as follows, numbered in alphabetical order (D.C. included in Maryland).

1. Ala	3.7	11. Ha	.9	21. Mass	5.8	31. NM	1.2	41. SD	.7
2. Alas	.4	12. Ida	.9	22. Mich	9.2	32. NY	17.7	42. Tenn	4.3
3. Ariz	2.4	13. Ill	11.2	23. Minn	4.0	33. NC	5.6	43. Tex	13.0
4. Ark	2.2	14. Ind	5.4	24. Miss	2.4	34. ND	.7	44. Ut	1.3
5. Cal	22.3	15. Ia	2.9	25. Mo	4.8	35. Oh	10.7	45. Vt	.5
6. Col	2.7	16. Kan	2.3	26. Mon	.8	36. Okla	2.8	46. Va	5.2
7. Conn	3.1	17. Ky	3.5	27. Neb	1.6	37. Ore	2.5	47. Wash	3.8
8. Del	.6	18. La	4.0	28. Nev	.7	38. Pa	11.8	48. W.Va	1.9
9. Fla	8.7	19. Me	1.1	29. NH	.9	39. RI	.9	49. Wis	4.7
10. Ga	5.1	20. Md	4.8	30. NJ	7.3	40. SC	2.9	50. Wy	.4

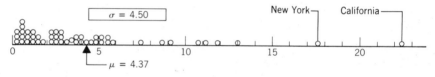

Population of state (millions)

a. Using the random digits from Appendix Table I, draw a random sample of $n = 5$ states. Calculate $\overline{X}$ to estimate μ.

b. On the graph above, mark the 5 observations and their mean $\overline{X}$. Is $\overline{X}$ closer to the target μ than the typical individual observation?

c. Estimate the total population of the country, and compare to the actual population (218 million).

6-29 a. Repeat the sampling experiment in Problem 6-28(a) several more times. Then graph the different values of $\overline{X}$ in order to see roughly how $\overline{X}$ varies from sample to sample.

b. If this sampling experiment was repeated *millions* of times, would the millions of values of $\overline{X}$ fall in any sort of pattern? If so, sketch it in the graph in part (a).

6-30 A small ski tram holding 8 skiers routinely waits until it is filled before going up. The resort owner is worried that the time required to fill it will exceed the patience of the skiers, in particular, will exceed 10 minutes.

Suppose that skiers arrive at random to board the tram. To be precise, denote the waiting time for the first skier by T_1, then the additional time for the second skier by T_2, and so on. Suppose that these successive waiting times $T_1, T_2, \ldots, T_8$ are independent, and that all have the following common distribution (in minutes):

Waiting Time t	0	1	2	3	4	5	6	7	8	9
$p(t)$	.33	.22	.14	.09	.07	.05	.04	.03	.02	.01

a. Simulate one loading of the ski tram. What is the waiting time required? Did it happen to exceed 10 minutes?

b. Using the theory of Section 6-3, calculate the probability that the total waiting time will exceed 10 minutes.

6-31 a. Simulate drawing a random sample of 5 observations from a normal population of women's heights, with $\mu = 65''$, $\sigma = 3''$. Calculate the sample mean and median.

*b. By pooling the answers from all the students in the class, have the instructor graph the sampling distribution of the sample mean and median. Which seems to be more accurate? Are the results similar to Figure 6-11?

*c. As an alternative to part **b**, carry out the simulation by computer. (We suggest somewhere between 10 and 1,000 samples be simulated.)

CHAPTER 6 SUMMARY

6-1 A *random* sample is one where all individuals in the population are equally likely to be sampled. Its great advantage is its fairness, which ensures *unbiased* estimation of the underlying population. Another important advantage is that sampling uncertainty can be evaluated.

In calculating the sample mean, an extreme observation tends to be diluted by more typical observations. Because of this averaging out, the sample mean doesn't vary as widely as the individual observations in the population.

6-2 To be specific, the sample mean $\overline{X}$ fluctuates around its target μ (the population mean) with a standard error of only $\sigma/\sqrt{n}$ (where σ is the standard deviation of the population, and n is the number of observations in the sample).

6-3 As sample size n increases, the sampling distribution of $\overline{X}$ concentrates more and more around its target μ. It also gets closer and closer to the normal shape. This normal approximation rule makes probability calculations very easy: We just standardize with $Z = (\overline{X} - \mu)/SE$, and refer to the standard normal table.

6-4 Proportions are very similar to means. The sample proportion P fluctuates normally about its target π, with a standard error of $\sqrt{\pi(1 - \pi)/n}$. This similarity of P to $\overline{X}$ is no accident. If the two categories such as Democrat-Republican are handled with a 0-1 variable, then it turns out that P is just a special case of $\overline{X}$.

*6-5 Keeping out the chips as they are drawn (sampling without replacement) is more efficient than sampling with replacement. Specifically, it reduces the standard error of $\overline{X}$ (or P) by the factor $\sqrt{(N - n)/(N - 1)}$, where N is the population size. (In large populations this "reduction factor" approaches 1, so it doesn't matter whether or not sampling is with replacement.)

*6-6 Monte Carlo (sampling simulation) confirms theoretical results like the normal approximation rule. Done quickly and cheaply on a computer, Monte Carlo is even more valuable in providing sampling distributions that are too difficult to derive theoretically.

REVIEW PROBLEMS

6-32 Match the symbol on the left with the phrase on the right:

μ	sample mean
$\overline{X}$	sample variance
σ^2	population proportion
s^2	population variance
π	sample proportion
P	population mean

6-33 Fill in the blank:

Suppose that in a certain election, the United States and California are alike in their proportion of Democrats, π, the only difference being that the United States is about 10 times as large a population. In order to get an equally reliable sampling estimate of π, the U.S. sample should be _____ as large as the California sample.

6-34 The diameters of the cylinders drilled into an engine block vary slightly, being normally distributed around the "target value" of 12.500 cm, with a standard deviation of .002 cm. If the diameter is within .003 cm of its target, it is acceptable.

    a. What proportion of the cylinders are acceptable?

    b. The foreman specifies that an engine block is acceptable if all four of its cylinders are acceptable. What is the probability that a block will be acceptable?

    c. His assistant feels that an engine block should be regarded as acceptable if the *average* diameter of its four cylinders is within .003 cm of its target. What proportion of the blocks satisfy this criterion?

    d. Who is right, the foreman or his assistant?

    e. What important assumption did you make in calculating:

        i. The answer to part **b**?

        ii. The answer to part **c**?

6-35 To get a cross section of public opinion, the Gallup poll typically interviews 1500 Americans. This poll in some ways is more reliable than simple random sampling. For example, it deliberately selects an equal number of men and women, rather than leaving the split to chance. (This is called *stratified sampling.*)

    In other ways, the Gallup poll is less expensive and a little less reliable than simple random sampling. For example, once a location has been randomly chosen, several people can be selected at random from it almost as easily as one person. Such *multistage sampling* makes it relatively easy to fill up the sample with 1500 people. Yet to the extent that later people in a given location tend to think like the first, they do not provide as much independent information as in a simple random sample, where each person is chosen entirely independently of the others.

    These two features—the increased reliability of stratified sampling and the decreased reliability of multistage sampling—tend to offset each other, so that the reliability of the sample is about the same as the reliability of a very simple random sample (VSRS).

    a. The Gallup poll interviews a fresh sample of 1500 adults several times a year—say 20 times. Roughly, what are the chances they will interview *you* sometime in the next 10 years? (Use the 1980 U.S. population of 230 million, with 160 million being adults over 18)

    b. As part **a** shows, the Gallup poll has a very small chance of ever picking any given voter. In view of this, can they make any legitimate claims to be representative, and provide a reliable picture of the *whole country*?

6-36 In the 1980 presidential election, 34.9 million voted Democratic and 43.2 million voted Republican (Carter versus Reagan, ignoring third parties as usual). The typical political poll randomly samples 1500 voters. What is the probability that such a poll would correctly forecast the election winner—that is, that a majority of the sample would be Republican?

6-37   American women have heights that are approximately normally distributed around a mean of 64 inches, with a standard deviation of 3 inches. Calculate and illustrate graphically each of the following:

    a.   If an American woman is drawn at random, what is the chance her height will exceed 66 inches (5 ft, 6 in.)?

    b.   If a sample of 25 are randomly drawn, what is the chance that the sample average will exceed 66 inches?

6-38   The 1200 tenants of a large apartment building have weights distributed as follows:

WEIGHT (POUNDS)	PROPORTION OF TENANTS
50	.20
100	.30
150	.40
200	.10

Each elevator in the building has a load limit of 2800 pounds. If 20 tenants crowd into one elevator, what is the probability it will be overloaded?

6-39   A population of potential jurors is 60% women. If a group of 30 potential jurors (called a "venire") is chosen at random, what is the probability that 10 or fewer would be women?

6-40   When polarized light passes through α-lactose sugar, it is rotated by an angle of exactly 90°. The *observed* angle, however, is somewhat in error; suppose it is normally distributed around 90° with a standard deviation of 1.2°. A sample of 4 independent observations is taken. What is the chance:

    a.   That the *first* observation exceeds 91°? (Since another sugar, D-xylose, rotates polarized light by 92°, the observer might then mistakenly think the α-lactose was D-xylose.)

    b.   That the *average* exceeds 91°?

    c.   That *all four* observations exceed 91°?
        [Before you calculate the chance, can you say how it will compare to parts **a** and **b**?]

6-41   At Las Vegas, roulette is played with a wheel that has 38 slots—20 losing slots and 18 winning slots. Your chances of losing your dollar are therefore 20/38, and of winning a dollar are 18/38.

    a.   In the very long run, what is the average loss per play?

    b.   What are a player's chances of ending up a net loser, if he plays a dollar:

        i. 5 times?

       ii. 25 times?

     iii. 125 times?

6-42 On March 1, a large greenhouse installs 1000 ultraviolet lamps. The manufacturer specifies that the length of life is normally distributed around a mean of 100 days, with a standard deviation of 25 days.

    a. What is the expected number that will have to be replaced by June 1 (92 days later)?

    b. What is the chance that more than 400 will have to be replaced by June 1?

6-43 An airline company notes that over a year's time, 15% of its passengers with reservations do not show up for the flight. So they oversell reservations by 10%. For example, for the Washington-London flight with 400 seats, they regularly sell 440 reservations.

    a. If the 440 passengers can be regarded every time as a random sample, what proportion of the flights will be overbooked—that is, will have more than 400 passengers with reservations show up?

    b. In what ways may the "random sample" assumption in part **a** be unrealistic? How would the realistic answer be different?

6-44 Recall that probability was defined as limiting relative frequency. That means, for example, that if a fair die is thrown a million times, the relative frequency of aces (or proportion P) will likely be very close to $\frac{1}{6}$. To be specific, calculate the probability that P will be within .001 of $\frac{1}{6}$.

Chapter

7

Point Estimation

Education is man's going forward from cocksure ignorance to thoughtful uncertainty.

DON CLARKS' SCRAPBOOK

7-1 Populations and Samples

In Table 7-1, we review the concepts of population and sample. It is essential to remember that the population mean μ and variance σ^2 are constants (though generally unknown). These are called population *parameters*.

By contrast, the sample mean $\overline{X}$ and sample variance s^2 are random variables. Each varies from sample to sample, according to its probability distribution. For example, the distribution of $\overline{X}$ was found to be approximately normal in (6-9). A random variable such as $\overline{X}$ or s^2, which is calculated from the observations in a sample, is given the technical name *sample statistic*. In Table 7-1 and throughout the rest of the text, we shall leave the *population gray* and make the *sample colored* in order to keep the distinction clear, just as we did in Chapter 6.

Now we can address the problem of statistical inference that we posed in Chapter 1: How can the population be estimated by the sample? Suppose, for example, that to estimate the mean family income μ in a certain region, we take a random sample of 100 incomes. Then the sample mean $\overline{X}$ surely is a reasonable estimator of μ. By the normal approximation rule (6-9), we know that $\overline{X}$ fluctuates about μ; sometimes it will be above μ, sometimes below. Even better than estimating μ with the single *point* estimate $\overline{X}$ would be to construct an *interval* estimate about $\overline{X}$ that is likely to bracket μ—a task we shall leave to Chapter 8.

TABLE **7-1**   **Review of Population versus Sample**

A RANDOM SAMPLE IS A RANDOM SUBSET OF THE POPULATION	
Relative frequencies f/n are used to compute	Probabilities p(x) are used to compute
$\overline{X}$ and s^2,	μ and σ^2,
which are examples of random statistics or estimators.	which are examples of fixed parameters or targets

For now, we ask whether another statistic, such as the sample median $\tilde{X}$, might do better than $\overline{X}$ as a point estimator. To answer such questions, we require criteria for judging a good estimator.

7-2 Desirable Properties of Estimators

A—NO BIAS

We already have noted that the sample mean $\overline{X}$ is, on average, exactly on its target μ. We therefore called $\overline{X}$ an *unbiased* estimator of μ.

To generalize, we consider any population parameter θ (Greek theta) and denote its estimator by U. If, on average, U is exactly on target as shown in Figure 7-1a, it is called an unbiased estimator. Formally, we state the definition.

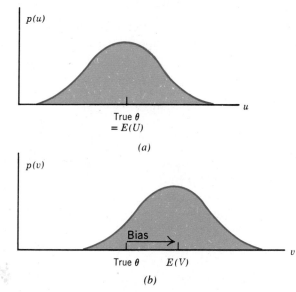

Figure 7-1
Comparison of (a) unbiased estimator, and (b) biased estimator

$$\boxed{\begin{array}{c} U \text{ is an unbiased estimator of } \theta \text{ if} \\ E(U) = \theta \end{array}} \qquad \begin{array}{c} (7\text{-}1) \\ \text{like } (6\text{-}5) \end{array}$$

Of course, an estimator V is called biased if $E(V)$ is different from θ. In fact, bias is defined as this difference.

$$\boxed{\text{Bias} \equiv E(V) - \theta} \qquad (7\text{-}2)$$

Bias is illustrated in Figure 7-1b. The distribution of V is off-target; since $E(V)$ exceeds θ, there will be a tendency for V to overestimate θ.

As we stressed already, to avoid bias we have to randomly sample the whole population. To show the difficulty we can encounter if we fail to follow this fundamental principle, we give an example of nonresponse bias.

EXAMPLE **7-1**

Let us give a concrete example of the teachers' questionnaire mentioned at the beginning of Chapter 6. A population of 10,000 teachers spend various numbers of hours X doing professional reading, according to Table 7-2a below.

The problem is, however, that the teachers are divided into two groups. The first is made up of those who would not respond to a questionnaire—many because they are too busy reading. The second group is made up of those who would respond, if they happened to be asked. This subpopulation who would respond is shown in Table 7-2b below.

a. What is the mean μ of the total population, and the mean μ_R of the subpopulation who would respond?

b. A random sample of 1000 mail questionnaires brought a response of 200 completed questionnaires, from which the average $\overline{X}$ was calculated. Is this an unbiased estimate of μ? If not, what is its bias?

TABLE **7-2** **Hypothetical Population, and Subpopulation Who Would Respond**

X = READING (HRS/MONTH)	(a) TOTAL POPULATION OF TEACHERS		(b) SUBPOPULATION WHO WOULD RESPOND	
	FREQUENCY f	REL FREQ. f/N	FREQUENCY f	REL. FREQ. f/N
0	2,000	.20	1,200	.60
20	5,000	.50	600	.30
40	3,000	.30	200	.10
	N = 10,000	1.00	N = 2,000	1.00

Solution

    a. Using the probabilities $p(x) = .20, .50, .30$, we obtain $\mu = 22$. Similarly, using the probabilities $p(x) = .60, .30, .10$, we obtain $\mu_R = 10$.

    b. The sample mean $\overline{X}$ has an obvious nonresponse bias, since heavy readers are less likely to respond. (Only 200 of the 3000 heavy readers would respond, whereas 1200 of the 2000 non-readers would respond.) Since heavy readers are seriously un-derrepresented, $\overline{X}$ will tend to be much too small.

        To calculate just how serious the bias is, note that $\overline{X}$ comes from a random sample drawn from the subpopulation who would respond. Since this population mean is $\mu_R = 10$, it follows that, according to (6-5),

$$E(\overline{X}) = \mu_R = 10$$

Recall that the target is the *whole* population mean $\mu = 22$, so that the bias is:

$$\text{Bias} = E(\overline{X}) - \mu \qquad\qquad \text{like (7-2)}$$
$$= 10 - 22 = -12 \qquad\qquad (7\text{-}3)$$

Thus the bias is indeed very large—an underestimate of 12 hours per month.

B—MINIMUM VARIANCE: THE EFFICIENCY OF UNBIASED ESTIMATORS

As well as being on target on the average, we also would like the distribution of an estimator to be highly concentrated—that is, to have a small variance. This is the notion of *efficiency*, shown in Figure 7-2. We describe the estimator U in panel (a) as more efficient than the estimator V in panel (b) because it has smaller variance. Formally, the relative efficiency of two estimators is defined as follows.

> For unbiased estimators,
>
> $$\text{Efficiency of } U \text{ compared to } V \equiv \frac{\text{var } V}{\text{var } U} \qquad (7\text{-}4)$$

    For example, when the population being sampled is symmetric, its center can be estimated without bias by either the sample mean $\overline{X}$ or median $\overset{\downarrow}{X}$. For some populations $\overline{X}$ is more efficient; for others, $\overset{\downarrow}{X}$ is more efficient. In the case of a normal parent population, for instance, it can be shown (in Appendix 7-2, for large samples) that $\overset{\downarrow}{X}$ has a standard error given approximately by:

$$\text{SE of } \overset{\downarrow}{X} \simeq 1.25 \, \sigma/\sqrt{n} \qquad (7\text{-}5)$$

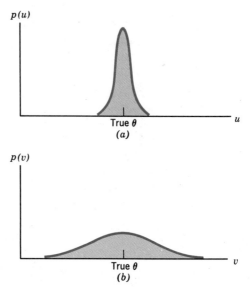

Figure 7-2
A comparison of (a) efficient estimator, and (b) inefficient estimator

Since $\overline{X}$ has a smaller standard error SE $= \sigma/\sqrt{n}$ [as given in (6-7)], it is more efficient. Specifically,

$$\text{Efficiency of } \overline{X} \text{ relative to } \overset{\downarrow}{X} \equiv \frac{\text{var } \overset{\downarrow}{X}}{\text{var } \overline{X}} \qquad\qquad \text{like} \quad (7\text{-}4)$$

$$= \left[\frac{\text{SE}(\overset{\downarrow}{X})}{\text{SE}(\overline{X})}\right]^2 \simeq \left[\frac{1.25\ \sigma/\sqrt{n}}{\sigma/\sqrt{n}}\right]^2$$

$$\simeq 1.56 = 156\% \qquad\qquad\qquad (7\text{-}6)$$

The greater efficiency of the sample mean is nicely confirmed by the Monte Carlo study in Figure 6-11—where the ratio of variances was also 1.56 ($= 1.26^2/1.01^2$). We conclude that, in estimating the center of a *normal* population, the sample mean $\overline{X}$ is about 56% more efficient than the sample median X. (In fact, it could be proved that the sample mean is more efficient than *every* other estimator of the center of a normal population.)

Of course, by increasing sample size n, we can reduce the variance of either the sample mean or median. This provides an alternative way of looking at the greater efficiency of the sample mean (in sampling from *normal* populations). The sample median will yield as accurate an estimate only if we take a larger sample (specifically, 56% larger). Hence the sample mean is more efficient because it costs less to sample. Note how the economic and statistical definitions of efficiency coincide.

Is the sample mean still efficient when sampling from *nonnormal* populations? An example will give the answer.

EXAMPLE **7-2**

One of the symmetric distributions with longer tails than the normal, is the Laplace distribution. In this case, it can be shown (in Appendix 7-2, again for large samples) that $\overset{\downarrow}{X}$ has a standard error given by

$$\text{SE of } \overset{\downarrow}{X} \simeq .71 \ \sigma/\sqrt{n} \tag{7-7}$$

Now what is the efficiency of $\overline{X}$ relative to $\overset{\downarrow}{X}$?

Solution

Since $\overline{X}$ always has SE $= \sigma/\sqrt{n}$, in this case we see that $\overset{\downarrow}{X}$ now has a smaller SE, and so $\overline{X}$ is now *less* efficient. Specifically:

$$\text{Efficiency of } \overline{X} \text{ relative to } \overset{\downarrow}{X} \equiv \frac{\text{var } \overset{\downarrow}{X}}{\text{var } \overline{X}} \qquad \text{like (7-4)}$$

$$= \left[\frac{\text{SE}(\overset{\downarrow}{X})}{\text{SE}(\overline{X})}\right]^2 \simeq \left[\frac{.71 \ \sigma/\sqrt{n}}{\sigma/\sqrt{n}}\right]^2 \tag{7-8}$$

$$\simeq .50 = 50\% \tag{7-9}$$

Thus the sample mean $\overline{X}$ has only about 50% of the efficiency of the sample median $\overset{\downarrow}{X}$ in the long-tailed Laplace distribution.

C—MINIMUM MEAN SQUARED ERROR: EFFICIENCY OF ANY ESTIMATOR

So far we have concluded that, in comparing unbiased estimators, we choose the one with minimum variance. But suppose we are comparing both biased and unbiased estimators, as in Figure 7-3. Now it may no longer be appropriate to select the estimator with least variance: W qualifies on that score, but is unsatisfactory because it is so badly biased. Nor do we necessarily pick the estimator with least bias: U has zero bias, but seems unsatisfactory because of its high variance. Instead, the estimator that seems to be closest to the target overall is V, because it has the best *combination* of small bias and small variance.

How can we make precise the notion of being "closest to the target overall"? We are interested in how an estimator, V let us say, is spread around its *true target* θ:

$$\boxed{\text{Mean squared error (MSE)} \equiv E(V - \theta)^2} \tag{7-10}$$
$$\text{like (4-34)}$$

This is similar to the variance, except that it is measured around the true target θ rather than around the (possibly biased) mean of the estimator. Then, as Appendix 7-4 proves, MSE does indeed turn out to be a combination of variance and bias:

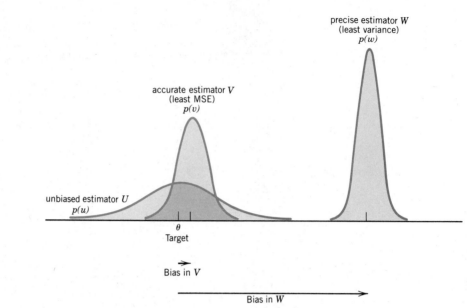

Figure 7-3
V is the estimator with the best combination of small bias and variance.

$$\boxed{\text{MSE} = (\text{variance of estimator}) + (\text{its bias})^2} \qquad (7\text{-}11)$$

We choose the estimator that minimizes this MSE.

This confirms two earlier conclusions: if two estimators with equal variance are compared (as in Figure 7-1), the one with less bias is preferred; and if two unbiased estimators are compared (as in Figure 7-2), the one with smaller variance is preferred. In fact, if two estimators are unbiased, it is evident from (7-11) that the MSE reduces to the variance. Thus, MSE may be regarded as a generalization of the variance concept, and leads to a generalized definition of the relative efficiency of two estimators:

$$\boxed{\begin{array}{c} \text{For any two estimators—whether biased or unbiased—} \\[4pt] \text{Efficiency of } U \text{ compared to } V \equiv \dfrac{\text{MSE}(V)}{\text{MSE}(U)} \end{array}} \qquad \begin{array}{l}(7\text{-}12) \\ \text{like } (7\text{-}4)\end{array}$$

To sum up, because it combines the two attractive properties of small bias and small variance, the concept of efficiency as defined in (7-12) becomes the single most important criterion for judging between estimators. An example will illustrate.

EXAMPLE **7-3**   In Example 7-1, recall Table 7-2 where we looked at a sample survey that suffered from a serious nonresponse bias. If we let V denote the biased estimator of μ, then we must recognize that V has variability as well as bias.

a. To measure how much V fluctuates around its target μ overall, calculate its MSE.

b. If the sample size was increased fivefold, how much would the MSE be reduced?

c. A second statistician takes a small sample survey of only $n = 20$ observations, persistently following up each questionnaire until he gets a response. Let this small but unbiased sample have a sample mean denoted by U. What is its MSE?

d. What is the efficiency of U relative to V?

e. In trying to publish his results, the second statistician was criticized for using a sample only 1/10 as large as the first. In fact, his sample size $n = 20$ was labeled "ridiculous." What defense might he offer?

Solution   **a.** Since V is the sample mean $\overline{X}$ for $n = 200$ observations drawn only from the subpopulation who would respond, this is the population whose moments are relevant:

Subpopulation who would respond (from Table 7-2b)

x	p(x)	xp(x)	$(x - \mu_R)$	$(x - \mu_R)^2$	$(x - \mu_R)^2 p(x)$
0	.60	0	-10	100	60
20	.30	6	$+10$	100	30
40	.10	4	$+30$	900	90
		$\mu_R = 10$			$\sigma_R^2 = 180$

Thus we can confirm that the bias in $\overline{X}$ was:

$$\text{Bias} = 10 - 22 = -12 \qquad (7\text{-}3)\text{ repeated}$$

Also, from the subpopulation variance $\sigma_R^2 = 180$ in the table, we can deduce the variance of $\overline{X}$:

$$\text{var}(\overline{X}) = \frac{\sigma_R^2}{n} = \frac{180}{200} = .90 \qquad \text{like (6-6)}$$

Now we can put the bias and variance together to get the MSE:

$$\text{MSE} = \text{var} + \text{bias}^2 \qquad (7\text{-}11)\text{ repeated}$$
$$= .90 + 144 = 144.90$$

b. A sample five times as large would reduce the variance:

$$\text{var}(\overline{X}) = \frac{\sigma_R^2}{n} = \frac{180}{5 \times 200} = .18$$

Unfortunately it would not reduce the bias—the same sampling mistake would merely be repeated more often. Thus,

$$\begin{aligned} \text{MSE} &= \text{var} + \text{bias}^2 \\ &= .18 + 144 = 144.18 \end{aligned} \tag{7-13}$$

Since the predominant term is the bias, which is unaffected by sample size, the MSE was reduced hardly at all (from 144.90 to 144.18).

c. U is the sample mean $\overline{X}$ for $n = 20$ observations from the population of both respondents and nonrespondents. (Recall that by persistent follow-up, the statistician has got even the "nonrespondents" to reply.) It is this total population whose moments are now relevant:

Whole population (from Table 7-2a)

x	p(x)	xp(x)	(x − μ)	(x − μ)²	(x − μ)²p(x)
0	.20	0	− 22	484	96.8
20	.50	10	− 2	4	2.0
40	.30	12	18	324	97.2
		μ = 22			σ² = 196

Since this sample is drawn from the whole population, $\overline{X}$ is now unbiased:

$$E(\overline{X}) = \mu = 22 \tag{6-5 repeated}$$

and

$$\text{var}(\overline{X}) = \frac{\sigma^2}{n} = \frac{196}{20} = 9.8$$

Thus we can calculate the MSE:

$$\begin{aligned} \text{MSE} &= \text{var} + \text{bias}^2 \tag{7-11 repeated} \\ &= 9.8 + 0 = 9.8 \tag{7-14} \end{aligned}$$

d.   Efficiency of U compared to $V \equiv \dfrac{\text{MSE}(V)}{\text{MSE}(U)}$

$$= \frac{144.9}{9.8} = 14.8 = 1480\%$$

Thus the small but unbiased sample produces an estimator that is nearly 15 times as efficient!

e. His defense would simply be that his estimator is far better because it has a far smaller MSE—15 times smaller! Or, in less mathematical terms, "it's the quality of the sample that counts, not mere quantity." He might even point out the lesson of part **b:** Increasing the sample size (even by 5 times) without dealing with its bias would provide little practical improvement.

PROBLEMS

7-1 True or false? If false, correct it.

    a. μ is a random variable (varying from sample to sample), and is used to estimate the parameter $\bar{X}$.

    b. The sample proportion P is an unbiased estimator of the population proportion π.

    c. In sampling from a normal population, the sample median and sample mean are both efficient estimators of μ. The difference is that the sample median is biased, whereas the sample mean is unbiased.

7-2 Each of three guns is being tested by firing 12 shots at a target from a clamped position. Gun A was not clamped down hard enough, and wobbled. Gun B was clamped down in a position that pointed slightly to the left, due to a misaligned sight. Gun C was clamped down correctly.

    a. Which of the following patterns of shots belongs to gun A? gun B? gun C?

    b. Which guns are biased? Which gun has minimum variance? Which has the largest MSE? Which is most efficient? Which is least efficient?

7-3 A researcher gathers a random sample of 500 observations, and loses the records of the last 180. Thus he has only 320 observations from which to calculate his sample mean. What is the efficiency of this, relative to what he could have obtained from the whole sample?

7-4 What is the efficiency of the sample median relative to the sample mean in estimating the center of a normal population? [*Hint:* Recall from (7-6) that the efficiency of the mean relative to the median was 156%.]

7-5 **a.** Answer True or False; if false, correct it.

In both Problems 7-3 and 7-4 we have examples of estimates that are only 64% efficient. In Problem 7-3, this inefficiency was obvious, because 36% of the observations were lost in calculating $\overline{X}$. In Problem 7-4, the inefficiency was more subtle, because it was caused merely by using the sample median instead of the sample mean. However, in terms of results—producing an estimate with more variance than necessary—both inefficiencies are equally damaging.

b. In view of part **a,** what advice would you give to a researcher who spends $100,000 collecting data, and $100 analyzing it?

*7-6 Suppose a magazine's readership consists of two classes of people:

70% are poor, each earning $8,000 annually
30% are rich, each earning $40,000 annually

A questionnaire insert asks, among other things, for the reader's income. But each rich person is twice as likely to answer as each poor person, so that the sample mean is a biased estimator of the population mean. If a sample of 10,000 replies is obtained,

a. How much is the bias of $\overline{X}$?

b. What is the MSE of $\overline{X}$?

c. It was recommended instead to take a *random* sample of the readership to eliminate the bias, with a sample size of only $n = 25$ to keep down the cost. How much more (or less) efficient would this sample mean be?

*7-7 In a certain county, suppose that the electorate is 80% urban and 20% rural; 70% of the urban voters, and only 25% of the rural voters, vote for D in preference to R. In a certain straw vote conducted by a small-town newspaper editor, a rural voter has 6 times the chance of being selected as an urban voter. This bias in the sampling will cause the sample proportion to be a biased estimator of the population proportion in favor of R.

a. How much is this bias?

b. Is the bias large enough to cause the average sample to be wrong (in the sense that the average sample "elects" a different candidate from the one the population elects)?

*7-8 In Problem 7-7, if we consider the viewpoint of the newspaper editor, realistically we cannot suppose that the population proportion π favoring D is known. However, we can suppose that the 80%–20% urban–rural split in the population is known—through census figures, for example. Suppose that the editor then obtains the following data from a biased sample of 700 voters:

| | | VOTE | |
LOCATION	FOR D	FOR R	TOTALS
Urban	210	92	302
Rural	80	318	398
Totals	290	410	700

The simple-minded and biased estimate of π is the sample proportion $290/700 = 41\%$. Calculate an *unbiased* estimate of π. Incidentally, a technique like this, which is based on several population strata (urban, rural) whose proportions are known and allowed for, is an example of *stratified sampling*.

7-9 **a.** Two polling agencies took random samples to estimate the proportion π of Americans who favored legalization of marijuana. The first poll showed proportion $P_1 = 60/200 = 30\%$ in favor. The second larger poll showed a proportion $P_2 = 240/1000 = 24\%$ in favor. To get an overall estimate, the simple average $P^* = 27\%$ was taken. What is the variance of this estimate? [Hint: $\text{var } P \simeq P(1-P)/n$]

b. The first poll is clearly less reliable than the second. So it was proposed to just throw the first away, and use the estimate $P_2 = 24\%$. What is the variance of this estimate? What then is its efficiency relative to P^*?

c. The best estimate of all, of course, would count each *observation* equally (not each *sample* equally). That is, take the overall proportion in favor, $P = (60 + 240)/(200 + 1000) = 25\%$. What is the variance of this estimate? Then what is its efficiency relative to P^*?

d. True or False? If false, correct it:
It is important to know the reliability of your sources. For example, if the information in an unreliable source is given too much weight, it can damage the overall estimate.

*7-10 (Requires Section 6-5) A random sample of 1000 students is to be polled from a population of 5000 students. Judging on the grounds of efficiency, is it better to sample with replacement, or without replacement? How much better?

7-3 Robust Estimation

A—THE SAMPLE MEAN COMPARED TO THE MEDIAN

It is important from time to time to recall just why we sample from a population: We are ignorant about the population, and want the sample to provide information quickly and inexpensively. Usually it is the population center that we want to estimate, but we must remember that we are generally ignorant about all other aspects too—the population spread and population shape, for example.

To estimate the population center, should we use the sample mean $\overline{X}$ or the sample median $\overset{\downarrow}{X}$? We have seen that $\overline{X}$ is better if we know the population shape is normal, while $\overset{\downarrow}{X}$ is better if we know the population shape is Laplace (Example 7-2). But, typically we *don't* know the population shape—so which should we choose?

Or is there a better alternative than either $\overline{X}$ or $\overset{\downarrow}{X}$? After all, there are many possible populations besides the normal and Laplace, and many possible estimators besides $\overline{X}$ and $\overset{\downarrow}{X}$ that can be devised. What we need is a *robust* estimator—one that works well in a wide variety of populations.

To illustrate the diversity of population shapes that may occur in practice, Figure 7-4 shows a typical population that has longer and thicker tails than the normal. These thick tails occasionally produce a far-out observation, a so-called *outlier*, as shown by the right-hand dot in the colored random sample. Such outliers have far too great a leverage on $\overline{X}$, the center of gravity of the sample, and make it an unreliable estimator of the population center: As Figure 7-4 shows, $\overline{X}$ is pretty far off target.

While $\overline{X}$ uses too *many* observations (all of them, including the problem outliers), the sample median $\overset{\downarrow}{X}$ would use too *few* (only the middle one). Might not some sort of compromise produce a better estimator in this population?

B—THE TRIMMED MEAN

One compromise between the median and the mean is to average just the *middle half* of the observations. By trimming 25% of the observations from each end, the outliers are removed; accordingly it is called the *trimmed mean*, or more precisely, the 25% trimmed mean. (In trimming 25% from each end, we use round numbers. For example, in a sample of $n = 45$ observations, we would trim ¼ of 45 ≃ 11 observations from each end.)

Of course, other degrees of trimming are possible, too. The percentage trimmed from each end can be written as a subscript; for example, $\overline{X}_{.10}$ is the 10% trimmed mean. An example will illustrate.

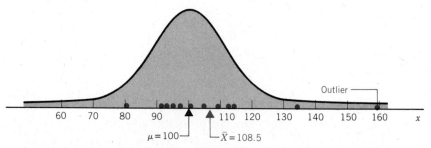

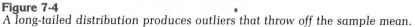

Figure 7-4
A long-tailed distribution produces outliers that throw off the sample mean.

EXAMPLE **7-4**   A random sample of n = 12 observations was drawn and ordered as follows:

81	106
94	110
95	113
95	114
97	135
102	160

Calculate the trimmed mean, for the following degrees of trimming from each end:

a.   0%
b.   8%
c.   25%
d.   42%

Solution   **a.**   Zero trim gives us just the ordinary mean $\overline{X}$:

$$\overline{X}_0 = \overline{X} = \frac{81 + 94 + \cdots + 135 + 160}{12} = \frac{1302}{12} = 108.5$$

b.   For 8% trim, we take 8% of 12 = 1 observation from each end. Thus:

$$\overline{X}_{.08} = \frac{94 + \cdots + 135}{10} = \frac{1061}{10} = 106.1$$

c.   For 25% trim, we take 25% of 12 = 3 observations from each end. Thus:

$$\overline{X}_{.25} = \frac{95 + 97 + 102 + 106 + 110 + 113}{6} = \frac{623}{6} = 103.8$$

d.   For 42% trim, we take 42% of 12 = 5 observations from each end. That leaves just the middle two observations, whose midpoint of course is the median $\overset{\downarrow}{X}$:

$$\overline{X}_{.42} = \overset{\downarrow}{X} = \frac{102 + 106}{2} = 104.0$$

Remarks   This sample, in fact, is the one shown in Figure 7-4, where the center (target) is known to be 100. We note that the large outlier X = 160 does indeed exert so much leverage that it makes the sample mean unduly large ($\overline{X}_0$ = 108.5). Trimming off this outlier gives a better estimate ($\overline{X}_{.08}$ = 106.1), closer to the target of 100. Further trimming cuts off the next outlier, and improves the estimate further ($\overline{X}_{.25}$ = 103.8). But if we trim as much as possible, we go a little too far: We get right down to the sample median, which is not quite as accurate ($\overline{X}_{.42}$ = 104.0).

Of course, the one random sample in Example 7-4, like one play at Las Vegas, does not prove much. Repeated sampling (Monte Carlo) is necessary. Such sampling experiments have confirmed that for populations with the sort of long tails shown in Figure 7-4, the mean involves too little trimming, and the median too much. 25% trimming provides about the best estimate, and works well in a wide variety of populations. The 25% trimmed mean thus provides a good robust estimate of the population center.

*C—THE BIWEIGHTED MEAN ($\overline{X}_b$)

A closer look at the 25% trimmed mean in Example 7-4(c) shows an interesting feature: The observations in the outside quarters don't count at all, while the observations in the middle half all count equally. When we graph these weights in panel (a) of Figure 7-5, we note the sudden drop at the upper and lower quartiles. This "black and white" weighting makes the computation simple, but lacks subtlety. Why not *gradually* weight the observations as in panel (b), with the least weight on the problem outliers (as suggested by Mosteller and Tukey, 1977)? Specifically, the weighting formula for any observation X is:

$$w(X) = (1 - Z^2)^2 \quad \text{if } |Z| \le 1 \qquad (7\text{-}15)$$
$$= 0 \qquad\qquad \text{if } |Z| > 1 \qquad (7\text{-}16)$$

where Z is a kind of standardized X value—standardized with the median $\overset{\downarrow}{X}$ and the interquartile range IQR:

$$Z = \frac{X - \overset{\downarrow}{X}}{3(\text{IQR})} \qquad\qquad (7\text{-}17)$$
$$\text{like (4-23)}$$

To see how this formula works, suppose X deviates from the median $\overset{\downarrow}{X}$ by more than 3 times the IQR; then (7-17) gives $|Z| > 1$, and so from (7-16), $w(X) = 0$. (Any observation that is so far out is given a zero weight.) On the other hand, if X is within this broad range, then $|Z| \le 1$ and X is given a positive weight according to (7-15). And the closer X is to the median $\overset{\downarrow}{X}$, the larger this weight will be. (In other words, $w(X)$ reaches a maximum at the median $\overset{\downarrow}{X}$.)

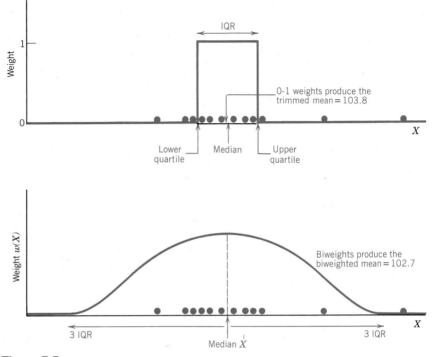

Figure 7-5
Two weighting functions that reduce the influence of the outliers, for the sample of Figure 7-4. (a) Weights for the 25% trimmed mean (b) More gradual biweights.

Because $w(X)$ in (7-15) has Z squared twice, it is called the *bisquare weight*, or more briefly, the *biweight*. To make the formula complete, we finally use the biweights to calculate the mean:

$$\text{Biweighted mean } \overline{X}_b \equiv \frac{\Sigma X w(X)}{\Sigma w(X)} \qquad (7\text{-}18)$$

This weight formula is familiar. For example, if the weights were frequency weights f, instead of biweights $w(X)$, then (7-18) would reduce to the ordinary mean $\overline{X} = \Sigma X f / \Sigma f = \Sigma X f / n$.

For the sample in Figure 7-4, the computer calculated the biweighted mean:

$$\overline{X}_b = 102.7$$

This is slightly closer than any of the other estimates to the target $\mu = 100$. And so we have illustrated two important points that are generally true:

1. In many populations, the biweighted mean $\overline{X}_b$ is slightly better than the other proposed estimates.
2. But $\overline{X}_b$ involves a lot more calculation. By hand, it is not worth it. By computer, however, the extra calculation is so trivial that $\overline{X}_b$ is well worth it.

   In fact, there is another and *even more important* advantage to the biweighted mean:
3. $\overline{X}_b$ is easily generalized to cover more complicated situations, such as estimating the relation between several variables in Chapter 11.

   These points are so important that they are well worth emphasizing:

> **The computer has made feasible many robust estimates like the biweighted mean $\overline{X}_b$ that provide insurance against possible outliers, even in complicated situations.**   (7-19)

*D—THE BIWEIGHTED MEAN ITERATED ($\overline{X}_{bi}$)

One of the great advantages of computers is that they can rapidly make the same calculations over and over. We can thus calculate and recalculate an estimate, improving it slightly each time—a technique called *iteration*. Let's see how it works for the biweighted mean $\overline{X}_b$.

How could $\overline{X}_b$ possibly be improved? The weighting function (7-17) is centered at the median $\overset{\vee}{X}$—roughly, the observed center of the data. But once $\overline{X}_b$ has been calculated, then it becomes a better estimate of the center. So we send the computer through all the calculations again. Here are the details:

We start now with $\overline{X}_b$ instead of $\overset{\vee}{X}$ in defining weights:

$$Z = \frac{X - \overline{X}_b}{3(\text{IQR})}$$ (7-20)
                                                  like (7-17)

This produces in (7-18) an improved value of $\overline{X}_b$. And so we keep repeating: Each time we get an improved value of $\overline{X}_b$, we plug it into (7-20) and get an even better value of $\overline{X}_b$ out of (7-18).

When no further real improvement occurs, the computer prints the final answer, and calls it the *biweighted mean iterated*, $\overline{X}_{bi}$. Not only is it a bit better than the original $\overline{X}_b$; even more important, it is particularly adaptable to more general statistical problems like regression, as we will see in Chapter 11.

*E—ROBUST MEASURE OF SPREAD: THE IQR

Having derived various robust measures of center, it is now time to briefly consider measures of spread. When we examine the standard deviation s, we see that, like the sample mean, it is unfortunately sensitive to outliers. In fact, since s involves *squared* deviations (which are *very* large for outliers), s is affected by outliers even more than $\overline{X}$ is.

In Chapter 2 we noted that the IQR was much more resistant to outliers than was s. It is therefore no surprise that the IQR is one of the most robust estimators of the population spread.

[In fact, it is because of its robustness that the IQR is used not only in estimating population *spread*, but also in defining two robust estimates of population *center*: Recall that in Figure 7-5, the IQR is the width of the weighting function for the trimmed mean $\overline{X}_{.25}$ in panel (a), and is used again in defining the biweighted mean $\overline{X}_b$ in panel (b).]

PROBLEMS

7-11 In each of the samples below, assume that the parent population is distributed symmetrically around a central value c. Which is the most efficient estimate of c—the sample mean, median, or 25% trimmed mean? Explain why it is preferred, then calculate it.

a. A chemist takes the following 12 measurements:

8.9	7.2	8.5	8.3	7.3	7.8
7.6	7.5	8.6	7.9	9.4	7.9

Assume that these observations differ only because of a normally distributed measurement error.

b. On New Year's day, suppose a random sample of 5 forecasters gave the following predictions of the unemployment rate in the coming year:

8.0%	7.4%	7.4%	8.3%	9.5%

Suppose also that the whole population of forecasters contains a few who make really extreme predictions, so that the distribution has long tails.

c. An anthropologist measured the width (in centimeters) of 9 skulls from a certain tribe. Unfortunately, his handwriting was so poor that the keypuncher occasionally had to guess at the digits as they were entered into the computer, as follows:

15.3	16.2	15.5	18.7
13.1	15.1	15.0	14.2	15.0

7-12 To compare two competing methods of filtering air, an engineer tries them both out on a random sample of 12 different days. Each day he measures the pollution index of the air after each filtering method, and calculates the difference A-B, obtaining these 12 numbers:

$$\begin{array}{cccccc}
105 & 105 & 125 & 97 & 99 & 103 \\
97 & 113 & 102 & 90 & 101 & -25
\end{array}$$

a. Find the median $\overset{\downarrow}{X}$, the mean $\overline{X}$, and the trimmed mean $\overline{X}_{.25}$.

b. Graph the data as 12 dots on the X-axis, and include the three statistics in part **a**.

c. This data was drawn from a population symmetrically distributed around the center $c = 100$. Which of your answers in part **a** happened to be closest to this target?

7-13 Repeat Problem 7-12 for the following sample of 5:

$$91 \quad 104 \quad 140 \quad 108 \quad 104$$

*7-14 Calculate a robust measure of *spread* for each of Problems 7-12 and 7-13.

*7-15 Calculate the biweighted mean for the sample in Problem 7-13.

*7-16 (Monte Carlo) The data in Problem 7-12 and Example 7-4 comes from a *mixed population:*

  80% of the population are normally distributed around a mean $\mu = 100$ with standard deviation $\sigma = 10$. The remaining 20% are more widespread, with $\sigma = 50$ (but still the same mean, $\mu = 100$).

a. From this mixture, simulate a random sample of $n = 12$ observations. (*Hint:* To simulate each observation, first draw a random digit from Appendix Table I: if the digit is 1, ... , 8, this means that you are sampling from the majority population with $\sigma = 10$; if the digit is 9 or 0, then you are sampling from the minority population with $\sigma = 50$. Then draw a random normal number Z from Appendix Table II, and calculate $X = \mu + Z\sigma$ to get the desired observation. Repeat 12 times to get all 12 observations. Incidentally, if you start at the top of Tables I and II, you will get the sample that appears in Problem 7-12.)

b. For this sample, find $\overset{\downarrow}{X}$, $\overline{X}$, and $\overline{X}_{.25}$. Which estimate happens to be closest to the target $\mu = 100$? Which is furthest?

c. Imagine parts **a** and **b** being repeated for many, many samples. Which of these estimators would you guess gives the best overall performance (minimum MSE)? And the worst?

***7-4 Consistency**

A—CONSISTENCY: EVENTUALLY ON TARGET

Like efficiency, consistency is one of the desirable properties of estimators. But consistency is more abstract, because it is defined as a limit: A consistent estimator is one that concentrates in a narrower and narrower band

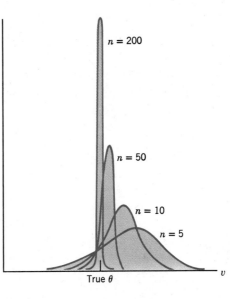

Figure 7-6
A consistent estimator, showing how the distribution of V concentrates on its target θ as n increases

around its target as sample size increases indefinitely. This is sketched in Figure 7-6, and made more precise in Appendix 7-4.

One of the conditions that makes an estimator consistent is if its MSE approaches zero in the limit. In view of (7-11), this may be reexpressed as follows.

> One of the conditions that makes an estimator consistent is:
>
> **if its bias and variance *both* approach zero.** (7-21)

EXAMPLE **7-5**   **a.** Is $\overline{X}$ a consistent estimator of μ?

       b. Is P a consistent estimator of π?

       c. Is the estimator $\overline{X}$ in Example 7-1 (based on partial response) a consistent estimator of μ?

Solution   **a.** From the normal approximation rule (6-9), we know that

$$\text{Bias} = 0 \quad \text{for all } n$$

$$\text{var} = \frac{\sigma^2}{n} \rightarrow 0 \quad \text{as } n \rightarrow \infty$$

Thus (7-21) assures us that $\overline{X}$ is a consistent estimator of μ.

b. From the normal approximation rule for P given in (6-12), we similarly see that P is a consistent estimator of π.

c. Because of nonresponse bias, the estimator $\overline{X}$ concentrates on a point below the target μ, and so is inconsistent.

B—ASYMPTOTICALLY UNBIASED ESTIMATORS

Sometimes an estimator has a bias that fortunately tends to zero as sample size increases ($n \to \infty$). Then it is called *asymptotically unbiased*. If its variance also tends to zero, (7-21) then assures us it will be consistent. An example will illustrate.

EXAMPLE **7-6**   Consider the mean squared deviation:

$$\text{MSD} = \frac{1}{n} \Sigma(X - \overline{X})^2 \qquad\qquad (7\text{-}22)$$
$$(2\text{-}10)\,\text{repeated}$$

This is a biased estimator of the population variance σ^2. Specifically, on the average it will underestimate, as can be seen very easily in the case of $n = 1$. Then $\overline{X}$ coincides with the single observed X, so that (7-22) gives MSD = 0, no matter how large the population variance σ^2 may be.

However, if we inflate MSD by dividing by $n - 1$ instead of n, we obtain the sample variance:

$$s^2 \equiv \frac{1}{n - 1} \Sigma(X - \overline{X})^2 \qquad\qquad (7\text{-}23)$$
$$(2\text{-}11)\,\text{repeated}$$

It can be proved (Lindgren, 1976) that s^2 is inflated just enough to be perfectly unbiased. [And in the extreme case above, where $n = 1$, the zero divisor in (7-23) makes s^2 undefined. This provides a simple warning that σ^2 cannot be estimated with a single observation, since this would give us no idea whatsoever how spread out the population may be.]

If you were puzzled earlier by the divisor $n - 1$ used in defining s^2, you now can see why. It is to ensure that s^2 will be an *unbiased* estimator of the population variance.

a. Although we have seen that MSD is biased, is it nevertheless asymptotically unbiased?

b. Suppose we used an even larger divisor, $n + 1$, to obtain the following estimator:

$$s_*^2 \equiv \frac{1}{n + 1} \Sigma(X - \overline{X})^2 \qquad\qquad (7\text{-}24)$$

Is s_*^2 asymptotically unbiased?

Solution   **a.** Let us write MSD in terms of the unbiased s^2:

$$\text{MSD} = \left(\frac{n-1}{n}\right)s^2 = \left(1 - \frac{1}{n}\right)s^2 \qquad \text{like (2-22)}$$

Then to calculate the expected value, we can take out the constant:

$$E(\text{MSD}) = \left(1 - \frac{1}{n}\right)E(s^2)$$

Finally, since s^2 is unbiased estimator of σ^2,

$$E(\text{MSD}) = \left(1 - \frac{1}{n}\right)\sigma^2 = \sigma^2 - \left(\frac{1}{n}\right)\sigma^2$$

Since $1/n \to 0$, the last term—the bias—tends to zero so that the MSD is indeed asymptotically unbiased.

b. Similarly, we may write:

$$s_*^2 = \left(\frac{n-1}{n+1}\right)s^2 = \left(1 - \frac{2}{n+1}\right)s^2$$

And since $2/(n+1) \to 0$, this is also asymptotically unbiased.

Remarks   We have shown that both MSD and s_*^2 are asymptotically unbiased. So also is s^2 itself, of course. It could further be shown that all three estimators have variance that approaches zero, so that they are all consistent.

Which of the three estimators should we use? Since all three are consistent, we need a stronger criterion to make a final choice, such as unbiasedness or efficiency. While s^2 is the unbiased estimator, it turns out that s_*^2 is most efficient (for most populations, including the normal). Consequently some statisticians use s_*^2 instead of s^2 to estimate σ^2.

C—CONCLUSIONS

Although consistency is quite abstract in its definition (in Appendix 7-4), it often provides a useful preliminary criterion for sorting out estimators.

Nevertheless, to finally sort out the best estimator, a more powerful criterion such as efficiency is required—as we saw in Example 7-6. Another familiar example will illustrate: In estimating the center of a normal population, both the sample mean and median satisfy the consistency criterion. To choose between them, efficiency is the criterion that will finally select the winner (the sample mean).

*PROBLEMS

7-17 The population of American personal incomes is skewed to the right (as we saw in Figure 2-5, for men in 1975, for example). Which of the following will be consistent estimators of the population mean μ?

    a. From a random sample of n incomes, the sample mean? The sample median? The sample trimmed mean?

    b. Repeat part **a,** for a sample of n incomes drawn at random from the cities over one million.

7-18 When S successes occur in n trials, the sample proportion $P = S/n$ customarily is used as an estimator of the probability of success π. However, sometimes there are good reasons to use the estimator $P^* \equiv (S + 1)/(n + 2)$. Alternatively, P^* can be written as a linear combination of the familiar estimator P:

$$P^* = \frac{nP + 1}{n + 2} = \left(\frac{n}{n + 2}\right)P + \left(\frac{1}{n + 2}\right)$$

    a. What is the MSE of P? Is it consistent?

    b. What is the MSE of P^*? Is it consistent? (*Hint:* Calculate the mean and variance of P^*, in terms of the familiar mean and variance of P.)

    c. To decide which estimator is better, P or P^*, does consistency help? What criterion *would* help?

    d. Tabulate the efficiency of P relative to P^*, for example, when $n = 10$ and $\pi = 0, .1, .2, \ldots, .9, 1.0$.

    e. State some possible circumstances when you might prefer to use P^* instead of P to estimate π.

CHAPTER 7 SUMMARY

7-1 To estimate the center of a population, sometimes $\overline{X}$ is best (if the population shape is normal), sometimes $\tilde{X}$ is better (if the population shape is long-tailed). In this chapter, we look for even better alternatives, and criteria to judge them.

7-2 Bias measures how far the sampling distribution of an estimator is centered off target. When bias is combined with variance, we obtain the overall error MSE (mean squared error) that measures how efficient an estimator is.

7-3 A robust estimator is one that is reasonably efficient in sampling from a wide variety of populations—especially those long-tailed populations that tend to produce troublesome outliers in the sample. Trimming off these outliers produces the robust trimmed mean. Or the computer can trim more subtly, producing the very robust biweighted mean.

*7-4   A consistent estimator is one that eventually is on target. (Not only on target on average, but the *whole distribution* gets squeezed onto the target, as the sample size n increases infinitely.) Thus persistent sampling bias, for example, makes an estimator inconsistent.

REVIEW PROBLEMS

7-19   Assuming as usual that samples are random, answer True or False; if False, correct it.

    a.   Samples are used for making inferences about the population from which the sample is drawn.

    b.   If we double the sample size, we halve the standard error of $\overline{X}$, and consequently double its accuracy in estimating the population mean.

    c.   The expected value of $\overline{X}$ is equal to the population mean (assuming the sample is random, with every member of the population equally likely to be drawn).

    d.   A robust estimator is one that has small MSE (is efficient) when the population shape is normal.

    e. $\overline{X}$ is a robust estimator of the population center, while s is a robust estimator of the population spread.

7-20   An estimator that has small variance (but may be biased) is called *precise*. An estimator that has small MSE is called *accurate*. To illustrate: A standard 100-gm mass was weighed many, many times, on a scale A, and the distribution of measurements is graphed below. Similarly, on scale B, and finally on scale C. To summarize the graphs:

$$
\begin{array}{lll}
\text{Scale A:} & \mu = 100.00, & \sigma = .05 \\
\text{Scale B:} & \mu = \ \ 99.98, & \sigma = .02 \\
\text{Scale C:} & \mu = 100.08, & \sigma = .01
\end{array}
$$

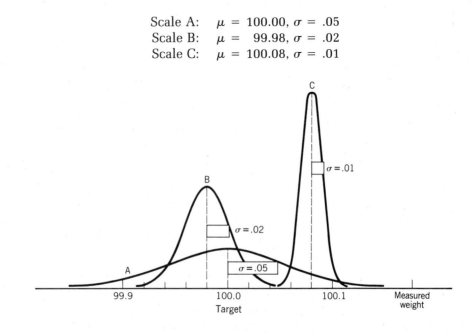

a. Which scale is most precise? Most accurate?

b. What is the relative efficiency of scale A relative to B? Of scale C relative to B? Do these answers agree with part **a**?

c. Which is more important: for an estimator to be precise or accurate?

7-21 **a.** Continuing Problem 7-20, since the scales were not perfect, it was decided in each case to weigh an object 25 times and take the average as the best estimate of the true weight. When used this way, which scale gives the most accurate $\overline{X}$?

b. Answer True or False; if False, correct it:

If a single measurement is taken, the random part (σ) and the systematic part (bias) are equally important.

When several measurements are averaged, the random part of the error gets averaged out, while the systematic part persists. Then it is particularly important to have little bias.

*7-22 A market survey of young business executives was undertaken to determine what sort of computer would suit a combination of their professional and personal needs. Since those with more children were thought to be more likely to buy a home computer, among other things the survey asked how many children they had.

Unfortunately, those with more children tended to have less time and inclination to reply to the survey, as the following table shows:

X = NUMBER OF CHILDREN OVER 5 YEARS OLD	TOTAL POPULATION (TARGET)		SUBPOPULATION WHO WOULD RESPOND		
	FREQUENCY f	REL. FREQUENCY f/N	FREQUENCY f	REL. FREQUENCY f/N	RESPONSE RATE
0	20,000	.40	6,200	.62	31%
1	12,000	.24	2,100	.21	17.5%
2	10,000	.20	1,200	.12	12%
3	6,000	.12	400	.04	6.7%
4	2,000	.04	100	.01	5%
	N = 50,000	1.00	N = 10,000	1.00	20%

Three types of sample survey were proposed:

i. *High volume*, with 1000 executives sampled, and with no followup. Therefore, their overall response rate would be 20% as given by the table, yielding 200 replies.

ii. *High quality*, with 25 executives sampled, and enough followup to get a 100% response rate.

iii. *Compromise*, with 100 executives sampled, and enough followup to get response rates of 60%, 40%, 30%, 30%, 30% in the five respective categories of X.

a. To estimate the mean number of children in the population, which of the three surveys would give an unbiased sample mean $\overline{X}$?

b. For each survey, find the MSE of $\overline{X}$. Then which survey would be most accurate?

*7-23 In the market survey of Problem 7-22, suppose the population could be classified into 4 categories, corresponding to the probability of their buying a personal computer within 2 years:

PROBABILITY OF BUYING π	TOTAL POPULATION (TARGET) FREQUENCY f	REL. FREQUENCY f/N	SUBPOPULATION WHO WOULD RESPOND FREQUENCY f	REL. FREQUENCY f/N	RESPONSE RATE
0	40,000	.80	7,000	.70	17.5%
10%	5,000	.10	1,000	.10	20%
25%	3,000	.06	1,000	.10	33%
50%	2,000	.04	1,000	.10	50%
	$N = 50,000$	1.00	$N = 10,000$	1.00	20%

a. To estimate the proportion π in the whole population who would buy a computer, which of the three surveys in Problem 7-22 would give an unbiased sample proportion P? (Assume the "compromise" survey has response rates of 40%, 50%, 70%, and 80% in the last column above.)

b. For each survey, find the MSE of P. Then which survey would be most accurate?

7-24 Suppose that a surveyor is trying to determine the area of a rectangular field, in which the measured length X and the measured width Y are independent random variables that fluctuate about the true values, according to the following probability distributions:

x	p(x)	y	p(y)
8	1/4	4	1/2
10	1/4	6	1/2
11	1/2		

The calculated area $A = XY$, of course, is a random variable, and is used to estimate the true area. If the true length and width are 10 and 5, respectively,

a. Is X an unbiased estimator of the true length?

b. Is Y an unbiased estimator of the true width?

c. Is A an unbiased estimator of the true area?
   (*Hint:* see Problem 5-36.)

7-25 A processor of sheet metal produces a large number of square plates, whose size must be cut within a specified tolerance. To measure the final product, a slightly worn guage is used: Its measurement error is normally distributed with a mean $\mu = 0$ and standard deviation $\sigma = .10$ inch. To improve the accuracy, and to protect against blunders, two independent measurements of a plate's length are taken with this gauge, say X_1 and X_2.

To find the area of a plate, the quality control foreman is in a dilemma:

   i.   Should he square first, and then average:

$$\frac{X_1{}^2 + X_2{}^2}{2}$$

   ii.   Should he average first, and then square:

$$\left(\frac{X_1 + X_2}{2}\right)^2$$

a.   Are methods **i** and **ii** really different, or are they just two different ways of saying the same thing? (*Hint:* Try a simulation. Suppose, for example, the two measured lengths are $X_1 = 5.9$ and $X_2 = 6.1$.)

b.   Which has less bias? [*Hint:* See equation (4-36).]

c.   As an alternative estimator of the area, what is the bias of X_1X_2? (*Hint:* See Problem 5-36.)

7-26   Suppose that two economists estimate μ (the average expenditure of American families on food), with two unbiased (and statistically independent) estimates U and V. The second economist is less careful than the first—the standard deviation of V is 3 times as large as the standard deviation of U. When asked how to combine U and V to get a publishable overall estimate, three proposals are made:

   i.
$$W_1 = \frac{1}{2} U + \frac{1}{2} V$$

   ii.
$$W_2 = \frac{3}{4} U + \frac{1}{4} V$$

   iii. $W_3 = 1U + 0V = U$

a.   Which are unbiased?

b.   Intuitively, which would you guess is the best estimator? The worst?

c.   Check out your guess in part **b** by making the appropriate calculations.

*d.   (Requires calculus) Find the optimal weights that give the most efficient unbiased estimator possible. [*Hint:* The general unbiased estimator is of the form $cU + (1 - c)V$, and its variance can be found from (5-22), in terms of c—call it $f(c)$. To find the minimum, set $f'(c) = 0$.]

*7-27   (Monte Carlo)

a.   Simulate a random sample of 5 observations from a normal population with $\mu = 0$, $\sigma = 1$. Calculate the mean, median, and 25% trimmed mean to estimate the population center. Which provides the best point estimate in your specific sample?

b. Which estimate would outperform the others if this sampling experiment were repeated many times?

*7-28 Repeat Problem 7-27, with one twist: The last observation inadvertently has its decimal point dropped (so that it is actually multiplied by 10, and can be viewed as coming from a population with $\mu = 0$ and $\sigma = 10$).

Interval Estimation

Reconnaissance is as important in the art of politics as it is in the art of war—or the art of love.   HENRY DURANT

8-1   A Single Mean

A—THEORY

In Chapter 7, we considered various *point estimators.* For example, we concluded that $\overline{X}$ was a good estimator of μ for populations that are approximately normal. Although *on average* $\overline{X}$ is on target, the specific sample mean $\overline{X}$ that we happen to observe is almost certain to be a bit high or a bit low. Accordingly, if we want to be reasonably confident that our inference is correct, we cannot claim that μ is precisely equal to the observed $\overline{X}$. Instead, we must construct an *interval estimate* or *confidence interval* of the form:

$$\mu = \overline{X} \pm \text{sampling error} \tag{8-1}$$

The crucial question is: How wide must this allowance for sampling error be? The answer, of course, will depend on how much $\overline{X}$ fluctuates, which we review in Figure 8-1.

First we must decide how confident we wish to be that our interval estimate is right—that it does indeed bracket μ. It is common to choose 95% confidence; in other words, we will use a technique that will give us, in the long run, a correct interval 19 times out of 20.

To get a confidence level of 95%, we select the smallest range under

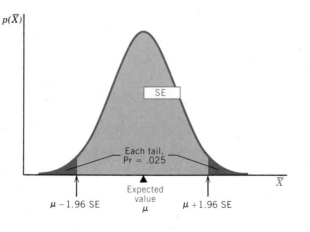

Figure 8-1
Normal distribution of the sample mean around the fixed but unknown parameter μ. 95% of the probability is contained within ± 1.96 standard errors.

the normal distribution of $\overline{X}$ that will just enclose a 95% probability. Obviously, this is the middle chunk, leaving $2\frac{1}{2}\%$ probability excluded in each tail. From Table IV, we find that this requires a z value of 1.96. That is, we must go above and below the mean by 1.96 standard errors (SE), as shown in Figure 8-1. In symbols:

$$\Pr(\mu - 1.96 \text{ SE} < \overline{X} < \mu + 1.96 \text{ SE}) = 95\% \qquad (8\text{-}2)$$

which is just the algebraic way of saying, "There is a 95% chance that the random variable $\overline{X}$ will fall between $\mu - 1.96$ SE and $\mu + 1.96$ SE."

The inequalities within the brackets may now be solved for μ—turned around—to obtain the equivalent statement:

$$\Pr(\overline{X} - 1.96 \text{ SE} < \mu < \overline{X} + 1.96 \text{ SE}) = 95\% \qquad (8\text{-}3)$$

We must be exceedingly careful not to misinterpret (8-3). μ has not changed its character in the course of this algebraic manipulation. It has not become a variable but has remained a population constant. Equation (8-3), like (8-2), is a probability statement about the random variable $\overline{X}$, or, more precisely, the *random interval* $\overline{X} - 1.96$ SE to $\overline{X} + 1.96$ SE. *It is this interval that varies, not μ.*

B—ILLUSTRATION

To appreciate the fundamental point that the confidence interval fluctuates while μ remains constant, consider an example. Suppose we wish to construct an interval estimate for μ, the mean height of the population of men on a large midwestern campus, on the basis of a random sample of $n = 10$ men. Moreover, to clearly illustrate what is going on, suppose

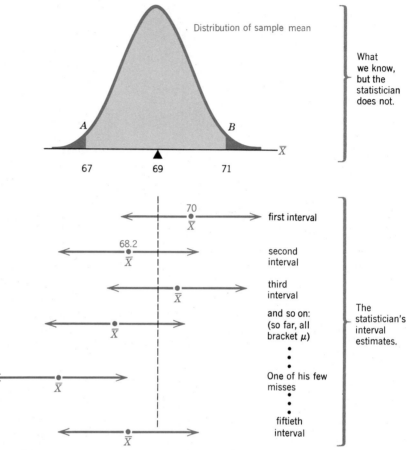

Figure 8-2
Constructing 50 interval estimates: About 95% correctly bracket the target μ.

that we have some supernatural knowledge of the population μ and σ.
Suppose that we know, for example, that:

$$\mu = 69$$
$$\sigma = 3.22$$

Then the standard error of $\overline{X}$ is:

$$\text{SE} = \frac{\sigma}{\sqrt{n}} \qquad\qquad (8\text{-}4)$$
$$\qquad\qquad\qquad (6\text{-}7)\,\text{repeated}$$
$$= \frac{3.22}{\sqrt{10}} = 1.02$$

Now let us observe what happens when the statistician (who does not

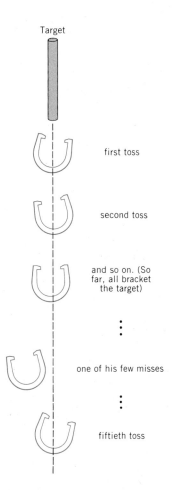

Target

first toss

second toss

and so on. (So far, all bracket the target)

one of his few misses

fiftieth toss

Figure 8-3
How an expert pitching 50 horsehoes is like the statistician in Figure 8-2: About 95% of his attempts correctly bracket the target.

have our supernatural knowledge of course) tries to estimate μ using (8-3). To appreciate the random nature of his task, suppose he makes 50 such interval estimates, each time from a different random sample of 10 men. Figure 8-2 shows his typical experience (compare to Figure 6-5).

First, in the normal distribution at the top we illustrate equation (8-2): $\overline{X}$ is distributed around $\mu = 69$, with a 95% probability that it lies as close as:

$$\pm 1.96\text{SE} = \pm 1.96(1.02) = \pm 2.0 \tag{8-5}$$

That is, there is a 95% probability that any $\overline{X}$ will fall in the range 67 to 71 inches.

But the statistician does not know this; he blindly takes his random sample, from which he computes the sample mean—let us say[1] $\overline{X} = 70$. From (8-3), he calculates the interval for μ:

$$\overline{X} \pm 1.96\text{SE} \tag{8-6}$$
$$= 70 \pm 1.96\,(1.02)$$
$$= 70 \pm 2$$
$$= 68 \text{ to } 72 \tag{8-7}$$

This interval estimate for μ is the first one shown in Figure 8-2. In his first effort, the statistician is right; μ is enclosed in this interval.

In his second sample, suppose he happens to draw a shorter group of individuals, and duly computes $\overline{X}$ to be 68.2 inches. From a similar evaluation of (8-6), he comes up with his second interval estimate shown in the diagram, and so on. If he continues in this way to construct 50 interval estimates, about 95% of them will bracket the target μ. Only about 2 or 3 will miss the mark.

We can easily see why the statistician is correct so often. For each interval estimate, he is simply adding and subtracting 2 inches from the sample mean; and this is the same ± 2 inches that was calculated in (8-5) and defines the 95% range AB around μ. Thus, if and only if he observes a sample mean within this range AB, will his interval estimate bracket μ. And this happens 95% of the time, in the long run.

In practice, of course, a statistician would not take many samples—he only takes one. And once this interval estimate is made, he is either right or wrong; this interval brackets μ or it does not. But the important point to recognize is that the statistician is using a method with a 95% chance of success; in the long run, 95% of the intervals he constructs in this way will bracket μ.

C—ANALOGY: PITCHING HORSE SHOES

Constructing 95% confidence intervals is like pitching horseshoes. In each case there is a fixed target, either the population μ or the stake. We are trying to bracket it with some chancy device, either the random interval or the horseshoe. This analogy is illustrated in Figure 8-3.

There are two important ways, however, that confidence intervals differ from pitching horseshoes. First, only *one* confidence interval is customarily constructed. Second, the target μ is not visible like a horseshoe stake. Thus, whereas the horseshoe player always knows the score (and specifically, whether or not the last toss bracketed the stake), the statistician does not. He continues to "throw in the dark," without knowing whether

---

[1] Strictly speaking, a specific realized value such as this should be denoted by the lower-case letter $\bar{x}$ to distinguish it from the potential value (random variable) denoted by $\overline{X}$. From now on, however, we will not bother to distinguish the realized value from the potential value, and will use $\overline{X}$ to refer to either. And for certain other variables, such as s, in order to conform to common usage we will always use a lower-case letter.

or not a specific interval estimate has bracketed μ. All he has to go on is the statistical theory that assures him that, in the long run, he will succeed 95% of the time.

D—REVIEW

To review, we briefly emphasize the main points:

1. The population parameter μ is constant, and remains constant. It is the interval estimate that is a random variable, because its center $\overline{X}$ is a random variable.
2. To appreciate where it came from, the confidence interval (8-6) may be written as

$$\mu = \overline{X} \pm z_{.025}\ \text{SE} \qquad (8\text{-}8)$$

where $z_{.025}$ is the value 1.96 obtained from Appendix Table IV—that is, the z value that cuts off $2\frac{1}{2}\%$ in the upper tail (and by symmetry, also $2\frac{1}{2}\%$ in the lower tail). Equation (8-8) is extremely useful as the prototype for all confidence intervals that we will study. When we substitute (8-4) into (8-8), we obtain another very useful form:

$$\boxed{\begin{array}{c}\textbf{95\% confidence interval}\\[4pt] \mu = \overline{X} \pm z_{.025}\ \dfrac{\sigma}{\sqrt{n}}\end{array}} \qquad (8\text{-}9)$$

3. As sample size is increased, $\overline{X}$ has a smaller standard error $\sigma/\sqrt{n}$, and the confidence interval becomes narrower. This increased accuracy is the value of increased sample size.
4. Suppose we wish to be more confident, for example, 99% confident. Then the range must be large enough to encompass 99% of the probability. Since this leaves .005 in each tail, the formula (8-9) would use $z_{.005} = 2.58$ (found in Table IV). Thus the confidence interval becomes wider—that is, vaguer. This is exactly what we would expect. The more certain we want to be about a statement, the more vague we must make it.

An example will illustrate how easily confidence intervals can be calculated.

EXAMPLE **8-1**   Twenty-five marks were randomly sampled from a very large class having a standard deviation of 13. If these 25 marks had a mean of 58, find a 95% confidence interval for the mean mark of the whole class.

Solution   Substitute $n = 25$, $\sigma = 13$, and $\overline{X} = 58$ into (8-9):

$$\mu = 58 \pm 1.96 \frac{13}{\sqrt{25}}$$

$$= 58 \pm 5$$

That is, the 95% confidence interval for μ is

$$53 < \mu < 63$$

Since confidence intervals are an ideal way to allow for sampling fluctuation, we will develop many other confidence intervals in this chapter
and indeed throughout the whole book. To make it easy to refer to them,
we have listed them all inside the back cover in a guide called WHERE
TO FIND IT. Here we list the commonest statistical problems that occur
in practice, and how to deal with each, including the appropriate confidence interval and examples of how problems can be solved.

E—WHEN SHOULD WE USE THE MEAN?

In this chapter and throughout the book, following the prototype (8-9) we
will build many other confidence intervals around the sample mean $\overline{X}$.
These confidence intervals are ideal if the parent populations are normal,
or nearly normal. (If the underlying population is not normal, we may be
able to get a narrower confidence interval by centering it on one of the
robust estimates discussed in Section 7-3.)

A confidence interval for the mean is also ideal whenever a total is
required, because a total is closely related to the mean (as we saw in
Chapter 2). For example, if a real estate board reports that their 2000
residential sales during March had a mean of $52,000, a median of $46,000,
and a mode of $45,000, it is the *mean* that gives the total value (2000 ×
$52,000 = $104,000,000).

PROBLEMS

8-1 Make the correct choice in each square bracket:

  a. The sample mean $[\overline{X}, \mu]$ is an unbiased estimate of the population
    mean $[\overline{X}, \mu]$.

  b. $\overline{X}$ fluctuates from sample to sample with a standard deviation equal
    to $[\sigma/n, \sigma/\sqrt{n}]$, which is also called the [standard error SE, population
    standard deviation].

  c. If we make an allowance of about $[\sqrt{n}, 2]$ standard errors on either
    side of $\overline{X}$, we obtain an interval wide enough that it has a 95% chance

of covering the target μ. This is called the 95% confidence interval for [$\overline{X}$, μ].

d. A statistician who constructed a thousand of these 95% confidence intervals over his lifetime would miss the target [practically never, about 50 times, about 950 times]. [And fortunately he would know, But unfortunately he would not know] just which times these were.

e. For greater confidence such as 99%, the confidence interval must be made [narrower, wider].

8-2 **a.** Suppose that you took a random sample of 100 accounts in a large department-store chain, and found that the mean balance due was $74. If you knew that the standard deviation of all balances due is $86, find the 95% confidence interval for the population mean.

b. If there were 243,000 accounts altogether, find a 95% confidence interval for the total balance due. Briefly explain to the vice president the meaning of your answer.

c. Suppose that the skeptical vice president undertook a complete accounting of the whole population of balances due, and found a total of $19,714,000. What would you say?

8-3 A random sample of 4 infants from a hospital nursery gave the following birth weights (kilograms):

$$3.1 \quad 2.8 \quad 3.6 \quad 3.7$$

a. If $\sigma = .4$, find the 95% confidence interval for the population mean birth weight.

b. Find the 99% confidence interval.

8-4 To determine the average age of its customers, a large manufacturer of men's clothing took a random sample of 50 customers and found $\overline{X} = 36$. If you knew $\sigma = 12$:

a. Find a 95% confidence interval for the mean age μ of all the customers.

b. Suppose you need to make the 95% confidence interval narrower— to be specific, ± 2 years. How large a sample is now required?

c. In part (b) we found out what sample size n was required to achieve a specified sampling error allowance $e = 2$. Show that the general formula for n in terms of any specified e and σ is:

$$n = \left(1.96 \frac{\sigma}{e}\right)^2 \qquad (8\text{-}10)$$

d. How large should n be if you need to reduce the sampling allowance to ± 1?

e. To achieve 16 times the accuracy (i.e., to reduce *e* to one-sixteenth of its former value) how much larger must the sample be?

8-5 A random survey of 225 of the 2700 institutions of higher learning in the U.S. was conducted in 1976 by the Carnegie Commission. The survey showed a mean enrolment of 3700 students per institution, with a standard deviation of 6000 students. Construct a 95% confidence interval for the total student enrolment in all 2700 institutions. (Since the sample is so large, the population standard deviation σ can be safely approximated with the sample standard deviation $s = 6000$.)

8-2 Small Sample *t*

In the previous section it was assumed, quite unrealistically, that a statistician knows the true population standard deviation σ. In this section, we consider the practical case where σ is unknown.

With σ unknown, the statistician wishing to evaluate the confidence interval (8-9) must use some estimate of σ—with the most obvious candidate being the *sample* standard deviation *s*. (Note that *s*, along with $\overline{X}$, can always be calculated from the sample data.) But the use of *s* introduces an additional source of unreliability, especially if the sample is small. To retain 95% confidence, we must therefore widen the interval. We do so by replacing the $z_{.025}$ value taken from the standard normal distribution with a larger $t_{.025}$ value taken from a similar distribution called *Student's t* distribution. (The close relationship between the standard normal and the *t* distribution is described in detail in Figure 9-2.)

When we substitute *s* and the compensating $t_{.025}$ into (8-9), we obtain:

> **95% confidence interval for the population mean**
>
> $$\mu = \overline{X} \pm t_{.025} \frac{s}{\sqrt{n}}$$ (8-11)

The value $t_{.025}$ is listed in the shaded column of Appendix Table V, and is tabulated according to the degrees of freedom (d.f.):

> **d.f. $\equiv$ amount of information used in calculating s^2**   (8-12)
> **$\equiv$ divisor in s^2**   like (2-16)

To explain this, recall from Section 2-3 that we can calculate the variance s^2 only if our sample size *n* exceeds 1. That is, in calculating s^2, there are essentially only $(n - 1)$ pieces of information (degrees of freedom):

> **d.f. $= n - 1$**   (8-13)

And this was the divisor in calculating s^2, you recall.

For example, if the sample size is n = 4, we read down Table V to d.f. = 3 (in the left-hand column), which gives $t_{.025} = 3.18$ to use in (8-11). (Note that this $t_{.025}$ is indeed larger than $z_{.025} = 1.96$.)

In practice, when do we use the normal z table, and when the t table? (Incidentally, these are the two most useful tables in statistics. So useful, in fact, that we repeat them inside the front cover, for easy reference.) If σ is known, the normal z value in (8-9) is appropriate. If σ has to be estimated with s, the t value in (8-11) is appropriate—regardless of sample size. However, if the sample size is large, the normal z is an accurate enough approximation[2] to the t. So in practice, t is used only for *small samples when σ is unknown*. An example will illustrate.

EXAMPLE **8-2**   From a large class, a random sample of 4 grades were drawn: 64, 66, 89, and 77. Calculate a 95% confidence interval for the whole class mean μ.

Solution   Since n = 4, d.f. = 3; and so from Table V inside the front cover, $t_{.025} = 3.18$. In Table 8-1, we calculate $\overline{X} = 74$ and $s^2 = 132.7$. When all these are substituted into (8-11), we obtain:

$$\mu = 74 \pm 3.18 \frac{\sqrt{132.7}}{\sqrt{4}}$$

$$= 74 \pm 18 \qquad (8\text{-}14)$$

That is, with 95% confidence we conclude that the mean grade of the whole class is between 56 and 92. This is a pretty vague interval because the sample is so small.

TABLE **8-1**   **Analysis of One Sample**

OBSERVED GRADE X	$(X - \overline{X})$	$(X - \overline{X})^2$
64	−10	100
66	−8	64
89	15	225
77	3	9
$\overline{X} = \dfrac{296}{4} = 74$	$0\sqrt{}$	$s^2 = \dfrac{398}{3} = 132.7$

To sum up, we note that (8-11) is of the following form:

$$\boxed{\mu = \overline{X} \pm t_{.025} \text{ (estimated standard error)}} \qquad (8\text{-}15)$$

---

[2]This may be verified from Table V. For example, a 95% confidence interval with d.f. = 60 should use $t_{.025} = 2.00$; but using $z_{.025} = 1.96$ is a very good approximation.

This equation is just like (8-8), the only difference being that the *estimated* standard error in (8-15) requires using the larger t value instead of the z value.

PROBLEMS

8-6 Answer True or False; if False, correct it:

If σ is unknown, then we must use (8-11) instead of (8-9). This involves replacing σ with its estimator s, an additional source of unreliability; and to allow for this, $z_{.025}$ is replaced by the larger $t_{.025}$ value in order to keep the confidence level at 95%.

8-7 A random sample of 5 women had the following cholesterol levels in their blood (grams per liter):

$$3.0 \quad 1.8 \quad 2.1 \quad 2.7 \quad 1.4$$

a. Calculate a 95% confidence interval for the mean cholesterol level for the whole population of women.

b. Construct a 99% confidence interval.

8-8 The reaction times of 30 randomly selected drivers were found to have a mean of .83 second and standard deviation of .20 second.

a. Find a 95% confidence interval for the mean reaction time of the whole population of drivers.

b. One more driver was randomly selected, and found to have a reaction time of .98 second. Is this surprising? Explain.

8-9 Fifty bags of coffee filled from bulk stock in a large supermarket were randomly sampled from the 800 bags on the shelf. Although the listed weight was "16 oz.," the actual weights were as follows:

WEIGHT (OZ.) (CELL MIDPOINT)	NO. OF BAGS
16.0	2
16.1	16
16.2	22
16.3	10

Calculate a 95% confidence interval for the mean weight of all 800 bags on the shelf.

8-10 A random sample of 40 cars were clocked as they passed by a checkpoint, and their speeds were as follows (km per hour):

49	83	58	65	68	60	76	86	74	53
71	74	65	72	64	42	62	62	58	82
78	64	55	87	56	50	71	58	57	75
58	86	64	56	45	73	54	86	70	73

Construct a 95% confidence interval for the average speed of all cars passing by that checkpoint. (*Hint:* You may find it reduces your work to group the observations into cells of width 5.)

8-11 **a.** Take a sample of five random normal numbers from Appendix Table II. Then find a 95% confidence interval for the mean of the population. Assume $\sigma = 1$ is known.

  b. Repeat part **a**, if σ is not known, and has to be estimated with s.

  c. Since you know $\mu = 0$, you can see whether or not your confidence intervals are correct. Have the instructor check the proportion of the confidence intervals that are correct in the class. What proportion would be correct in a very large class (in the long run)?

  d. Answer True or False; if False, correct it:
  The confidence intervals in part **b** based on the t distribution are always wider than those in part **a** based on the normal z distribution. This represents the cost of not knowing σ.

8-3 Two Means, Independent Samples

A—IF POPULATION VARIANCES ARE KNOWN (IN THEORY)

Two population means are commonly compared by forming their difference:

$$(\mu_1 - \mu_2) \tag{8-16}$$

This difference is the population target to be estimated. A reasonable estimate of this is the corresponding difference in *sample* means:

$$(\overline{X}_1 - \overline{X}_2)$$

Using a familiar argument, we could develop the appropriate confidence interval around the estimate:

$$(\mu_1 - \mu_2) = (\overline{X}_1 - \overline{X}_2) \pm z_{.025}\ SE \tag{8-17}$$

like (8-8)

In Appendix 8-3 we derive the formula for the SE; substituting it into (8-17) yields:

95% confidence interval for independent samples, when two populations have different and known variances

$$(\mu_1 - \mu_2) = (\overline{X}_1 - \overline{X}_2) \pm z_{.025}\sqrt{\frac{\sigma_1^2}{n_1} + \frac{\sigma_2^2}{n_2}} \tag{8-18}$$

When σ_1 and σ_2 are known to have a common value, say σ, this 95% confidence interval reduces to:

$$(\mu_1 - \mu_2) = (\overline{X}_1 - \overline{X}_2) \pm z_{.025}\, \sigma \sqrt{\frac{1}{n_1} + \frac{1}{n_2}} \qquad (8\text{-}19)$$

B—WHEN POPULATION VARIANCES ARE UNKNOWN

In practice, the population σ is not known, and has to be replaced with an estimate, customarily denoted by s_p. Then $z_{.025}$ has to be replaced with the broader value $t_{.025}$, and so (8-19) becomes:

95% confidence interval, when two populations have a common unknown variance:

$$(\mu_1 - \mu_2) = (\overline{X}_1 - \overline{X}_2) \pm t_{.025}\, s_p \sqrt{\frac{1}{n_1} + \frac{1}{n_2}} \qquad (8\text{-}20)$$

How do we derive the estimate s_p? Since both populations have the same variance σ^2, it is appropriate to pool the information from both samples to estimate it. So our estimate is called the *pooled* variance s_p^2. We add up all the squared deviations from both samples, and then divide by the total d.f. in both samples, $(n_1 - 1) + (n_2 - 1)$. That is:

$$s_p^2 = \frac{\sum(X_1 - \overline{X}_1)^2 + \sum(X_2 - \overline{X}_2)^2}{(n_1 - 1) + (n_2 - 1)} \qquad (8\text{-}21)$$

where X_1 (or X_2) represents the typical observation in the first (or second) sample. To complete (8-20), we need the d.f. for t. As suggested by (8-12), this is just the divisor used in calculating s_p^2:

$$\text{d.f.} = (n_1 - 1) + (n_2 - 1) \qquad (8\text{-}22)$$

Unless specified otherwise, we shall assume throughout this book that population variances are about equal, so that the pooling in (8-21) is appropriate. An example will illustrate.

EXAMPLE 8-3   From a large class, a sample of 4 grades were drawn: 64, 66, 89, and 77. From a second large class, an independent sample of 3 grades were drawn: 56, 71, and 53. Calculate the 95% confidence interval for the difference between the two class means, $\mu_1 - \mu_2$.

TABLE **8-2** **Analysis of Two Independent Samples**

	CLASS 1			CLASS 2	
X_1	$(X_1 - \overline{X}_1)$	$(X_1 - \overline{X}_1)^2$	X_2	$(X_2 - \overline{X}_2)$	$(X_2 - \overline{X}_2)^2$
64	-10	100	56	-4	16
66	-8	64	71	11	121
89	15	225	53	-7	49
77	3	9			
$\overline{X}_1 = \dfrac{296}{4}$ $= 74.0$	$0\checkmark$	398	$\overline{X}_2 = \dfrac{180}{3}$ $= 60.0$	$0\checkmark$	186

Solution

In Table 8-2 we calculate the sample means and the squared deviations. Thus:

$$s_p^2 = \frac{398 + 186}{3 + 2} = \frac{584}{5} = 117 \qquad \text{like (8-21)}$$

In the divisor we see d.f. $= 5$, so that from Table V, $t_{.025} = 2.57$. Substituting into (8-20), we obtain:

$$(\mu_1 - \mu_2) = (74.0 - 60.0) \pm 2.57 \sqrt{117} \sqrt{\frac{1}{4} + \frac{1}{3}}$$

$$= 14 \pm 21 \qquad\qquad (8\text{-}23)$$

$$= -7 \text{ to } +35$$

Thus with 95% confidence we conclude that the average of the first class (μ_1) may be 7 marks below the average of the second class (μ_2). Or μ_1 may be 35 marks above μ_2—or anywhere in between. There is a very large sampling allowance in this example because the samples were so small.

We emphasize that the confidence intervals in this section require the two samples to be *independent* as in Example 8-3. As another example, consider just one class of students, examined at two different times—say fall and spring terms. The fall population and spring population of grades could then be sampled and compared using (8-20)—provided, of course, that the two samples are independently drawn. (They would not be independently drawn, for example, if we made a point of canvassing the same students twice.)

PROBLEMS

8-12 To determine the effectiveness of a feed supplement for cattle, a pilot study randomly divided 8 steers into a group of 4 steers to be fed the supplement, and another group of 4 steers to be fed exactly the same diet except for the supplement. The weight gains (lbs) over a period of 6 months were as follows:

SUPPLEMENT GROUP	CONTROL GROUP
330	290
360	320
400	340
350	370

Construct a 95% confidence interval for the effect of the supplement.

8-13 In a large American university in 1969, the men and women professors were sampled independently, yielding the annual salaries given below (in thousands of dollars. From Katz, 1973):

WOMEN	MEN
9	16
12	19
8	12
10	11
16	22

    a. Calculate a 95% confidence interval for the mean salary difference between men and women.

    b. How well does this show the university's discrimination against women?

8-14 Repeat Problem 8-13 using the complete data given in Problems 2-1 and 2-2—independent random samples with the following characteristics:

WOMEN	MEN
$n = 5$	$n = 25$
$\overline{X} = 11.0$	$\overline{X} = 16.0$
$\Sigma(X - \overline{X})^2 = 40$	$\Sigma(X - \overline{X})^2 = 786$

8-15 In the hospital-birth example in Section 1-2, recall that a sample of 14 women were kept away from their infants (the *control* procedure), while a sample of 14 others were given extensive early contact with their infants (the *treatment*). Construct a 95% confidence interval for the difference that the treatment makes, for each of the following ways in which it was measured (Klaus and others, 1972):

    a. For each mother, a score (between 0 and 6) was assigned, indicating her interest and care during a physical exam of her infant. The score statistics turned out as follows:

	CONTROL MOTHERS	TREATMENT MOTHERS
Number n	14	14
Mean $\overline{X}$	3.43	5.36
Range	5	3
$\Sigma(X - \overline{X})^2$	33.3	11.2

b. At the same time an alternative score (between 0 and 12) was assigned to each mother, and produced the following distribution:

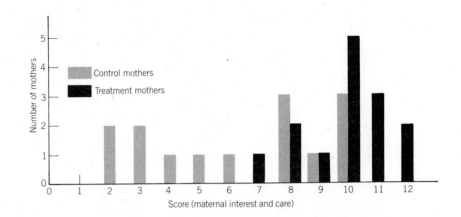

8-16 One of the pioneering controlled randomized experiments in medicine (clinical trials) was carried out in 1948 to evaluate the new penicillin drug streptomycin (Medical Research Council, 1948). A total of 107 tubercular patients were randomly assigned: 52 to be given bed rest only (control), and 55 to be also given streptomycin (treatment). The degree of improvement after 6 months was measured for each patient; if we code it between $+2$ ("considerable improvement") and -3 ("death"), the results were as follows:

IMPROVEMENT AFTER 6 MONTHS	TREATED PATIENTS	CONTROL PATIENTS
$+2$	28	4
$+1$	10	13
0	2	3
-1	5	12
-2	6	6
-3	4	14
	55	52

How well does this establish the superiority of streptomycin?

8-4 Two Means, Paired Samples

In the previous section we considered *independent* samples, for instance, a sample of students' grades in the fall was compared to a fresh sample of students' grades in the spring. In this section we will consider *dependent* or *paired* samples.

A—ANALYSIS BASED ON INDIVIDUAL DIFFERENCES

Suppose a comparison of fall and spring grades is done using the *same* students both times. Then the paired grades (spring X_1 and fall X_2) for each of the students can be set out, as in Table 8-3.

The natural first step is to see how each student changed; that is, calculate the difference $D = X_1 - X_2$, for each student. Once these differences are calculated, then the original data, having served its purpose, can be discarded.

We proceed to treat the four differences D now as a *single sample*, and analyze them just as we analyze any other single sample (e.g., the sample in Table 8-1). First, we calculate the average difference $\bar{D}$. Then we use this *sample* $\bar{D}$ appropriately to construct a confidence interval for the average *population* difference Δ, obtaining:

$$\boxed{\begin{array}{c} \textbf{95\% confidence interval for paired samples} \\[2mm] \Delta = \bar{D} \pm t_{.025}\, \dfrac{s_D}{\sqrt{n}} \end{array}} \qquad (8\text{-}24)$$

For our example in Table 8-3, we calculate $\bar{D} = 14$ and $s_D = \sqrt{15.3}$. In the divisor of s_D we saw d.f. $= n - 1 = 3$, so that $t_{.025} = 3.18$. Substituting into (8-24), we obtain

$$\Delta = 14 \pm 3.18\, \frac{\sqrt{15.3}}{\sqrt{4}}$$

$$= 14 \pm 6 \qquad (8\text{-}25)$$

$$= 8 \text{ to } 20$$

TABLE **8-3**   **Analysis of two Paired Samples**

	OBSERVED GRADES		DIFFERENCE		
STUDENT	X_1 (SPRING)	X_2 (FALL)	$D = X_1 - X_2$	$(D - \bar{D})$	$(D - \bar{D})^2$
Trimble	64	54	10	−4	16
Wilde	66	54	12	−2	4
Giannos	89	70	19	5	25
Ames	77	62	15	1	1
			$\bar{D} = \dfrac{56}{4}$	$0\checkmark$	$s_D^2 = \dfrac{46}{3}$
			$= 14.0$		$= 15.3$

Since the mean of the differences (Δ) exactly equals[3] the difference of the means ($\mu_1 - \mu_2$), we can re-express (8-24) as:

$$(\mu_1 - \mu_2) = (\bar{X}_1 - \bar{X}_2) \pm t_{.025} \frac{s_D}{\sqrt{n}} \tag{8-26}$$

In this form, it is clear that in (8-26) we are estimating the same parameter (the difference in two population means) as we did in the earlier formula (8-20). But the matched-pair approach (8-26) is better, because it has a smaller sampling allowance.

B—THE ADVANTAGE OF PAIRED SAMPLES

We have just seen how pairing samples reduces the sampling allowance (specifically, ± 6 in (8-25) compared to the ± 21 in (8-23) when the samples were independently drawn). The reason for this is intuitively clear. Pairing achieves a match that keeps many of the extraneous variables constant. In using the same four students, we kept sex, IQ, and many other factors exactly the same in both samples. We therefore had more leverage on the problem at hand—the difference in fall versus spring grades.

To summarize the chapter so far, we finish with an example that illustrates the important formulas for means.

EXAMPLE **8-4**

To measure the effect of a fitness campaign, a ski club randomly sampled 5 members before the campaign, and another 5 after. The weights were as follows (along with the person's initials):

Before: J.H. 168, K.L. 195, M.M. 155, T.R. 183, M.T. 169
After: L.W. 183, V.G. 177, E.P. 148, J.C. 162, M.W. 180

a. Calculate a 95% confidence interval for:
   i. The mean weight before the campaign.
   ii. The mean weight after the campaign.
   iii. The mean weight loss during the campaign.

b. It was decided that a better sampling design would be to measure the same people after, as before. In scrambled order, their figures were:

   After: K.L. 197, M.T. 163, T.R. 180, M.M. 150, J.H. 160

---

[3]This is also true for samples:

$$\bar{D} = \bar{X}_1 - \bar{X}_2$$

For example, for the data in Table 8-3, $\bar{D} = 14$, while $\bar{X}_1 - \bar{X}_2 = 74 - 60 = 14$, too.

On the basis of these people, calculate a 95% confidence interval for the mean weight loss during the campaign.

Solution

a.

	BEFORE			AFTER		
X_1	$(X_1 - \overline{X}_1)$	$(X_1 - \overline{X}_1)^2$	X_2	$(X_2 - \overline{X}_2)$	$(X_2 - \overline{X}_2)^2$
168	− 6	36	183	13	169
195	21	441	177	7	49
155	− 19	361	148	− 22	484
183	9	81	162	− 8	64
169	− 5	25	180	+ 10	100

$$\overline{X}_1 = \frac{870}{5} \qquad 0\checkmark \qquad 944 \qquad\qquad \overline{X}_2 = \frac{850}{5} \qquad 0\checkmark \qquad 866$$
$$= 174 \qquad\qquad\qquad\qquad\qquad\qquad = 170$$

   i. $\mu_1 = 174 \pm 2.78 \dfrac{\sqrt{944/4}}{\sqrt{5}}$ like (8-11)

          $= 174 \pm 19$

   ii. $\mu_2 = 170 \pm 2.78 \dfrac{\sqrt{886/4}}{\sqrt{5}}$ like (8-11)

          $= 170 \pm 18$

   iii. $\mu_1 - \mu_2 = (174 - 170) \pm$ like (8-20)

$$2.31 \sqrt{\frac{944 + 866}{4 + 4}} \sqrt{\frac{1}{5} + \frac{1}{5}}$$

          $= 4 \pm 22$

b. We must be sure to list the people in the same matched order so that it is meaningful to calculate the individual weight losses:

PERSON	X_1	X_2	$D = X_1 - X_2$	$(D - \overline{D})$	$(D - \overline{D})^2$
J.H.	168	160	+ 8	4	16
K.L.	195	197	− 2	−6	36
M.M.	155	150	+ 5	1	1
T.R.	183	180	+ 3	−1	1
M.T.	169	163	+ 6	2	4

$$\overline{D} = \frac{20}{5} \qquad 0\checkmark \qquad s_D^2 = \frac{58}{4}$$
$$\overline{D} = 4 \qquad\qquad\qquad\qquad = 14.5$$

$$\Delta = 4 \pm 2.78 \frac{\sqrt{14.5}}{\sqrt{5}}$$

$$= 4 \pm 5 \qquad\qquad\qquad \text{like (8-24)}$$

Thus the paired samples give a much more precise interval estimate for the weight loss ($\pm$ 5 vs. $\pm$ 22).

Pairing is obviously a desirable feature to design into any experiment, where feasible. If pairing cannot be achieved by using the same individual twice, we should look for other ways. For example, we might use pairs of twins—ideally identical twins. This would keep genetic and environmental factors constant. Of course, to decide which person within the pair is to be given the treatment, and which is to be left as the control, we would have to be fair and unbiased—that is, do it at random (with the flip of a coin, for example).

A further subtlety involved in pairing may be illustrated by the agricultural experiments that have historically played such an important role in statistics. If a plot of land is used in two successive years, many conditions in the second year will be different—rainfall, temperature, and so on. Therefore, it is much more effective to take two adjacent plots in the *same* year (the *split-plot* design).

PROBLEMS

8-17 Five people selected at random had their breathing capacity measured before and after a certain treatment, obtaining the data below.

BEFORE (X)	AFTER (Y)
2750	2850
2360	2380
2950	2930
2830	2860
2260	2330

Construct a 95% confidence interval for the mean increase in breathing capacity in the whole population.

8-18 To determine which of two seeds was better, a state agricultural station chose 7 two-acre plots of land randomly within the state. Each plot was split in half, and a coin was tossed to determine in an unbiased way which half would be sown with seed A, and which with seed B. The yields, in bushels, were as follows:

COUNTY	SEED A	SEED B
B	82	88
R	68	66
T	109	121
S	95	106
A	112	116
M	76	79
C	81	89

Which seed do you think is better? To back up your answer, construct a 95% confidence interval.

8-19 How much does an interesting environment affect the actual physical development of the brain? To answer this, for rats at least, Rosenzweig and others (1964) took 10 litters of purebred rats. From each litter, one rat was selected at random for the treatment group, and one for the control group. The two groups were treated the same, except that the treated rats lived altogether in a cage with interesting playthings, while the control rats lived in bare isolation. After a month, every rat was killed and its cortex (highly developed part of the brain) was weighed, with the following results (in centigrams) for the 10 pairs of littermates:

TREATMENT	CONTROL
68	65
65	62
66	64
66	65
67	65
66	64
66	59
64	63
69	65
63	58

a. Construct an appropriate 95% confidence interval.

b. State your answer to part **a** in a sentence or two that would be intelligible to a layman.

8-5 Proportions

A—LARGE SAMPLE FORMULA

Confidence intervals for proportions are very similar to means. We simply use the appropriate form of the normal approximation rule (6-12), and so obtain the 95% confidence interval for π:

$$\pi = P \pm 1.96 \sqrt{\frac{\pi(1 - \pi)}{n}}$$

For the unknown π that appears under the square root, we can substitute the sample P. (This is a strategy we have used before, when we substituted s for σ in the confidence interval for μ. This approximation introduces another source of error, which fortunately tends to zero as sample size n increases.) Thus:

> **95% confidence interval for the proportion, for large n**
>
> $$\pi = P \pm 1.96 \sqrt{\frac{P(1 - P)}{n}}$$

(8-27)

For this to be a good approximation, the sample size n ought to be large enough so that at least 5 successes and 5 failures turn up. As an example, the voter poll in Chapter 1 used this formula.

B—GRAPHICAL METHOD, LARGE OR SMALL SAMPLES

There is a graphical way to find an interval estimate for π, and it is very easy to see how it works for both large and small sample sizes. (The more complicated question—*why* it works—is covered in Appendix 8-5.) For example, suppose we observe 16 Republicans in a sample of 20 voters in a Kansas City suburb.

We first calculate $P = 16/20 = .80$. Then in Figure 8-4 we read up the vertical line passing through $P = .80$, noting where it cuts the curves labelled $n = 20$ (highlighted in white). These two points of intersection define the confidence interval for π, shown in color:

$$.55 < \pi < .95 \tag{8-28}$$

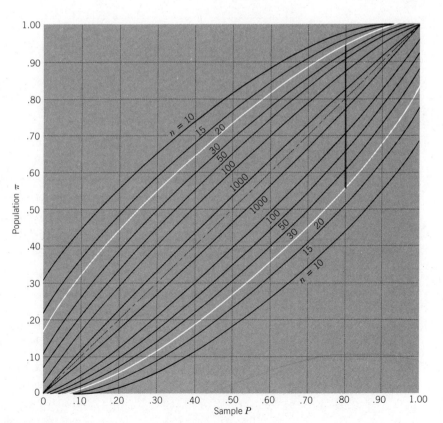

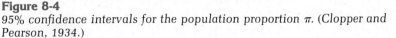

Figure 8-4
95% confidence intervals for the population proportion π. (Clopper and Pearson, 1934.)

This confidence interval is not symmetric about the estimate $P = .80$; the sampling allowance below is $.80 - .55 = .25$, while the allowance above is only $.95 - .80 = .15$. Why is it longer below? The answer may be seen by examining a more extreme case, where $P = 1.00$. Then the sampling allowance would have to be *entirely* below P, since π cannot exceed 1.00. (From Figure 8-4, we find $.83 < \pi < 1.00$.)

Although specifically designed for proportions, Figure 8-4 has illustrated two interesting features about confidence intervals that apply in wider contexts too: Confidence intervals are not always given by a formula, nor are they always defined symmetrically around the point estimate.

C—DIFFERENCE IN TWO PROPORTIONS, LARGE SAMPLES

Just as we derived the confidence interval to compare two means, we could similarly derive the confidence interval to compare two population proportions:

95% confidence interval for the difference in proportions, for large n_1 and n_2, and independent samples

$$(\pi_1 - \pi_2) = (P_1 - P_2) \pm 1.96 \sqrt{\frac{P_1(1 - P_1)}{n_1} + \frac{P_2(1 - P_2)}{n_2}}$$

(8-29)

like (8-18)

PROBLEMS

8-20 The Gallup Poll has sampled Americans over the past few years, in a random sample (or its equivalent) of about 1500 every time. The proportion who favor the legalization of marijuana has fluctuated as follows:

1969	12%
1977	28%
1980	25%

a. Construct a 95% confidence interval for the population proportion who favor legalization, in each of the three years.

b. Find a 95% confidence interval for the change in this proportion from 1969 to 1977, and also from 1977 to 1980.

8-21 In a random sample of tires produced by a large European multinational firm, 10% did not meet proposed new standards of blowout resistance. Construct a 95% confidence interval for the proportion π (in the whole population of tires) that would not meet the standards, if the sample is:

   a. $n = 10$.

   b. $n = 25$.

   c. $n = 50$.

   d. $n = 200$.

8-22 Two machines are used to produce the same good. In 400 articles produced by machine A, 16 were substandard. In the same length of time, the second machine produced 600 articles, and 60 were substandard. Construct 95% confidence intervals for:

   a. π_1, the true proportion of substandard articles from the first machine.

   b. π_2, the true proportion of substandard articles from the second machine.

   c. The difference between the two proportions $(\pi_1 - \pi_2)$.

8-23 Economists have long realized that GNP alone does not measure total welfare of a country. Less tangible factors are important too, such as leisure and freedom from pollution and crime. To get some idea of how these other factors vary among countries, in the 1970's a worldwide poll was undertaken (Gallup, 1976). To throw light on the issue of crime, the question was asked: "Are you afraid to walk the neighboring streets at night?" The replies were as follows:

	U.S.	JAPAN	LATIN AMERICA
Yes	40%	33%	57%
No	56%	63%	42%
No opinion	4%	4%	1%

Assuming each country's poll was equivalent in accuracy to a simple random sample of $n = 300$ people, find a 95% confidence interval for the difference in the proportions answering "yes":

   a. between the U.S. and Japan

   b. between the U.S. and Latin America

8-24 Repeat Problem 8-23, assuming that "No Opinion" replies were discarded, and thus not included in the samples.

8-25 In 1954 a large-scale experiment was carried out to test the effectiveness of a new polio vaccine. Among 740,000 children selected from grade 2 classes throughout the United States, 400,000 volunteered. Half of the volunteers were randomly selected for the vaccine shot; the remaining half were given a placebo shot of salt water. The results were as follows (taken, with rounding, from Meier, 1977):

GROUP	NUMBER OF CHILDREN	NUMBER OF CASES OF POLIO
Vaccinated	200,000	57
Placebo (control)	200,000	142
Refused to volunteer	340,000	157

a. For each of the three groups, calculate the polio rate (cases per 100,000).

b. Estimate the reduction in the polio rate that vaccination produces, including a 95% confidence interval.

c. Suppose *all* the volunteers had been vaccinated, leaving the refusals as the control group:

  i. Before analyzing the data, criticize this procedure.

  ii. What kind of data would you have obtained? Would it have given the correct answer to question (b)?

8-6 One-sided Confidence Intervals

There are occasions when, in order to establish a point, we wish to make a statement that a population value is at *least as large* as a certain value. The appropriate technique is then a one-sided confidence interval, which puts the 5% error allowance all in one tail:

> **95% confidence interval (one-sided)**
>
> $$\mu > \overline{X} - t_{.05} \frac{s}{\sqrt{n}}$$
>
> (8-30)
> like (8-11)

To illustrate this, we rework the data of Example 8-2.

EXAMPLE **8-5**

From a large class, a random sample of 4 grades had $\overline{X} = 74$ and $s = 11.5$. Calculate the 95% one-sided confidence interval to show how high the whole class mean is.

Solution

Since d.f. $= n - 1 = 3$, from Table V we have $t_{.05} = 2.35$. Also substitute $n = 4$, $\overline{X} = 74$, and $s = 11.5$ into (8-30):

$$\mu > 74 - 2.35 \frac{11.5}{\sqrt{4}} \qquad \text{like (8-30)}$$

$$\mu > 74 - 14 = 60$$

That is, with 95% confidence we can conclude that the mean grade of the whole class is greater than 60. This is better than the lower bound of 56 found earlier in the two-sided confidence interval (8-14). We must pay a price, however: The one-sided confidence interval has no upper bound at all.

Of course, if we want to state that a population value is *below* a certain figure, we would use a one-sided confidence interval of the following form:

$$\mu < \overline{X} + t_{.05} \frac{s}{\sqrt{n}} \qquad (8\text{-}31)$$

Any two-sided confidence interval may be similarly adjusted to give a one-sided confidence interval. For example, for two samples:

$$(\mu_1 - \mu_2) < (\overline{X}_1 - \overline{X}_2) + t_{.05}\, s_p \sqrt{\frac{1}{n_1} + \frac{1}{n_2}} \qquad \begin{array}{l}(8\text{-}32) \\ \text{like}\,(8\text{-}20)\end{array}$$

PROBLEMS

8-26 A manufacturing process has produced millions of lightbulbs with a mean life $\mu = 14{,}000$ hours, and $\sigma = 2{,}000$ hours. A new process produced a sample of 25 bulbs with $\overline{X} = 14{,}740$ (but assume σ remains unchanged at 2,000).

    a. To state how good the new process is, calculate a one-sided 95% confidence interval.

    b. Does the evidence indicate that the new process is better than the old?

8-27 A random sample of ten men professors' salaries gave a mean of 16 (thousand dollars, annually); a random sample of five women professors' salaries gave a mean of only 11. The pooled variance s_p^2 was 11.7. Calculate a one-sided 95% confidence interval to show how much men's mean salary is higher than women's.

8-28 In evaluating a dam, suppose government officials want to estimate μ, the mean annual irrigation benefit per acre. They therefore take a random sample of 25 one-acre plots and find that the benefit averages $8.10, with a standard deviation of $2.40.

    To promote the dam, they want to make a statement of the form "μ is at least as large as. . . ." Yet to avoid political embarrassment, they want 99% confidence in this statement. What value should they put into the blank?

8-29 Construct a one-sided 95% confidence interval for Problems 8-15a, 8-17, and 8-25b.

*8-7 The Jackknife

A—INTRODUCTION

So far, we have constructed confidence intervals to estimate population means. But how do we construct a confidence interval to estimate population variances, or percentiles, or even more complicated parameters? A statistical technique has been recently developed for the computer that will handle any of these confidence intervals. It is so handy for cutting through a wide variety of problems, in fact, that the technique is given the name *jackknife*.[4]

The usefulness of the jackknife is truly remarkable: Once an estimate has been calculated from the sample—*any estimate whatsoever*—the jackknife will provide a confidence interval around it. To illustrate, let us consider a typical problem that cannot be handled by our formulas so far.

EXAMPLE 8-6

The total U.S. population in 1950 was 151 million, and by 1975 had grown to 213 million, so that the true factor of increase Φ was:

$$\Phi = \frac{213}{151} = 1.41 \tag{8-33}$$

That is, the population had increased by 41%.

Now suppose we do not know Φ. How can we estimate it from a random sample of states such as Table 8-4?

Solution

Our estimate must be based on what we know from the sample. In those 6 states, the total population grew from $\Sigma Y = 7.6$ in 1950, to $\Sigma Z = 10.4$ in 1975. Thus the natural estimate of the factor of increase (I) is

$$\text{Ratio estimator:} \quad I = \frac{\Sigma Z}{\Sigma Y} \tag{8-34}$$

$$= \frac{10.4}{7.6} = 1.37 \tag{8-35}$$

TABLE 8-4   **Increased population (millions) in a random sample of 6 states.**

	1950 POPULATION	1975 POPULATION
STATE	Y	Z
RI	.8	.9
VT	.4	.5
ME	.9	1.1
MD	2.7	4.5
MON	.6	.7
OKLA	2.2	2.7

---

[4]In estimating a population variance, an alternative to jackknifing is described in Wonnacott (1984, IM).

Once we have an estimate such as I in (8-35), how can we construct a confidence interval around it when we don't even have a formula for its standard error (SE)? That's the crucial question we will solve with the jackknife.

B—NAIVE ANALYSIS: SPLITTING THE DATA INTO PARTS

We have already described how the standard error (SE) could be estimated by Monte Carlo: Take many samples, and for each sample calculate the estimate (such as $\overline{X}$, or I in this case). In seeing how the estimate fluctuates from sample to sample, we could calculate the required SE.

In Table 8-4, unfortunately, we have just the one sample. However, with a little ingenuity, we can split this sample of 6 states into 3 subsamples of 2 each:

TABLE **8-5** **Splitting the Total Sample into 3 Subsamples provides 3 "Sample Values" of _I_**

SAMPLE 1	Y	Z	SAMPLE 2	Y	Z	SAMPLE 3	Y	Z
RI	.8	.9	ME	.9	1.1	MON	.6	.7
VT	.4	.5	MD	2.7	4.5	OKLA	2.2	2.7
	1.2	1.4		3.6	5.6		2.8	3.4

$$I = \frac{1.4}{1.2} = 1.17 \qquad I = \frac{5.6}{3.6} = 1.56 \qquad I = \frac{3.4}{2.8} = 1.21$$

The splitting of a given sample into several subsamples has given us the repetition we need to see how I fluctuates: Specifically, we see that it varies between 1.17 and 1.56. More formally, in Table 8-6 we calculate the standard deviation of the three values of I (using the techniques and notation of Chapter 2).

Note that we treat the three different values of I (in the first column of Table 8-6) just like $n = 3$ different observations of X. Continuing to do so, we construct a standard 95% confidence interval:

TABLE **8-6** **Calculating the Mean and Standard Deviation of I (called _X_ here)**

ESTIMATE _I_ = X	X − $\overline{X}$	(X − $\overline{X}$)²
1.17	−.14	.020
1.56	+.25	.062
1.21	−.10	.010
$\overline{X} = \dfrac{3.94}{3}$		$s^2 = \dfrac{.092}{2}$
$= 1.31$		$s = \sqrt{.046}$

$$\mu = \overline{X} \pm t_{.025} \frac{s}{\sqrt{n}} \qquad \text{(8-11) repeated}$$

$$= 1.31 \pm 4.30 \frac{\sqrt{.046}}{\sqrt{3}} \qquad \text{(8-36)}$$

$$\mu = 1.31 \pm .53$$

Of course, μ represents the expected value of X—that is, the expected value of the sample increase I, which is the *population* increase[5] Φ. Thus:

$$\boxed{\Phi = 1.31 \pm .53} \qquad \text{(8-37)}$$

We note with satisfaction that this CI does indeed cover the true value $\Phi = 1.41$ in (8-33). And it has given us the main idea: Split a sample into subsamples in order to get a picture of the sampling fluctuation. Can this approach be improved?

C—SEARCH FOR A BETTER SPLIT

Why split the given 6 observations into 3 samples of 2 each, rather than 2 samples of 3 each—or, for that matter, into 6 samples of 1 each? This last suggestion would certainly give us a better t value. [With df = 5, $t_{.025}$ would be 2.57 instead of 4.30 in (8-36).] Unfortunately, however, it would introduce another problem—each sample would be so small that it would produce a very unreliable estimate.

What we would really like is lots of samples, *and* each one quite large. Since we only have 6 observations to work with, however, it looks as if we're asking for the impossible. But here is how we can eat our cake and have it, too:

In the left-hand box in Table 8-7, take all the observations in the original sample except the first and calculate I from this. Call it X_{-1} (since it *excludes* observation 1). Similarly, we calculate $X_{-2}, X_{-3}, \ldots$, the estimates obtained by deleting the 2nd, 3rd, ... observations, respectively.

But we must pay a price for using the same observations over and over. Since X_{-1} and X_{-2}, for example, share most of their data in common, they will be alike. (1.40 and 1.38 are indeed close.) Similarly, all the values $X_{-1} \cdots X_{-6}$ are pretty close together [i.e., close to the overall value $X_{\text{all}} = 1.37$ calculated from all the data in (8-35)]. In other words, because they are based on much the same data, $X_{-1} \cdots X_{-6}$ turn out to be statistically dependent.

In order to use formula (8-11), however, we require independence (just as observations are independent in a random sample). Or at least we need

---

[5]This is not *exactly* true, because of bias. But no matter: We are just using this method as an introduction to the jackknife, which will turn out to be unbiased itself.

TABLE **8-7** **Deleting One Observation at a Time Allows us to Construct Six "Sample Values" of I**

DELETING FIRST			DELETING SECOND		
STATE	Y	Z	STATE	Y	Z
~~RI~~	~~.8~~	~~.9~~	RI	.8	.9
VT	.4	.5	~~VT~~	~~.4~~	~~.5~~
ME	.9	1.1	M E	.9	1.1
MD	2.7	4.5	MD	2.7	4.5
MON	.6	.7	MON	.6	.7
OKLA	2.2	2.7	OKLA	2.2	2.7
	6.8	9.5		7.2	9.9

$$I = \frac{9.5}{6.8} = 1.40$$

Call this X_{-1}

$$I = \frac{9.9}{7.2} = 1.38$$

Call this X_{-2}

Similarly,

$$X_{-3} = 1.39 \quad X_{-4} = 1.20 \quad X_{-5} = 1.39 \quad X_{-6} = 1.43$$

components that are spread out enough so that they *act as if* they were independent. To spread them out, therefore, we define new values, for example:

$$X_{(1)} \equiv X_{all} + (n - 1)(X_{all} - X_{-1}) \tag{8-38}$$
$$= 1.37 + 5(1.37 - 1.40)$$
$$= 1.37 + 5(-.03) \tag{8-39}$$
$$= 1.22$$

In $X_{(1)}$ we have created a new value out of X_{-1} that is *five times* as far away from the central value X_{all}. [The difference between X_{-1} and X_{all} in (8-39) was .03, and was multiplied by 5. This brought $X_{(1)}$ down a total of .15 units below X_{all}, to 1.22.] In other words, $X_{(1)}$ is like X_{-1} except that its deviation from X_{all} is magnified 5 times—or in general $(n - 1)$ times.

Similarly, we spread out each of the other sample values (magnify its deviation from X_{all}):

$$X_{(2)} = 1.37 + 5(1.37 - 1.38) = 1.32$$
$$X_{(3)} = 1.37 + 5(1.37 - 1.39) = 1.27$$

Similarly, $X_{(4)} = 2.22$, $X_{(5)} = 1.27$, $X_{(6)} = 1.07$. It can be proved that these new values are indeed spread out enough that they do act as if they were independent. Therefore, they are called *pseudovalues* that make up a *pseudosample*, which we can treat just like a random sample of independent observations. We first calculate the sample moments in Table 8-8, and then substitute them into the confidence interval:

TABLE **8-8**   **Calculating the Mean and Standard Deviation of the Pseudovalues**

$X_{(i)}$	$X - \bar{X}$	$(X - \bar{X})^2$
1.22	$-.18$	.0324
1.32	$-.08$	.0064
1.27	$-.13$	.0169
2.22	$+.82$	.6724
1.27	$-.13$	.0169
1.07	$-.33$	.1089
$\bar{X} = \dfrac{8.37}{6}$		$s^2 = \dfrac{.8539}{5}$
$= 1.40$		$s = \sqrt{.171}$

$$\mu = \bar{X} \pm t_{.025} \frac{s}{\sqrt{n}} \qquad \text{(8-11) repeated}$$

$$= 1.40 \pm 2.57 \frac{\sqrt{.171}}{\sqrt{6}} \qquad \text{(8-40)}$$

$$\boxed{\Phi = 1.40 \pm .43} \qquad \text{(8-40)}$$

This is the jackknife confidence interval we desired: The factor of growth is between .97 and 1.83. This interval is better than the naive interval (8-37) in two ways:

1.   It is more accurate. (The allowance is $\pm$ .43 instead of $\pm$ .53.)

2.   It is standard. (It does not arbitrarily split the data into, say, 3 samples of 2 each, rather than 2 samples of 3 each.)

For these benefits, the cost is slight: The computations are increased n-fold—a trivial matter now for a computer.

D—JACKKNIFE IN GENERAL

The jackknife is an all-purpose technique for robust confidence intervals; its great virtue lies in its almost universal applicability. Once the estimate has been calculated from the sample, the jackknife provides the confidence interval around it, in 4 steps:

1.   Denote the estimate (based on all the data) by X_{all}.
2.   Calculate the estimated "sample value" X_{-1} based on all the data except the 1st observation. Similarly calculate X_{-2}, X_{-3}, and so on.
3.   Calculate the "pseudovalue" $X_{(1)}$, which is just the X_{-1} spread further from X_{all}:

$$X_{(1)} \equiv X_{all} + (n - 1)(X_{all} - X_{-1}) \qquad \text{(8-38) repeated}$$

$$(8\text{-}41)$$

Similarly calculate $X_{(2)}$, $X_{(3)}$, and so on. They will be spread far enough apart to act as if they were independent. Thus they constitute a "pseudosample" that acts like a random sample.

4. For the pseudosample, calculate the mean $\overline{X}$ and standard deviation s, and substitute them into the 95% confidence interval:

$$\text{Population parameter} = \overline{X} \pm t_{.025} \frac{s}{\sqrt{n}} \qquad (8\text{-}42)$$

Along with the confidence interval, jackknifing provides a bonus: If the original estimate is slightly biased (but asymptotically unbiased), jackknifing will often eliminate the bias (Wonnacott, 1984 IM).

*PROBLEMS

8-30 The population of 196 selected U.S. cities totalled 22,919,000 in 1920, and we would like to estimate by what factor Φ they grew in 1930 (from Cochran, 1977). A random sample of 4 cities was therefore drawn, and showed the following growth (population in thousands):

1920	1930
243	291
87	105
30	111
71	79

a. Calculate the estimate I of the true growth factor Φ (simple ratio estimate, without jackknife).

b. Using the jackknife, calculate a 95% confidence interval for Φ.

c. Since the total population was 22.9 millions in 1920, estimate the total population in 1930 by:

    i. Using the naive estimate in part **a.**

    ii. Using the confidence interval in part **b.** How is this better?

8-31 To test its effectiveness against scrub oak in a certain county, a herbicide was tried on 5 plots randomly chosen in the county. Each was of size 400 square meters, and had the following number of trees:

PLOT NUMBER	NUMBER OF TREES	NUMBER KILLED
1	71	48
2	49	24
3	80	58
4	110	66
5	97	62

252

a. In the 5 plots altogether, what was the proportion killed (kill rate)?

b. To get a 95% confidence interval, the jackknife provided the following pseudosample:

$$X_{all} = .634$$

i	X_{-i}	$X_{(i)}$
1	.625	.670
2	.654	.554
3	.612	.722
4	.646	.586
5	.632	.642

Verify the first line of the table. Then finish the jackknifing to obtain the 95% confidence interval.

c. In words, explain the parameter being estimated in part **b**.

8-32 In Problem 8-31, suppose instead that only 3 plots were randomly sampled, with the following results:

i	NUMBER OF TREES	NUMBER KILLED
1	26	16
2	65	51
3	128	98

Find the 95% confidence interval for the kill rate in the whole county.

*8-33 A pharmaceutical company produces 2.4 million capsules of vitamin C each day, and takes a small random sample of 4 capsules to make sure there is not too much variability in the dose. On a typical day, the 4 readings were as follows:

Dose of vitamin C (mg):   1008, 1098, 1002, 1116.

a. Calculate the unbiased estimate (s^2) of the variance of the whole day's production (σ^2).

b. Jackknife the estimate to get a robust 95% confidence interval for σ^2.

c. Suppose the director of quality-control inadvertently calculates the MSD. What is it? Is it biased?

d. Suppose he then jackknifes the MSD to obtain a 95% confidence interval. What is it? How is it different from the estimate in part (b)?

e. Answer True or False; if False, correct it:
At the same time that it provides a confidence for σ^2, the jackknife provides a bonus in part **c**: The jackknifed estimate coincides with the *unbiased* estimate.

CHAPTER 8 SUMMARY

8-1 From the normal tables, we find that 95% of the time, $\overline{X}$ will be within 1.96 standard errors of μ. This yields the 95% confidence interval (8-9):

$$\mu = \overline{X} \pm 1.96 \frac{\sigma}{\sqrt{n}}$$

8-2 In practice, σ is unknown in the formula above, and has to be estimated with s. To allow for the additional uncertainty, we replace the value $z_{.025} = 1.96$ with the wider value $t_{.025}$.

8-3 To compare two population means, for independent samples we take the difference of the two sample means, $\overline{X}_1 - \overline{X}_2$. To construct the confidence interval, we pool all the squared deviations in both samples to calculate the pooled variance s_p^2.

8-4 To compare two population means, for *dependent* (paired) samples we start by calculating the *individual* differences D. Their average $\overline{D}$ is then calculated, and the confidence interval built around it.

Paired samples are more efficient than independent samples, since pairing keeps many of the extraneous variables constant. Pairing should therefore be used whenever feasible.

8-5 Confidence intervals for proportions are like means, for large samples at least. For a small sample, a graphical alternative is best, reading the confidence interval off a 95% confidence band.

8-6 Any 95% confidence interval can be recast into a one-sided form by putting all of the 5% error allowance into one tail. This makes the claim stronger at one end (while being completely vague at the other end).

*8-7 In situations where the sampling distribution of an estimator is not known—and this occurs very often in practice—robust confidence intervals can be constructed with the jackknife.

REVIEW PROBLEMS

8-34 In a city of 20,000 households, 3% of the households were randomly sampled by a large survey research service. The average number of cars per household was 1.2, while the standard deviation was 0.8. As well, 15% of the households had no cars at all. For the whole city, construct a 95% confidence interval for:

   a. The average number of cars per household.
   b. The percentage of households with no car.

8-35 Suppose a 10% random sample from a stand of 380 trees gave the following distribution of volumes:

VOLUME	FREQUENCY
20	8
40	16
60	12
80	2
Total	38

Calculate a 95% confidence interval for:

a. The mean volume.

b. The total volume.

8-36 Soon after he took office in 1963, President Johnson was approved by 160 out of a sample of 200 Americans. With growing disillusionment over his Vietnam policy, by 1968 he was approved by only 70 out of a sample of 200 Americans.

a. What is the 95% confidence interval for the percentage of all Americans who approved of Johnson in 1963? in 1968?

b. What is the 95% confidence interval for the change?

8-37 In 1977, a sample of five men and five women was drawn from the population of American college graduates, and their incomes recorded (thousands of dollars, annually):

MEN	WOMEN
18	7
32	15
22	22
23	15
20	11

Calculate a 95% confidence interval to show how much men's mean income is higher than women's.

8-38 To determine the difference in gasoline A and B, four cars chosen at random were driven over the same route twice, once with gasoline A and once with gasoline B. Calculate a 95% confidence interval for the population difference, based on the mileages of four cars:

GAS A	GAS B
23	20
17	16
16	14
20	18

8-39 How seriously does alcohol affect prenatal brain development? To study this issue (Jones and others, 1974), six women were found who had been chronic alcoholics during pregnancy. The resulting children were tested at age 7, producing a sample of six IQ scores with a mean of 78, and $\Sigma(X - \bar{X})^2 = 1805$.

A control group of 46 women were found who on the whole were similar in many respects (same average age, education, marital status, etc.)—but without alcoholism in their pregnancy. Their 46 children had IQ scores that averaged 99, and $\Sigma(X - \bar{X})^2 = 11,520$.

a. If this had been a randomized controlled study, what would be the 95% confidence interval for the difference in IQ that prenatal alcoholism makes?

b. Since this in fact was an observational study, how should you modify your claim in part **a?**

8-40 A "Union Shop" clause in a contract requires every worker to join the union soon after starting to work for the company. In 1973 there were 31 states that permitted the Union Shop, and 19 states (mostly southern) that had earlier passed "Right-to-Work" laws that outlawed the Union Shop and certain other practices. A random sample of 5 states from each group showed the following average hourly wage within the state:

STATES WITH UNION SHOP, ETC.	STATES WITH RIGHT-TO-WORK
$4.00	$3.50
3.10	3.60
3.60	3.20
4.20	3.90
$4.60	$2.80

On the basis of these figures, a friend claims that the Right-to-Work laws are costing the average worker 50¢ per hour. Do you think this claim should be modified? If so, how?

8-41 An analysis was carried out (Gilbert and others, 1977) on 44 research papers that used randomized clinical trials to compare an innovative treatment (I) with a standard treatment (S), in surgery and anaesthesia. In 23 of the papers, I was preferred to S (and in the other 21 papers, S was preferred to I).

a. Assuming the 44 papers constitute a random sample from the population of all research papers in this field, construct a 95% confidence interval for the population proportion where I is preferred to S.

b. Do you agree with their interpretation?

"... When assessed by randomized clinical trials, innovations are successful only about half the time. Since innovations brought to the stage of randomized trials are usually expected by the innovators to be sure winners, we see that ... the value of the innovation needs empirical checking."

8-42 Lack of experimental control (e.g., failure to randomly assign the treatment and control) may affect the degree of enthusiasm with which a new medical treatment is reported. To test this hypothesis, 38 studies of a certain operation were classified as follows (Gilbert and others, 1977):

DEGREE OF CONTROL	REPORTED EFFECTIVE-NESS OF OPERATION	
	MODERATE OR MARKED	NONE
Well Controlled	3	3
Uncontrolled	31	1

a.   Assuming these 38 studies constitute a random sample, construct an appropriate 95% confidence interval. [Although (8-29) is only a rough approximation for small samples, it's the best you have, so use it.]

b.   Do you agree with the following interpretations reported by the authors?

". . . Nothing improves the performance of an innovation as much as the lack of controls."

". . . weakly controlled trials . . . may make proper studies more difficult to mount, as physicians become less and less inclined, for ethical reasons, to subject the issue to a carefully controlled trial lest the 'benefits' of a seemingly proven useful therapy be withheld from some patients in the study."

*8-43   A sample survey of homeowners, correctly carried out, concluded: "We estimate that 13% have recently insulated their homes (spending at least $500 over the past 3 years). In fact, the 95% confidence interval is .13 ± .03."

Imagine this survey was independently repeated over and over (same sample size and same population) and a 95% confidence interval calculated each time. Then how often would the following occur? Fill in the blanks with the appropriate answer (95%, more than 95%, or less than 95%):

a.   _____ of such intervals will cover the midpoint .13.

b.   _____ of such intervals will cover the population proportion.

c.   _____ of such intervals will completely cover the given interval (.10 to .16).

d.   _____ of such intervals will partially cover (overlap) the given interval (.10 to .16).

Chapter

Hypothesis Testing

There are no whole truths: all truths are half-truths.

A. N. WHITEHEAD

9-1 Hypothesis Testing Using Confidence Intervals[1]

A—A MODERN APPROACH[2]

A statistical *hypothesis* is simply a *claim about a population* that can be put to a test by drawing a random sample. A typical hypothesis, for example, is that the Democrats have a majority in a population of voters, such as the Kansas City suburb analyzed in (8-28). In fact, this hypothesis can be rejected, because the sample of 20 voters showed (at a 95% level of confidence) that between 55% and 95% of the population were Republicans; the Republicans, rather than the Democrats, had the majority.

This voter example illustrates how we can use a confidence interval to test an hypothesis. Another example will show this in more detail. (Recall that in the format of this book, a numbered Example is a problem for you to work on first; the solution in the text should only be consulted later as a check.)

---

[1]*Reminder to instructors:* This chapter is designed so that you may spend as much or as little time on hypothesis testing as you judge appropriate. At the end of any section you may skip ahead to the next chapters and thus have more time to cover ANOVA and regression.

[2]Actually, this is not all that new—in one form or another it has appeared ever since Neyman and Pearson formalized hypothesis testing 50 years ago. But since this approach does not yet commonly appear in introductory texts, we shall refer to it as a "new" or "modern" approach.

EXAMPLE **9-1**   In a large American university, 10 men and 5 women professors were independently sampled in 1969, yielding the annual salaries given below (in thousands of dollars. Same source as Problem 2-1):

MEN (X_1)		WOMEN (X_2)
12	20	9
11	14	12
19	17	8
16	14	10
22	15	16
$\overline{X}_1 = 16$		$\overline{X}_2 = 11$

These sample means give a rough estimate of the underlying population means μ_1 and μ_2. Perhaps they can be used to settle an argument: A husband claims that there is no difference between men's salaries (μ_1) and women's salaries (μ_2). In other words, if we denote the difference as $\Delta = \mu_1 - \mu_2$, he claims that

$$\Delta = 0 \tag{9-1}$$

His wife, however, claims that the difference is as large as 7 thousand dollars:

$$\Delta = 7 \tag{9-2}$$

Settle this argument by constructing a 95% confidence interval.

Solution   The 95% confidence interval is, from (8-20),

$$\Delta = (\overline{X}_1 - \overline{X}_2) \pm t_{.025}\, s_p \sqrt{\frac{1}{n_1} + \frac{1}{n_2}}$$

$$= (16 - 11) \pm 2.16 \sqrt{\frac{152}{13}} \sqrt{\frac{1}{10} + \frac{1}{5}}$$

$$= 5.0 \pm 2.16\,(1.87) \tag{9-3}$$

$$= 5.0 \pm 4.0 \tag{9-4}$$

Thus, with 95% confidence, Δ is estimated to be between 1.0 and 9.0. Thus the claim $\Delta = 0$ (the husband's hypothesis) seems implausible, because it falls outside the confidence interval. On the other hand the claim $\Delta = 7$ (the wife's hypothesis) seems more plausible, because it falls within the confidence interval.

In general, any hypothesis that lies outside the confidence interval may be judged implausible—that is, can be *rejected*. On the other hand, any hypothesis that lies within the confidence interval may be judged plausible or *acceptable*.[3] Thus:

> **A confidence interval may be regarded as just the set of acceptable hypotheses.** (9-5)

If a 95% confidence interval is being used, it would be natural to speak of the hypothesis as being tested at a 95% *confidence level*. In conforming to tradition, however, we usually speak of testing at an *error level* of 5%.

Thus, to return to our example, we formally conclude from (9-4) that, with a 5% chance of error (a "5% error level"), we can reject the hypothesis of no difference ($\Delta = 0$). In other words, we have collected enough evidence (and, consequently, have a small enough sampling allowance) so that we can discern a difference between men's and women's salaries. We therefore call the difference *statistically discernible* at the 5% error level.

Our formal statistical language must not obscure the important common-sense aspects of this problem, of course. Although we have shown (at the 5% error level) that men's and women's salaries are different, we have *not* shown that discrimination necessarily exists. There are many alternative explanations. For example, men may have a longer period of education than women, on average. What we really should do then is compare men and women of the *same qualifications*. (This will in fact be done later, using *multiple regression*, in Problem 14-26.)

EXAMPLE **9-2**

Suppose the confidence interval (9-4) had been based on a smaller sample and, consequently, had been vaguer. (Note how smaller sample sizes n_1 and n_2 increase the size of the sampling allowance.) Specifically, suppose we calculated the confidence interval to be:

$$\Delta = 5 \pm 8 \tag{9-6}$$
$$-3 < \Delta < 13 \tag{9-7}$$

Are the following interpretations true or false? If false, correct them:

a. Since the hypothesis $\Delta = 0$ falls within the interval (9-7), it cannot be rejected.

b. The true (population) difference may well be 0. That is, the population of men's salaries may be the same as the women's on average. The difference in *sample* means ($\overline{X}_1 - \overline{X}_2 = 5$) may represent only random fluctuation, and therefore cannot be used to show that a real difference exists in the population means.

---

[3]We use *acceptable* simply as a convenient shorthand that avoids the potentially confusing double negative in the more precise term *not rejected*.

c. The plausible population differences within the interval (9-7) include both negative and positive values; that is, we cannot even decide whether men's salaries on the whole are better or worse than women's.

d. In (9-6), we see that the sampling allowance ($\pm$ 8) overwhelms the estimated difference (5). Whenever there is this much sampling "fog," we call the difference *statistically indiscernible*.

Solution Each of these statements is correct, and illustrates that *samples that are too small produce statistical indiscernibility* (that is, keep us from a statistically discernible conclusion).

In summary, once a confidence interval has been calculated, it can be used immediately to test any hypothesis. For example, to use our Kansas City suburb example again, the claims that 60%, 70%, or 80% of the voters are Republican are acceptable hypotheses, because they fall within the confidence interval (8-28). But the hypotheses that only 30%, 40%, or 50% are Republican can be rejected because these hypotheses fall outside the confidence interval.

B—THE TRADITIONAL APPROACH

The hypothesis $\Delta = 0$ in (9-1) is of particular interest. Since it represents no difference whatsoever, it is called the *null hypothesis* H_0. In rejecting it because it lies outside the confidence interval (9-4), we established the important claim that there was indeed a difference between men's and women's income. Such a result is traditionally called *statistically significant* at the 5% *significance level*.

There is a problem with the term "statistically significant." It is a technical phrase that simply means enough data has been collected to establish that a difference does exist. It does *not* mean the difference is necessarily important. For example, if we had taken huge samples from nearly identical populations, the 95% confidence interval, instead of (9-4), might have been:

$$\Delta = .0005 \pm .0004$$

This difference is so miniscule that we could dismiss it as being of no human or scientific significance, even though it is just as statistically significant as (9-4). In other words, *statistical* significance is a technical term, with a far different meaning than *ordinary* significance.

There is also a problem with the term "5% significance level." It sounds like the higher this value (say 10% rather than 5%) the better the test. But precisely the reverse is true. (Our level of confidence would only be 90%, rather than 95%.)

Unfortunately, but understandably, many people tend to confuse statistical significance with ordinary significance.[4] To reduce the confusion, we prefer the word discernible instead of significant. In conclusion, therefore:

> **"Statistically *significant* at the 5% *significance* level" is the traditional phrase typically encountered in the scientific literature. It means exactly the same thing as our "statistically *discernible* at the 5% *error* level."**   (9-8)

PROBLEMS

9-1   For each of Problems 8-36 to 8-39, state whether the difference is statistically discernible at the 5% level.

9-2   (This problem will be analyzed two ways. Here we will use the familiar technique of confidence intervals. Later, in Example 9-3, we will develop a new technique.)

A manufacturing process has produced millions of TV tubes with a mean life $\mu = 1200$ hours, and standard deviation $\sigma = 300$ hours. A new process is tried on a sample of 100 tubes, producing a sample average $\overline{X} = 1265$ hours. (Assume σ remains unchanged at 300 hours.) Will this new process produce a long-run average that is better than the null hypothesis $\mu = 1200$?

    a.   Specifically, is the sample mean $\overline{X} = 1265$ statistically discernible (i.e., discernibly better than the H_0 value of 1200) at the 5% error level? Answer by seeing whether the one-sided 95% confidence interval excludes $\mu = 1200$.   *μ > 1215.8*

    b.   Repeat, for the 1% error level.   *μ > 1195.7*

    c.   Repeat, for the .1% error level.   *μ > 1176.2*

---

[4]To make matters worse, some writers simply use the single word "significant." While they may mean "statistically significant," their readers often interpret this to mean "ordinarily significant" or "important."

As an example of this, an article in a highly respected journal used the words "highly significant" to refer to relatively minor findings uncovered in a huge sample of 400,000 people. (For example, one such finding was that the youngest child in a family had an IQ that was 1 or 2 points lower on average than the second youngest child.)

From this example, it may be concluded that a large sample is like a large magnifying glass that allows us to discern the smallest molehill (smallest difference between two populations). It is unwise to use a phrase (such as "significant" or "highly significant") that makes this molehill sound like a mountain.

9-2 p-value (one-sided)

A—WHAT IS A p-VALUE?

In Section 9-1 we developed a simple way to test *any* hypothesis, by examining whether or not it falls within the confidence interval. Now we take a new perspective by concentrating on just one hypothesis, the null hypothesis H_0. We will calculate just how much (or how little) it is supported by the data.

EXAMPLE **9-3**

A traditional manufacturing process has produced millions of TV tubes, with a mean life $\mu = 1200$ hours and a standard deviation $\sigma = 300$ hours. A new process, recommended by the engineering department as better, produces a sample of 100 tubes, with an average $\overline{X} = 1265$. Although this sample makes the new process look better, is this just a sampling fluke? Is it possible that the new process is really no better than the old, and we have just turned up a misleading sample?

To give this problem more structure, we state the null hypothesis: the new process would produce a population that is no different from the old—that is, H_0: $\mu = 1200$, which is sometimes abbreviated:

$$\mu_0 = 1200 \tag{9-9}$$

The claim of the engineering department that the new process is better is called the *alternative hypothesis,*[5] H_1: $\mu > 1200$, which may be abbreviated to

$$\mu_1 > 1200 \tag{9-10}$$

How consistent is the sample $\overline{X} = 1265$ with the null hypothesis $\mu_0 = 1200$? Specifically, if the null hypothesis were true, what is the probability that $\overline{X}$ would be as high as 1265?

Solution

In Figure 9-1 we show the hypothetical distribution of $\overline{X}$, if H_0 is true. (Here, for the first time, we show the convention of drawing hypothetical distributions in ghostly white.) By the Normal Approximation Rule, this distribution is normal, with mean $\mu_0 = 1200$, and standard error (SE) $= \sigma/\sqrt{n} = 300/\sqrt{100} = 30$. We use these to standardize the observed value $\overline{X} = 1265$:

---

[5]In practice, the null and alternative hypotheses often are determined as follows: H_1 is the claim that we want to prove (9-10). Then H_0 is really everything else, $\mu \le 1200$. But we need not use all of this; instead, in (9-9), we use just the boundary point, $\mu = 1200$. Any other μ below 1200 would be easier to distinguish from H_1 since it is "further away"; we therefore need not bother with it.

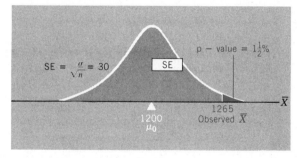

Figure 9-1
p-value $\equiv$ Pr($\overline{X}$ would be as large as the value actually observed, if H_0 is true).

$$Z = \frac{\overline{X} - \mu_0}{SE} \qquad \text{(9-11)}$$
$$\text{like (6-11)}$$
$$= \frac{1265 - 1200}{30} = 2.17$$

Thus $\qquad$ Pr($\overline{X} \geq 1265$) = Pr($Z \geq 2.17$) = .015 $\qquad$ (9-12)

What does this mean in plain English? If the new process were no better (that is, if H_0 were true), there would be only $1\frac{1}{2}\%$ probability of observing $\overline{X}$ as large as 1265. This $1\frac{1}{2}\%$ is therefore called the *p-value* for H_0. (It is also called the *one-sided* p-value. Like the one-sided confidence interval, it requires calculating probability just on one side.)

The p-value summarizes very clearly how much agreement there is between the data and H_0. In this example the data provided very little support for H_0; but if $\overline{X}$ had been observed closer to H_0 in Figure 9-1, the p-value would have been larger.

In general, for any hypothesis being tested, we define the p-value as:[6]

$$\text{p-value} \equiv \text{Pr}\left(\begin{array}{c}\text{The sample value would be as large}\\\text{as the value actually observed,}\\\text{if } H_0 \text{ is true}\end{array}\right) \qquad \text{(9-13)}$$

---

[6]For brevity, we use the term "as large as" to mean "*at least* as large as."
The p-value in Figure 9-1 is calculated in the right-hand tail, because the alternative hypothesis is on the right side ($\mu > 1200$). On the other hand, if the alternative hypothesis were on the left side ($\mu < 1200$), then the p-value would be calculated in the left-hand tail:

$$\text{p-value} = \text{Pr}\left(\begin{array}{c}\text{the sample value would be as small}\\\text{as the value actually observed,}\\\text{if } H_0 \text{ is true}\end{array}\right)$$

The p-value is an excellent way to *summarize what the data says*[7] *about the credibility of* H_0.

B—USING THE t DISTRIBUTION

We have seen how $\overline{X}$ was standardized so that the standard normal table could be used. The key statistic we evaluated was

$$Z = \frac{\overline{X} - \mu_0}{\sigma/\sqrt{n}}$$

        (9-14)
        like (9-11)

Usually σ is unknown, and has to be estimated with the *sample* standard deviation s. Then the statistic is called t instead of Z:

$$t = \frac{\overline{X} - \mu_0}{s/\sqrt{n}}$$

        (9-15)

Since $\overline{X}$ fluctuates around μ_0, Z fluctuates around 0. Similarly t fluctuates around 0—but with wider variability, as already noted in Chapter 8. (We no longer know the exact value σ, but instead have to use an estimate s, with its inevitable uncertainty.) The resulting wider distribution of t is shown in Figure 9-2.

There are many t distributions, one for each sample size; hence, one for each d.f. (degrees of freedom). In Figure 9-2 we see that the larger the sample size, the less spread out is the t distribution. (The larger the sample, the more reliable is the estimate s, and consequently the less variable is t.) Eventually, as sample size approaches infinity, s^2 estimates σ^2 dead on, and then the t distribution coincides with the normal z. This is reflected in the t table. Each of the distributions in Figure 9-2 corresponds to a row in Table V. In the last row, where d.f. = ∞, the t and z values become identical.

But in all other cases, t values are larger than z values. For example, in Figure 9-2, $z_{.025}$ is 1.96 while the corresponding $t_{.025}$ with 4 d.f. is 2.78. The use of t, when the sample size is small and σ has to be estimated with s, is easily illustrated with an example.

---

[7]Of course, the data is not the only thing to be considered if we want to make a final judgment on the credibility of H_0; common sense, or what sometimes is more formally called "personal prior probability," must be considered, too, especially when the sample is small and hence unreliable. For example, if you drew a coin from your pocket, flipped it 10 times, and found that it showed 8 heads, the p-value for H_0 (fair coin) would be .05 (from the binomial Table IIIc). But obviously it would be inappropriate to conclude from this that the coin was unfair. We know that practically all coins are fair; thus our common sense tells us that our sample result (of eight heads) was just "the luck of the draw," and we discount it accordingly.

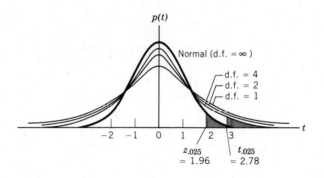

Figure 9-2
The standard normal distribution and the t distributions compared.

EXAMPLE **9-4**

A sample of $n = 5$ grades gave $\overline{X} = 65$, and $s = 11.6$. Suppose a claim is made that the population mean is only 50. What is the p-value for this hypothesis?

Solution

From (9-15) we calculate:

$$t = \frac{65 - 50}{11.6/\sqrt{5}} = 2.89 \qquad (9\text{-}16)$$

In calculating s, we used d.f. $= n - 1 = 4$. We therefore scan along the fourth row of Table V, and find that the observed t value of 2.89 lies beyond $t_{.025} = 2.78$. As Figure 9-3 shows, this means the tail probability is smaller than .025; that is:

$$\text{p-value} < .025 \qquad (9\text{-}17)$$

Since the p-value is a measure of the credibility of H_0, such a low value leads us to conclude that H_0 is an implausible hypothesis. In

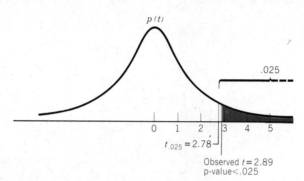

Figure 9-3
p-value found using t in Table V.

other words: If H_0 were true (that is, if the population mean $= 50$) there would be less than $2\frac{1}{2}$ chances in a hundred of getting a sample mean as high as the 65 actually observed.

We can easily generalize the use of t to cover other hypothesis tests. In (9-15), the numerator is the difference between the estimated value and the hypothetical value being tested (null hypothesis). The denominator is the estimated standard error. The generalization of (9-15) is therefore:

$$t = \frac{\text{estimate} - \text{null hypothesis}}{\text{standard error}} \qquad (9\text{-}18)$$

Often the null hypothesis is 0, in which case (9-18) takes on a very simple form:

$$t = \frac{\text{estimate}}{\text{standard error}} \qquad (9\text{-}19)$$

This makes good intuitive sense: The t ratio simply measures how large the estimate is, relative to its standard error.

Our earlier advice on when to use t is worth repeating. Use t when σ is unknown, and unreliably estimated in a small sample (n less than 30, 60, or 120, depending on the accuracy required). Otherwise, use z as a reasonable approximation, so that you can use the more detailed z values in Table IV, rather than the t values in Table V.

Finally, to illustrate the diversity of distributions for which the p-value can be calculated, we finish with two more examples—one involving a difference in means and the other involving a proportion.

EXAMPLE **9-5**   For the mean difference in men's and women's salaries (thousands of dollars, annually), we calculated the following 95% confidence interval (in Problem 8-27):

$$(\mu_1 - \mu_2) > (\overline{X}_1 - \overline{X}_2) - t_{.05} \, SE$$
$$> 5.0 - 1.77 \, (1.87) \qquad (9\text{-}20)$$
$$(\mu_1 - \mu_2) > 1.7 \qquad (9\text{-}21)$$

where there were 13 d.f. in calculating SE, and therefore 13 d.f. for the t distribution.

The null hypothesis (H_0: $\mu_1 - \mu_2 = 0$) is implausible, because it lies outside the confidence interval (9-21). To express just how little credibility the data attributes to H_0, calculate its p-value.

Solution   The null hypothesis is $\mu_1 - \mu_2 = 0$, so that (9-19) is appropriate:

$$t = \frac{\text{estimate}}{\text{SE}} = \frac{(\overline{X}_1 - \overline{X}_2)}{\text{SE}}$$

Of course, the values we need here are the ones already calculated for the confidence interval (9-20). Thus:

$$t = \frac{5.0}{1.87} = 2.67$$

Since d.f. = 13, we scan along the thirteenth row of Table V, and find that the observed t value of 2.67 lies beyond $t_{.010} = 2.65$. Thus:

$$\text{p-value} < .010 \qquad\qquad (9\text{-}22)$$

Accordingly, we conclude that H_0 has very little credibility indeed.

EXAMPLE **9-6**   To investigate whether black children a generation ago showed racial awareness and antiblack prejudice, a group of 252 black children was studied (Clark and Clark, 1958). Each child was told to choose a doll to play with from among a group of four dolls, two white and two nonwhite. A white doll was chosen by 169 of the 252 children.

What is the p-value for the null hypothesis that the children ignore color? (The alternative hypothesis is that the children are prejudiced against black—that is, favored white.)

Solution   First we formulate the problem mathematically. Suppose the 252 children can be regarded as a random sample from a large population of black children. (This is quite an assumption, so the final p-value we calculate should be interpreted cautiously.) In any case, the null hypothesis is that the population proportion choosing a white doll is 50–50; that is:

$$\pi_0 = .50 \qquad\qquad (9\text{-}23)$$

while the alternative hypothesis is

$$\pi_1 > .50$$

The observed sample proportion is $P = 169/252 = .67$. Its standard error is given by (6-23) as $\sqrt{\pi(1 - \pi)/n}$ and we use the null value $\pi = .50$, since p-value is always based on the null hypothesis.

(Remember that p-value was defined as the probability of . . . , assuming H_0 is true.) Thus:

$$t = \frac{\text{estimate} - \text{null hypothesis}}{\text{standard error}} \qquad \text{(9-18) repeated}$$

$$= \frac{.67 - .50}{\sqrt{.50(.50)/252}} = 5.40 \qquad (9\text{-}24)$$

Since the normal z distribution (Table IV) can be used in this large sample case instead of t, we have:

$$\text{p-value} = \Pr(Z \geq 5.40)$$
$$= .00000003$$

Since the p-value is miniscule, there is almost no credibility for the null hypothesis. Thus, to the extent that this sample reflects the properties of a random sample, it can be concluded that a generation ago, black children were prejudiced in favor of white.

PROBLEMS

9-3   In Boston in 1968, Dr. Benjamin Spock, a famous pediatrician and activist against the Vietnamese war, was tried for conspiracy to violate the Selective Service Act. The judge who tried Dr. Spock had an interesting record: Of the 700 people the judge had selected for jury duty in his previous few trials, only 15% were women (a simplified version of Zeisel and Kalven, 1972). Yet in the city as a whole, about 29% of the eligible jurors were women.

To evaluate the judge's fairness in selecting women:

a.   State the null hypothesis, in words and symbols.

b.   Calculate the p-value for H_0.

9-4   A watch-making firm wanted to test the relative effectiveness of two new training programs for its employees. Of 500 workers hired in 1985, 25 were assigned to the first program, and 25 to the second (and the rest to the old program). Then the productivity of the two samples was compared, and yielded a t value of 2.4, and hence a p-value of .01. Answer True or False; if False, correct it:

a.   To eliminate bias, the best 50 workers should have been chosen for the test, and then allowed to choose which new training program they preferred.

b.   The null hypothesis is that the two new training programs give equally high productivity on average; that is, $\overline{X}_1 = \overline{X}_2$.

c. If we repeated the experiment, and if H_0 were true, the probability is .01 of getting a t value as large as the one observed.

9-5 A random sample of 6 students in a phys-ed class had their pulse rate (beats per minute) measured before and after the 50-yard dash, with the following results:

BEFORE	AFTER	D	$(D-\bar{D})^2$
74	83	9	49
87	96	9	49
74	97	23	49
96	110	14	4
103	130	27	121
82	96	14	4

$$\bar{x} = \frac{516}{6} = 96 \quad \bar{x}_2 = \frac{612}{6} = 102 \quad \bar{x}_D = 16 \quad \frac{276}{5} = 55.2$$

a. Calculate the one-sided 95% confidence interval for the mean increase in pulse rate.

b. State the null hypothesis, in words and symbols. Then calculate its p-value.

9-6 How much do seat belts help? To answer this, a study was undertaken of cars that had been equipped with seat belts (lap-and-shoulder belts) and that had subsequently been involved in accidents. A random sample of 10,000 occupants showed the following injury rates (reconstructed from U.S. Department of Transportation, 1981):

SEVERE OR FATAL INJURY	SEAT BELT WORN YES	NO	TOTALS
Yes	3	119	122
No	829	9049	9878
Totals	832	9168	10,000

a. State H_0 in words and symbols.

b. To what extent does the data prove that seat belts help?

c. Would the study have been more effective if it had included not only cars with seat belts, but also those without? Why?

9-3 Classical Hypothesis Tests

A—WHAT IS A CLASSICAL TEST?

Suppose we have the same data as in Example 9-3. Recall that the traditional manufacturing process had produced a population of millions of TV tubes, with a mean life $\mu = 1200$ hours and a standard deviation σ

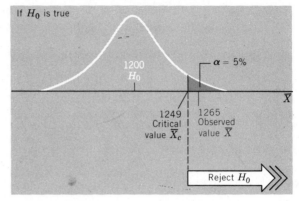

Figure 9-4
A classical test at level $\alpha = 5\%$.

$= 300$ hours. To apply a classical hypothesis test of whether a new process is better, we will proceed in three steps—the first two before any data is collected:

1. The null hypothesis (H_0: $\mu = 1200$) is formally stated. At the same time, we set the sample size (such as $n = 100$), and the error level of the test (such as 5%)[8] hereafter referred to as α.
2. We now assume temporarily that the null hypothesis is true—just as we did in calculating the p-value. And we ask, what can we expect of a sample mean drawn from this sort of world? Its specific distribution is again shown in Figure 9-4, just as it was in Figure 9-1. But there is one important difference in these two diagrams: Whereas in Figure 9-1 the shaded p-value was calculated from the observed $\overline{X}$, in Figure 9-4 the shaded area is arbitrarily set at $\alpha = 5\%$. This defines the critical range for rejecting the null hypothesis (shown as the big arrow). All this is done before any data is observed.
3. The sample is now taken. If the observed $\overline{X}$ falls in the rejection region in Figure 9-4, then it is judged sufficiently in conflict with the null hypothesis to reject H_0. Otherwise H_0 is not rejected.

The critical value $\overline{X}_c = 1249$ for this test was calculated by noting from Appendix Table IV that a 5% tail is cut off the normal distribution by a critical Z value of $z_{.05} = 1.64$; that is:

$$\text{Critical } Z = \frac{\overline{X}_c - \mu_0}{\sigma/\sqrt{n}} = 1.64$$

---

[8]This arbitrary choice of 5% is equivalent to the arbitrary choice of a confidence level at 95%.

Note in Figure 9-4 that we are cutting off a probability only in one tail. Accordingly, this is a "one-tailed test," consistent with the one-tailed alternative hypothesis (and equivalent to a one-sided confidence interval). Under certain circumstances a two-tailed test may be required, but this discussion is deferred to Section 9-6.

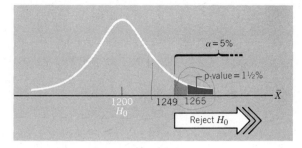

Figure 9-5
Classical hypothesis testing and p-value.

$$\frac{\overline{X}_c - 1200}{300/\sqrt{100}} = 1.64 \qquad \begin{array}{c}(9\text{-}25)\\ \text{like}\,(9\text{-}14)\end{array}$$

$$\overline{X}_c = 1249 \qquad (9\text{-}26)$$

In our example, the observed $\overline{X} = 1265$ is beyond this critical value, thus leading us to reject H_0 at the 5% error level.

There is another way of looking at this testing procedure. If we get an observed $\overline{X}$ exceeding 1249, there are two explanations:

1. H_0 is true, but we have been exceedingly unlucky and got a very improbable sample $\overline{X}$. (We're born to be losers; even when we bet with odds of 19 to 1 in our favor, we still lose).
2. H_0 is not true after all. Thus it is no surprise that the observed $\overline{X}$ was so high.

We opt for the more plausible second explanation. But we are left in some doubt; it is just possible that the first explanation is the correct one. For this reason we qualify our conclusion to be "at the 5% error level."

B—CLASSICAL HYPOTHESIS TESTING AND p-VALUE

For the example above, a comparison of classical testing and p-value is set out in Figure 9-5. Since the p-value [$1\frac{1}{2}$% from (9-12)] is less than α = 5%, the observed $\overline{X}$ is correspondingly in the rejection region; that is:

$$\boxed{\text{Reject } H_0 \text{ if its p-value} \le \alpha} \qquad (9\text{-}27)$$

To restate this, we recall that the p-value is a measure of the credibility of H_0. If this credibility falls below α, then H_0 is rejected.[9]

---

[9]Figure 9-5 provides another useful interpretation of p-value. Note that if we had happened to set the level of the test at the p-value of $1\frac{1}{2}$% (rather than 5%), it would have been just barely possible to reject H_0. Accordingly:

p-value is the lowest that we could push the level α
and still be able (barely) to reject H_0. $\qquad$ (9-28)

Applied statisticians increasingly prefer p-values to classical testing, because classical tests involve setting α arbitrarily (usually at 5%). Rather than introduce such an arbitrary element, it is often preferable just to quote the p-value, leaving the reader to pass his own judgment on H_0. [By determining whatever level of α he deems appropriate for his purposes, the reader may reach his own decision using (9-27).]

C—TYPE I AND TYPE II ERRORS

In the decision-making process we run the risk of committing two distinct kinds of error. The first is shown in Figure 9-6a (a reproduction of Figure 9-4), which shows what the world looks like if H_0 is true. In this event, there is a 5% chance that we will observe $\overline{X}$ in the shaded region, and thus erroneously reject the true H_0. Rejecting H_0 when it is true is called a type I error, with its probability of course being α, the error level of the test. (Now we see that when we use the term "error level of a test" we mean, more precisely, "the type I error level of a test.")

But suppose the null hypothesis is false—that is, the alternative hypothesis H_1 is true—and, to be specific, suppose $\mu = 1240$. Then we are living in a different sort of world. Now $\overline{X}$ is distributed around $\mu = 1240$,

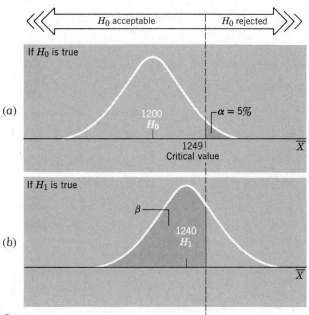

Figure 9-6
The two kinds of error that can occur in a classical test. (a) If H_0 is true, then α = probability of erring (by rejecting the true hypothesis H_0). (b) If H_1 is true, then β = probability of erring (by judging that the false hypothesis H_0 is acceptable).

TABLE **9-1** **Possible Results of an Hypothesis Test (Based on Figure 9-6)**

	DECISION	
STATE OF THE WORLD	H_0 ACCEPTABLE	H_0 REJECTED
If H_0 is true	Correct decision. Probability $= 1 - \alpha$ $\quad\quad = $ confidence level	Type I error. Probability $= \alpha$ $\quad\quad = $ level of the test
If H_0 is false	Type II error. Probability $= \beta$	Correct decision. Probability $= 1 - \beta$ $\quad\quad = $ power of the test

as shown in Figure 9-6b. The correct decision in this case would be to reject the false null hypothesis H_0. An error would occur if $\overline{X}$ were to fall in the H_0 acceptance region. Such acceptance of H_0 when it is false is called a *type II error*. Its probability is called β, and is shown as the shaded area in Figure 9-6b.

Table 9-1 summarizes the dilemma of hypothesis testing: The state of the real world, whether H_0 is true or false, is unknown. If a decision to reject or not reject must be made[10] in the face of this uncertainty, we have to take the risk of one error or another.

PROBLEMS

9-7 In a sample of 750 American men in 1974, 45% smoked; in an independent sample of 750 women, 36% smoked (*Gallup Opinion Index*, June 1974).

    a. Construct a one-sided 95% confidence interval for the difference between men and women in the population.

    b. Calculate the p-value for the null hypothesis of no difference.

    c. At the error level $\alpha = .05$, is the difference between 45% and 36% statistically discernible; that is, can H_0 be rejected? Answer two ways, making sure that both answers agree:

        i. Is H_0 excluded from the 95% confidence interval?

        ii. Does the p-value for H_0 fall below .05?

9-8 In a sample of 1500 Americans in 1971, 42% smoked; in an independent sample of 1500 Americans a year later, 43% smoked. For the increase over the year, repeat Problem 9-7.

9-9 (Acceptance Sampling) The prospective purchaser of several new shipments of waterproof gloves hopes they are as good as the old shipments, which had a 10% rate of defective pairs. But he fears that they may be

---

[10]Of course, other more complicated decision rules may be used. For example, the statistician may decide to suspend judgment if the observed $\overline{X}$ is in the region around 1250 (say $1240 < \overline{X} < 1260$). If he observes an ambiguous $\overline{X}$ in this range, he then would undertake a second stage of sampling—which might yield a clear-cut decision, or might lead to further stages (i.e., "sequential sampling").

worse. So for each shipment, he takes a random sample of 100 pairs and counts the proportion P of defective pairs so that he can run a classical test. The level of this test was determined by such relevant factors as the cost of a bad shipment, the cost of an alternative supplier, and the new shipper's reputation. When all these factors were taken into account, suppose the appropriate level of this test was $\alpha = .09$.

a. State the null and alternative hypotheses, in words and symbols.

b. What is the critical value of P? That is, how large must P be in order to reject the null hypothesis (i.e., reject the shipment)?

c. Suppose that for 6 shipments, the values of P turned out to be 12%, 25%, 8%, 16%, 24%, and 21%. Which of these shipments should be rejected?

9-10 In Problem 9-9, suppose that the purchaser tries to get away with a small sample of only 10 pairs. Suppose that instead of setting $\alpha = .09$, he sets the rejection region to be $P \geq 20\%$ (i.e., he will reject the shipment if there are 2 or more defective pairs among the 10 pairs in the sample).

For this test, what is α? (*Hint:* Rather than using the normal approximation, the binomial distribution is easier and more accurate.)

*9-11 Consider the problem facing an air traffic controller at Chicago's O'Hare Airport. If a small irregular dot appears on the screen, approaching the flight path of a large jet, she must decide between:

H_0: All is well. It's only a bit of interference on the screen.
H_1: A collision with a small private plane is imminent.

Fill in the blanks:

A "false alarm" is a type_____error, and its probability is denoted by_____. A "missed alarm" is a type_____error, and its probability is denoted by_____. By making the equipment more sensitive and reliable, it is possible to reduce both_____and_____.

9-12 In manufacturing machine bolts, the quality control engineer finds that it takes a sample of $n = 100$ to detect an accidental change of .5 millimeter in the mean length of the manufactured bolt. Suppose he wants more precision, to detect a change of only .1 millimeter, with the same α and β. How much larger must his sample be? (*Hint:* It is easy, if you rephrase it in terms of confidence intervals. To make a confidence interval 5 times as precise, how much larger must the sample be?)

*9-13 Consider the classical test shown in Figure 9-6.

a. If the observed $\overline{X}$ is 1245, do you reject H_0?

b. Suppose that you had designed this new process, and you were convinced that it was better than the old; specifically, on the basis of sound engineering principles, you really believed H_1: $\mu = 1240$. Common sense suggests that H_0 should be rejected in favor of H_1. In this conflict of common sense with classical testing, which path would you follow? Why?

c. Suppose that sample size is doubled to 200, while you keep $\alpha = 5\%$ and you continue to observe $\overline{X}$ to be 1245. What would be the classical test decision now? Would it, therefore, be true to say that your problem in part **a** may have been inadequate sample size?

d. Suppose now that your sample size n was increased to a million, and in that huge sample you observed $\overline{X} = 1201$. An improvement of only one unit over the old process is of no economic significance (i.e., does not justify retooling, etc.); but is it statistically significant (discernible)? Is it therefore true to say that a sufficiently large sample size may provide the grounds for rejecting any specific H_0—no matter how nearly true it may be—simply because it is not *exactly* true?

*9-4 Classical Tests Reconsidered

A—REDUCING α AND β

In panel (a) of Figure 9-7 we show again the two error probabilities of Figure 9-6: α, if H_0 is true; and β, if H_1 is true. In panel (b), we illustrate how decreasing α (by moving the critical point to the right to, say, 1270) will at the same time increase β. That is, pushing down α will raise β. Figure 9-8 illustrates this trade-off between conflicting objectives.

There is an interesting legal analogy. In a murder trial, the jury is being asked to decide between H_0, the hypothesis that the accused is innocent, and the alternative H_1, that he is guilty. A type I error is committed if an innocent man is condemned, while a type II error occurs if a guilty man is set free. The judge's admonition to the jury that "guilt must be proved beyond a reasonable doubt" means that α should be kept very small.

There have been many legal reforms (e.g., such as limiting the power of the police to obtain a confession) that have been designed to reduce α, the probability that an innocent man will be condemned. But these same reforms have increased β, the probability that a guilty man will evade punishment. There is no way of pushing α down to 0 (ensuring absolutely against convicting an innocent man) without letting β rise to 1 (letting every defendant go free and making the trial meaningless).

The only way that one error can be reduced without increasing the other is by gathering better evidence. For example, Figure 9-7c shows how increasing the sample size n makes the sampling distributions more accurate. Hence β can be made smaller without increasing α.

B—SOME PITFALLS IN CLASSICAL TESTING

Figure 9-7 can be used to illustrate some of the difficulties that may be encountered in applying a classical reject-or-accept hypothesis test at an

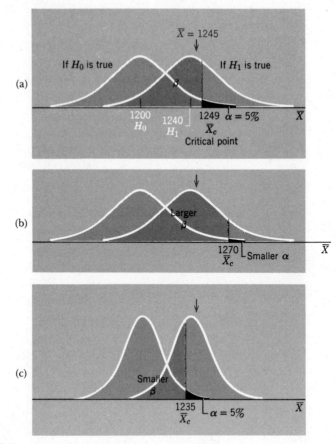

Figure 9-7
(a) Hypothesis test of Figure 9-6 showing α and β. (b) How a reduction in α increases β, other things being equal. (c) How an increase in sample size allows one error probability (β) to be reduced, without increasing the other (α).

arbitrary level α. In panel (a), suppose we have observed a sample $\overline{X} = 1245$. This is not quite extreme enough to allow rejection of H_0 at level $\alpha = 5\%$, so $\mu_0 = 1200$ is accepted. But if we had set $\alpha = 10\%$, then H_0 would have been rejected. This illustrates once again how an arbitrary specification of α leads to an arbitrary decision.

There is an even deeper problem. Accepting $\mu_0 = 1200$ (at level $\alpha =$

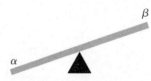

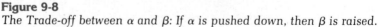

Figure 9-8
The Trade-off between α and β: If α is pushed down, then β is raised.

5%) is a disaster, if we had prior grounds for believing H_1 is true—that is, for expecting that the new process would yield a mean $\mu_1 = 1240$. In this case, our prior belief in the new process is strongly supported by the sample observation of 1245. Yet we have used this *confirming* sample result (in this classical hypothesis test) to *reverse* our original view. That is, we have used an observation of 1245 to judge in favor of a population value of 1200, and against a population value of 1240. In this case our decision makes no sense at all. This serves as a warning of the serious problem that may exist in a classical test if a small sample is used to accept a null hypothesis.[11] Accordingly, it is wise to stop short of explicitly saying "accept H_0." Instead, we use the more reserved phrase "H_0 is acceptable,"[12] or "H_0 is not rejected."[13]

At the other extreme, a huge sample size may lead us into another kind of error [the one encountered in Problem 9-13(d)]. This is the error of rejecting an H_0 that, although essentially true, is not exactly true This difficulty arises because a huge sample may reduce the standard error to the point where even a miniscule difference becomes statistically discernible; that is, H_0 is rejected, although it is practically true.

Both these pitfalls of classical testing can be avoided by simply quoting the p-value for H_0 instead. Then no arbitrary judgment need be made about α; instead, the reader can make his own decision on whether or not to reject H_0.

C—WHY IS CLASSICAL TESTING EVER USED?

In light of all these accumulated reservations about a classical accept-or-reject hypothesis test, why do we take the time to study it? There are three reasons:

1. It is helpful in guiding the student through the classical statistical literature.
2. It is helpful in clarifying certain theoretical issues, like type I and type II errors.

---

[11]In Figure 9-7, it was the small sample in panel (a) that misled us into judging H_0 acceptable. By contrast, the large sample shown in panel (c) would allow us to reject H_0.

[12]That is, we conclude that H_0 is one of many acceptable hypotheses (without passing judgment on which of these, if any, should actually be accepted).

[13]Thus we see that, although classical testing provides a rationale for rejecting H_0, it provides no formal rationale for accepting H_0. The null hypothesis may sometimes be uninteresting, and one that we neither believe nor wish to establish; it is selected because of its simplicity. In such cases, it is the alternative H_1 that we are trying to establish, and we prove H_1 by rejecting H_0. We can see now why statistics sometimes is called *the science of disproof*. H_0 cannot be proved, and H_1 is proved by disproving (rejecting) H_0. It follows that if we wish to prove some proposition, we often will call it H_1 and set up the contrary hypothesis H_0 as the "straw man" we hope to destroy. And, of course, if H_0 is only such a straw man, then it becomes absurd to accept it in the face of a small sample result that really supports H_1.

3. A classical hypothesis test may be preferred to the calculation of p-value if the test level α can be determined rationally, and if many samples are to be classified. [In industrial quality control, for example, samples of output are periodically taken to see whether production is still properly in control—i.e., to see whether H_0 ("no problems") is still acceptable, or whether it should be rejected.]

To illustrate this last point, consider again the familiar Example 9-3; but now suppose we are considering five new production processes, rather than just one. If a sample of 100 tubes is taken in each case, suppose the results are as shown below.

NEW PROCESS	$\overline{X}$	IS PROCESS REALLY BETTER THAN THE OLD PROCESS (WHERE $\mu = 1200$)?
1	1265	?
2	1240	?
3	1280	?
4	1150	?
5	1210	?

We now have two options. We can calculate five p-values for these five processes, just as we calculated the p-value for the first process in Example 9-3. But this will involve a lot more work than using a classical testing approach, which requires only that we specify α (at say 5%) and then calculate one single figure, the critical value $\overline{X}_c = 1249$ derived in (9-26). Then all five of the sample values can immediately be evaluated without any further calculation. (Note that H_0 is rejected only for processes 1 and 3; in other words, these are the only two that may be judged superior to the old method.)

But if a classical hypothesis test is to be used in this way, the level α should not be arbitrarily set. Instead, α should be determined rationally, on the basis of two considerations:

1. *Prior belief.* To again use our example, the less confidence we have in the engineering department that assured us these new processes are better, the smaller we will set α (i.e., the stronger the evidence we will require to judge in their favor). So we need to answer questions such as: Are the engineers' votes divided? How often have they been wrong before?

2. *Losses involved in making a wrong decision.* The greater the costs of a type I error (i.e., needlessly retooling for a new process that is actually no better), the smaller we will set α, the probability of making that sort of error. Similarly, the greater the cost of a type II error (i.e., failing to switch to a new process that is actually better), the smaller we will set β. The problem is that with limited resources, we cannot make both α and β smaller (recall Figure 9-7b). But in setting their level, we must obviously take into account the relative costs of the two errors.

It is possible to take account of prior beliefs and losses in a more formal way. Called *Bayesian decision analysis*, this theory is developed in Chapter 20.

*PROBLEMS

9-14   Answer True or False. If False, correct it.

a.   Comparing hypothesis testing to a jury trial, we may say that the type I error is like the error of condemning a guilty man.

b.   Suppose, in a certain test, that the p-value turns out to be .013. Then H_0 would be acceptable at the 5% level and also at the 1% level.

c.   There are two disadvantages in arbitrarily specifying α at 5%, in a classical accept-or-reject test:
If the sample is very small, we may find H_0 acceptable even when H_0 is quite false.
If the sample is very large, we may reject H_0 even when H_0 is approximately correct and hence a good working hypothesis.

9-15   In the jury trial analogy, what would be the effect on α and β of reintroducing capital punishment on a large scale?

9-16   Consider a very simple example in order to keep the philosophical issues clear. Suppose that you are gambling with a single die, and lose whenever the die shows 1 dot (ace). After 100 throws, you notice that you have suffered a few too many losses—20 aces. This makes you suspect that your opponent is using a loaded die; specifically, you begin to wonder whether this is one of the crooked dice recently advertised as giving aces one-quarter of the time.

a.   Find the critical proportion of aces beyond which you would reject H_0 at the 5% level.

b.   Illustrate this test with a diagram similar to Figure 9-6. From this diagram, roughly estimate α and β.

c.   With your observation of 20 aces, what is your decision? Suppose that you are playing against a strange character you have just met on a Mississippi steamboat. A friend passes you a note indicating that this stranger cheated him at poker last night; and you are playing for a great deal of money. Are you happy with your decision?

d.   If you double α, use the diagram in part **b** to roughly estimate what happens to β.

*9-17   In Problem 9-10, suppose the alternative hypothesis is that the shipment is 30% defective.

a.   State this in symbols.

b.   Calculate β.

*9-18   Repeat the problem above, for Problem 9-9 in place of 9-10. Then, is it fair to say that an increased sample size reduced both α and β?

***9-5   Operating Characteristics Curve (OCC)**

Now that we have defined and discussed the type II error probability β at some length, it is time to actually calculate it.

EXAMPLE **9-7**   Consider the test shown in Figure 9-6, where the cutoff value $\overline{X}_c$ is 1249, and the alternative hypothesis is $\mu_1 = 1240$.
   Calculate β (the probability that when H_1 is true, we will make the error of finding H_0 acceptable). For your solution, recall that $\overline{X}$ was based on a sample of n = 100 observations, and the population standard deviation was $\sigma = 300$.

Solution   According to Figure 9-6b (where H_1 is true), the error of accepting H_0 occurs when $\overline{X}$ falls below the cutoff value $\overline{X}_c = 1249$. To find the probability of this, we standardize the critical value $X_c = 1249$, seeing how far it deviates from the mean $\mu_1 = 1240$:

$$Z = \frac{\overline{X}_c - \mu_1}{SE}$$

$$= \frac{1249 - 1240}{300/\sqrt{100}} = .30$$

Thus, $Pr(\overline{X} < 1249) = Pr(Z < .30)$
$$= 1 - .382 = .618$$
$$\beta = 62\% \tag{9-29}$$

In many situations, it is unrealistic to pin down the alternative hypothesis to one specific value. For example, a more realistic version of Example 9-7 would include several possible alternative hypotheses, as follows.

EXAMPLE **9-8**   Continuing Example 9-7, calculate the type II error probability β for several more alternative hypotheses:

   a. $\mu_1 = 1280$
   b. $\mu_1 = 1320$

Solution   We use exactly the same method as in Example 9-7:

   a.
$$Z = \frac{1249 - 1280}{300/\sqrt{100}} = -1.03$$

Thus

$$Pr(\overline{X} < 1249) = Pr(Z < -1.03) = .152$$
$$\beta = 15\%$$

b.

$$Z = \frac{1249 - 1320}{300/\sqrt{100}} = -2.37$$

Thus

$$Pr(\overline{X} < 1249) = Pr(Z < -2.37) = .009$$
$$\beta = 1\%$$

The values of β in Examples 9-7 and 9-8 (corresponding to $\mu_1 = 1240$, 1280, and 1320) are shown in Figure 9-9 as the three shaded areas (to the left of the cutoff point, where we accept H_0). Of course, an even more realistic alternative hypothesis would include *any* number larger than $\mu_0 = 1200$; that is,

$$\mu_1 > 1200 \qquad\qquad (9\text{-}10)\,\text{repeated}$$

In other words,

$$\mu_1 = 1201$$
$$\text{or } \mu_1 = 1202$$
$$\text{or } \mu_1 = 1203$$
$$\vdots$$

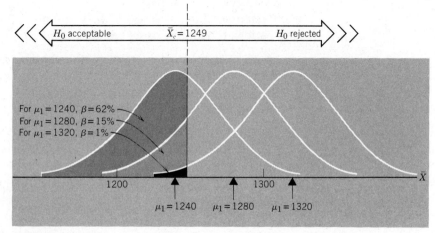

Figure 9-9
Calculation of β, the probability of type II error, for a composite H_1.

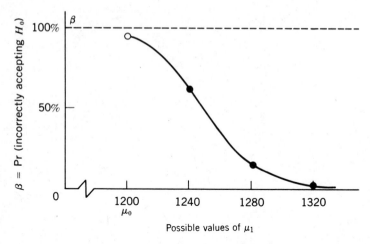

Figure 9-10
The β function, or Operating Characteristics Curve (OCC).

An hypothesis like this, composed of many possibilities, is called a *composite* hypothesis. For each possible μ_1, we can calculate the corresponding β—as we already have for μ_1 = 1240, 1280, and 1320. When all the values of β are graphed, we obtain the β *function* in Figure 9-10— more commonly called the *operating characteristics curve* (OCC).

Figure 9-10 shows very graphically that as the alternative hypothesis μ_1 gets further away from μ_0 = 1200, the probability of error β drops. That is, the bigger the difference between μ_0 and μ_1, the less chance of confusing the two.

*PROBLEMS

9-19   A certain type of seed has always grown to a mean height of 8.5 inches, with a standard deviation of 1 inch. A sample of 100 seeds is grown under new enriched conditions to see whether the mean height might be improved.

   a.   At the 5% level, calculate the cutoff value $\overline{X}_c$ above which H_0 should be rejected.

   b.   If the sample of 100 seeds actually turns out to have a mean height $\overline{X}$ = 8.8 inches, do you reject H_0?

   c.   Roughly graph the OCC for this test. (*Hint:*   Calculate β for a rough grid of values, say μ_1 = 8.6, 8.8, 9.0. Then sketch the curve joining these values.)

   d.   What would be the approximate chance of failing to detect a mean height improvement, if the sample of 100 seeds were to come from a population whose mean was 8.9 inches?

9-20   **a.**   For the acceptance sampling in Problem 9-9, sketch the OCC.

    b. Repeat part **a,** for Problem 9-10.

    c. By comparing part **a** with part **b,** note how a larger sample gives a better OCC.

9-21 A large electronics firm buys printed circuits in lots of 50,000 each, and in the past has found the proportion that are substandard is $\pi = 2\%$ (The null hypothesis). To protect themselves against a possible deterioration of quality, they decide to sample each lot: If more than 25 in a sample of 1,000 are defective, they will reject the lot and return it.

    a. Calculate the level α of this test.

    b. Calculate β, the probability of the type II error of this test, if the proportion of defectives is:

        i. $\pi = 2.5\%$ **iii.** $\pi = 3.5\%$

        ii. $\pi = 3.0\%$ **iv.** $\pi = 4.0\%$

    c. Sketch the graph of β—the OCC curve.

    d. Instead of graphing β (the probability of mistakenly accepting the lot) we could have graphed $1-\beta$ (the probability of correctly rejecting the lot). If we call this the *power* curve, what relation does it have to the OCC curve? Graph this power curve.

    e. Suppose 50 shoddy lots were shipped, each having 3% defective. About how many would be detected as shoddy, and rejected?

*9-6 Two-sided Tests

So far we have concentrated on the one-sided test, in which the alternative hypothesis, and consequently the rejection region and p-value, are just on one side. The one-sided test (like the one-sided confidence interval) is appropriate when there is a one-sided claim to be made, such as, "more than," "less than," "better than," "worse than," "at least," and so on.

However, there are occasions when it is more appropriate to use a two-sided test (like a two-sided confidence interval). These occasions often may be recognized by symmetrical claims such as, "different from," "changed for better or worse," "unequal," and so on. This section discusses the minor modifications required.

A—TWO-SIDED p-VALUE[14]

An example will show how easy it is to accommodate a two-sided hypothesis.

---

[14]Since the one-sided p-value is commonest, it is often called simply the p-value (as in Section 9-2). To avoid confusion, therefore, whenever we mean a two-sided p-value, we specifically refer to it as "two-sided."

EXAMPLE **9-9**   Consider again the testing of TV tubes in Example 9-3. Suppose that the null hypothesis remains as:

$$\mu_0 = 1200$$

But now change the alternative hypothesis by supposing that our engineers cannot advocate the new process as better but must concede that it may be worse. Then the alternative hypothesis would be:

$$\mu_1 > 1200 \quad \text{or} \quad \mu_1 < 1200$$

That is:

$$\mu_1 \neq 1200 \tag{9-30}$$

In other words, we now are testing whether the new process is *different* (whereas in Example 9-3, we were testing whether it was *better*). Thus, even before we collect any data, we can agree that a value of $\overline{X}$ well below 1200 would be just as strong evidence against H_0 as a value of $\overline{X}$ well above 1200; that is, what counts is how far away $\overline{X}$ is, on *either side*.

If the sample mean $\overline{X} = 1265$, what is the two-sided p-value? That is, what is the probability that $\overline{X}$ would be this distant (in either direction) from the null hypothesis $\mu_0 = 1200$? (For your solution, recall that $\overline{X}$ was based on a sample of $n = 100$ observations, and the population standard deviation was $\sigma = 300$.)

Solution   As usual, we measure how far away $\overline{X}$ is from the null hypothesis μ_0 by standardizing:

$$Z = \frac{\overline{X} - \mu_0}{SE}$$

$$= \frac{1265 - 1200}{300/\sqrt{100}} = 2.17 \qquad \text{like (9-11)}$$

Thus, $\Pr(\overline{X} > 1265) = \Pr(Z > 2.17) = .015$

This is the chance that we would observe $\overline{X}$ this far above μ_0, as shown in the right-hand tail of Figure 9-11. But there is the same probability (.015) that we would observe $\overline{X}$ this far below μ_0, as shown in the left-hand tail. The two-sided p-value is therefore .030.

In general, whenever the alternative hypothesis is two-sided, it is appropriate to calculate the two-sided p-value for H_0. As we have just seen

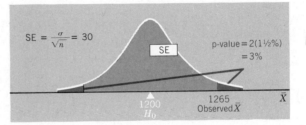

Figure 9-11
Two-sided p-value ≡ Pr ($\overline{X}$ would be as extreme as the value actually observed, if H_0 is true). Compare with Figure 9-1.

in Figure 9-11, whenever the sampling distribution is symmetric, the two-sided p-value is just twice the one-sided p-value.

B—CLASSICAL TESTS AND CONFIDENCE INTERVALS

It is easy to modify a one-sided classical test to a two-sided classical test; we merely reject H_0 if the *two*-sided p-value falls below the specified level α. For instance, in Example 9-9 where the two-sided p-value was only 3%, H_0 could be rejected at the 5% level.

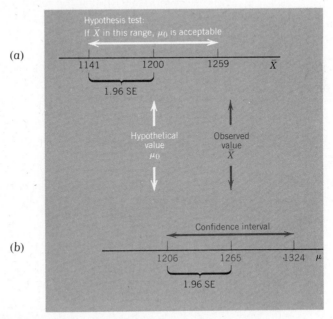

Figure 9-12
The equivalence of two-sided classical testing and confidence intervals. (a) The classical test at level $\alpha = 5\%$. Since $\overline{X}$ falls beyond the cutoff points, we can reject H_0. (b) The 95% confidence interval. Since μ_0 falls outside the confidence limits we can reject H_0.

In a repetitive situation where we need to test many samples, instead of calculating the many p-values we could calculate once and for all a classical rejection region—a two-sided region that cuts off $2\frac{1}{2}\%$ in each tail, for a total of $\alpha = 5\%$. Figure 9-12a illustrates this for the TV-tube data of Example 9-9. We note that an observed $\overline{X} = 1265$ would mean H_0 is rejected.

Whenever a two-sided test is appropriate, an ordinary two-sided confidence interval also is appropriate, as shown in Figure 9-12b. Since the null hypothesis $\mu_0 = 1200$ lies outside this confidence interval, it can be rejected. It is clear why the confidence interval and the test are exactly equivalent: In both cases, we simply check whether the magnitude of the difference $|\overline{X} - \mu_0|$ exceeds $1.96\,SE$ (standard errors). These two approaches differ only because the classical test uses the null hypothesis μ_0 as its reference point, whereas the confidence interval uses the observed $\overline{X}$ as its reference point. Thus, (9-5) finally is confirmed.

*PROBLEMS

9-22 Answer True or False; if False, correct it.

a. If the alternative hypothesis is two-sided, then the p-value, classical test, and confidence interval should be two-sided too.

b. To decide whether the probability π of a die coming up ace is fair, suppose we are testing the hypothesis

$$H_0:\ \pi = 1/6 \quad \text{against} \quad H_1:\ \pi < 1/6$$

Then we should use a two-sided test, rejecting H_0 when π turns out to be large.

9-23 In a Gallup poll of 1500 Americans in 1975, 45% answered "yes" to the question, "Is there any area right around here—that is, within a mile—where you would be afraid to walk alone at night?" In an earlier poll of 1500 Americans in 1972, only 42% had answered "yes."

a. Construct a two-sided confidence interval for the change in proportion who are afraid.

b. Calculate the two-sided p-value for the null hypothesis of no change.

c. At the error level $\alpha = 5\%$, is the increase from 42% to 45% statistically discernible? That is, can H_0 be rejected? Answer in two ways, making sure that both answers agree:

    i. Is H_0 excluded from the 95% confidence interval?

    ii. Does the p-value for H_0 fall below 5%?

9-24 Repeat Problem 9-23, with the appropriate changes in wording, for the following income data sampled from U.S. physicians in 1979 (net income after expenses, but before taxes):

	GENERAL PRACTITIONERS	PSYCHIATRISTS
Sample size n	200	100
Mean income $\overline{X}$	$62,000	$62,600
Standard deviation s	$32,000	$35,000

Pooled standard deviation s_p = $33,000

CHAPTER 9 SUMMARY

9-1 The values inside a 95% confidence interval are called acceptable hypotheses (at the 5% error level), while the values outside are called rejected hypotheses. The no-difference hypothesis is called the null hypothesis H_0; when it is rejected, a difference is established and so the result is called statistically discernible (significant).

9-2 The p-value is defined as the chance of getting a value of $\overline{X}$ as large as the one actually observed, if H_0 were true. That is, it is the tail area (of the distribution centered on μ_0) beyond the observed value of $\overline{X}$. The p-value therefore measures the credibility of H_0.

9-3 For a classical test, we reject H_0 if its p-value (credibility) falls below a specified error level α (usually 5%). Alternatively, a classical test can be set up with extreme values of $\overline{X}$ forming a "rejection region," which is particularly useful for repetitive situations such as industrial quality control: Whenever $\overline{X}$ falls into this extreme region, H_0 is rejected.

*9-4 Classical tests may be misleading—especially if H_0 is accepted on the basis of a small sample, or H_0 is rejected on the basis of a very large sample.

*9-5 When the alternative hypothesis H_1 is true, the luck of the draw may erroneously lead us to conclude that H_0 is acceptable. The chance of this "type II error" is called β, and the graph of β is called the OCC (operating characteristics curve).

*9-6 The two-sided p-value is double the one-sided p-value, and is appropriate whenever the alternative hypothesis is two-sided. An hypothesis test using the two-sided p-value is equivalent to a test using a two-sided confidence interval.

REVIEW PROBLEMS

9-25 A professor was evaluated on a five-point scale by 15 students randomly selected from one class, and 40 students from another, with the following results:

| | FREQUENCIES | |
SCALE VALUE	10:30 AM CLASS	3:30 PM CLASS
1 (terrible)	0	4
2	0	7
3	4	13
4	7	9
5 (excellent)	4	7
	$n = 15$	$n = 40$

a. Graph the frequency distributions of the two samples.

b. Calculate the two sample means and standard deviations, and show them on the graph.

9-26 In Problem 9-25, the registrar wanted to know whether the difference in the two samples was just a sampling fluke, or reflected a real difference in the classes. In fact, she suspected it showed a real difference, because it was customary for late classes to be restless and give their professors rather low ratings.

Answer her as thoroughly as you can. (*Hint:* We suggest you calculate the one-sided confidence interval, and p-value for H_0.)

9-27 Suppose that a scientist concludes that a difference in sample means is "statistically significant (discernible) at the 1% level." Answer True or False; if False, correct it.

a. There is at least a 99% chance that there is a real difference in the population means.

b. The p-value for H_0 (population means are exactly equal) is 1% or less.

c. If there were no difference in the population means, the chance of getting such a difference (or more) in the sample means is 1% or less.

d. The scientist's conclusion is sound evidence that a difference in population means exists. Yet in itself it gives no evidence whatever that this difference is large enough to be of practical importance. This illustrates that statistical significance and practical significance are two entirely different concepts.

9-28 In a 1965 court case (Kaye, 1982), the defense argued that the grand jury had racially discriminated. Although 25% of the men in the jurisdiction were black, only 17% (177 out of 1050) of the men called for duty had been black. The null hypothesis of a fair draw (selection of jurors) may be stated as H_0: The 1050 men called for jury duty constituted a *random* sample of men drawn from the jurisdiction.

a. Calculate the p-value for H_0.

b. The judge said the difference between 17% and 25% was not sufficient to establish that discrimination had occurred. Do you agree?

*9-29 Suppose that, in Problem 9-28, the judge instead had ruled in favor of the defense—that is, had ruled that discrimination *had* occurred. Then the prosecution could have come up with this counter argument: Only literate males were eligible, and the literacy rate for black men (48%) was different from white men (83%).

a. Of the eligible population, what proportion is black?

b. Now let us define the null hypothesis as H_0: The 1050 men called for jury duty constituted a random sample from the eligible population. Calculate the p-value for H_0.

c. Do you think there was racial discrimination? What kind?

d. What do you think was the real problem—the selection of jurors, the eligibility rule, or something else?

Chapter

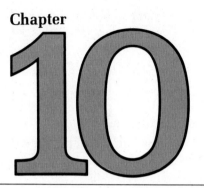

Analysis of Variance (ANOVA)

Variability is the law of life.

SIR WILLIAM OSLER

10-1 One-factor Analysis of Variance

A—INTRODUCTION

In Chapters 8 and 9, we made inferences about one population mean, and then compared two means. Now we will compare several means.

As an illustration, suppose that three machines are to be compared. Because these machines are operated by people, and because of other, inexplicable reasons, output per hour is subject to chance fluctuation. In the hope of "averaging out" and thus reducing the effect of chance fluctuation, a random sample of 5 different hours is obtained from each machine and set out in Table 10-1, where each sample mean is then calculated. (The part of the table below this can be ignored for now.)

Some explicit notation will be useful in this chapter. The ith sample mean is denoted $\overline{X}_i$. The typical t^{th} observation within this sample requires another subscript, so we write it X_{it}. If there are c samples (or columns) and n observations in each sample, then we may write our data as follows:

$$X_{it} \quad \begin{aligned} i &= 1,2,\ldots,c \\ t &= 1,2,\ldots,n \end{aligned}$$

B—VARIATION BETWEEN SAMPLES

The first question is "Are the machines really different?" That is, are the sample means $\overline{X}_1, \overline{X}_2,$ and $\overline{X}_3$ in Table 10-1 different because of differences

TABLE **10-1** **Sample Outputs of Three Machines**

	MACHINE 1	MACHINE 2	MACHINE 3	
	47	55	54	
	53	54	50	
	49	58	51	
	50	61	51	
	46	52	49	
$\overline{X}_i$	$\overline{X}_1 = 49$	$\overline{X}_2 = 56$	$\overline{X}_3 = 51$	$\overline{\overline{X}} = 52$
$(\overline{X}_i - \overline{\overline{X}})$	-3	4	-1	$\Sigma(\overline{X}_i - \overline{\overline{X}}) = 0\checkmark$
$(\overline{X}_i - \overline{\overline{X}})^2$	9	16	1	$\Sigma(\overline{X}_i - \overline{\overline{X}})^2 = 26$

in the underlying population means μ_1, μ_2, and μ_3 (where μ_1 represents the life-time performance of the first machine, etc.)? Or may these differences in the $\overline{X}$'s be reasonably attributed to chance fluctuations alone? To illustrate, suppose that we collect three samples from just one machine, as shown in Table 10-2. As expected, sampling fluctuations cause small differences in the $\overline{X}$'s in the bottom row even though the μ's in this case are identical. So the question may be rephrased, "Are the differences in the $\overline{X}$'s of Table 10-1 of the same order as those of Table 10-2 (and thus attributable to chance fluctuation), or are they large enough to indicate a difference in the underlying μ's?" The latter explanation seems more plausible; but how do we develop a formal test?

As usual, the hypothesis of "no difference" in the population means is called the null hypothesis:

$$H_0: \ \mu_1 = \mu_2 = \mu_3 \tag{10-1}$$

A test of this hypothesis first requires a numerical measure of how much the sample means $\overline{X}_i$ differ. We therefore calculate their variance at the bottom of Table 10-1. (This first required a calculation of $\overline{\overline{X}}$, the *grand mean* of the $\overline{X}_i$). Thus we obtain:

$$s_{\overline{X}}^2 = \frac{26}{2} = 13 \tag{10-2}$$

TABLE **10-2** **Three Sample Outputs of the Same Machine**

	SAMPLE 1	SAMPLE 2	SAMPLE 3	
	49	52	55	
	55	51	51	
	51	55	52	
	52	58	52	
	48	49	50	
	$\overline{X}_1 = 51$	$\overline{X}_2 = 53$	$\overline{X}_3 = 52$	

The variance formula we used here, of course, is (2-11)—where $\overline{X}_i$ is substituted for X, and c (the number of sample means or columns) is substituted for n, so that:

$$s_{\overline{X}}^2 = \frac{1}{c-1} \Sigma(\overline{X}_i - \overline{\overline{X}})^2 \tag{10-3}$$

C—VARIATION WITHIN SAMPLES

The variance between machines that we have just calculated does not tell the whole story. For example, consider the data of Table 10-3, which has the same $s_{\overline{X}}^2$ as Table 10-1, yet more erratic machines that produce large chance fluctuations within each column. The implications of this are shown in Figure 10-1. In panel (b), the machines are so erratic that all samples could be drawn from the same population. That is, the differences in sample means may be explained by chance. On the other hand, the same differences in sample means can hardly be explained by chance in panel (a), because the machines in this case are not so erratic.

We now have our standard of comparison. In panel (a) we conclude that the μ's are different—reject H_0—because the variance in sample means ($s_{\overline{X}}^2$) is large relative to the chance fluctuation.

How can we measure this chance fluctuation? Intuitively, it seems to be the spread (or variance) of observed values within each sample. Thus we compute the squared deviations within the first sample in Table 10-1:

$$\Sigma (X_{1t} - \overline{X}_1)^2 = (47 - 49)^2 + (53 - 49)^2 + \cdots = 30$$

Similarly, we compute the squared deviations within the second and third samples, and add them all up. Then we divide by the total d.f. in all three samples ($n - 1 = 4$ in each). We thus obtain the pooled variance s_p^2 (just as we did for the two-sample case in Chapter 8):

$$s_p^2 = \frac{30 + 50 + 14}{4 + 4 + 4} = \frac{94}{12} = 7.83 \tag{10-4}$$

TABLE **10-3** **Sample Outputs of Three Erratic Machines**

MACHINE 1	MACHINE 2	MACHINE 3
50	48	57
42	57	59
53	65	48
45	59	46
55	51	45
$\overline{X}_1 = 49$	$\overline{X}_2 = 56$	$\overline{X}_3 = 51$

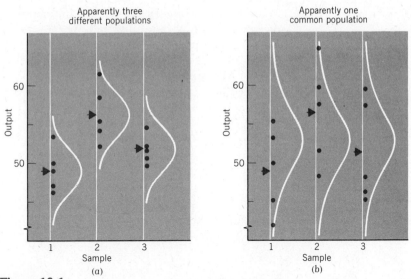

Figure 10-1
(a) Outputs of 3 relatively predictable machines (data from Table 10-1) (b) Outputs of 3 erratic machines (data from Table 10-3). They have the same sample means $\overline{X}$, hence the same $s_{\overline{X}}^2$, as the machines in (a).

The generalization is easy to see. When there are c columns of data, each having n observations, then:

$$s_p^2 = \frac{\Sigma(X_{1t} - \overline{X}_1)^2 + \Sigma(X_{2t} - \overline{X}_2)^2 + \cdots}{c(n - 1)}$$

$$\text{(10-5)}$$
$$\text{like (8-21)}$$

D—THE F TEST

The key question now can be stated. Is $s_{\overline{X}}^2$ large relative to s_p^2? That is, is the ratio $s_{\overline{X}}^2/s_p^2$ large? It is customary to examine a slightly modified ratio, called F in honor of the renowned English statistician Sir Ronald Fisher (1890–1962):

$$\boxed{F = \frac{ns_{\overline{X}}^2}{s_p^2}} \qquad \text{(10-6)}$$

Here n has been introduced into the numerator to make it equal on average to the denominator (when H_0 is true)—that is, to make the F ratio fluctuate[1] around 1.

---

[1]Why does F fluctuate around 1? The three samples are drawn from three populations, and when H_0 is true, these populations are identical. We could then regard the three samples as all coming from this one common population. And so there would be two alternative ways of estimating its variance σ^2:

(Con't)

If H_0 is not true (and the μ's are not the same), then $ns_{\bar{X}}^2$ will be relatively large compared to s_p^2, and the F ratio in (10-6) will tend to be much greater than 1. The larger is F, therefore, the less credible is the null hypothesis.

To measure the credibility of H_0 numerically, we find its p-value—in this case, the probability in the tail of the F distribution beyond the observed value. We get this p-value from Appendix Table VI, which lists the critical values of the F distribution when H_0 is true (just like Table V lists the critical values of t). In Table VI, the F distribution depends on the degrees of freedom in the numerator variance $(c - 1)$, and in the denominator variance $[c(n - 1)]$. This is written briefly as:

$$\text{d.f.} = (c - 1) \text{ and } c(n - 1) \qquad (10\text{-}7)$$

An example is the easiest way to demonstrate the actual calculation of the p-value.

EXAMPLE **10-1**

For the data in Table 10-1, we have already calculated how much variance there is *between* the 3 sample means:

$$s_{\bar{X}}^2 = 13 \qquad \qquad (10\text{-}2) \text{ repeated}$$

and how much residual variance there is *within* the 3 samples (of 5 observations each):

$$s_p^2 = 7.83 \qquad \qquad (10\text{-}4) \text{ repeated}$$

a. Calculate the F ratio.
b. Calculate the degrees of freedom for F.
c. Find the p-value for H_0 (no difference in population means).

---

1. Average the variance within each of the three samples. This is s_p^2 in the denominator of (10-6).
2. Or, infer σ^2 from $s_{\bar{X}}^2$, the observed variance between the sample means. Recall how the variance of the sample means ($\sigma_{\bar{X}}^2$, which we have so far been referring to as SE^2) is related to the variance of the population (σ^2) as follows:

$$\sigma_{\bar{X}}^2 = \frac{\sigma^2}{n} \qquad \qquad \text{like (6-6)}$$

Thus:

$$\sigma^2 = n\sigma_{\bar{X}}^2$$

This suggests estimating σ^2 with $ns_{\bar{X}}^2$, which is recognized as the numerator of (10-6).

If H_0 is true, we could estimate σ^2 by either of these methods; since the two will be about equal, their ratio F will fluctuate around 1.

But if H_0 is not true, then the numerator in (10-6) will blow up because the difference in population means will result in a large spread in the sample means (large $s_{\bar{X}}^2$). At the same time, the denominator s_p^2 will still estimate σ^2. Then the F ratio will be large.

Solution   **a.**

$$F = \frac{ns_{\bar{X}}^2}{s_p^2} \qquad \text{(10-6) repeated}$$

$$= \frac{5(13)}{7.83} = 8.3$$

b.
$$\begin{aligned} \text{d.f.} &= (c - 1) \text{ and } c(n - 1) \qquad \text{(10-7) repeated} \\ &= (3 - 1) \text{ and } 3(5 - 1) \\ &= 2 \text{ and } 12 \qquad\qquad\qquad\qquad \text{(10-8)} \end{aligned}$$

c. We look up Table VI where d.f. = 2 and 12, and find five critical values listed in a column that we scan down—till we find that the observed F value of 8.3 lies beyond $F_{.01} = 6.93$. As Figure 10-2 shows, we conclude that:

$$\text{p-value} < .01 \qquad\qquad \text{(10-9)}$$

This means that if H_0 were true, there is less than a 1% chance of getting sample means that differ so much. Accordingly we reject H_0 and conclude that the 3 machines in Table 10-1 are different.

Let us see what the F test shows for the other cases in Tables 10-2 and 10-3.

EXAMPLE **10-2**   **a.** Calculate the p-value for H_0 using the data in Table 10-2, which showed 3 samples from the same machine. In that case $s_{\bar{X}}^2$ was only 1.0, while s_p^2 was 7.83.

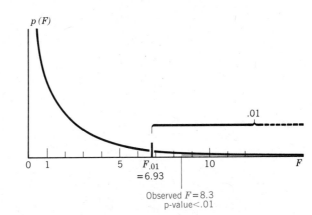

Figure 10-2
The p-value found using F in Table VI (compare to Figure 9-3).

b. Calculate the p-value for H_0, using the data in Table 10-3, which showed 3 erratic machines. Thus s_p^2 took on the large value 39.0, while $s_{\bar{X}}^2$ was 13.0.

Solution   **a.**

$$F = \frac{5(1)}{7.83} = .64 \tag{10-10}$$

Using d.f. = 2 and 12 again, we find the observed F value of .64 falls far short of the first critical value $F_{.25} = 1.56$. Therefore,

$$\text{p-value} \geqslant .25$$

Since the p-value is much greater than .25, H_0 is very credible. This is the correct conclusion, of course, since we generated these 3 samples in Table 10-2 from the same machine.

b.

$$F = \frac{5(13)}{39} = 1.67 \tag{10-11}$$

Using d.f. = 2 and 12 still, we find the observed F value of 1.67 lies just beyond $F_{.25} = 1.56$. Therefore:

$$\text{p-value is just less than .25}$$

This credibility level for H_0 is sufficiently high that we do not reject H_0. The large difference in sample means may well have occurred because each machine is erratic, not because of a difference in machines.

Of course, the F distribution shown in Figure 10-2 is only one of many. As Figure 10-3 illustrates, there is a different F distribution for every

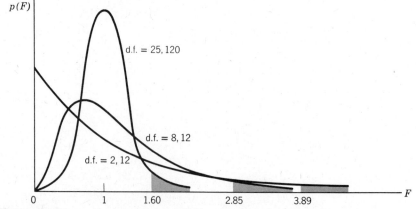

Figure 10-3
Some typical F distributions, with various d.f. in the numerator and denominator. Note how the 5% critical point (beyond which H_0 is customarily rejected) moves left toward 1 as d.f. increase.

...nbination of d.f. in the numerator and denominator (like the different ...stributions in Figure 9-2).

HE ANOVA TABLE

...alculations we have done so far can be followed more easily if we ...arize them in a table of standard form, called the ANOVA table. As ...10-4 shows, it is mostly a bookkeeping arrangement, with the first ...owing calculations of the numerator of the F ratio, and the second ...denominator. Panel (a) gives the ANOVA Table in general sym- ...while panel (b) gives the numerical results for the data in Table ...nally, panel (c) gives the same numerical results computed on ...B.

Table, General

...IATION ...QUARES, SS)	d.f.	VARIANCE (MEAN SQUARE, MS)	F RATIO
$\sum_{1}^{c} (\overline{X}_i - \overline{\overline{X}})^2$	$(c - 1)$	$MS_{cols} = SS_{cols}/(c-1)$ $= ns_{\overline{X}}^2$	$F = \dfrac{MS_{cols}}{MS_{res}}$ $= \dfrac{ns_{\overline{X}}^2}{s_p^2}$
$\sum_{t=1} (X_{it} - \overline{X}_i)^2$	$c(n - 1)$	$MS_{res} = SS_{res}/c(n - 1)$ $= s_p^2$	
Total $SS_{tot} = \sum_{i=1}^{c} \sum_{t=1}^{n} (X_{it} - \overline{\overline{X}})^2$	$nc - 1$		

TABLE **10-4b** **ANOVA Table, for Observations Given in Table 10-1**

SOURCE	SS	d.f.	MS	F RATIO	p-VALUE
Between machines	130	2	65	$\dfrac{65}{7.83} = 8.3$	$p < .01$
Residual	94	12	7.83		
Total	224√	14√			

TABLE **10-4c** **Corresponding Computer Output**

```
ANALYSIS OF VARIANCE
SOURCE     DF       SS        MS        F
FACTOR      2    130.00    65.00     8.30
ERROR      12     94.00     7.83
TOTAL      14    224.00
```

In addition, the ANOVA table provides a handy check on our calculations. For example, in Table 10-4b consider the sums of squares (SS, also called *variations*) in the second column. At the bottom we list the total SS, which is obtained by ignoring the column-by-column structure of the data. That is, we just take all 15 numbers within Table 10-1 and calculate how much each deviates from the overall mean $\overline{\overline{X}}$. All the resulting deviations $(X - \overline{\overline{X}})$ are then squared and summed, to produce the total variation SS_{tot}. To indicate that we must sum over the whole table, both rows and columns, we write the Σ sign twice:

$$SS_{tot} = \Sigma\Sigma(X - \overline{\overline{X}})^2 \qquad (10\text{-}12)$$
$$= (47 - 52)^2 + (53 - 52)^2 + \cdots + (49 - 52)^2$$
$$= 25 + 1 + \cdots + 9 = 224 \qquad (10\text{-}13)$$

Then we note that the variations (SS) in the second column add up to this total; that is:

$$
\begin{array}{ccccc}
SS_{tot} & = & SS_{cols} & + & SS_{res} \\
\text{Total} & = & \text{Between-column} & + & \text{Residual} \\
\text{variation} & & \text{variation} & & \text{variation}
\end{array}
\qquad (10\text{-}14)
$$

This breakdown of the total variation is so important that it is proved in general in Appendix 10-1. In fact, this is what the name ANOVA is all about: ANalysis Of VAriation.

Just as the SS in the second column of Table 10-4b add up, so do the d.f. in the next column, which we indicate with a check ($\checkmark$).

When each SS is divided by its appropriate d.f., we obtain the Mean Square (MS, also called the *variance*) in Table 10-4. The variance between columns is *explained* by the fact that the columns in Table 10-1 may come from different parent populations (machines that perform differently). The residual variance within columns is *unexplained* because it is the random or chance variation that cannot be systematically explained (by differences in machines). Thus F is sometimes referred to as the variance ratio:

$$F = \frac{\text{Explained variance}}{\text{Unexplained variance}} \qquad (10\text{-}15)$$

This suggests a possible means of strengthening the F test. Suppose, for example, that these three machines are sensitive to differences in the men operating them. Why not introduce this human factor explicitly into the analysis? If some of the previously unexplained variation could thus be explained by differences in operator, the denominator of (10-15) would be reduced. With the larger F value that would result, we would have a more powerful test of the machines (i.e., we would be in a stronger position to reject H_0). Thus, our ability to detect whether one factor (machine) is important would be strengthened by introducing another factor (operator)

TABLE **10-5** **Modification of ANOVA Table 10-4 for Unequal Sample Sizes** $n_1, n_2, \ldots, n_i, \ldots$

SOURCE OF VARIATION	VARIATION (SUM OF SQUARES, SS)	d.f.	VARIANCE (MEAN SQUARE, MS)	F RATIO
Between columns (machines), due to differences between column means $\overline{X}_i$	$SS_{cols} = \sum\limits_{i=1}^{c} n_i(\overline{X}_i - \overline{\overline{X}})^2$	$(c - 1)$	$MS_{cols} = SS_{cols}/(c - 1)$ $= ns_{\overline{X}}^2$	$F = \dfrac{MS_{cols}}{MS_{res}}$ $= \dfrac{ns_{\overline{X}}^2}{s_p^2}$
Residual, due to differences between observations X_{it} and column means $\overline{X}_i$	$SS_{res} = \sum\limits_{i=1}^{c} \sum\limits_{t=1}^{n_i} (X_{it} - \overline{X}_i)^2$	$\sum\limits_{i=1}^{c} (n_i - 1)$	$MS_{res} = SS_{res}/\Sigma(n_i - 1)$ $= s_p^2$	
Total	$SS_{tot} = \sum\limits_{i=1}^{c} \sum\limits_{t=1}^{n_i} (X_{it} - \overline{\overline{X}})^2$	$\sum\limits_{i=1}^{c} n_i - 1$		

Where $\overline{\overline{X}} \equiv$ the grand average of all the X_{it}:

$$\overline{\overline{X}} \equiv \frac{\Sigma\Sigma X_{it}}{\Sigma n_i} = \frac{\Sigma n_i \overline{X}_i}{\Sigma n_i}$$

to help explain variance. In fact, we shall carry out precisely this calculation in the next section, two-factor ANOVA.

*F—UNEQUAL SAMPLE SIZES

The most efficient way to collect observations is to make all samples the same size n. However, when this is not feasible, it still is possible to modify the ANOVA calculations. Table 10-5 provides the necessary modifications of Table 10-4 to take into account different sample sizes n_1, n_2, and so on. Note especially that the definition of $\overline{\overline{X}}$ is no longer the simple average of $\overline{X}_i$, but rather a *weighted* average with weights n_i.

PROBLEMS

For Problems 10-1 to 10-3, calculate the ANOVA table, including the p-value for the null hypothesis.

10-1 Twelve plots of land are divided randomly into 3 groups. Fertilizers A and B are applied to the first two groups, while the third group is a control C with no fertilizer. The yields were as follows:

A	B	C
75	74	60
70	78	64
66	72	65
69	68	55

10-2   In a large American university in 1969, the male and female professors were sampled independently, yielding the following annual salaries (in thousands of dollars, rounded. From Katz, 1974):

MEN	WOMEN
12	9
11	12
19	8
16	10
22	16

10-3   Suppose a sample of American homeowners reported the following home values (in thousands of dollars) by region:

REGION	POPULATION	SAMPLE OF HOME VALUES					$\overline{X}$	$\Sigma(X - \overline{X})^2$
Northeast	19,000,000	69	54	36	45	81	57	1314
Northcentral	22,000,000	51	96	36	39	33	51	2718
South	24,000,000	39	30	48	36	57	42	450
West	15,000,000	39	114	51	75	51	66	3564

10-4   In 1977, 50 full-time working women were sampled from each of three educational categories, and their incomes (in thousands of dollars, annually) were reported as follows:

YEARS OF SCHOOL COMPLETED	$\overline{X}$	$\Sigma(X - \overline{X})^2$
Elementary school (8 years)	7.8	1835
High school (12 years)	9.7	2442
College (16 years)	14.0	4707

   a.   Calculate the ANOVA table.

   b.   Is it fair to say that completing high school and college nearly doubles income? Why?

*10-5   **a.**   (Monte Carlo) Simulate 5 observations from a normal population with $\mu = 50$, $\sigma = 10$, and call it sample A. Then simulate a second and third sample (B and C) from the same population. From the array of 15 observations, evaluate the ANOVA table to test whether these three samples are from the same population.

   b.   Have the instructor record from every student the F statistic found in part **a.** Graph the resulting distribution of F. Does the graph resemble Figure 10-2? What proportion of the F values exceed $F_{.05} = 3.89$? What proportion would exceed this value if the class were very large?

*10-6   A sample of 11 American families in 1971 reported the following annual incomes (in thousands of dollars) by region.

NORTHEAST	NORTH CENTRAL	SOUTH	WEST
8	13	7	7
14	9	14	7
		8	16
		7	

Construct the ANOVA table, including the p-value for H_0.

*10-7 When all n_i are equal to a common value n, verify that the ANOVA Table 10-5 reduces to the ANOVA Table 10-4. Specifically, show that $\overline{\overline{X}}$, SS_{cols}, residual d.f., and total d.f. all reduce to the simpler form of Table 10-4.

10-2 Two-factor Analysis of Variance

A—THE ANOVA TABLE

We have suggested that the F test on the differences in machines given in (10-15) would be strengthened if the unexplained variance could be reduced by taking other factors into account. Suppose, for example, that the sample outputs given in Table 10-1 were produced by five different operators—with each operator producing one of the sample observations on each machine. This data, reorganized according to a two-way classification (by machine *and* operator), is shown in Table 10-6. It is necessary to complicate our notation somewhat. We now are interested in the average of each operator (each row average $\overline{X}_{\cdot j}$) as well as the average of each machine (each column average $\overline{X}_{i\cdot}$).[2]

Now the picture is clarified; some operators are efficient (the first and fourth), and some are not. The machines are not that erratic, after all; there is just a wide difference in the efficiency of operators. We shall now show that an explicit adjustment for this will, as we hoped earlier, reduce the unexplained or chance variance in the denominator of (10-15). Since the numerator will remain unchanged, the F ratio will be larger as a consequence, with the final question being: Will this allow us to reject H_0? To sum up, it appears that another factor (difference in operators) was responsible for a lot of extraneous noise[3] in our simple one-way

---

[2]The dot indicates the subscript over which summation occurs. For example:

$$\overline{X}_{i\cdot} = \frac{\sum\limits_{j=1}^{r} X_{ij}}{r}$$

[3]As another example of noise removal, suppose three varieties of seed are to be investigated, each used on five plots of land. A naive procedure is to just take 15 plots of land, and randomly select five for seed A, five for seed B, and five for seed C. A better procedure is to find five blocks of land that can each be split into three plots, one for each seed variety. Then two-factor ANOVA can be used, the factors being seed and block (analogous to machine and operator in Table 11-6). This experimental design is called *blocking*—not only in agricultural applications, but more generally as well.

analysis in the previous section. By removing this noise, we hope to get a much more powerful test of the machines.

The analysis begins with calculating the familiar column (machine) variation in Table 10-6. Similarly we calculate the row (operator) variation.

The residual calculation is the trickiest feature. In one-factor ANOVA we calculated residual variation by looking at the spread of n observed values within a category—that is, within a whole column in Table 10-1. But in the two-factor Table 10-6, we have split our observations row-wise, as well as column-wise; this leaves us with only one observation within each category. Thus, for example, there is only a single observation (61) of how much output is produced on machine 2 by operator 1. Variation no longer can be computed within that cell. What should we do?

Well, if there were no random residual, how would we predict the output on machine 2 by operator 1, for example? The second machine is relatively good (from the next-to-lowest margin in Table 10-6, we see it is 4 units above average). And the first operator is also relatively good (from the second-last margin on the right, we see she is 3 units above average). Adding these two components to the overall grand mean $\overline{\overline{X}} = 52$ provides the following prediction of her output on this machine (a predicted value is indicated with a hat):

$$\text{Predicted } \hat{X}_{21} = 52 + 4 + 3 = 59 \qquad (10\text{-}16)$$

Comparing this to the observed $X_{21} = 61$, we calculate the residual to be $61 - 59 = 2$.

TABLE **10-6** **Sample of Production X_{ij} of Three Different Machines (as given in Table 10-1) but Now Arranged According to Operator**

GIVEN DATA, WITH AVERAGES CALCULATED

OPERATOR	MACHINE $i = 1$	2	3	Operator Means $\overline{X}_{\cdot j}$
$j = 1$	53	61	51	55
2	47	55	51	51
3	46	52	49	49
4	50	58	54	54
5	49	54	50	51
Machine Means $\overline{X}_{i\cdot}$	49	56	51	$\overline{\overline{X}} = 52$

OPERATOR VARIATION (ROWS)

$(\overline{X}_{\cdot j} - \overline{\overline{X}})$	$(\overline{X}_{\cdot j} - \overline{\overline{X}})^2$
3	9
-1	1
-3	9
2	4
-1	1
$0\checkmark$	24×3 $= 72$

Number of columns c

MACHINE VARIATION (COLUMNS)

$(\overline{X}_{i\cdot} - \overline{\overline{X}})$	-3	4	-1	$0\checkmark$
$(\overline{X}_{i\cdot} - \overline{\overline{X}})^2$	9	16	1	26×5 $= 130$

Number of rows r

In general, the predicted value $\hat{X}_{ij}$ is similarly:

$$\hat{X}_{ij} = \overline{\overline{X}} + (\overline{X}_{i\cdot} - \overline{\overline{X}}) \qquad (\overline{X}_{\cdot j} - \overline{\overline{X}})$$

$$= \text{overall mean} + \text{machine adjustment} + \text{operator adjustment}$$

$$(10\text{-}17)$$

And the residual is calculated by comparing this predicted value with the observed value X_{ij}:

$$\text{Residual} = X_{ij} - \hat{X}_{ij}$$

In Table 10-7 the residual is calculated for all cells; then it is squared and summed as usual, to find the residual variation of 22.

The three variations—row, column, and residual—are then summarized in Table 10-8. Panel (a) is general, and panel (b) is specific to the data of Table 10-6.

Note that the unexplained variation (94) earlier in Table 10-4b has been broken down in Table 10-8b into a major component that is explained by operators (72) and what has become a minor unexplained variation (22). Thus all the component sources of variation still must sum to the total variation in the last line, just as they did earlier in one-way ANOVA. (Appendix 10-2 gives the formal proof):

$$\boxed{SS_{tot} = SS_{cols} + SS_{rows} + SS_{res}} \qquad (10\text{-}18)$$

$$\text{like } (10\text{-}14)$$

B—THE F TEST

Now that we have broken the total variation down into its components, we can test whether there is a discernible difference in machines. We also

TABLE **10-7** **Calculation of Residuals from Table 10-6**

	FITTED VALUES $\hat{X}_{ij} = \overline{\overline{X}} + (\overline{X}_{i\cdot} - \overline{\overline{X}}) + (\overline{X}_{\cdot j} - \overline{\overline{X}})$				RESIDUALS $X_{ij} - \hat{X}_{ij}$		
		i				i	
j	1	2	3	j	1	2	3
1	52	59	54	1	1	2	−3
2	48	55	50	2	−1	0	1
3	46	53	48	3	0	−1	1
4	51	58	53	4	−1	0	1
5	48	55	50	5	1	−1	0

Sum of squares $= 1^2 + 2^2 + \cdots + 0^2$

$= 22$

TABLE **10-8a**   **Two-Way ANOVA, General**

SOURCE	VARIATION (SUM OF SQUARES, SS)	d.f.	VARIANCE (MEAN SQUARE, MS)	F
Between columns (machines), due to differences between column means $\overline{X}_{i\cdot}$	$SS_{cols} = r \sum_{i=1}^{c} (\overline{X}_{i\cdot} - \overline{\overline{X}})^2$	$c - 1$	$MS_{cols} = \dfrac{SS_{cols}}{c - 1}$	$\dfrac{MS_{cols}}{MS_{res}}$
Between rows (operators), due to differences between row means $\overline{X}_{\cdot j}$	$SS_{rows} = c \sum_{j=1}^{r} (\overline{X}_{\cdot j} - \overline{\overline{X}})^2$	$r - 1$	$MS_{rows} = \dfrac{SS_{rows}}{r - 1}$	$\dfrac{MS_{rows}}{MS_{res}}$
Residuals due to differences between the actual observations X_{ij} and the fitted values $\hat{X}_{ij} = \overline{X}_{i\cdot} + \overline{X}_{\cdot j} - \overline{\overline{X}}$	$SS_{res} =$ $\sum_{i=1}^{c} \sum_{j=1}^{r} (X_{ij} - \overline{X}_{i\cdot} - \overline{X}_{\cdot j} + \overline{\overline{X}})^2$	$(c - 1)$ $\times (r - 1)$	$MS_{res} = \dfrac{SS_{res}}{(c - 1)(r - 1)}$	
Total	$SS_{tot} = \sum_{i=1}^{c} \sum_{j=1}^{r} (X_{ij} - \overline{\overline{X}})^2$	$cr - 1$		

TABLE **10-8b**   **Two-Way ANOVA, for Observations Given in Table 10-6**

SOURCE	SS	d.f.	MS	F	p-VALUE
Between machines	130	2	65	23.6	$p < .001$
Between operators	72	4	18	6.5	$p < .05$
Residual (unexplained)	22	8	2.75		
Total	224$\checkmark$	14$\checkmark$			

can test whether there is a discernible difference in operators. In either test, the extraneous influence of the other factor will be taken into account.

On the one hand, we test for differences in machines by constructing the F ratio:

$$F = \frac{\text{variance explained by machines}}{\text{unexplained variance}}$$

(10-19)
like (10-15)

Specifically, from Table 10-8b, we have:

$$F = \frac{65}{2.75} = 23.6$$

If H_0 is true, this has an F distribution. We therefore look up Table VI, with d.f. = 2 and 8, and find the observed F value of 23.6 lies beyond $F_{.001} = 18.5$. Therefore:

$$p\text{-value} < .001$$

This is stronger evidence against H_0 than we got from one-factor ANOVA in (10-9). The numerator has remained unchanged, but the chance variation in the denominator is much smaller, since the effect of the operators' differences has been netted out. Thus, our statistical leverage on H_0 has been increased.

Similarly, we can get a powerful test of the null hypothesis that the operators perform equally well. Once again, F is the ratio of explained to unexplained variance; but this time, of course, the numerator is the variance between operators. Thus:

$$F = \frac{\text{variance explained by operators}}{\text{unexplained variance}} = \frac{18}{2.75} = 6.5 \qquad (10\text{-}20)$$

We look up Table VI, with d.f. = 4 and 8, and find the observed F value of 6.5 lies beyond $F_{.05} = 3.84$. Therefore:

$$\text{p-value} < .05$$

Accordingly, we could conclude that at the 5% level, there is a discernible (significant) difference in operators.

C—INTERACTION

So far we have assumed the simple *additive model*, where the effects of the two factors are simply added together as in (10-16). That is, we have assumed there is *no interaction* between the two factors—as would occur, for example, if certain operators did unusually well on certain machines. Such interaction would require several observations per cell and a more complex model.

The problem of interaction also occurs in regression. To appreciate it in both contexts we defer it until Chapter 14 (or see Wonnacott, 1984 *IM*).

PROBLEMS

10-8 To refine the experimental design of Problem 10-1, suppose that the 12 plots of land are on 4 blocks (3 plots on each). Since you suspect that there may be a difference in fertility between blocks, you retabulate the data in Problem 10-1, according to fertilizer *and* block, as follows:

| | FERTILIZER | | |
BLOCK	A	B	C
1	69	72	60
2	75	74	64
3	70	78	65
4	66	68	55

Incidentally, this is a classic experimental design called the *randomized block* design.

Calculate the ANOVA table, including the p-value for the two null hypotheses.

10-9 Three men work on an identical task of packing boxes. The number of boxes packed by each in three selected hours is shown in the table below.

HOUR	MAN A	B	C
11–12 A.M.	24	19	20
1–2 P.M.	23	17	14
4–5 P.M.	25	21	17

Calculate the ANOVA table, including the p-value for the two null hypotheses.

10-10 Three methods of estimating inventory value were used at the end of each month, for a six-month period, thus producing the following 18 estimates:

METHOD	J	F	M	A	M	J	AVERAGES
Method 1	14	12	16	15	10	11	13
Method 2	18	16	17	19	13	13	16
Method 3	16	14	12	14	13	9	13
Averages	16	14	15	16	12	11	14

Part of the ANOVA table is given below. Finish it.

SOURCE	VARIATION
Between months	66
Between methods	36
Residual	
Total	124

*10-11 Any number of factors can be studied by ANOVA, not just one or two. For example, suppose that Problem 10-8 was extended to also investigate two different seed varieties—X and Y. The data would then start out like this:

BLOCK 1	SEED	FERTILIZER A	B	C
	X	69	72	60
	Y	73	78	65

BLOCK 2	SEED	FERTILIZER A	B	C
	X	75	74	64
	Y	80	77	68

⋮

BLOCK 4

If we assume as usual the simple additive model, the ANOVA table would then start out like this:

SOURCE	VARIATION	d.f.	VARIANCE	F	p-VALUE
Fertilizers	608	2			
Seeds	183	1			
Blocks	261	3			
Residual					
Total	1152				

Complete the ANOVA table.

10-3 Confidence Intervals

A—A SINGLE CONFIDENCE INTERVAL

Thus far, we have concentrated on testing the null hypothesis of no difference between means. Often it is even more useful to find a confidence interval for the difference by using the very familiar formula:

$$(\mu_1 - \mu_2) = (\overline{X}_1 - \overline{X}_2) \pm t_{.025}s\sqrt{\frac{1}{n_1} + \frac{1}{n_2}} \qquad \text{(10-21)} \\ \text{like (8-20)}$$

where

μ_1, μ_2 = the two population means to be compared;

$\overline{X}_1, \overline{X}_2$ = the two corresponding sample means (for example, the first two column means);

n_1, n_2 = the number of observations over which we average to get $\overline{X}_1$ and $\overline{X}_2$;

s^2 = the residual variance in the ANOVA table (also called MS_{res});

$t_{.025}$ = the d.f. associated with s^2 in the ANOVA table.

This is just like the original formula (8-20), except for the more reliable value of s that contains as much information (d.f.) as possible about the unknown σ.

Although we used subscripts 1 and 2, this formula works equally well, of course, to compare *any* specific pair of means. An example will illustrate.

EXAMPLE **10-3**   Construct a 95% confidence interval for the difference between the second and third machines in Table 10-1 (whose ANOVA calculations were given in Table 10-4b).

Solution   We substitute the appropriate quantities into (10-21). We are careful to use $s^2 = 7.83$ from the ANOVA Table 10-4b, and its corresponding d.f. = 12 for the $t_{.025}$ value. Thus:

$$(\mu_2 - \mu_3) = (56 - 51) \pm 2.18 \sqrt{7.83} \sqrt{\frac{1}{5} + \frac{1}{5}}$$

$$= 5.0 \pm 3.9 \approx 5 \pm 4 \tag{10-22}$$

At the 95% level of confidence, we can therefore conclude that the second machine is from 1 to 9 units more productive than the third machine.

B—SIMULTANEOUS CONFIDENCE INTERVALS

In Example 10-3 there are two other contrasts we could make—between μ_1 and μ_2, and between μ_1 and μ_3. Then all three confidence intervals would be:

$$\left.\begin{array}{l} \mu_1 - \mu_2 = -7.0 \pm 3.9 \\ \mu_1 - \mu_3 = -2.0 \pm 3.9 \\ \mu_2 - \mu_3 = +5.0 \pm 3.9 \end{array}\right\} \tag{10-23}$$

Since *each* of these statements has a 5% chance of being wrong, however, the chance that the *whole system* will somewhere be wrong is nearly 15%. If we want full 95% confidence that this whole system of statements is correct, we need to increase the confidence of each statement. The easiest way to do this (Scheffé, 1959) is to use a broader confidence allowance. We replace $t_{.025}$ in (10-21) with the larger allowance:

$$\boxed{\sqrt{(k - 1)F_{.05}}} \tag{10-24}$$

where

$k =$ the number of means to be compared, for example, the number of columns c. (Alternatively, if we wish, k may refer to the number of rows r in two-factor ANOVA.)

$F_{.05} =$ the critical value of F that leaves 5% in the upper tail. Its d.f. are given in the ANOVA table—the same d.f. we used in looking up F to find the p-value.

When we substitute (10-24), then (10-21) becomes:

With 95% confidence, all the following intervals are simultaneously true:

$$(\mu_1 - \mu_2) = (\bar{X}_1 - \bar{X}_2) \pm \sqrt{(k-1)F_{.05}}\, s\sqrt{\frac{1}{n_1} + \frac{1}{n_2}}$$

$$(\mu_1 - \mu_3) = (\bar{X}_1 - \bar{X}_3) \pm \sqrt{(k-1)F_{.05}}\, s\sqrt{\frac{1}{n_1} + \frac{1}{n_3}}$$

$$\cdot$$
$$\cdot$$
$$\cdot$$

$$(\mu_{k-1} - \mu_k) = (\bar{X}_{k-1} - \bar{X}_k) \pm \sqrt{(k-1)F_{.05}}\, s\sqrt{\frac{1}{n_{k-1}} + \frac{1}{n_k}} \qquad (10\text{-}25)$$

F ratio

MS of residual

An example will illustrate.

EXAMPLE **10-4** Construct simultaneous 95% confidence intervals for *all* the differences between machines in Table 10-1 (whose ANOVA calculations were given in Table 10-4*b*).

Solution The $t_{.05}$ value used in Example 10-3 is now replaced with $\sqrt{(k-1)F_{.05}}$. Here k = the number of machines to be compared ($c = 3$), and the d.f. for $F_{.05}$ are 2 and 12 as already calculated in the ANOVA Table 10-4*b*. Thus from Table VI, $F_{.05} = 3.89$, and so (10-25) becomes:

$$\mu_1 - \mu_2 = (49 - 56) \pm \sqrt{(3-1)3.89}\,\sqrt{7.83}\,\sqrt{\frac{1}{5} + \frac{1}{5}}$$

That is,

Similarly,
$$\left.\begin{array}{l} \mu_1 - \mu_2 = -7.0 \pm 4.9 \\ \mu_1 - \mu_3 = -2.0 \pm 4.9 \\ \mu_2 - \mu_3 = +5.0 \pm 4.9 \end{array}\right\} \qquad (10\text{-}26)$$

These allowances in (10-26) are vaguer than they were in (10-23)—the price we pay for *simultaneous* confidence in the whole set of statements.

PROBLEMS

10-12 Construct 95% simultaneous confidence intervals for each of Problems 10-1, 10-2, 10-8, and 10-9.

10-13   An executive moving from Hartford to Atlanta was interested in comparing home values in the South with the Northeast. Using the data in Problem 10-3, construct the appropriate confidence interval:

   a.   That has 95% confidence, by itself.

   b.   That can be included with all other comparisons, at the 95% simultaneous confidence level.

10-14   A sociologist sampled 25 incomes from each of 8 groups, and then tested H_0 with a one-factor ANOVA F test. He then proposed to compare every possible pair using the 95% confidence interval (10-21) based on t. (The value of s_p^2 was obtained by pooling all 8 groups.)

   a.   How many such pairs are there to be compared?

   b.   Then what is the expected number that will be wrong?

   c.   If he wanted 95% certainty that *all* these confidence intervals will be right, by how much should he increase their width?

CHAPTER 10 SUMMARY

10-1   When more than two populations are to be compared, we need an extension of the two-sample t test. This is provided by the F test, which compares the variance explained by differences *between* the sample means, with the unexplained variance *within* the samples. The ANOVA table provides an orderly way to calculate F, step by step, to test whether the factor is statistically discernible.

10-2   A second factor can be introduced—not only for its own intrinsic interest, but also to sharpen the analysis of the first factor. For this two-factor ANOVA, we consider just the simplest model (the additive model—leaving the more complex interactive model to Chapter 14).

10-3   ANOVA can be used to construct confidence intervals as well as test hypotheses. To compare all possible differences among a set of means, we simply broaden the ordinary confidence interval (using $F_{.05}$ instead of $t_{.025}$). This gives us 95% simultaneous confidence that *all* comparisons are true.

REVIEW PROBLEMS

10-15   To compare 3 varieties of potatoes, an experiment was conducted by assigning each variety at random to 3 equal-size plots at each of 3 different soil types. The following yields, in bushels per plot, were recorded:

	VARIETY OF POTATO		
SOIL	A	B	C
Sand	21	20	16
Clay	16	18	11
Loam	23	31	24

a. Construct the ANOVA table.

b. Calculate the 95% simultaneous confidence intervals for the differences in the 3 varieties.

*c. The botanist who developed variety B remarked that he had worked 10 years to find something that grew well in a loam soil. As you glance at the data, do you think he succeeded? In the light of this information, what would you say about your analysis in parts **a** and **b**?

10-16 In 1980, a Gallup poll asked Americans to give their opinion of their country on a numerical scale from $+5$ (very favorable) to -5 (very unfavorable). The following table gives the relative frequency of replies for the four regions of the country:

REPLY	RELATIVE FREQUENCY OF REPLY (%)			
	EAST	MIDWEST	SOUTH	WEST
$+5$ (favorable)	67%	63%	77%	54%
$+4$	17	17	11	17
$+3$	8	11	5	16
$+2$	3	4	3	5
$+1$	3	2	2	4
0 (don't know)	1	1	2	2
-1	0	1	0	0
-2	0	0	0	0
-3	0	0	0	1
-4	0	1	0	0
-5 (unfavorable)	1	0	0	1
Total	100%	100%	100%	100%
Sample size (approx)	400	400	400	400
Mean	4.31	4.21	4.52	3.92
Standard deviation	1.42	1.45	1.08	1.70
Sum of squared deviations	806	834	468	1149

9. 1379

a. Verify the 3 summary statistics at the bottom of the first column (East region).

b. Construct the ANOVA table.

c. What is the null hypothesis, in words? How credible is it? Are the differences between regions statistically discernible at the 5% level?

d. What region has the most favorable view of the country? Which has the least favorable? How big is the difference?

e. Construct 95% simultaneous confidence intervals for all differences.

REVIEW PROBLEMS, CHAPTERS 6 to 10

10-17 The admissions record for a large Graduate School in 1975 was as follows (same data as Problem 1-8):

	MEN		WOMEN	
FACULTY	NUMBER OF APPLICANTS	NUMBER ADMITTED	NUMBER OF APPLICANTS	NUMBER ADMITTED
Arts	2300	700	3200	900
Science	6000	3000	1100	600
Total	8300	3700	4300	1500

a. Construct a 95% confidence interval for the difference in admission rates between men and women for:

  i. Arts

  ii. Science

  iii. the whole school

b. What is the underlying population being discussed in part **a?**

c. In words, summarize the evidence of sex discrimination.

10-18 Answer True or False; if False, correct it:

a. Problem 10-17 illustrates a very important principle: Confidence intervals are a useful way to express the uncertainty of an estimate. Yet in observational studies, getting an *unbiased estimate* in the first place is even more important.

b. In observational studies, one very effective way to remove the bias of extraneous factors is multiple regression (which we will study in Chapter 13).

c. For randomized controlled experiments that are free of bias, confidence intervals are correct 95% of the time. That is, the interval of the form:

$$\text{Parameter} = \text{estimate} \pm \text{SE}$$

covers the true parameter 95% of the time.

10-19 Suppose that a 95% confidence interval for a population mean was calculated to be $\mu = 170 \pm 20$. Answer True or False; if False, correct it:

a. Any hypothesis in the interval $150 < \mu < 190$ is called the null hypothesis, while any hypothesis outside this interval is called the alternate hypothesis.

b. The population mean is a random variable with expectation 170 and standard deviation 20.

c. If this sampling experiment were repeated many times, and if each time a confidence interval were similarly constructed, 95% of these confidence intervals would cover $\mu = 170$.

10-20 A psychologist runs 6 people through a pilot experiment, and collects the following data on heart rate:

| | HEART RATE | |
SUBJECT	BEFORE EXPERIMENT	AFTER EXPERIMENT
Smith	71	84
Jones	67	72
Gunther	71	70
Wilson	78	85
Pestritto	64	71
Seaforth	69	80

Calculate a 95% confidence interval for the effect of the experiment on heart rate.

10-21 A controlled study to investigate (among other things such as oxygen damage to the retina) the relationship between oxygen therapy and infant mortality was carried out (Schank and Lehman, 1954).

All infants with birth weight between 1000 and 1850 grams and less than 12 hours old on admission to the nursery were assigned at random either to high oxygen concentration or to low oxygen concentration treatments. Except for the type of oxygen therapy, the two groups were cared for in the same manner. The results were as follows:

| CONCENTRATION OF OXYGEN | THREE-MONTH SURVIVAL | | |
	DIED	SURVIVED	TOTALS
High	9	36	45
Low	12	28	40
Totals	21	64	85

Calculate the appropriate confidence interval, and give a verbal interpretation.

10-22 Do men and women agree on the ideal number of children in a family? To answer this question, a Gallup poll of 750 men and 750 women in 1980 gave the following reply:

IDEAL NUMBER OF CHILDREN	RELATIVE FREQUENCY OF REPLY	
	MEN	WOMEN
0	1%	1%
1	3%	3%
2	52%	49%
3	21%	20%
4	11%	13%
5	2%	2%
6	2%	3%
No opinion	8%	9%
Total	100%	100%

SUMMARY STATISTICS OF THOSE WITH AN OPINION		
Mean	2.57	2.65
Median	2	2
Range	6	6
IQR	1	1
$\Sigma(X - \overline{X})^2 f$	755	880

a. Verify the very first summary statistic ($\overline{X}_1 = 2.57$).

b. Of those who have an opinion, what difference is there between men and women in the whole population? Construct a 95% confidence interval.

c. Is the difference statistically discernible (at the 5% level)?

10-23 A ski lift is designed with a load limit of 18,000 lbs. It claims a capacity of 100 persons. If the weights of all the people using the lift have a mean of 175 lbs and a standard deviation of 30 lbs, what is the probability that a group of 100 persons will exceed the load limit?

10-24 A large survey research service conducts polls to estimate various percentages in the American population. Enough voters are included in each poll to make the standard error (standard deviation of P) about 4%. In 100 polls last year, about how many were off by more than 5%:

a. If each poll was unbiased?

b. If each poll was biased by 10 percentage points?

10-25 In the hospital birth example of Section 1-2, recall that the 28 women were divided into treatment and control groups according to whether they arrived at the hospital for their delivery on an even- or odd-numbered day. For all practical purposes, this should be equivalent to randomization—which we shall call the null hypothesis H_0.

To see whether the data support H_0, calculate its p-value for each of the following sets of observations:

a. The number of baby boys turns out to be 6 out of 14 in the treatment group, and 8 out of 14 in the control group.

   b. The number of married mothers was 4 out of 14 in the treatment group, and 5 out of 14 in the control group.

10-26 For each of two educational levels, a random sample of 3 American household incomes was drawn (in thousands of dollars, 1978):

EDUCATION OF HOUSEHOLDER	
8 YEARS OR LESS	**9 TO 12 YEARS**
10	18
12	11
8	19

Calculate a 95% confidence interval for the mean difference in income:

   a. With the ordinary confidence interval (from Chapter 8, using the t table).

   b. With the simultaneous confidence interval (from Chapter 10, using the F table).

   c. In parts **a** and **b**, did you get the same answer? Did you make the same assumptions? What were these assumptions?

10-27 Repeat Problem 10-26, calculating p-values for H_0 instead of confidence intervals.

10-28

EDUCATION OF HOUSEHOLDER			
8 YEARS OR LESS (ELEMENTARY)	**9 TO 12 YEARS** (HIGH SCHOOL)	**13 TO 15** (COLLEGE)	**16 OR MORE** (COLLEGE COMPLETED)
10	18	15	40
12	11	32	25
8	19	10	16

   a. The data above is just an extension of the earlier data in Problems 10-26 and 10-27. Would you guess the p-value for H_0 will be the same as earlier? Why?
   Calculate the ANOVA table and p-value to verify your guess.

   b. Would you guess the 95% simultaneous confidence intervals for all comparisons will be the same width as earlier? Why?
   Calculate them to verify your guess.

10-29 Based on your experience in Problems 10-26 and 10-28, answer True or False; if False, correct it:

   a. The traditional comparison of 2 independent samples (using the t table) can be extended to c independent samples (using the F table); this extension is called one-factor ANOVA.

   b. In other words, F in ANOVA gives exactly the same answer as t for 2 samples, and then goes on to give answers for any number of samples.

10-30 A household income was randomly drawn from each of 4 U.S. regions and 2 races (in thousands of dollars, 1978):

| | RACE | |
REGION	WHITE	BLACK
Northeast	18.7	11.7
Northcentral	19.0	13.6
South	17.1	9.9
West	18.8	10.8

Repeat the same questions as in Problems 10-26, 10-27, and 10-29, for the mean difference between *races.*

10-31 Referring to Problem 10-28, is it fair to say that completing college more than doubles income, compared to the lowest educational level? Why?

10-32 Each week a hospital administrator monitors the number of deaths in the intensive care unit in order to be warned if anything starts to go wrong. Over the past year when everything was working well, records showed that 7% of the patients died.

At present, approximately 90 patients pass through intensive care each week. The administrator decides that if the number of deaths ever exceed 10 in a given week, an investigation is necessary.

a. In a year when everything is working well, about how many needless investigations (false alarms) will result from this rule?

b. Suppose a serious uncorrected mistake by the hospital staff causes 6 needless deaths in a certain week. What is the chance that this will fail to trigger an investigation (i.e., what is the chance of a missed alarm?) What is the customary symbol for this probability?

Part

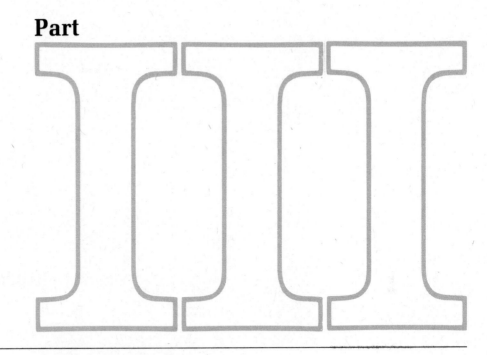

REGRESSION: RELATING TWO OR MORE VARIABLES

Chapter 11

Fitt[...]

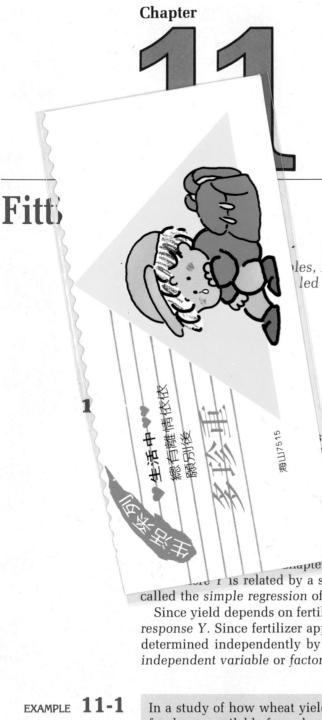

[...]les, least squares is the most simple: by [...]led into the most complicated

K. F. GAUSS, 1809

[...] study more than an isolated single variable [...]We usually want to look at how one variable [...]hat statisticians call *regression*.

[...]heat yield depends on the amount of fer-[...]ield Y that follows from various amounts [...]to Figure 11-1 might be observed. From [...]e amount of fertilizer does affect yield. [...]le to describe *how*, by fitting a curve [...]apters 11 and 12, we will stick to the simplest [...] Y is related by a straight line to just one variable X. This is called the *simple regression* of Y against X.

Since yield depends on fertilizer, it is called the *dependent variable* or *response* Y. Since fertilizer application is not dependent on yield, but is determined independently by the experimenter, it is referred to as the *independent variable* or *factor* or *regressor* X.

EXAMPLE **11-1** In a study of how wheat yield depends on fertilizer, suppose that funds are available for only seven experimental observations. So X is set at seven different levels, with one observation Y in each case, as shown in Table 11-1.

TABLE **11-1**

Fertilizer and Yield Observations

X FERTILIZER (LB/ACRE)	Y YIELD (BU/ACRE)
100	40
200	50
300	50
400	70
500	65
600	65
700	80

a. Graph these points, and roughly fit a line by eye.

b. Use this line to predict yield Y, if fertilizer application X is 400 pounds.

Solution

a. In Figure 11-2 we fit a line by eye to the scatter of observations.

b. With a fertilizer application of $X = 400$ pounds, the predicted yield (denoted with a hat) is the height $\hat{Y} \simeq 60$ bushels. This is the point on the fitted line in Figure 11-2 above the X value of 400. The deviation d of the actual value Y from its predicted value $\hat{Y}$ is of particular interest. Roughly speaking, in fitting the line by eye we have tried to keep this pattern of vertical deviations as small as possible.

How good is a rough fit by eye, such as we used in Example 11-1? In Figure 11-3, we note in panel (a) that if all the points were exactly in a line, then the line could be fitted by eye perfectly accurately. (Since all

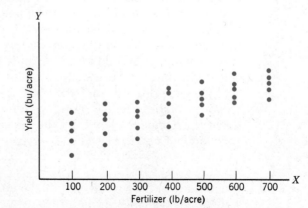

Figure 11-1
Observed relation of wheat yield to fertilizer application on 35 experimental plots.

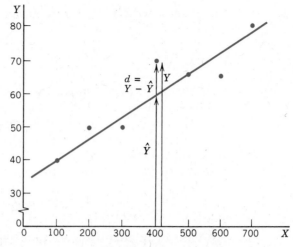

Figure 11-2
Regression line fitted by eye to the data of Table 11-1.

the points would be on this line, every one of the deviations would be zero.) But as we progress to the highly scattered case in panel (c), we need to find another method—a method that is more objective, and is easily computerized. The following section, therefore, sets forth algebraic formulas for fitting a line.

11-2 Ordinary Least Squares (OLS)

A—THE LEAST SQUARES CRITERION

Our objective is to fit a line whose equation is of the form

$$\hat{Y} = a + bX \tag{11-1}$$

That is, we must find a formula to calculate the slope b and intercept a. (A review of these concepts is given in Appendix 11-1.)

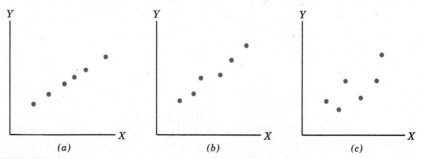

Figure 11-3
Various degrees of scatter.

The question is: How do we select a and b so that we minimize the pattern of vertical Y deviations (prediction errors) in Figure 11-2, where

$$\text{deviation } d = Y - \hat{Y}$$

On first thought, we might try to minimize Σd, the sum of all the deviations. However, because some of the points are above the line and others below, some deviations d will be positive and others negative; to the extent they cancel, they will make the total (Σd) deceptively near zero. To avoid this problem, we could first take the *absolute value* of each deviation, and then minimize their sum:

$$\text{minimize } \Sigma|d| = \Sigma|Y - \hat{Y}|$$

A familiar alternative is to *square* each deviation, and then minimize the sum of all of these:

$$\boxed{\text{minimize } \Sigma d^2 = \Sigma(Y - \hat{Y})^2} \qquad \text{(11-2)}$$
$$\text{like (2-10)}$$

This is called the criterion of *ordinary least squares* (OLS), and it selects a unique line called the OLS line.

B—THE LEAST SQUARES FORMULAS

Recall the line to be fitted is

$$\hat{Y} = a + bX \qquad \text{(11-1) repeated}$$

The OLS slope b is calculated from the following formula (derived in Appendix 11-2):

$$b = \frac{\Sigma(X - \overline{X})(Y - \overline{Y})}{\Sigma(X - \overline{X})^2} \qquad \text{(11-3)}$$

The deviations $(X - \overline{X})$ and $(Y - \overline{Y})$ will appear so often that it is worthwhile abbreviating them. Let

$$\left. \begin{aligned} x &\equiv X - \overline{X} \\ y &\equiv Y - \overline{Y} \end{aligned} \right\} \qquad \text{(11-4)}$$

The small x (or y) notation provides a reminder that the deviations x are typically smaller numbers than the original values X. (This is apparent in the first four columns of Table 11-2.) With this notation, the formula for b can now be simplified:

TABLE **11-2** **Fitting the Least Squares Line to the Data of Example 11-1**

DATA		DEVIATION FORM		PRODUCTS	
		$x = X - \overline{X}$ $= X - 400$	$y = Y - \overline{Y}$ $= Y - 60$		
X	Y			xy	x^2
100	40	−300	−20	6000	90,000
200	50	−200	−10	2000	40,000
300	50	−100	−10	1000	10,000
400	70	0	10	0	0
500	65	100	5	500	10,000
600	65	200	5	1000	40,000
700	80	300	20	6000	90,000
$\overline{X} = 400$	$\overline{Y} = 60$	$\Sigma x = 0\checkmark$	$\Sigma y = 0\checkmark$	$\Sigma xy = 16,500$	$\Sigma x^2 = 280,000$

$$b = \frac{\Sigma xy}{\Sigma x^2}$$

(11-5)

Once b is calculated, the intercept a can then be found from another simple formula (also derived in Appendix 11-2):

$$a = \overline{Y} - b\overline{X}$$

(11-6)

For the data in Example 11-1, the calculations for a and b are laid out in Table 11-2. We calculate Σxy and Σx^2, and substitute them into (11-5):

$$b = \frac{\Sigma xy}{\Sigma x^2} = \frac{16,500}{280,000} = .059$$

Then we use this slope b (along with $\overline{X}$ and $\overline{Y}$ calculated in the first two columns of Table 11-2) to calculate the intercept a from (11-6):

$$a = \overline{Y} - b\overline{X} = 60 - .059(400) = 36.4$$

Plugging these estimated values a and b into (11-1) yields the equation of the OLS line:

$$\hat{Y} = 36.4 + .059X$$

(11-7)

From its graph in Figure 11-4, we can see how closely this OLS line resembles the fit by eye in Figure 11-2. And it easily gives an estimate of yield for any desired fertilizer application. For example, if $X = 400$ lbs. of fertilizer,

$$\hat{Y} = 36.4 + .059(400) = 60 \text{ bushels} \qquad \text{like (11-7)}$$

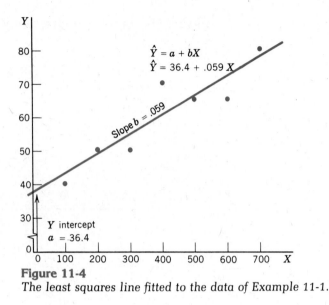

Figure 11-4
The least squares line fitted to the data of Example 11-1.

C—MEANING OF THE SLOPE b

By definition, the slope of a line is the change in height Y, when we move to the right by one unit in the X direction. That is,

> Slope b = change in Y that accompanies a unit change in X (11-8)

This is illustrated geometrically in Appendix 11-1. It is so important that we give an alternative algebraic derivation here. Suppose in (11-7), for example, that fertilizer X was increased by 1 unit, from 75 to 76 pounds, say. Then to calculate the increase in yield Y:

$$\text{Initial } Y = 36.4 + .059(75)$$
$$\text{New } Y = 36.4 + .059(75 + 1)$$
$$= 36.4 + .059(75) + .059$$
$$= \text{initial } Y + .059$$

That is, Y has increased by .059 as X has increased by 1—which is the slope b. And so (11-8) is established.

D—RANDOMIZED EXPERIMENTS VS. OBSERVATIONAL STUDIES

If the values of X are assigned at random, then we can make a stronger statement than (11-8): b then estimates the increase in Y *caused* by a unit increase in X.

If, however, the relation of Y to X occurs in an uncontrolled observational study, then we cannot necessarily conclude anything about causation. In that case, as emphasized in Chapter 1, the increase in Y that accompanies a unit change in X would include not only the effect of X, but also *the effect of any extraneous variables* that might be changing simultaneously.

PROBLEMS

11-1   Suppose that 4 levels of fertilizer were randomly assigned to 4 plots, resulting in the following yields of corn:

FERTILIZER X (CWT/ACRE)	YIELD Y (BU/ACRE)
1	70
2	70
4	80
5	100

a.   Calculate the regression line of yield against fertilizer.

b.   Graph the 4 points and the regression line. Check that the line fits the data reasonably well.

c.   Use the regression equation to predict:

   i.   The yield if 3 cwt/acre (hundredweight per acre) of fertilizer is applied.

   ii.   The yield if 4 cwt/acre of fertilizer is applied.

   iii.   The increase in yield for every 1 cwt/acre increase in fertilizer. Show these on the graph.

11-2   Suppose that a random sample of 5 families had the following annual incomes and savings (in thousands of dollars):

FAMILY	INCOME X	SAVING S
A	16	1.2
B	22	2.4
C	18	2.0
D	12	1.4
E	12	.6

Calculate the regression line of S on X.

11-3   During the 1950s, radioactive waste leaked from a storage area near Hanford, Washington, into the Columbia River nearby. For nine counties downstream in Oregon, an index of exposure X was calculated (based on distance from Hanford, and distance of the average citizen from the river, etc.). Also, the cancer mortality Y was calculated (deaths per 100,000 person-years, 1959–1964). The data was as follows (from Fadeley, 1965; via Anderson and Sclove, 1978):

COUNTY	RADIOACTIVE EXPOSURE X	CANCER MORTALITY Y
Clatsop	8.3	210
Columbia	6.4	180
Gilliam	3.4	130
Hood River	3.8	170
Morrow	2.6	130
Portland	11.6	210
Sherman	1.2	120
Umatilla	2.5	150
Wasco	1.6	140

From this data, summary statistics were computed:

$$\overline{X} = 4.6 \qquad \overline{Y} = 160$$

$$\Sigma x^2 = 97.0 \qquad \Sigma y^2 = 9400 \qquad \Sigma xy = 876$$

a. Calculate the regression line for predicting Y from X.

b. Estimate the cancer mortality if X were 5.0. And if X were 0.

c. Graph the nine counties, and your answers in parts **a** and **b**.

d. To what extent does this data prove the harmfulness of radioactive exposure?

*11-3 Advantages of OLS—a More Robust Version

A—ADVANTAGES OF OLS

Ordinary least squares (OLS) is the basis for developing many of the other formulas that appear in Chapters 11 through 15. Since it is used so extensively, we should ask just why it is chosen as the criterion for fitting a line (that is, selecting a and b). For example, why didn't we instead minimize the *absolute* deviations? There are several good reasons:

1. OLS leads to relatively simple formulas for calculating a and b, as we have just seen in Section 11-2.
2. OLS is closely related to ANOVA, as we will see in Chapter 14.
3. The squaring that occurs in (11-2) is like the squaring that occurred in defining the variance s^2 in Chapter 2. The analogy goes further: In the special case where Y has no relation to X (i.e., the slope b is zero), then the OLS fit (11-1) becomes simply

$$\hat{Y} = a$$

But noting (11-6) and the zero value of b in this special case, a becomes $\overline{Y}$; thus,

$$\hat{Y} = \overline{Y} \qquad (11\text{-}9)$$

In other words, the best prediction we can make of Y is just the sample mean $\overline{Y}$. The sample mean is therefore a special case of OLS.

4. Just as we have seen that the sample mean is unbiased and often efficient, so too is OLS (as we will show in Section 12-2).

5. OLS is very flexible and easy to modify. Specifically, for the OLS criterion we added up all the squared deviations, giving an equal weight to each. But we could give a different weight to each, in which case we would have *weighted least squares* (WLS).

For example, in Figure 11-4, suppose the second observation had a Y value of 80 rather than 50. Such an outlier would have a great influence: The fitted regression line in that figure would become dramatically different from the one shown. To appropriately reduce the effect of this outlier, we could give it a smaller weight—that is, use weighted least squares.

In applying WLS, the crucial question of course is: How do we determine the weights to be used? We will next outline one of the many practical answers now available.

B—A MORE ROBUST FORM OF REGRESSION—BIWEIGHTED LEAST SQUARES

In Section 7-3, we showed how to get a robust estimate of a population center, by giving outlier observations relatively little weight. Exactly the same idea (Mosteller and Tukey, 1977) can be used in fitting a line.

First, to determine whether or not an observation Y is an outlier, we measure its distance from the OLS fitted value $\hat{Y}$; that is, we measure each deviation as in Figure 11-2:

$$d = Y - \hat{Y} \qquad (11\text{-}10)$$

To see how *relatively* large this is, we convert it into a kind of standardized deviation:

$$u = \frac{Y - \hat{Y}}{3S} \qquad \begin{array}{l}(11\text{-}11)\\ \text{like } (7\text{-}17)\end{array}$$

where S is some overall measure of deviation or spread. (For example, S could be the interquartile range of all the deviations d.) Then we compute the *biweights*:

$$\begin{array}{ll} w = (1 - u^2)^2 & \text{if } u \leq 1 \qquad (11\text{-}12)\\ = 0 & \text{if } u > 1 \qquad \text{like } (7\text{-}15)\end{array}$$

Notice how the second equation is simply stating that an observation will be given zero weight if it is an outlier—that is, if its deviation d in (11-10) is so large that it makes the u value in (11-11) greater than 1. Moreover, the first equation above describes how we weight all other observations: An observation will be given less weight the further it is from the regression line—that is, the greater is its deviation d and hence its u value.

The weights determined in (11-12) are now used in all the regression formulas:

$$b = \frac{\Sigma wxy}{\Sigma wx^2}$$

$$\qquad (11\text{-}13)$$
like (11-5)

$$a = \overline{Y} - b\overline{X}$$

$$\qquad (11\text{-}14)$$
like (11-6)

where

$$\overline{Y} = \frac{\Sigma wY}{\Sigma w} \quad \text{and} \quad \overline{X} = \frac{\Sigma wX}{\Sigma w} \qquad \text{like (7-18)}$$

This new value of a and b gives us a more robust line. With it, we can start all over again [recalculating the deviations in (11-10)] and thus get an even better line. And so on.

To summarize:

1. We start by fitting the OLS line. [If you like, you can think of this as a special application of weighted least squares where all the weights are equal to 1 in (11-13) and (11-14). Of course, in this special case, these WLS formulas are exactly the same as the OLS formulas.]
2. We then measure $Y - \hat{Y}$, the deviation of each observation from the fitted line. These determine the set of weights (11-12) we then use in (11-13) and (11-14) to fit a more robust line.
3. With our new fitted line, we can then repeat step 2 as many times as we like. For example, the computer can keep iterating until there is no substantial improvement.

*PROBLEMS

11-4 Answer True or False; if False, correct it:

a. OLS may be regarded as the extension of a familiar technique—fitting a mean to a sample as in Chapter 2—and so shares many of the virtues of the sample mean.

b. For example, OLS is efficient if the underlying population is normal, or has extremely long tails.

c. The OLS formulas for the slope b and intercept a are relatively difficult to compute.

d. Weighted least squares (WLS) is a very flexible modification of OLS. It just requires that each deviation be weighted. (The greater the deviation, the less weight it is given.)

e. If outliers are a problem, for example, we can increase their influence by giving them large weights, such as biweights (a close analogy to the biweights used in Chapter 7 to smoothly trim the mean).

CHAPTER 11 SUMMARY

11-1 To show how a response Y is related to a factor (regressor) X, a line can be fitted called the regression line, $\hat{Y} = a + bX$.

11-2 Using the criterion of ordinary least squares (OLS), the slope b and intercept a can be calculated from simple formulas. The slope b gives the change in Y that accompanies a unit change in X.

*11-3 The OLS formulas can be easily modified to create biweighted least squares, a more robust form of regression. To give possible outliers much less weight, this modification uses the same kind of biweighting we used in estimating the mean in Section 7-3.

REVIEW PROBLEMS

11-5 Suppose a random sample of 4 pharmaceutical firms had the following profits and research expenditures:

PROFIT, P (THOUSANDS OF DOLLARS)	RESEARCH EXPENDITURE, R (THOUSANDS OF DOLLARS)
50	40
60	40
40	30
50	50

a. Fit a regression line of P against R.

b. Graph the data and the fitted line.

c. How well does this regression line show how research generates profits for pharmaceutical firms?

11-6 Each of the 300 dots in this figure represents a pair of heights: a son's height Y plotted against his father's height X. Note that very few sons are exactly the same height as their father (i.e., very few dots are on the 45° line where $Y = X$). Instead, the dots are scattered quite widely around the 45° line, in a pattern very similar to a classic study carried out about 1900 by a geneticist (Galton and Lee, 1903).

a. Suppose we had to predict the height of the son, if we know the father is 6 feet. The relevant father-son pairs are those in the shaded vertical strip above 6 feet.

Roughly guessing by eye, mark the average son's height Y in this strip with a small bar. This average height of course represents the best prediction.

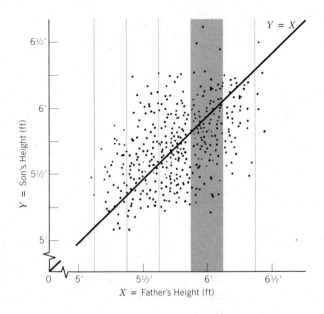

b. Repeat (*a*) for all the other fathers' heights (from 5¼, 5¾, up to 6¼ ft).

c. The predicting bars lie roughly on a line. Draw in this line, by eye.

d. Calculate the regression line of Y on X from the following summary statistics (in feet):

$$\overline{X} = 5.8 \qquad\qquad \overline{Y} = 5.8$$
$$\Sigma x^2 = 18.3 \qquad\qquad \Sigma y^2 = 19.0$$
$$\Sigma xy = 9.1$$

e. Graph the line in d. How does it compare to the line in c?

11-7 In the early 1900s, six provinces in Bavaria had recorded their infant mortality (deaths per 1000 live births) and bottle-feeding (percentage of infants bottle-fed) as follows (Knodel, 1977):

PROVINCE	MORTALITY (DEATHS PER 1000)	BOTTLE-FEEDING (%)
Mittelfranken	250	40
Niederbayern	320	70
Oberfranken	170	10
Oberpfalz	300	40
Schwaben	270	60
Unterfranken	190	20
Mean	250	40%

a. Calculate the appropriate regression line to predict the infant mortality of two other provinces, Oberbayern and Pfalz, which had bottle-feeding rates of 63% and 15%.

   If the actual infant mortalities were 290 and 168, calculate the prediction errors.

b. To what extent does this data prove the benefits of bottle-feeding?

11-8 Suppose the data in Figure 11-1 was collected by an agricultural research station in Kansas, in order to make recommendations for fertilizer application within the state. Consider two ways the data might have been obtained:

    i. Thirty-five farmers are chosen at random from the state, and each one reports his average yield Y and fertilizer application X for the current crop of wheat.

    ii. Thirty-five one-acre plots of land are chosen at random from the state; the seven levels of fertilizer are assigned at random to these plots, and applied by the agricultural station. Although each farmer's permission must be obtained, of course, he is not told what level of fertilizer was assigned, and is instructed to care for his plot in the usual fashion.

a. Which of these ways is better? Why?

b. Can you suggest an even better way to collect the data?

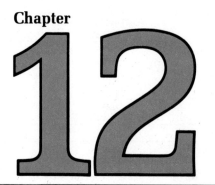

Chapter 12

Simple Regression

Models are to be used, but not to be believed.   HENRI THEIL

So far, our treatment of a sample scatter of observations has been descriptive: We have just summarized it with a fitted line. Now we wish to make inferences about the parent population from which this sample was drawn. To do so, we must build a statistical model that allows us to construct confidence intervals and test hypotheses.

12-1 The Regression Model

A—SIMPLIFYING ASSUMPTIONS

Suppose in Figure 12-1a that we set fertilizer at level X_1 for many, many plots. The resulting yields will not all be the same, of course; the weather might be better for some plots, the soil might be better for others, and so on. Thus we would get a distribution (or population) of Y values, appropriately called the probability distribution of Y_1 given X_1, or $p(Y_1/X_1)$. There will similarly be a distribution of Y_2 at X_2, and so forth. We can therefore visualize a whole set of Y populations such as those shown in Figure 12-1a. There would obviously be great problems in analyzing populations as peculiar as these. To keep the problem manageable, therefore, we make three assumptions about the regularity of these Y distributions— as shown in Figure 12-1b:

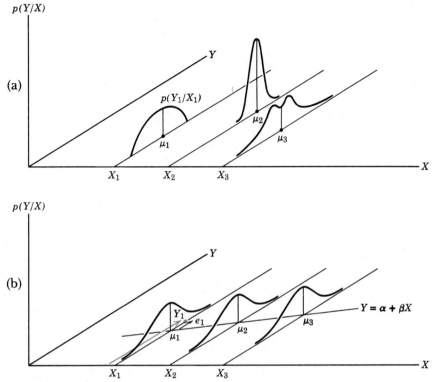

Figure 12-1
(a) General populations of Y, given X. (b) The special form of the populations of Y assumed in simple linear regression.

1. *Homogeneous Variance:* All the Y distributions have the same spread. Formally, this means that the probability distribution $p(Y_1/X_1)$ has the same variance σ^2 as $p(Y_2/X_2)$ and so on.

2. *Linearity:* The mean of each of the Y distributions lies on a straight line, known as the true (population) regression line:

$$E(Y) = \mu = \alpha + \beta X \qquad (12\text{-}1)$$

The population parameters α and β (alpha and beta) specify the line, and are to be estimated from sample information.

3. *Independence:* The random variables $Y_1, Y_2, \ldots$ are statistically independent. For example, if Y_1 happens to be large, there is no reason to expect Y_2 to be large (or small); that is, Y_2 is statistically unrelated to Y_1.

These three assumptions may be written more concisely as:

> The random variables Y_1, Y_2, ... are independent, with
>
> $$\text{Mean} = \alpha + \beta X$$
> $$\text{Variance} = \sigma^2$$
> $\hfill$ (12-2)

On occasion, it is useful to describe the deviation of Y from its expected value as the error or disturbance term e, so that the model may alternatively be written as:

> $$Y = \alpha + \beta X + e$$
>
> where e_1, e_2, ... are independent errors, with
>
> $$\text{Mean} = 0$$
> $$\text{Variance} = \sigma^2$$
> $\hfill$ (12-3)

For example, in Figure 12-1b the first observed value Y_1 is shown in color, along with the corresponding error term e_1.

B—THE NATURE OF THE ERROR TERM

Now let us consider in more detail the purely random part of Y, the error or disturbance term e. Why does it exist? Why doesn't a precise and exact value of Y follow, once the value of X is given? The random error may be regarded as the sum of two components:

1. *Measurement error.* There are various reasons why Y may be measured incorrectly. In measuring crop yield, there may be an error resulting from sloppy harvesting or inaccurate weighing. If the example is a study of the consumption of families at various income levels, the measurement error in consumption might consist of budget and reporting inaccuracies.
2. *Inherent variability* occurs inevitably in biological and social phenomena. Even if there were no measurement error, repetition of an experiment using exactly the same amount of fertilizer would result in somewhat different yields. The differences could be reduced by tighter experimental control—for example, by holding constant soil conditions, amount of water, and so on. But complete control is impossible—for example, seeds cannot be exactly duplicated.

C—ESTIMATING α AND β

Suppose the true (population) regression, $Y = \alpha + \beta X$, is the black line shown in Figure 12-2. This is unknown to the statistician, who must estimate it as best he can by observing X and Y. If at the first level X_1, the

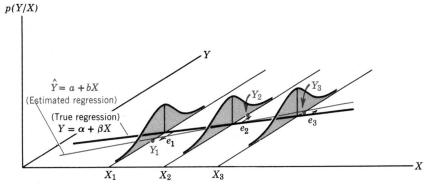

Figure 12-2
True (population) regression and estimated (sample) regression.

random error e_1 happens to take on a negative value as shown in the diagram, he will observe Y_1 below the true line. Similarly, if the random errors e_2 and e_3 happen to take on positive values, he will observe Y_2 and Y_3 above the true line.

Now the statistician applies the least squares formulas to the only information he has—the sample points Y_1, Y_2, and Y_3. This produces an estimated line $\hat{Y} = a + bX$, which we color blue in Figure 12-2.

Figure 12-2 is a crucial diagram. Before proceeding, you should be sure that you can clearly distinguish between: (1) the true regression and its surrounding e distribution; since these are population values and cannot be observed they are shown in black. (2) the Y observations and the resulting fitted regression line; since these are sample values, they are known to the statistician and colored blue.

Unless the statistician is very lucky indeed, it is obvious that his estimated line will not exactly coincide with the true population line. The best he can hope for is that it will be reasonably close to the target. In the next section we develop this idea of "closeness" in more detail.

PROBLEMS

12-1 (Monte Carlo). Suppose the true (long-run average) relation of corn yield Y to fertilizer X is given by the line

$$Y = 2.40 + .30X$$

where Y is measured in tons of corn per acre, and X is measured in hundreds of pounds of fertilizer per acre. In other words, the population parameters are $\alpha = 2.40$ and $\beta = .30$.

a. First, play the role of nature. Graph the line for $0 \leq X \leq 12$. Suppose the yield varies about its expected value on this line, with a standard deviation $\sigma = 1$, and a distribution that is normal. Simulate a sample of five such yields, one each for $X = 2, 4, 6, 8, 10$. (*Hint:* First calculate the five mean values from the given line. Then add to each a random normal error e from Appendix Table II.)

b. Now play the role of the statistician. Calculate the least squares line that best fits the sample. Graph the sample and the fitted line.

c. Finally, let the instructor be the evaluator. Have him graph several of the lines found in part **b.** Then have him tabulate all the values of b, and graph their relative frequency distribution. How good are the estimates b? Specifically, find the expected value and standard error of b, approximately.

d. Why are the expected value and standard error in part **c** only approximate? Suggest how you could get better answers.

12-2   Sampling Variability

A—SAMPLING DISTRIBUTION OF b

How close does the estimated line come to the true population line? Specifically, how is the slope estimate b distributed around its target β? (Because of its greater importance, we shall concentrate for the rest of this chapter on the slope b, rather than the intercept a.)

While the sampling distribution of b can be approximately derived using Monte Carlo methods, statisticians have been able to do even better. They have theoretically derived its distribution, just as they derived the distribution of the sample mean $\overline{X}$. The moments of b, for example, are derived in part C.

Once again they found a sampling distribution that is approximately normal, regardless of whether or not the shape of the parent population— shaded in gray in Figure 12-2—is normal. (As one might expect, if the parent population is highly skewed, or otherwise non-normal, then it will take a larger sample size for b to reach approximate normality.) Specifically:

Normal Approximation Rule for Regression
The slope estimate b is approximately normally distributed, with

$$\text{Expected value of } b = \beta$$

$$\text{Standard error of } b = \frac{\sigma}{\sqrt{\Sigma x^2}}$$

(12-4)
like (6-9)

Here σ represents the standard deviation of the Y observations about the population line, and each small x as usual represents the deviation of X from the mean $\overline{X}$. The normal distribution of b is shown in Figure 12-3.

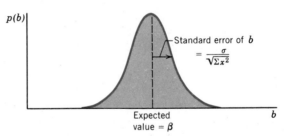

Figure 12-3
The sampling distribution of the estimate b.

B—HOW TO REDUCE THE STANDARD ERROR (SE) OF *b*

Although (12-4) may be the easiest formula to calculate the SE, the easiest formula for *understanding* the SE is obtained by re-expressing it:

$$\text{SE of } b = \frac{\sigma}{\sqrt{n\left(\frac{\Sigma x^2}{n}\right)}} \qquad (12\text{-}5)$$

Now recall that $\Sigma x^2/n$ is just the MSD, which is approximately s_X^2, the variance of X. Thus (12-5) can be written as

$$\boxed{\text{SE of } b \simeq \frac{\sigma}{\sqrt{n}} \cdot \frac{1}{s_X}} \qquad \begin{array}{c}(12\text{-}6)\\ \text{like } (6\text{-}7)\end{array}$$

In this form, we can see there are three ways the SE can be reduced to produce a more accurate estimate *b*:

1. By reducing σ, the inherent variability of the Y observations.
2. By increasing n, the sample size.
3. By increasing s_X, the spread of the X values, which are determined by the experimenter. (Recall in our example how the experimenter fixed the fertilizer levels.)

This third point is particularly interesting: Because increasing s_X improves *b*, s_X is called the *leverage* of the X values on *b*.

   To illustrate, Figure 12-4 shows why increasing the leverage s_X increases the reliability of the slope *b*. In panel (a), the X values are bunched together with very little spread (s_X is small). This means that the small part of the line being investigated is obscured by the error *e*, making the slope estimate *b* very unreliable. In this specific instance, our estimate has been pulled badly out of line by the errors—in particular by the one indicated by the arrow.

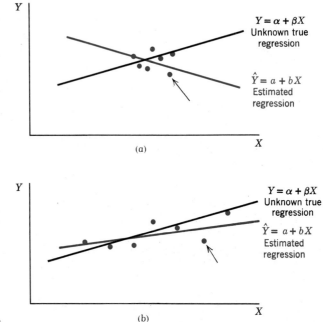

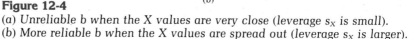

Figure 12-4
(a) Unreliable b when the X values are very close (leverage s_X is small).
(b) More reliable b when the X values are spread out (leverage s_X is larger).

By contrast, in panel (b) we show how the X values have more leverage when they are more spread out (that is, when s_X is larger). Even though the errors e remain the same, the estimate b is much more reliable.

As a concrete example, suppose we wish to examine how sensitive U.S. imports Y are to the international value of the dollar X. A much more reliable estimate should be possible using the recent periods when the dollar has been floating (and taking on a range of values) than in earlier periods when currencies were fixed and only allowed to fluctuate within very narrow limits.

*C—THE MOMENTS OF THE SLOPE b: DERIVATION

Although (11-5) may be the easiest formula for calculating b, the easiest formula for understanding the nature of b is the equivalent formula in Appendix 11-2 that uses Y values in their original, rather than deviation, form:

$$b = \frac{\Sigma xY}{\Sigma x^2}$$

(12-7)
(11-18A) repeated

To appreciate what this means, we write out explicitly the sum in the numerator:

$$b = \frac{1}{\Sigma x^2} [x_1 Y_1 + x_2 Y_2 + \cdots + x_n Y_n] \qquad (12\text{-}8)$$

That is,
$$b = w_1 Y_1 + w_2 Y_2 + \cdots + w_n Y_n \qquad (12\text{-}9)$$

where the coefficients or weights are:

$$w_1 = \frac{x_1}{\Sigma x^2} \qquad w_2 = \frac{x_2}{\Sigma x^2} \cdots \qquad (12\text{-}10)$$

Since each x is a constant (in our example, a specific level of fertilizer fixed by the experimenter), each w coefficient is also a constant. Thus, from (12-9) we establish the important conclusion:

> **b is a weighted sum (linear combination) of the random variables $Y_1, Y_2, \ldots, Y_n$** $\qquad (12\text{-}11)$

Hence, by (5-17) we may write:

$$E(b) = w_1 E(Y_1) + w_2 E(Y_2) + \cdots + w_n E(Y_n) \qquad (12\text{-}12)$$

Moreover, noting that the variables $Y_1, Y_2, \ldots$ are assumed to be independent, it follows from (5-22) that:

$$\text{var}(b) = w_1^2 \text{ var } Y_1 + w_2^2 \text{ var } Y_2 + \cdots + w_n^2 \text{ var } Y_n \qquad (12\text{-}13)$$

Now we just have to work out the details. For the mean of b, we substitute (12-1) into (12-12):

$$\begin{aligned} E(b) &= w_1(\alpha + \beta X_1) + w_2(\alpha + \beta X_2) + \cdots + w_n(\alpha + \beta X_n) \\ &= \alpha (w_1 + w_2 + \cdots + w_n) \\ &\quad + \beta (w_1 X_1 + w_2 X_2 + \cdots + w_n X_n) \end{aligned} \qquad (12\text{-}14)$$

To simplify the first set of brackets, we substitute the value of each of the w coefficients given in (12-10):

$$(w_1 + w_2 + \cdots + w_n) = \frac{x_1 + x_2 + \cdots + x_n}{\Sigma x^2}$$

The sum of all the positive and negative deviations x in the numerator is zero, according to (2-9). Thus

$$(w_1 + w_2 + \cdots + w_n) = \frac{0}{\Sigma x^2} = 0 \qquad (12\text{-}15)$$

To simplify the last set of brackets in (12-14), we again substitute the value of each of the w coefficients given in (12-10), and also express each $X = x + \overline{X}$ (since $x \equiv X - \overline{X}$):

$$
(w_1X_1 + w_2X_2 + \cdots + w_nX_n)
$$

$$
= \frac{x_1(x_1 + \overline{X}) + x_2(x_2 + \overline{X}) + \cdots + x_n(x_n + \overline{X})}{\Sigma x^2}
$$

$$
= \frac{(x_1^2 + x_2^2 + \cdots + x_n^2) + (x_1 + x_2 + \cdots + x_n)\overline{X}}{\Sigma x^2}
$$

$$
= \frac{(\Sigma x^2) + 0\overline{X}}{\Sigma x^2} = 1 \tag{12-16}
$$

Consequently (12-14) can be written

$$
E(b) = \alpha(0) + \beta(1) = \beta \tag{12-17}
$$

Thus we have proved that b is an unbiased estimate of β.

To find the variance of b, we recall from (12-2) that each Y has the same variance σ^2. Thus (12-13) becomes:

$$
\begin{aligned}
\text{var}(b) &= w_1^2\sigma^2 + w_2^2\sigma^2 + \cdots + w_n^2\sigma^2 \\
&= (w_1^2 + w_2^2 + \cdots + w_n^2)\sigma^2 \tag{12-18}
\end{aligned}
$$

To simplify what is in brackets, we again write out the meaning of each w given in (12-10):

$$
(w_1^2 + w_2^2 + \cdots + w_n^2) = \left(\frac{x_1}{\Sigma x^2}\right)^2 + \left(\frac{x_2}{\Sigma x^2}\right)^2 + \cdots + \left(\frac{x_n}{\Sigma x^2}\right)^2
$$

$$
= \left(\frac{1}{\Sigma x^2}\right)^2 (x_1^2 + x_2^2 + \cdots + x_n^2)
$$

$$
= \left(\frac{1}{\Sigma x^2}\right)^2 (\Sigma x^2) = \left(\frac{1}{\Sigma x^2}\right)
$$

Substitute this into (12-18):

$$
\text{var}(b) = \left(\frac{1}{\Sigma x^2}\right)\sigma^2 \tag{12-19}
$$

That is,
$$
\text{SE} = \frac{\sigma}{\sqrt{\Sigma x^2}} \tag{12-4 proved}
$$

*D—THE GAUSS-MARKOV THEOREM

In (12-11) we saw that the OLS estimate b is a linear combination of $Y_1, Y_2, \ldots,$ and in (12-17) we saw that it was unbiased. If we look at other

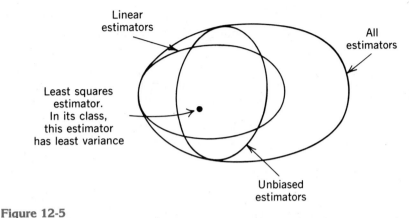

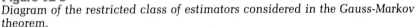

Figure 12-5
Diagram of the restricted class of estimators considered in the Gauss-Markov theorem.

possible estimates in this same class—unbiased linear estimates—it is impossible to find one that is more efficient than OLS (i.e., that has smaller variance). This result, known as the Gauss-Markov Theorem, is illustrated in Figure 12-5; it is one more reason why OLS is popular.

PROBLEMS

12-2 At an agricultural research station investigating the effect of fertilizer on soybean yield, several proposals were made to change the experimental design. If we can assume that yield Y increases linearly with fertilizer application X (over the range of X values considered, at least), state how much the standard error of the slope b will change if they take:

a. Four times as many observations, spread over the same X range.

b. The same number of observations, spread over 4 times the former X range.

c. Half as many observations, spread over twice the former X range.

d. The same number of observations, spread over 5 times the former X range, and with each observation more accurate because of tighter experimental control (less variation from plot to plot of soil acidity, spacing between plants, etc.). In fact, suppose the variance σ^2 is reduced to one half its former value.

*12-3 Suppose that the experimenter of Example 11-1 was in a hurry to analyze his data, and so drew a line joining the first and last points. We shall denote the slope of this line by b_1.

a. Calculate b_1 for the data in Table 11-1.

b. Write out a formula for b_1 in terms of X_1, Y_1, X_7, Y_7.

c. Is b_1 a linear estimator?

   d. Is b_1 an unbiased estimator of β?

   e. Without doing any calculations, can you say which has a larger SE, b_1 or the OLS estimator b?

   f. Verify your answer in part **e** by actually calculating the SE of b_1 and of b for the data of Example 11-1. (Express your answer in terms of the unknown error variance σ^2.)

*12-4   In Problem 12-3, we considered the two extreme values, X_1 and X_7. Now let us consider an alternative pair of less extreme values—say X_2 and X_6. We shall denote the slope of this line by b_2. Like b_1, it can easily be shown to be linear and unbiased.

   a. Calculate the SE of b_2. How does it compare with the SE of b_1?

   b. Answer True or False; if False, correct it: b_1 has a smaller SE than does b_2, which illustrates a general principle: The more a pair of observations are spread out, the more "statistical leverage" they exert, hence the more efficient they are in estimating the sample slope b.

12-3   Confidence Intervals and Tests for β

A—ESTIMATING THE STANDARD ERROR OF b

With the expected value, standard error, and normality of b established, statistical inferences about β are now in order. But first there is one remaining problem: From (12-4), the standard error of b is $\sqrt{\sigma^2/\Sigma x^2}$, where σ^2 is the variance of the Y observations about the population line. But σ^2 is generally unknown, and must be estimated. A natural way to estimate σ^2 is to use the deviations of Y about the fitted line. Specifically, consider the mean squared deviations about the fitted line:

$$\frac{1}{n}\Sigma d^2 = \frac{1}{n}\Sigma(Y - \hat{Y})^2 \qquad \text{like (11-2)}$$

We make one small modification: Instead of n we use the divisor $n - 2$. [The reason is familiar: Two degrees of freedom (d.f.) have already been used up in calculating a and b in order to get the fitted line. This then leaves $(n - 2)$ d.f. to estimate the variance.][1] We therefore estimate σ^2 with the *residual variance* s^2, defined as

---

[1]A more detailed view may be helpful. If there were only $n = 2$ observed points, a least-squares line could be fitted—indeed it would turn out to provide a perfect fit. (Through *any* two points, a line can always be drawn that goes through them exactly.) Thus, although a and b would be determined easily enough, there would be no "information left over" to tell us anything about σ^2, the variance of the observations about the regression line. Only to the extent that n exceeds 2 can we get information about σ^2. That is, $n - 2$ d.f. remain when we use s^2 to estimate σ^2.

$$s^2 \equiv \frac{1}{n-2}\Sigma(Y - \hat{Y})^2$$

(12-20)
like (2-11)

When s is substituted for σ in (12-5), we obtain the estimated standard error (SE):

$$SE = \frac{s}{\sqrt{\Sigma x^2}}$$

(12-21)

With this in hand, statistical inferences can now be made.

B—CONFIDENCE INTERVALS

Using the same argument as in Chapter 8 earlier, we could easily show that the 95% confidence interval for β is

$$\beta = b \pm t_{.025}\ SE$$

(12-22)
like (8-15)

Substituting SE from (12-21) yields[2]

95% confidence interval for the slope[3]

$$\beta = b \pm t_{.025}\ \frac{s}{\sqrt{\Sigma x^2}}$$

(12-23)

where the degrees of freedom for t are the same as the divisor used in calculating s^2:

$$\text{d.f.} = n - 2$$

(12-24)

---

[2]We can see [from equation (12-9) for example] that b is a kind of *weighted mean* of the observed Y values—with weights in (12-10) proportional to x. It is therefore no surprise that everything we said about means earlier (in Section 8-1e) is still relevant: If the underlying populations in Figure 12-1 are not normal, the confidence interval for β would still have 95% confidence. But we could get a *narrower* confidence interval for β if we used the robust estimate derived in Section 11-3.

[3]The 95% confidence interval for the intercept α, although of less interest, may be similarly derived. It turns out to be (see Problem 12-19):

$$\alpha = (\bar{Y} - b\bar{X}) \pm t_{.025}\ s\sqrt{\frac{1}{n} + \frac{\bar{X}^2}{\Sigma x^2}}$$

(12-25)

EXAMPLE **12-1**

The slope relating wheat yield to fertilizer was found to be .059 bu/lb. Of course, this was based on a mere sample of $n = 7$ observations. If millions of observations had been collected, what would the resulting population slope be? Calculate a 95% confidence interval. (Use the statistics calculated in Table 11-2.)

Solution

TABLE **12-1**

Calculations for the Residual Variance s^2

X	Y	$\hat{Y} = 36.4 + .059X$	$Y - \hat{Y}$	$(Y - \hat{Y})^2$
100	40	42.3	−2.3	5.29
200	50	48.2	1.8	3.24
300	50	54.1	−4.1	16.81
400	70	60.0	10.0	100.00
500	65	65.9	−0.9	.81
600	65	71.8	−6.8	46.24
700	80	77.7	2.3	5.29

$$s^2 = \frac{177.68}{7 - 2}$$
$$= 35.5$$

We first use (12-20) to calculate s^2 in Table 12-1. The critical t value then has d.f. $= n - 2 = 7 - 2 = 5$ (the same as the divisor in s^2). From Appendix Table V, this $t_{.025}$ is found to be 2.57. Finally, note that Σx^2 was already calculated in Table 11-2. When these values are substituted into (12-23),

$$\beta = .059 \pm 2.57 \frac{\sqrt{35.5}}{\sqrt{280,000}}$$

$$= .059 \pm 2.57(.0113) \tag{12-26}$$

$$= .059 \pm .029 \tag{12-27}$$

$$.030 < \beta < .088 \tag{12-28}$$

C—TESTING HYPOTHESES

The hypothesis that X and Y are unrelated may be stated mathematically as $H_0: \beta = 0$. To test this hypothesis at the 5% error level, we merely note whether the value 0 is contained in the 95% confidence interval.

EXAMPLE **12-2**

In Example 12-1, test at the 5% level the null hypothesis that yield is unrelated to fertilizer.

Solution

Since $\beta = 0$ is excluded from the confidence interval (12-28), we reject this null hypothesis, and conclude that yield is indeed related to fertilizer.

Equivalently, we note that the estimate of .059 in (12-27) stands out beyond its sampling allowance, so we can conclude that it is statistically discernible.

D—p-VALUE

Rather than simply accept or reject, a more appropriate form for a test is the calculation of the p-value. We first calculate the t statistic:

$$t = \frac{b}{SE}$$

(12-29)
like (9-19)

Then we look up the probability in the tail beyond this observed value of t; this is the p-value.

EXAMPLE **12-3**

In Example 12-1, what is the p-value for the null hypothesis that yield does not increase with fertilizer?

Solution

In the confidence interval (12-26) we have already calculated b and its standard error, which we can now substitute into (12-29):

$$t = \frac{.059}{.0113} \approx 5.2$$

In Appendix Table V, we scan the row where d.f. = 5, and find the observed t value of 5.2 lies beyond $t_{.0025} = 4.77$. Thus

$$\text{p-value} < .0025$$

(12-30)

This provides so little credibility for H_0 that we could reject it, and conclude once again that yield does indeed increase with fertilizer.

In this example, we calculated the one-sided p-value (that is, the p-value in only one tail). But in other cases, it may be appropriate to calculate a two-sided p-value. Appendix 12-3 provides detail on this issue.

PROBLEMS

12-5 For Problems 11-1 and 11-2, construct a 95% confidence interval for the population slope β.

12-6 Suppose that a random sample of 4 families had the following annual incomes and savings:

FAMILY	INCOME X (THOUSANDS OF $)	SAVING S (THOUSANDS OF $)
A	22	2.0
B	18	2.0
C	17	1.6
D	27	3.2

a. Estimate the population regression line $S = \alpha + \beta X$.

b. Construct a 95% confidence interval for the slope β.

c. Graph the 4 points and the fitted line, and then indicate as well as you can the acceptable slopes given by the confidence interval in **b.**

12-7 Which of the following hypotheses is rejected by the data of Problem 12-6 at the 5% level?

a. $\beta = 0$ **c.** $\beta = .10$

b. $\beta = .05$ **d.** $\beta = .50$

12-8 In Problem 12-6, suppose the population marginal propensity to save (β) is known to be positive, if it is not 0. Accordingly, you decide on a one-sided test.

a. State the null and alternative hypothesis in symbols.

b. Calculate the p-value for H_o.

c. Construct a one-sided 95% confidence interval of the form, "β is at least as large as such and such."

d. At the 5% level, can we reject the null hypothesis $\beta = 0$? Test in two ways:

 i. Is the p-value less than 5%?

 ii. Is $\beta = 0$ excluded from the confidence interval?

12-9 Repeat Problem 12-8 using the data of Problem 11-2.

12-4 Predicting Y at a Given Level of X

So far we have considered broad aspects such as the position of the whole line (determined by α and β). In this section, we will consider the narrower, but often very practical, problem of predicting what Y will be, for a given level of X.

A—CONFIDENCE INTERVAL FOR μ_0, THE MEAN OF Y AT X_0

For a given value X_0 (say 550 lbs. of fertilizer), what is the confidence interval for μ_0, the corresponding mean value of Y_0 (wheat yield)? This is the interval that the chemical company may want to know, to describe the long-run (mean) performance of its fertilizer.

As we can see in panel (a) of Figure 12-6, the point μ_0 that we are trying to estimate is the point on the true regression line above X_0. The best point estimate is, of course, the point on the *estimated* regression line:

$$\hat{Y}_0 = a + bX_0 \tag{12-31}$$

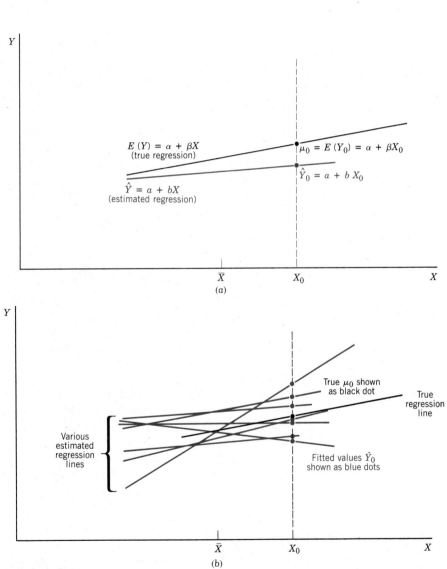

Figure 12-6
How the estimator $\hat{Y}_0$ is related to its target. (a) One single $\hat{Y}_0$ (b) A whole series of possible $\hat{Y}_0$ illustrates the sampling distribution of $\hat{Y}_0$

As a point estimate, this involves some error because of the sampling error in a and b. For example, panel (a) shows the estimate $\hat{Y}_0$ (on the blue sample line) lying below the target μ_0 (on the population line). In this case $\hat{Y}_0$ underestimates μ_0 because b underestimated β.

Panel (b) shows what happens if sampling is repeated. Each sample would give a different regression line, and hence a different estimate $\hat{Y}_0$. All these $\hat{Y}_0$ values would be scattered around the target μ_0, illustrating the sampling distribution of $\hat{Y}_0$. The standard error of $\hat{Y}_0$ is derived in Appendix 12-4; using this, we construct the confidence interval:

95% confidence interval for the mean (of Y_0) at level X_0

$$\mu_0 = (a + bX_0) \pm t_{.025}\, s \sqrt{\frac{1}{n} + \frac{(X_0 - \overline{X})^2}{\Sigma x^2}} \qquad (12\text{-}32)$$

The two terms under the square root reflect the uncertainty in estimating the height of the line α and the uncertainty in estimating the slope of the line β. This second source of uncertainty is seen in Figure 12-6 to be particularly serious if X_0 is far removed from the center of the data $\overline{X}$. Then the $(X_0 - \overline{X})^2$ appearing at the end of (12-32) becomes large, and the estimate of μ_0 becomes very unreliable.

B—PREDICTION INTERVAL FOR A SINGLE OBSERVATION Y_0

Now consider a different question: How widely would we hedge our estimate if we are making a single application of $X_0 = 550$ lbs. of fertilizer and wish to predict the one yield Y_0 that will result? This is the sort of interval that an individual farmer might want to know, in order to plan his budget for the coming year.

In predicting this single Y_0, we will face all the problems involved in estimating the mean μ_0; namely, we will have to recognize the sampling error involved in the estimates a and b. But now we have an additional problem because we are trying to estimate only one observed Y (with its inherent fluctuation) rather than the stable average of all the possible Y's. It is no surprise that this produces an interval similar to (12-32) except for an additional large term at the end, and the requirement of a normal population shape, as well [as proved in Appendix 12-4]:

---

[4]As n gets very large, the first two terms under the square root approach zero, leaving just the 1. Then the 95% prediction interval is approximately:

$$Y_0 \approx (a + bX_0) \pm t_{.025}\, s \qquad (12\text{-}34)$$

In view of this, s (which we have so far called the "residual standard deviation") is sometimes also called the *standard error of the estimate*.

> **95% Prediction Interval[4] for an *individual* Y_0 at level X_o**
>
> $$Y_o = (a + bX_o) \pm t_{.025}\, s \sqrt{\frac{1}{n} + \frac{(X_o - \overline{X})^2}{\Sigma x^2} + 1}$$ (12-33)

We call (12-33) a *prediction interval* (for a single observation Y_o) to contrast it with the earlier *confidence interval* (for the parameter μ_o).

C—COMPARISON OF THE TWO INTERVALS

An example will illustrate the confidence interval for μ_o and the wider prediction interval for the individual Y_o.

EXAMPLE **12-4**

In the fertilizer-yield example, we earlier calculated the following statistics:

$$n = 7 \qquad \overline{X} = 400$$
$$a = 36.4 \qquad \Sigma x^2 = 280{,}000$$
$$b = .059 \qquad s = \sqrt{35.5} = 5.96$$

a. If 550 pounds of fertilizer is to be applied over and over, find a 95% confidence interval for the long-run (mean) yield.

b. If the 550 pounds of fertilizer is to be applied to just one plot, find a 95% prediction interval for the one resulting yield.

Solution

a.

$$\mu_o = (a + bX_o) \pm t_{.025}\, s \sqrt{\frac{1}{n} + \frac{(X_o - \overline{X})^2}{\Sigma x^2}} \qquad \text{(12-32) repeated}$$

$$= (36.4 + .059 \times 550) \pm 2.57\,(5.96) \sqrt{\frac{1}{7} + \frac{(550 - 400)^2}{280{,}000}}$$

$$= 69 \pm 2.57\,(5.96)\sqrt{.223} \qquad\qquad (12\text{-}35)$$

$$= 69 \pm 7 \qquad\qquad (12\text{-}36)$$

b. For the prediction interval, we add 1 under the square root of (12-35):

$$Y_o = 69 \pm 2.57\,(5.96)\sqrt{.223 + 1}$$

$$= 69 \pm 17 \qquad\qquad (12\text{-}37)$$

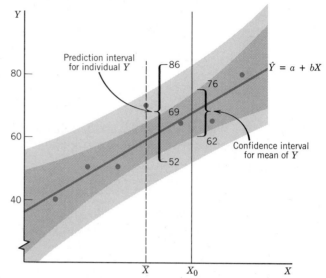

Figure 12-7
95% confidence interval for the mean μ_0 (dark blue), and wider 95% prediction interval for an individual Y_0 (light blue).

> This interval is more than twice as wide as (12-36), which shows how much more difficult it is to predict an *individual* Y observation than the *mean* of Y.

Figure 12-7 illustrates the two intervals calculated in this example. As well as at $X_0 = 550$, it gives the intervals at all possible values of X_0; the result is two colored bands. The dark inner band shows the relatively precise confidence interval for μ_0, while the light outer band shows the vaguer prediction interval for an individual Y_0. Note that both bands become more vague as X_0 moves away from the center of the data $\overline{X}$—a feature we first saw in Figure 12-6.

We also note that only the outer band captures all 7 original observations. It is this band that is wide enough to be 95% certain of capturing yet another observation Y_0.

PROBLEMS

12-10 For 48 successive months, the output of a hosiery mill X and the corresponding total cost Y was plotted below (Dean, 1941). The summary statistics follow.

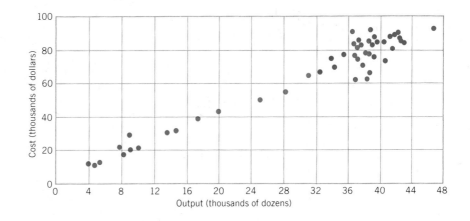

$$n = 48 \qquad \Sigma xy = 13{,}800$$
$$\overline{X} = 30 \qquad \Sigma x^2 = 6900$$
$$\overline{Y} = 63 \qquad \Sigma y^2 = 29{,}000$$

Regression equation $\hat{Y} = 3.0 + 2.0X$
Residual standard deviation, $s = 6$

We assume for now that these 48 points are a random sample from the population of many years' production.

a. Verify the regression equation, and graph it on the given figure. Does it fit the data pretty well? Is the typical residual about equal to $s = 6$?

b. If another month was sampled, guess the cost of production Y if the output X was:

    i. $X = 6$,   ii. $X = 18$,   iii. $X = 30$,   iv. $X = 42$.

In each case, include an interval wide enough so that you could be 95% sure it would turn out to be correct.

c. Graph the four intervals in part b, and sketch the band that joins them, as in Figure 12-7. Does this band contain about 95% of the individual points?

d. Graph the approximate band (12-34) given in footnote 4.

12-11 Repeat Problem 12-10 parts b and c, if we are now interested in predicting the long-run *average* of the total cost values Y at each given level of X.

*12-12 Circle the correct choice in each pair of brackets:

a. Suppose n is at least 25, and X_0 is no more than 2 standard deviations s_X away from the central value $\overline{X}$. Then the exact 95% prediction interval (12-33) for Y_o would be at most [4%, 10%] [*wider, narrower*] than the approximation (12-34)—as we saw in Problem 12-10.

b. The intuitive reason this approximation works so well is that we can safely ignore the minor source of uncertainty, which is [*the position of the population line, the inherent variability of the observations*].

c. This approximation is not only adequate for many purposes, but it is also easy to graph—because it is just a band of two [*divergent curves, parallel straight lines*] on either side of the regression line.

12-5 Extending the Model

A—EXTRAPOLATION

In predicting Y_o, the given level X_o may be *any* value. If X_o lies *among* the observed values $X_1, \ldots, X_n$, the process is called *interpolation*. If X_o lies *beyond* the observed values $X_1, \ldots, X_n$—either to the left or right— then the process is called *extrapolation*. The further we extrapolate beyond the observed X values, the greater the risk we run—for two reasons:

1. We emphasized in the previous section that both intervals (12-32) and (12-33) get wider as X_o moves further from the center $\overline{X}$. This is true, even if all the assumptions underlying our model hold exactly.

2. In practice, we must recognize that a mathematical model is never absolutely correct. Rather, it is a useful approximation. In particular, we cannot take seriously the assumption that the population means are strung out *exactly* in a straight line. If we consider the fertilizer example, it is likely that the true relation increases initially, but then bends down eventually as a "burning point" is approached, and the crop is overdosed. This is illustrated in Figure 12-8, which is an extension of Figure 11-2 with the scale appropriately changed. In the region of interest, from 0 to 700 pounds, the relation is *practically* a straight line, and no great harm is done in assuming the linear model. However, if the linear model is extrapolated far beyond this region of experimentation, the result becomes meaningless. In such cases, a nonlinear model should be considered (as in Chapter 14).

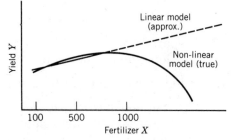

Figure 12-8
Comparison of linear and nonlinear models.

B—EXTENSION TO UNCONTROLLED OR RANDOM *X*

So far we have assumed that the independent variable X has been set at various *controlled* levels (for example, fertilizer application was set at 100, 200, ...). But in many cases—for example, in examining the effect of rainfall on yield—we must recognize that X (rainfall) is a random variable that is completely outside our control. Fortunately, this chapter remains valid whether X is a controlled variable *or* a random variable, provided the model (12-3) is appropriately interpreted, as follows: If X is random, then for any level of X, we still require the conditional expected value of the error e to be 0, and the conditional variance to be a constant value σ^2. In Chapter 15 we will say much more about this.

CHAPTER 12 SUMMARY

12-1 The actual observations must be recognized as just a sample from an underlying population. As usual in describing a population, we use a Greek letter for the slope β. This is the target being estimated by the sample slope b.

12-2 If sampling is random, then from sample to sample b fluctuates around β with a specified standard error—just as $\overline{X}$ fluctuated around μ earlier.

12-3 We can therefore construct a confidence interval for β, or calculate the p-value for $\beta = 0$. In doing so, we estimate the variance σ^2 around the population line with the residual variance s^2 around the sample line.

12-4 For any given level X_o, we predict the corresponding $\hat{Y}_o$ to be the point on the fitted regression line. To be 95% confident, an interval allowance must be made. This *prediction* interval for a single Y_o has to be wider than the *confidence* interval for the mean μ_o (the more stable average of all possible Y_o's).

12-5 We must be very cautious about extrapolating the fitted line beyond the given data.

　　The theory of this chapter applies whether X is a controlled variable (like the fertilizer application specified by the researcher) or a random variable (like rainfall).

REVIEW PROBLEMS

12-13 In 1970, a random sample of 50 American men aged 35 to 54 showed the following relation between annual income Y (in dollars) and education X (in years). (Reconstructed from the 1970 U.S. Census 1 in 10,000 sample):

$$\hat{Y} = 1200 + 800X$$

Average income was $\overline{Y} = \$10,000$ and average education was $\overline{X} = 11.0$ years, with $\Sigma x^2 = 900$. The residual standard deviation about the fitted line was $s = \$7300$.

a. Calculate a 95% confidence interval for the population slope.

use p-value

b. Is the relation of income to education statistically discernible at the 5% level?

c. Predict the income of a man who completed 2 years of high school ($X = 10$). Include an interval wide enough that you would bet on it at odds of 95 to 5.

d. Would it be fair to say that each year's education is worth $800. Why?

12-14 A random sample of 5 boys had their heights (in inches) measured at age 4 and again at age 18, with the following results:

BOYS' INITIALS	AGE 4	AGE 18
J.K.	40	68
A.M.	43	74
D.B.	40	70
I.S.	40	68
J.B.	42	70
Average	41	70

a. Suppose another boy is randomly drawn from the same population. If he was 42 inches tall at age 4, would you predict he would be 70 inches tall at age 18 (like the last boy in the sample)? Why?

b. Construct a 95% prediction interval for what his height will be at age 18.

12-15 A 95% confidence interval for a regression slope was calculated on the basis of 1000 observations: $\beta = .38 \pm .27$. Calculate the p-value for the null hypothesis that Y does not increase with X.

12-16 A class of 150 registered students wrote two tests, for which the grades were denoted X_1 and X_2. The instructor calculated the following summary statistics:

$$\overline{X}_1 = 60 \qquad \overline{X}_2 = 70$$

$$\Sigma(X_1 - \overline{X}_1)^2 = 36,000 \qquad \Sigma(X_2 - \overline{X}_2)^2 = 24,000$$

$$\Sigma(X_1 - \overline{X}_1)(X_2 - \overline{X}_2) = 15,000$$

Residual variance of regression of X_1 against X_2, $s^2 = 180$

Residual variance of regression of X_2 against X_1, $s^2 = 120$

The instructor then discovered that there was one more student, who was unregistered; worse yet, one of this student's grades (X_1) was lost, although the other grade was discovered ($X_2 = 55$). The dean told the instructor to estimate the missing grade X_1 as closely as possible.

a. Calculate the best estimate you can, including an interval that you have 95% confidence will contain the true grade.

b. What assumptions did you make implicitly?

12-17 In order to estimate this year's inventory (excess inventory), a tire company sampled 6 dealers, in each case getting inventory figures for both this year and last:

X = INVENTORY LAST YEAR	Y = INVENTORY THIS YEAR	$\hat{y}$	$(y-\hat{y})^2$
70	60	89.30	859.66
260	320	329.86	97.22
150	230	190.6	1552.36
100	120	127.3	575.64
20	50	26.02	575.64
60	60	76.66	277.56

$$3361.84$$

Summary statistics are $\overline{X} = 110$, $\overline{Y} = 140$.

$$\hat{y} = .7 + 1.266X$$

$$\Sigma x^2 = 36,400 \qquad \Sigma y^2 = 61,800 \qquad \Sigma xy = 46,100$$

$$s^2 = 840.46$$

a. Calculate the least squares line showing how this year's inventory Y is related to last year's X.

b. Suppose that a complete inventory of all dealers is available for last year (but not for this year). Suppose also that the mean inventory for last year was found to be $\mu_X = 180$ tires per dealer. On the graph below, we show this population mean μ_X and sketch the population scatter (although this scatter remains unknown to the company, because Y values are not available yet). On this graph, plot the six observed points, along with $\overline{X}$, $\overline{Y}$, and the estimated regression line.

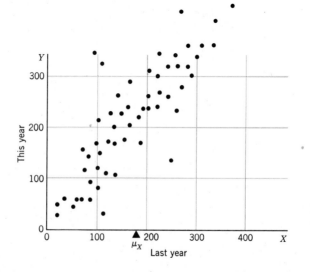

    c. Indicate on the graph how μ_Y should be estimated. Construct a 95% confidence interval for μ_Y.

    d. Construct a 95% confidence interval for μ_Y, if last year's data X had been unavailable or ignored (i.e., using only Y values).

    e. Comparing part **c** to **d,** state in words the value of exploiting prior knowledge about last year's inventory.

12-18   **a.** In Problem 12-17, for the fitted regression line $\hat{Y} = a + bX$, what was $\Sigma(Y - \hat{Y})^2$ (the prediction errors, squared and summed)?

    b. Suppose a vice president of the tire company suggests using the following prediction: $\hat{Y} = 10 + X$ (that is, this year is predicted to be just 10 more than last year). Can you say what $\Sigma(Y - \hat{Y})^2$ will now be, compared to (a)? Why? Then calculate $\Sigma(Y - \hat{Y})^2$.

12-19   **a.** Prove (12-25). [*Hint:* (12-32)]

    b. For the data in Problem 12-17, calculate a 95% confidence interval for α.

12-20   **a.** In Problem 5-14, part **a,** we calculated $E(Y/X)$, for four values of X. Graph these 4 points. Is the linearity assumption in (12-2) satisfied?

    b. Continue Problem 5-14: calculate var(Y/X) for each value of X. Is the assumption of constant variance in (12-2) satisfied?

Chapter

Multiple Regression

"The cause of lightning," Alice said very decidedly, for she felt quite sure about this, "is the thunder—no, no!" she hastily corrected herself, "I meant it the other way."

"It's too late to correct it," said the Red Queen. "When you've once said a thing, that fixes it, and you must take the consequences."

LEWIS CARROLL

13-1 Why Multiple Regression?

Multiple regression is the extension of simple regression, to take account of more than one independent variable X. It is obviously the appropriate technique when we want to investigate the effect on Y of several X variables simultaneously. Yet, even if we are interested in the effect of only one variable, it usually is wise to include the other variables in a multiple regression analysis, for two reasons:

1. In observational studies, it is essential to eliminate the bias of extraneous variables by including them as regressors—a point first emphasized in Chapter 1.
2. In both observational studies and randomized experiments, the introduction of additional variables can reduce the residual variance s^2, and hence improve confidence intervals and tests—just as the introduction of a second factor in ANOVA strengthened those tests.

EXAMPLE **13-1** Suppose the fertilizer and yield observations in Example 11-1 were taken at seven different plots across the state. If soil conditions and temperature were essentially the same in all these areas, we still might ask whether part of the fluctuation in Y can be explained by varying levels of rainfall in different areas. A better prediction of yield may be possible if *both* fertilizer and rainfall are examined. The observed levels of rainfall are therefore given in Table 13-1, along with the original observations of yield and fertilizer.

TABLE **13-1**

Observations of Yield, Fertilizer, and Rainfall

Y WHEAT YIELD (BU/ACRE)	X_1 FERTILIZER (LB/ACRE)	X_2 RAINFALL (INCHES)
40	100	10
50	200	20
50	300	10
70	400	30
65	500	20
65	600	20
80	700	30

a. On Figure 11-4 (p. 324), tag each point with its value of rainfall X_2. Then, considering just those points with low rainfall ($X_2 = 10$), fit a line by eye. Next, repeat for the points with moderate rainfall ($X_2 = 20$), and then for the points with high rainfall ($X_2 = 30$).

b. Now, if rainfall were kept constant, estimate the slope of yield on fertilizer. That is, estimate the increase in yield per pound of additional fertilizer.

c. If fertilizer were kept constant, estimate the increase in yield per inch of additional rainfall.

d. Roughly estimate the yield if fertilizer were 400 pounds, and rainfall were 10 inches.

Solution

a. The three fitted lines are shown in Figure 13-1.

b. The largest slope in Figure 13-1 is 10/200 = .05 (on the lowest line where $X_2 = 10$). The smallest slope is 10/300 = .03 (on the highest line where $X_2 = 30$). On average these slopes are about .04 bushels per pound of fertilizer.

c. Let us keep fertilizer constant—for example, at the center of the data, where $X_1 = 400$. The dashed line shows the vertical distance of about 15 bushels, between the line where rainfall $X_2 = 10$ and the line where $X_2 = 30$. Since this increase of 15 bushels comes from an increase of 20 inches of rain, rain increases yield by about 15/20 = .75 bushels per inch of rainfall.

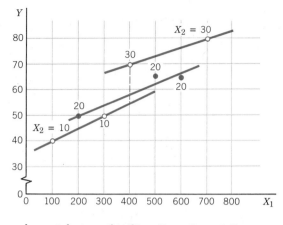

Figure 13-1
How yield depends on 2 factors (fertilizer X_1 and rainfall X_2).

d. On Figure 13-1 we go along the horizontal axis to $X_1 = 400$, and then go directly up to the line where $X_2 = 10$. At that point we read off the yield $\simeq 55$ bushels.

Figure 13-1 shows clearly why adding an additional rainfall variable X_2 gives a better idea of the effect of fertilizer X_1 on yield Y. When rainfall was not held constant and was ignored, we obtained the slope in Figure 11-4—which is steeper because high rainfall tends to accompany high fertilizer. Thus the slope in Figure 11-4 is thrown off because we erroneously attribute to fertilizer the effects of both fertilizer *and* rainfall.

This tendency for high rainfall to accompany high fertilizer may occur even if the two factors are determined in a randomized experiment. (Such an imbalance would be a fluke, like a coin tossed 10 times coming up heads 7 or 8 times.) In this case the error in the slope in Figure 11-4 would be a random error, which would be averaged out, were the sample size to increase indefinitely. However, in a finite (realistic) sample this error would not quite average out, and it is helpful to use multiple regression to remove it.

On the other hand, in an observational study of the kind we are considering here, this tendency for high rainfall and high fertilizer to occur together may be more than a fluke, and may persist undiminished even in very large samples. (To see why, suppose there is no randomized assignment of fertilizer and instead this decision is made by the owners of the land. Those with the most productive and profitable land—because it receives more rainfall—may be able to afford more fertilizer.) In this case the error in the slope of Figure 11-4 is bias, and would *never* average out, even in a very large sample. In such circumstances, it is *essential* to use multiple regression to remove this bias.

PROBLEMS

13-1 Suppose that a random sample of 5 families yielded the following data (Y and X_1 are measured in thousands of dollars annually):

FAMILY	SAVING Y	INCOME X_1	CHILDREN X_2
A	2.1	15	2
B	3.0	28	4
C	1.6	20	4
D	2.1	22	3
E	1.2	10	2

a. On the $(X_1 Y)$ plane, graph the 5 points (and tag each point with its value of X_2). Then fit a line by eye to just those points tagged with the lowest value of X_2. Repeat, for those points tagged with the highest value of X_2.

b. If X_2 is kept constant, estimate the change in Y for a unit increase in X_1.

c. If X_1 is kept constant, estimate the change in Y for a unit increase in X_2.

d. Estimate the saving Y for a family with an income $X_1 = 25$ thousand, and $X_2 = 4$ children.

13-2 **a.** Continuing Problem 13-1, calculate the simple regression of Y against X_1, and show it on the same graph.

b. How would you interpret the slope? Why is it less than in the multiple regression?

13-3 Suggest possible additional regressors that might be used to improve the multiple regression analysis of wheat yield.

13-2 The Regression Model and Its OLS Fit

A—THE GENERAL LINEAR MODEL (GLM)

Yield Y is now to be regressed against two or more factors (such as fertilizer X_1 and rainfall X_2). Let us suppose the relationship is of the form

$$E(Y) = \beta_0 + \beta_1 X_1 + \beta_2 X_2$$

$$\text{(13-1)}$$
$$\text{like (12-1)}$$

Geometrically this equation is a plane in the three-dimensional space shown in Figure 13-2. (The geometry of lines and planes is reviewed in Appendix 11-1.) For any given combination of factors (X_1, X_2), the expected Y is the point on the plane directly above, shown as a hollow dot.

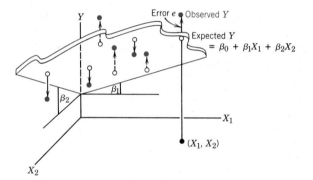

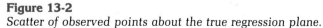

Figure 13-2
Scatter of observed points about the true regression plane.

The observed yield Y, shown as usual as a colored dot, will be somewhat different of course, and the difference is the random error. Thus any observed Y may be expressed as its expected value plus the random error e:

$$Y = \underbrace{\beta_0 + \beta_1 X_1 + \beta_2 X_2}_{E(Y)} + e \qquad (13\text{-}2)$$

like (12-3)

And we make the same assumptions about e as in Chapter 12.

β_1 is geometrically interpreted as the slope of the plane as we move in the X_1 direction keeping X_2 constant, sometimes called the marginal effect of fertilizer X_1 on yield Y. Similarly β_2 is the slope of the plane as we move in the X_2 direction keeping X_1 constant, called the marginal effect of X_2 on Y.

Of course, equation (13-2) can be extended to include more than two regressors:

$$Y = \beta_0 + \beta_1 X_1 + \beta_2 X_2 + \beta_3 X_3 + \cdots + e \qquad (13\text{-}3)$$

We will see (in Chapter 14) how ingenious choices of the regressors can make this model very useful indeed—of such general applicability that it is often called the *General Linear Model (GLM)*.

B—THE LEAST SQUARES FIT

As in simple regression, the problem is that the statistician does not know the true equation (13-2). Instead, she must fit an *estimated* equation of the form

$$\hat{Y} = b_0 + b_1 X_1 + b_2 X_2 \qquad (13\text{-}4)$$

like (11-1)

When we apply the usual least squares criterion for a good fit in multiple regression, we unfortunately do not get easy formulas for b_0, b_1, and b_2. Instead, we get 3 equations to solve, sometimes called *estimating equations* or *normal equations*:

$$\Sigma x_1 y = b_1 \Sigma x_1^2 + b_2 \Sigma x_1 x_2 \tag{13-5}$$

$$\Sigma x_2 y = b_1 \Sigma x_1 x_2 + b_2 \Sigma x_2^2 \tag{13-6}$$

$$b_0 = \overline{Y} - b_1 \overline{X}_1 - b_2 \overline{X}_2 \tag{13-7}$$

where once again we have used the convenient deviations:

$$\left. \begin{array}{l} x_1 \equiv X_1 - \overline{X}_1 \\ x_2 \equiv X_2 - \overline{X}_2 \\ y \equiv Y - \overline{Y} \end{array} \right\} \tag{13-8}$$

Equations (13-5) and (13-6) may be simultaneously solved for b_1 and b_2. (The algebraic solution of a set of simultaneous linear equations is reviewed in Appendix 13-2 at the back of the book.) Then equation (13-7) may be solved for b_0. For the data of Table 13-1, these calculations are shown in Table 13-2, and yield the fitted multiple regression equation

$$\hat{Y} = 28 + .038X_1 + .83X_2 \tag{13-9}$$

TABLE **13-2** Calculations for the Multiple Regression of Y Against X_1 and X_2

Y	X_1	X_2	$y = Y - \overline{Y}$	$x_1 = X_1 - \overline{X}_1$	$x_2 = X_2 - \overline{X}_2$	$x_1 y$	$x_2 y$	x_1^2	x_2^2	$x_1 x_2$
40	100	10	−20	−300	−10	6000	200	90,000	100	3000
50	200	20	−10	−200	0	2000	0	40,000	0	0
50	300	10	−10	−100	−10	1000	100	10,000	100	1000
70	400	30	10	0	10	0	100	0	100	0
65	500	20	5	100	0	500	0	10,000	0	0
65	600	20	5	200	0	1000	0	40,000	0	0
80	700	30	20	300	10	6000	200	90,000	100	3000
$\overline{Y} = 60$	$\overline{X}_1 = 400$	$\overline{X}_2 = 20$	$0\checkmark$	$0\checkmark$	$0\checkmark$	$\Sigma x_1 y = 16,500$	$\Sigma x_2 y = 600$	$\Sigma x_1^2 = 280,000$	$\Sigma x_2^2 = 400$	$\Sigma x_1 x_2 = 7000$

Estimating equations (13-5) and (13-6) $\begin{cases} 16,500 = 280,000 b_1 + 7000 b_2 \\ 600 = 7000 b_1 + 400 b_2 \end{cases}$

Solution $\begin{cases} b_1 = .0381 \\ b_2 = .833 \end{cases}$

From (13-7), $b_0 = 60 - .0381(400) - .833(20)$
$b_0 = 28.1$

TABLE **13-3** **Computer Solution for Wheat Yield Multiple Regression (Data in Table 13-1)**

```
          > READ'Y','X1','X2'
   DATA>      40 100   10
   DATA>      50 200   20
   DATA>      50 300   10
   DATA>      70 400   30
   DATA>      65 500   20
   DATA>      65 600   20
   DATA>      80 700   30

        > REGRESS'Y' ON 2 REGRESSORS'X1','X2'

   THE REGRESSION EQUATION IS
   Y = 28.1 + 0.0381 X1 + 0.833 X2
```

As we mentioned in the preface, statistical calculations are usually done on a computer using a standard program. This is particularly true of multiple regression, where the calculations generally become much too complicated to do by hand. (Imagine what Table 13-2 would look like with 100 rows of data and 5 regressors—a fairly typical situation.) Therefore in Table 13-3 we show a computer estimate of a regression equation such as (13-9).

We recommend that you use a computer yourself to calculate the homework problems. It is a good opportunity to meet the machine, since a standard multiple regression program is available in practically every computer center (or even in some sophisticated pocket calculators).

If a computer is not available, you can still calculate the regression in a tableau like Table 13-2. Or you can skip the calculation entirely, and go on to the rest of the problem where the interpretation of the regression is discussed—this is the part that requires the human touch. For instance, the next example will show how the multiple regression equation (13-9) nicely answers the same questions as Example 13-1.

EXAMPLE **13-2**   **a.** Graph the relation of yield Y to fertilizer X_1 given by (13-9), when rainfall has the constant value:

(**i**) $X_2 = 10$, (**ii**) $X_2 = 20$, (**iii**) $X_2 = 30$

b. Compare to the figure in Example 13-1

Solution   **a.** Substitute $X_2 = 10$ into (13-9):

$$\hat{Y} = 28 + .038X_1 + .83X_2 \qquad \text{(13-9) repeated}$$
$$\hat{Y} = 28 + .038X_1 + .83(10)$$
$$= 36.3 + .038X_1 \qquad\qquad \text{(13-10)}$$

This is a line with slope .038 and Y intercept 36.3. Similarly, when we substitute $X_2 = 20$ and then $X_2 = 30$ we obtain the lines

$$\hat{Y} = 44.6 + .038X_1 \qquad (13\text{-}11)$$
$$\hat{Y} = 52.9 + .038X_1 \qquad (13\text{-}12)$$

These are lines with the same slope .038, but higher and higher intercepts, as shown in Figure 13-3.

b. The three lines given by the multiple regression model are evenly spread because we had evenly spaced values of X_2 in part **a**. The lines are also parallel, because they all have the same slope $b = .038$.

In contrast, the three earlier lines fitted in Example 13-1 were not constrained to have the same spacing or slope.

Note that the three separate lines calculated earlier in Example 13-1 fit the data a little closer than the multiple regression estimated in (13-9). However, they are harder to summarize: In Example 13-1b, we calculated the average slope to be about .04, whereas in (13-9), we can immediately see the slope is .038. Similarly, in Example 13-1c, we calculated the average effect of one inch of rain to be about .75, whereas in (13-9) we can immediately see that it is .83.

Another important difference is that in Example 13-1 we had only 2 or 3 observations for each fitted line. However, in estimating the multiple regression (13-9), we used all 7 observations; this will put us in a stronger position to test hypotheses or construct confidence intervals.

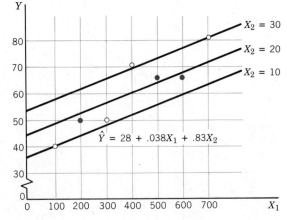

Figure 13-3
Multiple regression fits the data with parallel lines.

PROBLEMS

13-4   In Problem 13-1, the multiple regression of Y against X_1 and X_2 was computed to be

$$\hat{Y} = .77 + .148X_1 - .52X_2$$

    a.   Graph the relation of Y to X_1 when X_2 has the constant value:
        (i) $X_2 = 2$, **(ii)** $X_2 = 3$, **(iii)** $X_2 = 4$.
        Compare to the graph in Problem 13-1, part **a.**

    b.   If X_2 is kept constant, estimate the change in Y for a unit increase in X_1.

    c.   If X_1 is kept constant, estimate the change in Y for a unit increase in X_2.

    d.   Estimate the saving Y for a family with an income $X_1 = 25$ thousand, and $X_2 = 4$ children.

    *e.   Using a computer, or a tableau like Table 13-2, calculate the multiple regression coefficients and verify that they agree with the given equation.

13-5   In the midterm U.S. congressional elections (between Presidential elections), the party of the President usually loses seats in the House of Representatives. To measure this loss concretely, we take as our base the average congressional vote for the President's party over the previous 8 elections; the amount that the congressional vote drops in a given midterm election, relative to this base, will be our *standardized vote loss Y*.

    Y depends on several factors, two of which seem important and easily measurable: X_1 = Gallup poll rating of the President at the time of the midterm election (percent who approved of the way the President is handling his job) and X_2 = change over the previous year in the real disposable annual income per capita.

YEAR	$Y =$ STANDARDIZED VOTE LOSS	$X_1 =$ PRESIDENT'S GALLUP RATING	$X_2 =$ CHANGE IN REAL INCOME OVER PREVIOUS YEAR
1946	7.3%	32%	− \$40
1950	2.0	43	100
1954	2.3	65	− 10
1958	5.9	56	− 10
1962	− .8	67	60
1966	1.7	48	100

    From the above data (Tufte, 1974), the following multiple regression equation was computed:

$$\hat{Y} = 10.9 - .13X_1 - .034X_2$$

a. On the (X_1, Y) plane, graph the 6 points (and tag each point with its X_2 value). Then graph the grid of 4 lines you get from the regression equation by setting $X_2 = 100, 50, 0$, and -50.

b. If X_2 is kept constant, estimate the change in Y for a unit change in X_1.

c. If X_1 is kept constant, estimate the change in Y for a unit change in X_2.

d. Estimate the vote loss Y for a midterm election when $X_1 = 60\%$ approval, and $X_2 = \$50$ increase in real income.

*e. Using a computer or a tableau like Table 13-2, calculate the multiple regression coefficients, and verify that they agree with the given equation.

13-6 (Continuing Problem 13-5)

a. From the graph, find the fitted 1946 vote loss Y, given that $X_1 = 32\%$ and $X_2 = \$-40$. Confirm it exactly from the regression equation. Compared to the actual vote loss $Y = 7.3\%$, what is the error?

b. Now consider a real prediction. Put yourself back in time, just before the 1970 midterm election, when President Nixon's rating was $X_1 = 56\%$, and the change in real income was $X_2 = \$70$. From the graph, predict the 1970 vote loss. Confirm it exactly from the regression equation. It turns out that the actual vote loss Y was 1.0%; what therefore is the prediction error?

13-7 In the estimating equation (13-5), suppose it is known a priori that Y has no relation whatever to X_2; in other words, $b_2 = 0$. When you solve for b_1, what do you get?

13-3 Confidence Intervals and Statistical Tests

A—STANDARD ERROR

As in simple regression, the true relation of Y to any X is measured by the unknown population slope β; we estimate it with the sample slope b. Whereas the true β is fixed, the estimate b varies randomly from sample to sample, fluctuating around its target β with an approximately normal distribution. The standard deviation or standard error of b is customarily computed at the same time as b itself, as shown in Table 13-4. The meaning and use of the standard error are quite analogous to the simple regression case. For example, the standard error forms the basis for confidence intervals and tests.

B—CONFIDENCE INTERVALS AND p-VALUES

For each β coefficient, the formula for estimating it with a 95% confidence interval is of the standard form:

TABLE **13-4** **Computed Standard Errors and *t* Ratios (Continuation of Table 13-3)**

COLUMN	COEFFICIENT	ST. DEV. OF COEF.	T-RATIO = COEF/S.D.
	28.095	2.491	11.28
X1	0.038095	0.005832	6.53
X2	0.8333	0.1543	5.40

$$\boxed{\beta = b \pm t_{.025} \text{ SE}}$$ (13-13)
                                                     like (12-22)

When there are k regressors as well as the constant term, there are $(k + 1)$ coefficients to estimate, which leaves $n - (k + 1)$ d.f. That is,

$$\boxed{\text{d.f.} = n - k - 1}$$ (13-14)
                                                     like (12-24)

As usual, the observed t ratio to test $\beta = 0$ is

$$\boxed{t = \frac{b}{\text{SE}}}$$ (13-15)
                                                     like (12-29)

The p-value is then the tail area read from Appendix Table V. An example will illustrate.

EXAMPLE **13-3**   From the computer output in Table 13-4:

   a. Calculate a 95% confidence interval for the coefficient β_1.

   b. Calculate the p-value for the null hypothesis $\beta_1 = 0$ (fertilizer doesn't increase yield).

Solution   **a.** From (13-14), d.f. $= 7 - 2 - 1 = 4$, so that Appendix Table V gives $t_{.025} = 2.78$. Also substitute b_1 and its SE from Table 13-4; then (13-13) gives, for the fertilizer coefficient,

$$\beta_1 = .03810 \pm 2.78 \, (.00583)$$ (13-16)

$$= .038 \pm .016$$ (13-17)

b. In (13-16), b and its SE are given, so that we can easily form their ratio:

$$t = \frac{.03810}{.00583} = 6.53 \qquad (13\text{-}18)$$

Or, equivalently, this same t ratio can be read from the last column of Table 13-4. In any case, we again refer to Appendix Table V, scanning the row where d.f. = 4. We find that the observed t value of 6.53 lies beyond $t_{.0025} = 5.60$. Thus

$$\text{p-value} < .0025 \qquad (13\text{-}19)$$

With such little credibility, the null hypothesis can be rejected; we conclude that yield is indeed increased by fertilizer.

It is possible to summarize Example 13-3 (and similar calculations for the other regression coefficients) by arranging them in equation form as follows:

$$\text{YIELD} = 28 + .038 \text{ FERTILIZER} + .83 \text{ RAINFALL}$$

Standard error	.0058	.154
95% CI	± .016	± .43
t ratio	6.5	5.4
p-value	< .0025	< .005

$$(13\text{-}20)$$

C—A WARNING ABOUT DROPPING A REGRESSOR

In equation (13-20), both fertilizer and rainfall are kept in the model as statistically discernible (significant) regressors because their t values are large enough to allow H_o to be easily rejected in each case.

But now suppose we had a smaller sample and therefore weaker data; specifically, suppose that the standard error for rainfall was much larger—say .55 instead of .15. Then the t ratio would be $t = .83/.55 = 1.51$, so we could *not* reject H_o at the 5% level. If we use this evidence to actually accept H_o (no effect of rainfall) and thus drop rainfall as a regressor, *we may seriously bias the remaining coefficients*—as we emphasized in Section 13-1. As well, we may encounter the same philosophical difficulty we discussed earlier in Section 9-4. Since this argument is so important in regression analysis, let us briefly review it.

Although it is true that a t ratio of 1.51 for rainfall would be statistically indiscernible, this *would not prove* there is no relationship between rainfall and yield. It is easy to see why. We have strong biological grounds for believing that yield is positively related to rainfall. This belief would be confirmed by the positive coefficient $b = .83$. Thus our statistical evidence would be consistent with our prior belief, even though it would be weaker confirmation than we would like. To actually accept the null hypothesis $\beta = 0$, and conclude there is no relation, would be to contradict directly both the (strong) prior belief and the (weak) statistical evidence. We would be reversing a prior belief, even though the statistical evidence weakly confirmed it. And this would remain true for any positive t ratio—although, as t became smaller, our statistical confirmation would become weaker. Only if the coefficient were zero or negative would the statistical results contradict our prior belief.

To summarize: If we have strong prior grounds for believing that X is related positively to Y, X generally should not be dropped from the regression equation if it has the right sign. Instead, it should be retained along with the information in its confidence interval and p-value.

D—WHEN CAN A REGRESSOR BE DROPPED?

Continuing our example, suppose now our prior belief is that H_o is approximately true. Then our decision on whether or not to drop a variable would be quite different. For example, a weak observed relationship (such as $t = 1.51$) would be in some conflict with our prior expectation of no relationship. But it is a minor enough conflict that is easily explained by chance (p-value $\simeq .10$). Hence, resolving it in favor of our prior expectation and continuing to use H_o as a working hypothesis might be a reasonable judgment. In this case, the regressor could be dropped from the equation.

We conclude once again that classical statistical theory alone does not provide firm guidelines for accepting H_o; acceptance must be based also on extrastatistical judgment. Such prior belief plays a key role, not only in the initial specification of which regressors should be in the equation, but also in the decision about which ones should be dropped in light of the statistical evidence.

PROBLEMS

13-8 The following regression was calculated for a class of 66 students of nursing (Snedecor and Cochran, 1967):

$$\hat{Y} = 3.1 + .021X_1 + .075X_2 + .043X_3$$

Standard error	(.019)	.034)	(.018)
95% CI ( )	( )	( )	
t ratio ( )	( )	( )	
p-value ( )	( )	( )	

where Y = student's score on a theory examination
X_1 = student's rank (from the bottom) in high school
X_2 = student's verbal aptitude score
X_3 = a measure of the student's character

    a. Fill in the blanks.

    b. What assumptions were you making in part **a?** How reasonable are they?

    c. Which regressor gives the strongest evidence of being statistically discernible? (This also tends to be the regressor producing the largest change in student's score Y.)

    *d. In writing up a final report, would you keep the first regressor in the equation, or drop it? Why?

13-9 In Problem 13-5, the congressional vote loss of the President's party in midterm elections (Y) was related to the President's Gallup rating (X_1) and change in real income over the previous year (X_2). Specifically, the following regression was computed from $n = 6$ points:

$$\hat{Y} = 10.9 - .13X_1 - .034X_2$$
$$\text{Standard error} \quad (.046) \quad (.010)$$

Answer the same questions as in Problem 13-8.

13-10 Suppose that your roommate is a bright student, but that he has studied no economics, and little statistics. (Specifically, he understands only simple—but not multiple—regression.) In trying to explain what influences the U.S. price level, he has regressed U.S. prices on 100 different economic variables one at a time (i.e., in 100 simple regressions). Moreover, he apparently selected these variables in a completely haphazard way without any idea of potential cause-and-effect relations. He discovered 5 variables that were statistically discernible at the level $\alpha = 5\%$, and concluded that each of these has an influence on U.S. prices.

    a. Explain to him what reservations you have about his conclusion.

    b. If he had uncovered 20 statistically discernible variables, what reservations would you now have? How could he improve his analysis?

13-4 Regression Coefficients as Multiplication Factors

A—SIMPLE REGRESSION

The coefficients in a linear regression model have a very simple but important interpretation. Recall the simple regression model,

$$Y = a + bX \qquad\qquad (13\text{-}21)$$
$$\text{like } (11\text{-}1)$$

The coefficient b is the slope:

$$\frac{\Delta Y}{\Delta X} = b \qquad\qquad (13\text{-}22)$$

where ΔX is any change in X, and ΔY is the corresponding change in Y. We can rewrite (13-22) in another form:

$$\Delta Y = b \, \Delta X \qquad (13\text{-}23)$$

Since this is so important, we write it verbally:

$$\boxed{\textbf{Change in } Y = b \textbf{ (change in } X)} \qquad (13\text{-}24)$$

For example, consider the fertilizer-yield example,

$$Y = 36 + .06X \qquad \text{like (11-7)}$$

How much higher would yield Y be, if fertilizer X were 5 pounds higher? From (13-24) we find,

$$\text{Change in yield} = .06(5) = .30 \text{ bushel} \qquad (13\text{-}25)$$

Since any change that occurs in X is multiplied by b in order to find the corresponding change in Y, we can call b the *multiplication* factor.

Of course, when the change is $\Delta X = 1$, then (13-24) becomes: Change in $Y = b$; that is,

$$\boxed{b = \textbf{change in } Y \textbf{ that accompanies a unit change in } X} \qquad (13\text{-}26)$$
$$\text{like (11-8)}$$

Thus, for example, the coefficient $b = .06$ means that a change of .06 bushel in yield Y accompanies a 1-pound change in fertilizer X, as illustrated in Figure 13-4.

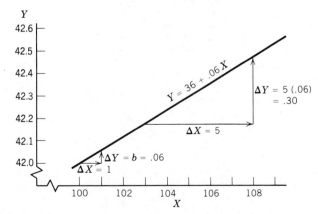

Figure 13-4
Interpretation of slope: b = *change in Y that accompanies a unit change in X.*

B—MULTIPLE REGRESSION: "OTHER THINGS BEING EQUAL"

Consider again the multiple regression model

$$Y = b_0 + b_1X_1 + b_2X_2 \qquad (13\text{-}27)$$
$$\text{like } (13\text{-}4)$$

If X_2 *remains constant*, it is still true that:

$$\Delta Y = b_1 \Delta X_1 \qquad (13\text{-}28)$$
$$\text{like } (13\text{-}23)$$

Equation (13-28) is so important that it is worthwhile giving a simple proof: If we keep X_2 constant, while we increase X_1 to $(X_1 + \Delta X_1)$, then from (13-27),

$$\text{initial } Y = b_0 + b_1X_1 + b_2X_2$$
$$\text{new } Y = b_0 + b_1 (X_1 + \Delta X_1) + b_2X_2$$

$$\text{difference: } \Delta Y = b_1 \Delta X_1 \qquad (13\text{-}28) \text{ proved}$$

Of course, we can easily generalize to the case of several regressors: (13-28) still describes how Y will be related to any one of the X regressors, *provided all the others remain constant*—that is, that "other things remain equal."

> **If one regressor, say X_1, changes while all the others remain constant, then**
> **change in Y = b_1 (change in X_1)** $\qquad (13\text{-}29)$

For example, suppose wheat yield Y is related to fertilizer X_1, rainfall X_2, and temperature X_3, as follows:

$$Y = 30 + .036X_1 + .81X_2 + .02X_3$$

How much would yield Y increase if fertilizer X_1 were increased 5 pounds, while X_2 and X_3 did not change? From (13-29),

$$\text{Change in yield} = .036(5) = .18 \text{ bushel} \qquad (13\text{-}30)$$

Thus the coefficient $b_1 = .036$ is a multiplication factor in multiple regression.

Again, when the change is $\Delta X_1 = 1$, then (13-29) becomes:

> **b_1 = change in Y that accompanies a unit change in the regressor X_1, if all the other regressors remain constant.** $\qquad (13\text{-}31)$

So far we have only seen what happens when one regressor changes. But what happens if all the regressors change simultaneously? Just as we proved (13-28), we could now show that the change in Y is just the sum of the individual changes:

$$\text{If } Y = b_0 + b_1 X_1 + b_2 X_2 + \cdots,$$
$$\text{then } \Delta Y = b_1 \Delta X_1 + b_2 \Delta X_2 + \cdots. \qquad (13\text{-}32)$$

To contrast the different uses of multiple and simple regression, we end with an example.

EXAMPLE **13-4**   The simple and multiple regressions of yield against fertilizer and rainfall (obtainable from Table 13-2) are as follows:

$$\text{YIELD} = 36 + .059\ \text{FERT} \qquad (13\text{-}33)$$
$$\text{YIELD} = 30 + 1.50\ \text{RAIN} \qquad (13\text{-}34)$$
$$\text{YIELD} = 28 + .038\ \text{FERT} + .83\ \text{RAIN} \qquad (13\text{-}35)$$

a.   If a farmer adds 100 more pounds of fertilizer per acre, how much can he expect his yield to increase?

b.   If he irrigates with 3 inches of water, how much can he expect his yield to increase? (Assume that water from irrigation and rainfall have the same effect.)

c.   If he simultaneously adds 100 more pounds of fertilizer per acre, and irrigates with 3 inches of water, how much can he expect his yield to increase?

d.   We have already remarked that high fertilizer application tends to be associated with high rainfall, in the data on which these three regression equations were calculated. If this same tendency persisted, how much more yield would you expect on an acre that has been getting 3 more inches of rainfall than another acre?

Solution   a.   The question is: What happens if he adds 100 pounds of fertilizer, with rainfall unchanged? The answer therefore is provided by the multiple regression equation. We multiply the 100 pounds by the multiple regression coefficient .038 (i.e., the magnification factor for fertilizer):

$$.038(100) = 3.8 \text{ bushels} \qquad \text{like (13-28)}$$

b.   The question is: What happens if he adds 3 inches of water, with fertilizer held constant? Again, the answer lies in the multiple regression equation. We multiply the 3 inches by the multiple regression coefficient .83 (i.e., the magnification factor for water):

$$.83(3) = 2.5 \text{ bushels}$$

c. From (13-32),

$$\begin{aligned}
\Delta Y &= .038(100) + .83(3) \\
&= 3.8 + 2.5 \\
&= 6.3 \text{ bushels}
\end{aligned} \tag{13-36}$$

d. Now we are not holding fertilizer constant, but letting it vary as rainfall varies, in the same pattern that produced (13-34) (i.e., a pattern where, for example, farmers that get more rainfall are more prosperous and can afford more fertilizer). So we do *not* want to use the coefficient of .83 in (13-35), because this shows how yield rises with rainfall alone (with fertilizer constant). Instead, we go back to (13-34), whose coefficient of 1.50 shows how yield rises with rainfall when fertilizer is changing too:

$$1.50(3) = 4.5 \text{ bushels}$$

To sum up: This is larger than our answer in part **b**—because the simple regression coefficient of 1.50 shows how yield is affected by rainfall *and* the associated fertilizer increase.

PROBLEMS

13-11 To determine the effect of various influences on land value in Florida, the sale price of residential lots in the Kissimmee River Basin was regressed on several factors. With a data base of $n = 316$ lots, the following multiple regression was calculated (Conner and others, 1973; via Anderson and Sclove, 1978):

$$\hat{Y} = 10.3 + 1.5X_1 - 1.1X_2 - 1.34X_3 + \cdots$$

where

$$\begin{aligned}
Y &= \text{price per front foot} \\
X_1 &= \text{year of sale } (X_1 = 1, \ldots, 5 \text{ for } 1966, \ldots, 1970) \\
X_2 &= \text{lot size (acres)} \\
X_3 &= \text{distance from the nearest paved road (miles)}
\end{aligned}$$

    a. Other things being equal, such as year of sale and distance from the nearest paved road, was the price (per front foot) of a 5 acre lot more or less than a 2 acre lot? How much?

    b. Other things being equal, how much higher was the price (per front foot) if the lot was ½ mile closer to the nearest paved road?

    c. Was the average selling price of a lot (per front foot) higher in 1970 than in 1966? How much?

13-12 A study of several hundred professors' salaries in a large American university in 1969 yielded the following multiple regression. (From Katz, 1973. These same professors were discussed in Problems 2-1 and 8-13. But now we are using more factors than just sex. In order to be brief, however, we do not write down all the factors that Katz included.)

$$\hat{S} = 230B + 18A + 100E + 490D + 190Y + 50T + \cdots$$

Standard error	(86)	(8)	(28)	(60)	(17)	(370)
95% CI	( )	( )	( )	( )	( )	( )
t ratio	( )	( )	( )	( )	( )	( )
p-value	( )	( )	( )	( )	( )	( )

where, for each professor,

    S = the annual salary (dollars)
    B = number of books written
    A = number of ordinary articles written
    E = number of excellent articles written
    D = number of Ph.D's supervised
    Y = number of years' experience
    T = teaching score as measured by student evaluations, severely rounded: the best half of the teachers were rounded up to 100% (i.e., 1): The worst half were rounded down to 0.

    a. Fill in the blanks below the equation.

    b. For someone who knows no statistics, briefly summarize the influences on professors' incomes, by indicating where strong evidence exists and where it does not.

    c. Answer True or False; if False, correct it.

        i. The coefficient of B is estimated to be 230. Other social scientists might collect other samples from the same population and calculate other estimates. The distribution of these estimates would be centered around the true population value of 230.

        ii. Other things being equal, we estimate that a professor who has written one or more books earns $230 more annually. Or, we might say that $230 estimates the value (in terms of a professor's salary) of writing one or more books.

        iii. Other things being equal, we estimate that a professor who is 1 year older earns $190 more annually. In other words, the annual salary increase averages $190.

    d. Similarly, interpret all the other coefficients for someone who knows no statistics.

13-13 Each year for 20 years, the average values of hay yield Y, temperature T, and rainfall R were recorded in England (Hooker, 1907; via Anderson, 1958), so that the following regressions could be calculated:

$$\hat{Y} = 40.4 - .208T$$
$$\text{Standard error} \quad (.112)$$

$$\hat{Y} = 12.2 + 3.22R$$
$$\text{Standard error} \quad (.57)$$

$$\hat{Y} = 9.14 + .0364T + 3.38R$$
$$\text{Standard error} \quad (.090) \quad (.70)$$

Estimate the average increase in yield from one year to the next:

a. If rainfall increases 3, and temperature remains the same.

b. If temperature increases 10, and rainfall remains the same.

c. If rainfall increases 3, and temperature increases 10.

d. If rainfall increases 3, and we don't know how much temperature changes (although we know it likely will drop, since wet seasons tend to be cold).

e. If rainfall increases 3, and temperature decreases 13.

f. If temperature increases 10, and we don't know how much rainfall changes (although we know it will likely fall, since hot seasons tend to be dry).

13-14 In Problem 13-13:

a. What yield would you predict if $T = 50$ and $R = 5$?

b. What yield would you predict if $T = 65$ and $R = 7$?

c. By how much has yield increased in part **b** over part **a**? Confirm this answer using (13-32).

13-15 Answer True or False; if False, correct it:

a. The simple regression equations in Problem 13-13 occasionally can be useful. For example, in the absence of any information on temperature, the second equation would correctly predict that a year with below-average rainfall would produce above-average yield.

b. In view of the positive multiple regression coefficient, however, it would improve the crop to irrigate.

*13-16 In Problem 13-13, parts **a** and **b,** put 95% confidence intervals on your answers, assuming the 20 years formed a random sample. What is the population being sampled?

*13-5 Simple and Multiple Regression Compared

A—DIRECT AND INDIRECT EFFECTS

The idea of regression coefficients being multiplication factors or multipliers is so useful that we will illustrate it with another example.

EXAMPLE **13-5** In a fertility survey of 4700 Fiji women (Kendall and O'Muircheartaigh, 1977), the following variables were measured for each woman:

$$AGE = \text{woman's present age, at time of the study} \quad X_1$$
$$EDUC = \text{woman's education, in years} \quad X_2$$
$$CHILDN = \text{number of children the woman has borne} \quad Y$$

From this data, two regression equations were calculated:

$$CHILDN = 3.4 + .059 \, AGE - .16 \, EDUC \qquad (13\text{-}37)$$

$$EDUC = 7.6 - .032 \, AGE \qquad (13\text{-}38)$$

a. For a woman who is 1 year older than another, calculate:

  i. The expected change in CHILDN, if EDUC is constant.
  ii. The expected change in EDUC.
  iii. The expected change in CHILDN, if EDUC is changing too.

b. What is the simple regression coefficient of CHILDN against AGE?

Solution **a.** A diagram will help to keep all these equations and questions straight. In Figure 13-5, regression *coefficients are shown as thin black arrows,* with the dependent *Y* variable on the right. Thus, the two arrows b_1 and b_2 leading to CHILDN are the multiple regression coefficients in (13-37), while the single downward arrow b is the simple regression coefficient in (13-38).

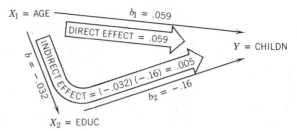

Figure 13-5
The direct and indirect effects of age on children.

Noting that the change in age is $\Delta AGE = 1$, we can now answer the questions:

i. Since EDUC is constant, the expected change for CHILDN is given by the multiple regression coefficient b_1 (i.e., the arrow pointing from AGE to CHILDN): There is an expected increase of .059 children.

ii. The expected change in EDUC is given by the regression coefficient b (i.e., the arrow pointing down from AGE to EDUC): There is an expected increase of $-.032$ years of education, that is, .032 *fewer* years of education.

iii. Since EDUC is changing simultaneously, the expected change in CHILDN is found from the *simple* regression of CHILDN against AGE, which is unfortunately not given. Let us see if we can calculate it from the regressions that *are* given. If we consider women 1 year older ($\Delta AGE = 1$), two effects must be considered:

    1. *The direct effect*: By itself, we have already seen in **i** that AGE produces an increase of .059 in CHILDN (black arrow b_1).

    2. *The indirect effect*: AGE affects CHILDN indirectly because, following the lower two black arrows, AGE affects EDUC, which in turn affects CHILDN. Specifically, in part **ii** we saw that the 1-year change in AGE produced a change of $-.032$ in EDUC, which in turn produces a change in CHILDN of:

$$\Delta CHILDN = b_2 \, \Delta EDUC$$
$$= -.16\,(-.032) = +.005 \qquad (13\text{-}39)$$

The broad white arrows in the diagram sum up these two effects:

> Direct effect of AGE on CHILDN $b_1 = .059$
>
> Indirect effect, via EDUC $bb_2 = (-.032)(-.16) = .005$
>
> ---
>
> Total effect $b_1 + bb_2 = .064$
>
> Notice that to calculate the indirect effect, we simply *follow the arrows*, multiplying the coefficients as we go along.
>
> **b.** When we calculate the simple regression coefficient of CHILDN against AGE, we are *not* holding EDUC constant; so we should be capturing both the direct and indirect effects estimated above. We therefore expect:
>
> Simple regression coefficient = Total effect = .064   (13-40)
>
> We could confirm that this is indeed correct by actually computing the simple regression coefficient from the data. (It turned out to be .064, just as we expected.)

We can generally prove (in Appendix 13-5) our discovery in Example 13-5 that the simple regression coefficient reflects the total effect[1]:

$$\boxed{\begin{aligned}\textbf{Total effect} &= \textbf{Direct effect} + \textbf{Indirect effect}\\ &= b_1 + bb_2\end{aligned}}$$
(13-41)

The advantage of decomposing the total effect (the simple regression coefficient) into its direct and indirect effects is that we can immediately see their relative importance. We can also identify the bias that will occur if we erroneously used the simple regression coefficient (the total effect) when we should be using the multiple regression coefficient (the direct effect), as we will next examine in detail.

B—BIAS FROM OMITTING IMPORTANT EXTRANEOUS REGRESSORS

Recall that in Example 13-5a we want to know how AGE affects CHILDN, other things (i.e., EDUC) being constant; in other words, we want to estimate how many more children a woman will have if she is 1 year older

---

[1]Equation (13-41)—and path analysis later in equation (13-44)—have been easier to state and understand in terms of cause-and-effect relationships. But remarkably this analysis still holds even if there is no such cause and effect. (For example, X_1 need not influence X_2 as black arrow b in Figure 13-5 suggests. It might be that X_2 influences X_1, or that some other factor influences both.) Whatever the relation, the regression coefficients still satisfy equation (13-41). This is a remarkably general result, and as Appendix 13-5 proves, is a consequence of the way that regression coefficients are calculated—algebraically, without any reference whatsoever to cause and effect.

than another woman with exactly the same education. Specifically, this estimate is given by the *multiple* regression coefficient $b_1 = .059$ that results from a regression of CHILDN on both AGE and EDUC. But suppose we erroneously omit the extraneous EDUC regressor. (We call EDUC extraneous here, because we are not interested in its effect—indeed we want to disallow this effect.) In ignoring the EDUC variable, we would then be running a *simple* regression of CHILDN against AGE alone and thereby capturing the *total* effect, which includes both the direct effect we want and the indirect effect that we don't want. In other words, the indirect effect would be bias.

To sum up: If we want to know how CHILDN is affected by AGE, *other things being constant*, we should run a *multiple* regression of CHILDN against both AGE and the extraneous variable EDUC, and then examine the top broad white arrow in Figure 13-5. If we erroneously ignore the extraneous regressor EDUC, then our simple regression of CHILDN against AGE will give us *both* of the white arrows; in other words, our estimate will be biased by the inclusion of the lower white arrow.

Of course, the world is full of extraneous regressors in which we have no interest. Obviously we cannot include them all. We only include an extraneous regressor like EDUC when it is *important*—that is:

i. It influences CHILDN (b_2 in Figure 13-5 is not zero).
ii. It is related to AGE (b is not zero).

In short, we include EDUC because it is related to *both* AGE and CHILDN. (If either b or b_2 were zero, then EDUC would become unimportant: The bias from excluding it—the lower white arrow—would disappear, and we could have left it out.)

In practice, there may be several important extraneous variables like EDUC. To avoid bias we should include them all. In other words:

> **The more we fail to include important extraneous regressors in an observational study, the more bias we risk (with the riskiest case being simple regression, where *all* such regressors are omitted).**

(13-42)

C—RANDOMIZATION REMOVES *ALL* BIAS

Equation (13-42) points up a real difficulty with observational studies: In many cases it is impossible to measure all the extraneous variables that we should be including.

For example, suppose that we carried out an observational study of how fertilizer affects yield, collecting data from various farmers across the state who chose to apply various amounts of fertilizer. Farmers who were more prosperous because they lived on richer land, for instance, might be the very ones who could afford the most fertilizer. Thus their increased yield would reflect not only fertilizer, but also the quality of their land. If such an extraneous influence isn't taken into account, it will introduce bias. But how do you measure the quality of land?

Fortunately, there is another way of removing bias—by randomization: In this example let the level of fertilizer be assigned to farmers at random, rather than through their own choice. Thus high levels of fertilizer would be assigned equally often, on average, to rich *and* poor land. This would break the tendency of the rich land to get the most fertilizer, and thus remove the reason for bias in the original observational study. In conclusion, we see that whenever it is feasible, random assignment of treatment (fertilizer) breaks the link between the treatment (the regressor whose direct effect we want) and *all* the extraneous variables—including the ones we cannot measure (or don't even know exist). Thus we confirm an important conclusion in Chapter 1:

Random assignment of treatment completely removes bias.

(13-43)
like (1-5)

*PROBLEMS

13-17

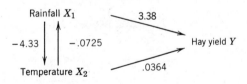

The diagonal arrows above show the multiple regression of Y against X_1 and X_2, and the vertical arrows show the simple regression of X_2 against X_1 and vice versa (same data as Problem 13-13).

a. What is the direct relation of rainfall to yield?

b. What is the total relation of rainfall to yield?

c. If an inch of irrigation was added every year, what effect do you estimate it would have?

d. What is the simple regression coefficient of yield against rainfall? And yield against temperature?

13-18

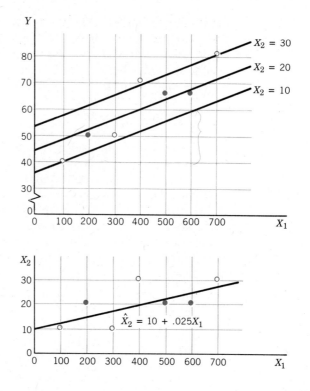

In the top figure, we repeat the data of Figure 13-3, which we suppose was gathered from an observational study. In the lower figure, we show explicitly the positive relation of fertilizer X_1 and rainfall X_2, including the fitted regression line.

a. Fill in the numerical regression coefficients in this figure (like Figure 13-5):

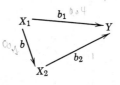

b. Calculate the indirect relation bb_2 and hence the total relation (simple regression of Y on X_1).

c. Graph the simple regression in the top figure, using the slope found in part **b**. (The Y-intercept is $a = 36$.) Does this simple regression fit the data well?

d. As the sample gets larger, would you expect the relation of fertilizer X_1 to rainfall X_2 to disappear? Would you therefore expect the indirect effect to disappear?

13-19 Repeat Problem 13-18 when fertilizer is assigned at *random* to 20 fields, producing the results below. (To draw the graph in part **c**, the Y-intercept is $a = 44$.)

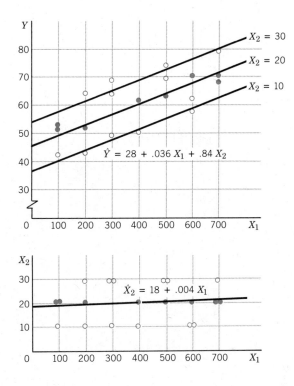

*13-6 Path Analysis

A *path analysis* is just a generalization of Figure 13-5: The first step is to lay out a sequence of variables from left to right typically in order of cause and effect. Each variable is then regressed against all the "previous" variables to its left that influence it. Using this information, we can then show how the total effect of one variable on another is the sum of its direct and indirect effects, just as we did in Figure 13-5. An example will show how easily this earlier procedure can be generalized.

EXAMPLE **13-6**   A more extensive study of the fertility (number of children, CHILDN) of the 4700 Fiji women in Example 13-5 included another influence: the regressor X_3 = AGEMAR = the woman's age at marriage. Then the new, appropriately ordered set of variables became the ones shown in Figure 13-6. Each variable was regressed against all the

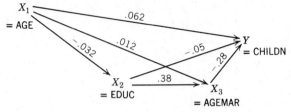

Figure 13-6
Path diagram of Fiji fertility.

previous variables to its left, with the following results (ignoring constants in each regression equation):

$$Y = .062X_1 - .05X_2 - .28X_3$$
$$X_3 = .012X_1 + .38X_2$$
$$X_2 = -.032X_1$$

Then all these regression coefficients were recorded on the path diagram in Figure 13-6.

Calculate the total effect on the number of children Y of:

a. Education X_2. (*Hint:* Just treat the triangle X_2X_3Y like the triangle in Figure 13-5.)

b. Present age X_1. (*Hint:* Extending the approach in part **a**, calculate *all* the paths between X_1 and Y, and sum them.)

c. Marriage age X_3.

Solution   **a.** From X_2 to Y there are two paths—the direct path from X_2 to Y and the indirect path via X_3. Thus, the triangle X_2X_3Y has exactly the same structure as the triangle in Figure 13-5, and the calculations are similar.

Direct effect of X_2 on Y		$= -.05$
Indirect effect via X_3	$(.38)(-.28)$	$= -.11$
	Total effect	$= -.16$

That is, a better educated woman has fewer children on average (.16 fewer children for each year of education she has received). Partly this is because more education *directly* reduces childbearing, and partly because of an *indirect effect:* More education tends to produce a later marriage age, which in turn results in fewer children.

b. From X_1 to Y there are so many paths that it is helpful to work them out systematically so that nothing is missed. As well as the direct path, there are now several indirect paths through the intervening variables X_3 and X_2; we consider each in turn:

Direct effect of X_1 on Y $= .062$

Indirect effect via X_3 $(.012)(-.28) = -.003$

Indirect effect via X_2: via X_2 alone $(-.032)(-.05) = +.002$

via X_2 and X_3 $(-.032)(.38)(-.28) = +.003$

_____

Total effect $= .064$

We note with satisfaction that this total effect agrees, as it should, with the answer of .064 given earlier in Example 13-5.

c. From X_3 to Y there is just one path, and so the answer is immediately obtained:

$$\text{Total effect of } X_3 = -.28$$

To sum up: The total effect of one variable (say X_1) on a later variable (say Y) is defined as the change occurring in Y when X_1 changes one unit—taking into account all the changes in the intervening variables between X_1 and Y. The total effect can be calculated from the network of direct effects using a natural extension of (13-41): As we follow each path from X_1 to Y, we multiply together all the coefficients we encounter. Then

> **Total effect of X_1 on Y = the sum of all paths (following the arrows from X_1 to Y)**

(13-44)
like (13-41)

*PROBLEMS

13-20 In the following path diagram, the variables are defined as in Figure 13-6. (The sample, however, is taken from the Indian subpopulation of Fiji, so the path coefficients are slightly different.)

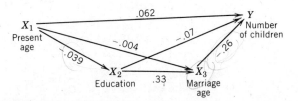

Calculate the total effect on Y (number of children) of each of the following:

a. The total effect of X_3.

b. The total effect of X_2.

c. The total effect of X_1.

13-21 In Problem 13-20, how many children would you expect of:

a. A woman whose marriage was 1 year later than the marriage of a woman of the same age and education?

b. A woman having 1 year more education than a woman of the same age, and the same age at marriage?

c. A woman having 1 year more education than a woman of the same age? (Age at marriage not known, but probably different, of course.)

d. A woman who is 1 year older than another woman? (Education and age at marriage not known, but probably different, of course.)

13-22 Suppose the demographer who did the Fiji study in Problem 13-20 had not measured X_3 (marriage age). Then his path diagram would look like this:

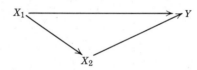

Suppose that otherwise his data was exactly the same, based on the same women. When he calculates the multiple regression coefficients (path coefficients), what will they be? [*Hint:* The path from X_1 to Y, for example, will be the sum of two former paths: the direct path, and the indirect path via X_3, which is now omitted.]

13-23 How can one explain the choice of occupation? A study of Blau and Duncan (1967, via Van de Geer, 1971) measured the following variables:

$$X_1 = \text{father's education level}$$
$$X_2 = \text{father's occupational status}$$
$$X_3 = \text{son's educational level}$$
$$X_4 = \text{occupational level of son's first job}$$
$$Y = \text{occupational level of son's second job}$$

Then the multiple regression of each variable on the previous ones was calculated as follows:

$$X_2 = .52X_1$$
$$X_3 = .31X_1 + .28X_2$$
$$X_4 = .02X_1 + .22X_2 + .43X_3$$
$$Y = -.01X_1 + .12X_2 + .40X_3 + .21X_4$$

a. Draw the path diagram that illustrates these equations.

b. Find the total effect on Y of:

    (i) X_3 **(ii)** X_2 **(iii)** X_1

c. If father's education level X_1 increased 3 units, what change would you expect in the occupational level of the son's second job Y if:

    i. The other variables were constant?

    ii. If nothing was known about the other variables (except that they probably changed in response to the change in X_1)?

CHAPTER 13 SUMMARY

13-1 To show how a response Y is related to two or more X factors, we can fit the multiple regression, $\hat{Y} = b_0 + b_1X_1 + b_2X_2 \cdots$. Each coefficient b shows how that particular X is related to Y, if all the other regressors were constant.

13-2 The underlying population regression as usual is denoted by Greek letters: $E(Y) = \beta_0 + \beta_1X_1 + \beta_2X_2 + \cdots$. Geometrically, this is a plane, and each coefficient β is the slope in the corresponding X direction.

    The estimated coefficients $b_0, b_1, b_2, \ldots$ can be calculated from the data, again using the criterion of least squares. For these calculations there are standard computer packages, which become invaluable when there are many regressors.

13-3 The computer package also calculates the standard error and t ratio for each b, from which we can calculate confidence intervals and p-values.

13-4 In the model $Y = b_0 + b_1X_1 + b_2X_2 + \cdots$, the change in Y is given by $\Delta Y = b_1\Delta X_1 + b_2\Delta X_2 + \cdots$. In particular, if only one X is changing, say X_1, then $\Delta Y = b_1\Delta X_1$. That is, the multiplication factor b_1 is applied to ΔX_1 to give its effect on Y.

*13-5 The total effect of a regressor on Y (the simple regression coefficient) can be expressed as the sum of the direct effect (the multiple regression coefficient) and the indirect effects. The indirect effects represent the bias in an observational study that omits important extraneous variables.

*13-6 A path diagram typically involves many variables listed in order of cause and effect, each variable being related to the previous ones by a multiple regression. The total effect of one variable on another can be obtained by summing along all the paths from one to the other.

REVIEW PROBLEMS

13-24 A multiple regression of lung capacity Y (in milliliters), as a function of age (in years) and height (in inches) and amount of smoking (in packs per day), was computed for a sample of 50 men randomly sampled from a large population of workers (Lefcoe and Wonnacott, 1974):

VARIABLE	COEFFICIENT	STANDARD ERROR	t RATIO	p-VALUE
x_1 AGE	− 39.0	9.1		
x_2 HEIT	+ 98.4	32.0		
x_3 SMOKIN	− 180.0	206.4		

a. Fill in the last two columns.

b. If the whole population were run through the computer, rather than a mere sample of 50, in what range would you expect to find each of the computed coefficients?

c. Other things being equal, what would you estimate is the effect of:

  i.   Smoking 1 pack per day?

  ii. Being 5 years older?

  iii. Smoking 2 packs per day and being 10 years older?

d. As far as lung capacity is concerned, the effect of smoking 1 pack per day is equivalent to aging how many years?

e. Your answer to part **d** is only approximate. What are the sources of error?

13-25 A simple regression of life expectancy Y (in years) on smoking X (in packs per day) for U.S. males (age 21) would yield the following approximate regression equation (U.S. Surgeon-General, 1979):

$$\hat{Y} = 70 - 5X$$

Is it fair to say that smoking cuts 5 years off the end of your life, for each pack smoked? If so, why? If not, how *would* you estimate the effect of smoking 1 pack per day?

13-26 The costs of air pollution (in terms of higher disease and mortality rates) was studied by two economists (Lave and Seskin, 1974). They introduced the problem as follows:

> "The cause of a disease is often difficult to establish. For chronic diseases, establishing a cause-and-effect relationship is especially difficult. Many studies show that populations exposed to urban air pollution have a shorter life expectancy and higher incidence of lung cancer, emphysema, and other chronic respiratory diseases. Yet it is a long step to assert that this observed association between air pollution and ill health is proof that air pollution causes ill health . . .
>
> "The situation is similar to the controversy as to whether cigarette smoking causes lung cancer. Both the cigarette smoking and air pollution controversies stem from the fact that many other possible causal factors are present. Urban dwellers live in more crowded conditions, get less exercise, and tend to live more tense lives. Since each of these factors is known to increase the morbidity and mortality

rates, care must be taken to control or account for each factor before drawing inferences . . ."

a. Before drawing inferences about causation, the authors suggest three extraneous factors that should be controlled or accounted for: crowding, lack of exercise, and tenseness. What other extraneous factors might be important?

b. Suggest how they might control or account for these extraneous factors.

13-27 The authors in Problem 13-26 continued:

. . . the basic problem is one of allowing for all causal variables. Accounting for confounding factors in observed data is one of the purposes of multivariate statistical analysis. Since we have some notion of the model and desire estimates of the direct contribution of each factor, an appropriate technique is multiple regression."

To carry out this multiple regression, they collected measurements on the following variables, in each of 117 SMSA's (Standard Metropolitan Statistical Areas) in 1960:

M = mortality rate (deaths per 10,000 people annually)
P = pollution (mean of the biweekly suspended particulate readings, in $\mu g/m^3$)
SP = sulphate pollution (the smallest of the biweekly sulphate readings, in $\mu g/m^3$)
D = density of population (people per square mile)
B = percentage of population that is nonwhite
E = percentage of population that is elderly (over 65)

The regression equation of mortality on pollution and the extraneous factors was:

$$M = 19.6 + .041P + .71SP + .001D + .41B + 6.87E$$
$$(t \text{ ratio}) \qquad (2.5) \quad (3.2) \quad (1.7) \quad (5.8) \quad (18.9)$$

a. For Pittsburgh, a typical SMSA, the levels of the various factors were: $P = 170, SP = 6.0, D = 790, B = 6.8$ and $E = 9.5$. What is Pittsburgh's predicted mortality rate? Pittsburgh's actual mortality rate was 103 (deaths per 10,000 annually). How accurate was the prediction?

b. For all 117 SMSA's, the average levels of the various factors were $\overline{P} = 120, \overline{SP} = 4.7, \overline{D} = 760, \overline{B} = 12.5$, and $\overline{E} = 8.4$. What was the average mortality rate?

c. Other things being equal, estimate how much the average mortality rate would change:

   i. If the proportion of elderly people was halved (from an average of 8.4 percent given in **b** down to 4.2 percent).

      ii. If the population density was halved.

     iii. If both forms of pollution (both P and SP) were halved.

  d. Suppose the 117 SMSA's are regarded as a sample from a large conceptual population. Then for the population coefficient of pollution P:

       i. What would be the 95% confidence interval?

      ii. What would be the p-value for H_0?

     iii. At the 5% level, would pollution be a statistically discernible factor?

13-28 **a.** In Problem 13-27, part **c,** we measured the effects of three factors in terms of relative changes. In this sense, which of the three factors is most important? Least important?

  b. Since the percentage of elderly people E was found to be very important, what might have happened if it had been omitted from the study?

  c. Note that variables such as occupational exposure, personal habits and smoking were omitted from the study. Would you agree with the authors that if these variables "are related to the level of air pollution, then our pollution estimates will be biased as indicators of causality." Then why do you suppose they were omitted?

13-29 The costs and control of crime have now become important questions addressed by statisticians. In each of the 121 Standard Metropolitan Statistical Areas (SMSA) of more than 250,000 inhabitants in the U.S. in 1970, the following variables were measured (Jacobs, 1979):

P = number of policemen per 100,000 inhabitants
D = disparity or inequality of income
I = mean family income annually (dollars)
S = number of small drug and liquor stores per 100,000 inhabitants
R = number of riots between 1960 and 1970
C = crime rate as measured by FBI crimes known to the police

Often a regression is carried out with *standardized* variables, denoted with a prime symbol (') and defined as

$$X' \equiv \frac{X - \overline{X}}{s_X} \qquad \text{like (4-23)}$$

To explain how the number of policemen P depended on the other variables, the following multiple regression was calculated:

$$P' = .34D' + .15I' + .44S' - .004R' + .27C' + \cdots$$

As indicated by the dots, the multiple regression also included other relevant variables (such as % unemployment, population size, etc.). The following simple regressions were also calculated:

$$P' = .63D', \qquad P' = .47R', \qquad P' = .41C'$$

a. Suppose we compared an SMSA that had 2 standard deviations more income disparity than another SMSA; that is, $\Delta D' = 2$. How many more police do you estimate it would have if the two SMSA's were:

   i.   Alike in all other respects?

   ii. Alike in all other respects except that the one with the greater income disparity also has a crime rate that is 1 standard deviation lower?

   iii. Just drawn at random from the 121 available?

b. Sue and Laurie are having an argument. Sue thinks the evidence shows that the number of police by 1970 was increased substantially in reaction to the riots in the 1960s. Do you agree or disagree? Why?

*13-30 Continuing Problem 13-29, the data showed that income I had a mean of $10,000 and standard deviation of $1200 approximately.

a. If the original variable I had been used instead of the standardized I' in the multiple regression, what would be its coefficient b?

b. Choose the correct alternative in each bracket: One of the advantages of standardizing is that the regression coefficient is [independent of, closely related to] the units in which the variable is measured. Then a small coefficient, for example, assures us that the impact of the variable itself is [small, large]—a property not found in the unstandardized form, as part **a** illustrated. Thus we can conclude that the variable with the smallest impact on number of police is [number of riots R, number of small stores S].

*13-31 Suppose that:

$$Y = a + b_1X_1 + b_2X_2 + b_3X_3$$

Prove that, if X_3 remains constant while X_1 and X_2 change:

$$\Delta Y = b_1\Delta X_1 + b_2\Delta X_2$$

Regression Extensions

Happy the man who has been able to learn the causes of things.

VIRGIL

In this chapter we will see how versatile a tool multiple regression can be, when it is given a few clever twists. It can be used on categorical data (such as yes-no responses) as well as the numerical data we have dealt with so far. And it can be used to fit curves as well as linear relations.

14-1   Dummy (0-1) Variables

A—PARALLEL LINES FOR TWO CATEGORIES

In Chapter 6 we introduced dummy variables for handling data that came in two categories (such as Democrat versus Republican, or treatment versus control). By associating numbers (0 and 1) with the two categories, a dummy variable ingeniously transformed the problem into a numerical one, and so made it amenable to all the standard statistical tools. (For example, standard errors and confidence intervals could be constructed.) Now we shall see how a dummy variable can be equally useful in regression analysis.

EXAMPLE **14-1**   A certain drug (drug A) is suspected of having the unfortunate side effect of raising blood pressure. To test this suspicion, 10 women were randomly sampled, 6 of whom took the drug once a day, and 4 of whom took no drug, and hence served as the control group. To transform "drug use" into a numerical variable, for each patient let:

$$D = \text{number of doses of this drug she} \qquad (14\text{-}1)$$
$$\text{takes daily} \qquad \text{like (6-17)}$$

that is,

$$D = 1 \text{ if she took the drug} \qquad (14\text{-}2)$$
$$= 0 \text{ if she did not (i.e., was a control)} \qquad \text{like (6-18)}$$

In the form (14-1), it is clear that D is a variable that can be run through a regression computer program like any other variable. Moreover, in the form (14-2) it is clear that D is a 0-1 variable that clearly distinguishes between the two groups of women; and this is the form most commonly used.

We want to investigate how this drug affects blood pressure Y, keeping constant extraneous influences such as age X. Following blood pressure Y in column 1 of Table 14-1, therefore, we type in as regressors both age X (rounded to the nearest decade in column 2) and drug use D (in column 3). The computer calculates the equation:

$$\hat{Y} = 69.5 + .44X + 4.7D \qquad (14\text{-}3)$$

which is of the general form:

$$\hat{Y} = a + bX + cD \qquad (14\text{-}4)$$

TABLE **14-1** **Computer Solution for Drug Trials Using a Dummy Variable _D_.**

```
        > READ'Y','X', 'D'
DATA>      85   30    0
DATA>      95   40    1
DATA>      90   40    1
DATA>      75   20    0
DATA>     100   60    1
DATA>      90   40    0
DATA>      90   50    0
DATA>      90   30    1
DATA>     100   60    1
DATA>      85   30    1


        > REGRESS'Y' ON 2 REGRESSORS'X','D'

THE REGRESSION EQUATION IS
Y = 69.5 + 0.442 X + 4.65 D
```

COLUMN	COEFFICIENT	ST. DEV. OF COEF.	T-RATIO = COEF/S.D.
	69.535	2.905	23.93
X	0.44186	0.07301	6.05
D	4.651	1.885	2.47

a. Graph the relation of Y to X given by (14-3), when
   (i) $D = 0$,
   (ii) $D = 1$.

b. What is the meaning of the coefficient of D? Answer using the graph in **a**. Also answer using the fundamental interpretation (13-31).

c. Construct a 95% confidence interval for the (true) population coefficient of D. Is it discernible at the 5% level?

Solution   a. Equation (14-4) takes on two forms, depending on whether D is zero or one.

If $D = 0$, the equation
for the control group is: $\hat{Y} = a + bX$ (14-5)

If $D = 1$, the equation
for the treatment group $\hat{Y} = a + bX + c$
is: $= (a + c) + bX$ (14-6)

Compared to (14-5), the line (14-6) is seen to be a line with the same slope b, but an intercept that is c units higher. That is, the line where $D = 1$ is parallel and c units higher. For the specific equation (14-3), the two lines are plotted in Figure 14-1.

b. On the graph, suppose we keep age constant, at $X = 55$ for example. The difference the drug makes is shown as the vertical distance between the lines, $c = 4.7$.

Alternatively, we could use the fundamental interpretation for any regression coefficient: The 4.7 coefficient of D is the change

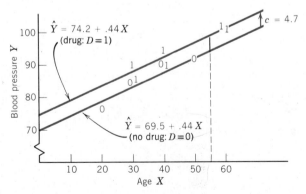

Figure 14-1
Graph of equation (14-3) relating blood pressure to age and treatment. (Each of the 10 patients is plotted as a 1 or a 0, depending on whether or not she was taking the drug.)

in Y that accompanies a unit change in D (while X remains constant). This unit change in D can only be from 0 to 1—that is, from no drug to drug. Thus there is an increase in blood pressure Y of 4.7 as we go from a woman without the drug ($D = 0$) to a woman of the same age with the drug ($D = 1$). This agrees with the interpretation in the previous paragraph.

c. Our answers about inference will be the standard regression answers: We construct the 95% confidence interval from the estimate of 4.7 and its standard error (i.e., standard deviation $\doteq$ 1.88 given in Table 14-1). For $t_{.025}$, we use d.f. $= n-k-1 = 10-2-1 = 7$. Thus:

$$\text{Population coefficient} = c \pm t_{.025}\ \text{SE} \qquad \text{like (13-13)}$$
$$= 4.7 \pm 2.36(1.88)$$
$$= 4.7 \pm 4.4 \qquad\qquad (14\text{-}7)$$

Since the estimate 4.7 exceeds the sampling allowance 4.4, the relationship between D and Y is discernible at the 5% level. (But we should be careful about concluding that D *causes* an increase in blood pressure Y, because the drug was not assigned at random to the women. Thus our multiple regression kept constant only one of the possible extraneous factors—age X.)

In conclusion, this example shows how a two-category factor can be nicely handled with a 0-1 regressor:

> **If D is a 0-1 variable in the regression model**
>
> $$\hat{Y} = a + bX + cD$$
>
> then relative to the reference line where $D = 0$,
> the line where $D = 1$ is parallel and c units higher.
> Furthermore, standard confidence intervals and tests
> can still be applied. $\qquad$ (14-8)

B—SEVERAL CATEGORIES

So far we have considered a factor that has only two categories—treatment and control. The dummy variable D measured the effect of the treatment ($D = 1$) relative to the control ($D = 0$).

What happens if a factor has three categories? For example, suppose we are clinically testing two drugs A and B against a control C, with a sample of 30 patients. For each patient, we again measure the response Y and extraneous variables X_1, X_2 . . . (such as age, weight, etc.). But now, in measuring drug use, we must use two dummy variables:

$$D_A = 1 \text{ if drug } A \text{ given; 0 otherwise} \qquad (14\text{-}9)$$

$$D_B = 1 \text{ if drug } B \text{ given; 0 otherwise} \qquad (14\text{-}10)$$

Thus, typical data would start out as follows (illustrating the three different ways drug could be assigned):

COMMENTS		DATA				
PERSON	DRUG	Y	D_A	D_B	X_1	X_2 · · ·
Koval	On drug A	.	1	0	. . .	. . .
Bellhouse	On drug B	.	0	1	. . .	. . .
Haq	Control C	.	0	0	. . .	. . .
.	.	.	.	.	. . .	. . .
.	.	.	.	.	. . .	. . .
.	.	.	.	.	. . .	. . .

$$(14\text{-}11)$$

Suppose that the data yielded the multiple regression equation:

$$\hat{Y} = 5 + 30D_A + 20D_B - 13X_1 + \cdots \qquad (14\text{-}12)$$

This means that the effect of drug A (relative to the reference category—control group C) is to raise Y by 30. The effect of drug B is to raise Y by 20. Thus drug A increases Y by 10 units more than does drug B.

Note that we need only 2 dummy variables to handle 3 categories—since one category (in our example, the control group) is the reference category against which the other two are measured. This is generally true for any number of categories: We need one less dummy than there are categories.

C—DIFFERENT SLOPES AS WELL AS INTERCEPTS

It is easy to extend our model to have different slopes as well as different intercepts. We just introduce one more regressor (XD—the variable X times D), which is called the *interaction* between X and D:

$$\hat{Y} = a + bX + cD + eXD \qquad (14\text{-}13)$$

By setting D equal to its two possible values 0 and 1, this yields the pair of equations:

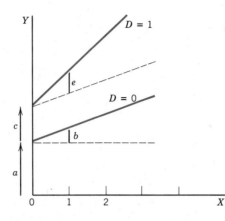

Figure 14-2
Interpretation of dummy variable regression with interaction: $\hat{Y} = a + bX + cD + eXD$.

If $D = 0$, the equation for the control group is:	$\hat{Y} = a + bX$
If $D = 1$, the equation for the treatment group is:	$\hat{Y} = a + bX + c + eX$ $= (a + c) + (b + e)X$

These equations show that the treatment group has not only an intercept that is greater by the amount c, but also a slope that is greater by the amount e, as illustrated in Figure 14-2.

PROBLEMS

14-1 A regression equation related personal income Y (annual, in $1000) to education E (in years) and geographical location, measured with dummies as follows:

$$D_S = 1 \text{ if in the south; } 0 \text{ otherwise}$$
$$D_W = 1 \text{ if in the west; } 0 \text{ otherwise}$$

The remaining region (northeast) is left as the reference region.

a. Suppose the fitted regression was

$$\hat{Y} = 4.5 + 0.5E - 1.0D_S + 1.5D_W$$

Graph the estimated income $\hat{Y}$ as a function of E, for each of the 3 regions, with E running from 8 to 16.

b. Redraw the graph in part **a,** according to the following interactive model:

$$\hat{Y} = 4.5 + 0.5E + 1.4D_S + 0.3D_W - 0.2D_SE + 0.1D_WE$$

14-2 When the data of Example 14-1 was fitted with an interactive model, the following regression was computed:

$$\hat{Y} = 67.5 + .50X + 8.0D - .09XD$$

Standard error   (.13)   (6.4)   (.16)

  a. Graph the two lines (for $D = 0$ and $D = 1$).

  b. Find a 95% confidence interval for the difference in the 2 population slopes.

  c. What is the p-value for the null hypothesis that there is no difference in the population slopes (Chow test)?

14-3 In a 1974 study of 1072 men, a multiple regression was calculated to show how lung function was related to several factors, including some hazardous occupations (Lefcoe and Wonnacott, 1974):

$$AIRCAP = 4500 - 39\ AGE - 9.0\ SMOK$$
(SE)   (1.8)   (2.2)
$$- 350\ CHEMW - 380\ FARMW - 180\ FIREW$$
                 (46)   (53)   (54)

where

AIRCAP	= air capacity (milliliters) that the worker can expire in one second
AGE	= age (years)
SMOK	= amount of current smoking (cigarettes per day)
CHEMW	= 1 if subject is a chemical worker, 0 if not
FARMW	= 1 if subject is a farm worker, 0 if not
FIREW	= 1 if subject is a firefighter, 0 if not

A fourth occupation, physician, served as the reference group, and so did not need a dummy. Assuming these 1072 people were a random sample,

  a. Calculate the 95% confidence interval for each coefficient.

     Fill in the blanks, and choose the correct word in square brackets:

  b. Other things being equal (things such as _____), chemical workers on average have AIRCAP values that are _____ milliliters [higher, lower] than physicians.

  c. Other things being equal, chemical workers on average have AIRCAP values that are _____ milliliters [higher,lower] than farmworkers.

  d. Other things being equal, on average a man who is 1 year older has an AIRCAP value that is _____ milliliters [higher,lower].

  e. Other things being equal, on average a man who smokes one pack

(20 cigarettes) a day has an AIRCAP value that is _____ milliliters [higher,lower].

    f. As far as AIRCAP is concerned, we estimate that smoking one pack a day is roughly equivalent to aging _____ years. But this estimate may be biased because of _____.

14-4   **a.** From the regression equation in Problem 14-3, fill in the estimated values of AIRCAP for the 8 cells in the following table (assuming AGE = 30):

DAILY SMOKING	OCCUPATION			
	PHYSICIANS	CHEMW	FARMW	FIREW
0				
20				

    b. Using part **a,** graph the values of AIRCAP as a function of daily smoking—4 lines for the 4 occupations.

14-5 In the same study as Problem 14-3, each worker's lungs were alternatively evaluated by whether or not he had bronchitis—using a dummy variable:

$$\text{BRONC} = 1 \text{ if worker has bronchitis, } 0 \text{ otherwise}$$

Then BRONC was regressed on the same factors:

$$\text{BRONC} = -.04 + .0021 \text{ AGE} + .0047 \text{ SMOK}$$
$$(.0009) \qquad (.0011)$$
$$+ .065 \text{ CHEMW} + .002 \text{ FARMW} - .032 \text{ FIREW}$$
$$(.024) \qquad\qquad (.027) \qquad\qquad (.027)$$

    This represents something new: a dummy variable for the response, as well as dummy variables as regressors. Nevertheless, the interpretation is still the same. For example, other things being equal, an increase of 10 years in age means an increase in BRONC of $10 \times .0021 = .021$—that is, an increase of 2.1 percentage points in the bronchitis rate (or in the *probability* of an individual having bronchitis).

    Answer the same questions as in Problem 14-3, substituting BRONC for AIRCAP (and for the units, substitute percentage points for milliliters).

14-6 In an observational study to determine the effect of a drug on blood pressure it was noticed that the treated group (taking the drug) tended to weigh more than the control group. Thus, when the treated group had higher blood pressure on average, was it because of the treatment or their weight? To untangle this knot, some regressions were computed, using the following variables:

$$\text{BP} = \text{blood pressure}$$
$$\text{WEIGHT} = \text{weight}$$
$$\text{D} = 1 \text{ if taking the drug, } 0 \text{ otherwise}$$

When the 15 patients' records were run through the computer, the following outputs were obtained:

D	WEIGHT	BP
0	180	81
0	150	75
0	210	83
0	140	74
0	160	72
0	160	80
0	150	78
0	200	80
0	160	74
1	190	85
1	240	102
1	200	95
1	180	86
1	190	100
1	220	90

```
THE REGRESSION EQUATION IS
BP = 54.0 + 10.6 D + 0.139 WEIGHT
```

		ST. DEV.	T-RATIO =
COLUMN	COEFFICIENT	OF COEF.	COEF/S.D.
	54.040	8.927	6.05
D	10.596	2.984	3.55
WEIGHT	0.13950	0.05248	2.66

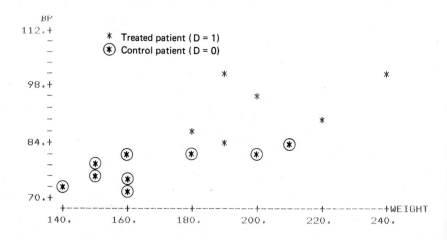

a. Verify that the computer has correctly graphed the first two patients. Then graph the regression equation.

b. How much higher on average would the blood pressure be:

  i. For someone of the same weight who is on the drug?

  ii. For someone on the same treatment who is 10 lbs. heavier?

c. On the computer graph, show your answers to part **b.**

d. How would the simple regression coefficient compare to the multiple regression coefficient for weight? Why?

14-7 **a.** Continuing Problem 14-6, sort out the values of BP into two groups according to whether they had taken the drug or not (treatment and control groups). Then calculate the confidence interval for the difference in two means, using (8-20).

b. Calculate the confidence interval for the simple regression coefficient of BP against D, using the following computer output:

```
THE REGRESSION EQUATION IS
BP = 77.4 + 15.6 D

                                 ST. DEV.     T-RATIO =
COLUMN          COEFFICIENT      OF COEF.     COEF/S.D.
                   77.444          1.784        43.42
D                  15.556          2.820         5.52
```

c. Which of the two confidence intervals above is a better measure of the effect of the drug? Is there an even better measure?

14-8 To generalize Problem 14-7, answer True or False; if False, correct it.

a. The confidence interval for 2 independent samples is equivalent to using simple regression on a 0-1 regressor.

b. However, it is much better to use *multiple* regression to avoid the bias that occurs in simple regression (or in calculating the difference in two means).

14-2 Analysis of Variance (ANOVA) by Regression

A—ONE-FACTOR ANOVA

In Example 14-1, we were mainly concerned with the effect on blood pressure of the dummy regressor (drug), and introduced the numerical regressor (age) primarily to keep this extraneous variable from biasing the estimate of the drug effect. Used in this way, multiple regression is sometimes called *analysis of covariance* (ANOCOVA, with a table similar to the ANOVA Table 10-4).

If the numerical regressors are omitted entirely so that dummy regressors alone are used, then a regression equation such as (14-12) reduces to a form like

$$\hat{Y} = 5 + 30D_A + 20D_B \tag{14-14}$$

A regression of this form is equivalent to the traditional one-factor ANOVA in Chapter 10. (There the single factor was machines; here it is drugs.)

B—TWO-FACTOR ANOVA

It is possible to use dummy variables to introduce a second factor, or more. For example, in the drug study, suppose we want to see the effect of another factor, sex. We can take care of it with one dummy variable:

$$M = 1 \text{ if male; } 0 \text{ if female} \qquad (14\text{-}15)$$

Then the fitted regression equation, instead of (14-14), would be something like

$$\hat{Y} = 5 + 30D_A + 20D_B + 15M \qquad (14\text{-}16)$$

One convenient way to graph (14-16) is like Figure 14-1—with M playing the former role of age X, and D_A and D_B playing the former role of the dummy D. This graph is shown in Figure 14-3a.

This dummy regression is equivalent to traditional two-factor ANOVA in Chapter 10. (There the two factors were machines and operators; here they are drugs and sex.) However, the dummy regression approach is much more flexible than the traditional ANOVA approach for two reasons:

1. It can easily handle any number of observations per cell, rather than the single observation per cell that we had in Table 10-8.
2. It can more easily handle interaction, as we will show next.

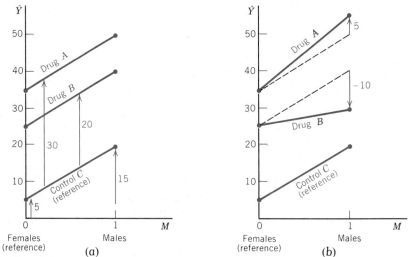

Figure 14-3
Graph of two-factor ANOVA. (a) Additive model. (b) Interactive model.

C—TWO-FACTOR ANOVA, INTERACTIVE MODEL

To extend (14-16) to a model with different slopes as well as different intercepts, we again introduce interaction terms. Then the estimated regression would be something like:

$$\hat{Y} = 5 + 30D_A + 20D_B + 15M + 5MD_A - 10MD_B \qquad (14\text{-}17)$$
$$\text{like (14-13)}$$

This relationship is graphed in panel (b) of Figure 14-3. In this interactive model, it is clear that drug A has a greater effect than B, especially for males. [Contrast this with the simple additive model in panel (a), where the effect of drug A was the same for males as females.] We therefore say that there is *positive interaction* or *synergism* between drug A and being male.

To generalize two-factor ANOVA, suppose the first factor has c categories, while the second has r categories. Then we can analyze the data with multiple regression as follows:

   i. For the first factor, use $c - 1$ dummies, $D_1, \ldots, D_{c-1}$

  ii. For the second factor, use $r - 1$ dummies, $M_1, \ldots, M_{r-1}$

 iii. Interaction terms like the last two terms in (14-17) can be introduced as desired. If a *full* interaction model is judged necessary, it would require $(c - 1)(r - 1)$ interaction terms of the form $D_1M_1, \ldots, D_{c-1}M_{r-1}$

Such a multiple regression can be used not only to answer such standard ANOVA questions as "Does sex matter?" "Does drug matter?" but also more subtle ANOVA questions like "What sort of interaction—if any—is there between sex and drug?" To answer this, we would test to see whether the interaction coefficients are discernibly different from zero.

PROBLEMS

14-9 Twelve plots of land are divided randomly into three groups. The first is held as a control group C, while fertilizers A and B are applied to the other two groups. Yield is observed to be:

C	A	B
60	75	74
64	70	78
65	66	72
55	69	68

The regression of yield Y was computed to be:

$$\hat{Y} = 61 + 9D_A + 12D_B$$

where the dummies are defined as

$$D_A = 1 \text{ if fertilizer } A \text{ used, } 0 \text{ otherwise}$$
$$D_B = 1 \text{ if fertilizer } B \text{ used, } 0 \text{ otherwise}$$

a. What is the estimated yield $\hat{Y}$ for:

   i. The control C?   **ii.** Fertilizer A?   **iii.** Fertilizer B?

b. Calculate the average yield for each of the 3 columns. How do they compare with the answers in part **a**?

c. Is this called one-factor or two-factor ANOVA? What is the "response"? What are the "factor(s)"?

*__d.__ Using a computer, verify the given multiple regression.

14-10 Two new drugs, A and B, along with a control C, were tried out on men and women to estimate their effect on blood pressure Y. Since the numbers of men and women were unequal, it was decided that a multiple regression would be easier than traditional ANOVA. The following dummies were defined:

$$D_A = 1 \text{ if patient on drug } A, 0 \text{ otherwise}$$
$$D_B = 1 \text{ if patient on drug } B, 0 \text{ otherwise}$$
$$M = 1 \text{ if male, } 0 \text{ otherwise.}$$

Two regression models were fitted:

a. Additive $\hat{Y} = 65 + 5D_A - 10D_B + 10M$

b. Interactive $\hat{Y} = 68 + 6D_A - 15D_B + 5M - 2D_A M + 9D_B M$

Graph each, as in Figure 14-3.

14-11

Fitted Response $\hat{Y}$

	SEX	
DRUG	FEMALE	MALE
A		
B		
Control C		

a. Fill in the table above for each of the two models in Problem 14-10, and check that your answers agree with the graphs there.

b. Make the correct choice in each bracket: In the [additive, interactive] model, the improvement of males over females is the same for all drugs. Then it is equally true that the improvement of drug A—or drug B—over drug C is the same for [both sexes, all treatments].

14-12 The strength of yarn produced by 4 machines varied according to the type of raw material used. In fact, a sample of several hundred pieces of yarn gave the following fitted responses:

Yarn Strengths

MATERIAL	MACHINE			
	I	II	III	IV
A	14	12	15	12
B	18			
C	19			

a. Fill in the blank cells, assuming the model is additive.

b. Make the correct choice in each bracket: Consider now a general 2-way table, with $r \times c$ cells. Suppose the fitted response has been filled in, in the top row and left-hand column. To fill in the remaining [rc, $(r - 1)(c - 1)$] cells of the table, it is enough to additionally know either (1) [$(r - 1)(c - 1)$, $rc - 1$] interaction coefficients; or (2) that all the interaction coefficients are zero—that is, the model is [additive, interactive].

14-3 Simplest Nonlinear Regression

Straight lines and planes are called *linear* functions. They are characterized by a simple equation (such as $Y = a + bX$) where the independent variable appears just as X, rather than in some more complicated nonlinear way such as X^2, $\sqrt{X}$, or $1/X$. In this section we will look at some nonlinear functions.

A—EXAMPLE

Let us reconsider how wheat yield Y depends on fertilizer X. Suppose that vast amounts of the fertilizer we are testing begin to burn the crop, causing the yield to fall. An appropriate model might therefore be a second-degree equation (parabola) as shown in Figure 14-4 of the form:

$$\hat{Y} = b_0 + b_1 X + b_2 X^2 \tag{14-18}$$

To find the equation that best fits the data, we simply define new variables X_1 and X_2 as:

$$\left. \begin{array}{l} X_1 \equiv X \\ X_2 \equiv X^2 \end{array} \right\} \tag{14-19}$$

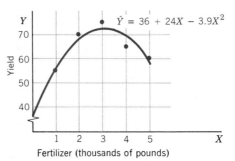

Figure 14-4
How yield depends on fertilizer—a parabolic relation.

Then (14-18) becomes the ordinary multiple regression:

$$\hat{Y} = b_0 + b_1 X_1 + b_2 X_2 \tag{14-20}$$

This is of the standard form that can be run through a computer, as Table 14-2 shows. The given data Y and X—the dots in Figure 14-4—were read into the computer, and the values of X were then squared to give a new variable, 'X SQ.' The computed regression was then:

$$\hat{Y} = 36 + 24X - 3.9X^2 \tag{14-21}$$

This is the least squares regression graphed in Figure 14-4. It shows a nice fit to the data. (It is the best fit, of course, in the sense of least squared error.)

Of course, the computer output can be used for standard statistical tests too. For example, suppose we wish to test whether the parabolic model

TABLE **14-2**   **Computer Solution for Parabolic Fit Using Multiple Regression (data and graph of output are in Figure 14-4).**

Y	X	X SQ
55	1	1
70	2	4
75	3	9
65	4	16
60	5	25

> REGRESS 'Y' ON 2 REGRESSORS 'X', 'X SQ'

THE REGRESSION EQUATION IS
Y = 36.0 + 24.1 X - 3.93 X SQ

COLUMN	COEFFICIENT	ST. DEV. OF COEF.	T-RATIO = COEF/S.D.
	36.000	8.502	4.23
X	24.071	6.479	3.72
X SQ	-3.929	1.059	-3.71

is really necessary—that is, whether it is necessary to have X^2 appear in this equation. The answer lies in prior knowledge and the statistical discernibility (significance) of the regressor X^2. Our prior expectation that this was a downward bending parabola is confirmed with the very negative t value of -3.71 in Table 14-2; so we retain X^2.

B—GENERALIZATION

As well as parabolas, there are many other nonlinear functions that can be fitted using multiple regression on redefined variables. For example:

1. General Polynomials: $\hat{Y} = b_0 + b_1 X + b_2 X^2 + b_3 X^3 + \cdots$

2. Reciprocals: $\qquad\qquad \hat{Y} = b_0 + \dfrac{b_1}{X}$

    In this case, letting $X_1 = 1/X$, we obtain the linear regression,

    $$\hat{Y} = b_0 + b_1 X_1$$

3. Annual Cycles: $\qquad \hat{Y} = b_0 + b_1 \sin\left(\dfrac{2\pi X}{12}\right) + b_2 \cos\left(\dfrac{2\pi X}{12}\right)$

    In this case, letting $X_1 = \sin(2\pi X/12)$ and $X_2 = \cos(2\pi X/12)$, we obtain the linear regression

    $$\hat{Y} = b_0 + b_1 X_1 + b_2 X_2$$

This method—simply redefining variables so that the regression becomes a linear one—works in all the equations above because the only nonlinearity is in the X variables. The coefficients b_0, b_1, and b_2 still appear in the same linear way as they did in Chapter 13. In other words, in each of these models, Y is a linear combination of the b coefficients. (Nowhere do terms like $\sqrt{b}$ or e^{3b} or $b_1 b_2$ appear.)

PROBLEMS

14-13 To see how tomato yield depends on irrigation, various amounts of irrigation (I) were assigned at random to 24 experimental plots, with the resulting 24 yields Y used to fit the following least squares regression:

$$Y = 42 + 12\,I - 1.5\,I^2$$
$$\text{(SE)} \qquad (.8) \qquad (.4)$$

a. Consider the null hypothesis H_0: Expected yield Y increases *linearly* with irrigation I. Before you had a look at the data, how credible would you find H_0?

b. To find out how much credibility the data gives to H_0, calculate its p-value. Does this answer confirm **a**, or not?

c. Graph the fitted parabola. Where does the maximum yield occur? Is this the appropriate level to irrigate?

d. What is the estimated increase in Y if I is increased one unit, from $I = 2$ to $I = 3$?

14-14 In testing the efficiency of a chemical plant producing nitric acid from ammonia, the stack losses and several related variables were measured for $n = 17$ different days. Then the following regression equation was fitted by least squares (Daniel and Wold, 1971):

$$Y = 1.4 + .07X_1 + .05X_2 + .0025X_1X_2$$

where Y = stack loss (% of ammonia lost)
X_1 = air flow (deviation from average flow)
X_2 = temperature of cooling water (deviation from average temperature)

a. Graph the relation of Y to X_1 when:

  i. $X_2 = -4$ **ii.** $X_2 = 0$ **iii.** $X_2 = 4$ **iv.** $X_2 = 8$

b. When air flow is 4 units above average, and the temperature is 6 units above average, what would you predict the stack loss to be? Show this on the graph too.

c. Choose the correct alternative: This model is called [additive, interactive] because the effect of X_1 [depends, does not depend] on the level of X_2.

  It was [easy, difficult] to estimate the coefficients from the 17 data points, because [nonlinearity requires a special program to minimize the sum of squared deviations $\Sigma(Y - \hat{Y})^2$, we could use multiple regression of Y on the three regressors X_1 and X_2 and X_1X_2].

*14-4   Nonlinearity Removed by Logs

In Section 14-3 we saw how equations that were nonlinear in the X variables could still be handled with multiple regression, as long as they were linear in the b coefficients. Now we will look at some more subtle models where the nonlinearity involves the b coefficients too. For example,

$$Y = e^{bX} \qquad \qquad (14\text{-}22)$$
$$\text{or} \quad Y = b_0 X^{b_1}$$

To find the least squares estimates, we will first have to transform the equation so that it becomes linear in the b coefficients. An extended example will illustrate.

A—A SIMPLE GROWTH MODEL

Panel (a) of Figure 14-5 shows the U.S. population during a period of sustained growth (1850–1900). It curves up in a way that suggests *exponential* growth (*constant percentage* growth, like compound interest or unrestrained biological growth). To verify this, in panel (b) we plot the same values of P on a *log scale*, which is explicitly designed to transform exponential curves into straight lines. Since the points in panel (b) lie very close to a straight line, we conclude that the appropriate model is indeed an exponential one:

$$P = Ae^{bX} \tag{14-23}$$

where A and b are the estimated coefficients, P is the population (in millions), X is time (in years, since 1850), and $e \simeq 2.718$ is the base for natural logarithms. One advantage of using the base e is that, as we will show in Example 14-3, it gives b a very straightforward and useful interpretation:

> In the growth model $P = Ae^{bX}$,
>
> $b \simeq$ **the annual growth rate** $\tag{14-24}$

To determine A and b, we first transform the exponential model (14-23) into the standard linear regression model by taking logs:

$$\log P = \log A + bX \tag{14-25}$$

Then we simply define:

$$Y \equiv \log P \tag{14-26}$$
$$a \equiv \log A \tag{14-27}$$

Substituting these new variables into (14-25) yields a standard linear regression:

$$Y = a + bX \tag{14-28}$$

Table 14-3 shows how a and b were computed. The given data P and X—the dots in Figure 14-5—were read in, and $Y = \log P$ was computed. Then the computer could find the simple regression of Y against X:

$$Y = \log P = 3.24 + .022X \tag{14-29}$$

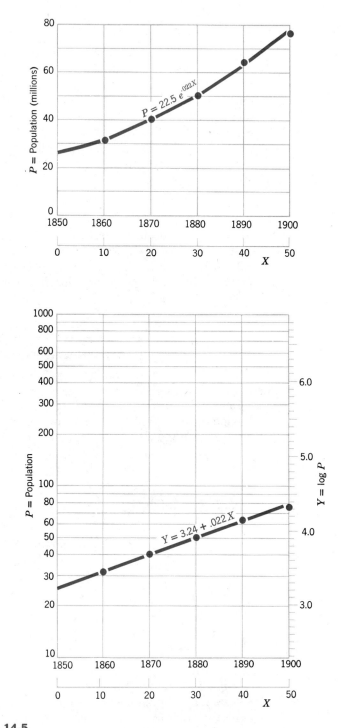

Figure 14-5
(a) U.S. population growth, and its exponential fit. (b) Same curve plotted on log paper.

TABLE **14-3** **Computer Solution for Exponential Fit of Population Growth (Data and graph are in Figure 14-5).**

```
> LOGE OF 'P' INTO 'Y'

> PRINT'P','X','Y'
        P       X           Y

      31.4     10      3.44681
      39.8     20      3.68387
      50.2     30      3.91601
      63.0     40      4.14313
      76.0     50      4.33073

> REGRESS'Y' ON 1 REGRESSOR'X'

THE REGRESSION EQUATION IS
Y = 3.24 + 0.0223 X

                                    ST. DEV.     T-RATIO =
   COLUMN        COEFFICIENT        OF COEF.     COEF/S.D.
                   3.23598          0.01825        177.30
   X             0.0222712          0.0005503       40.47
```

Taking exponentials (reversing our original log transformation):

$$P = e^{3.24}\, e^{.022X}$$

That is,

$$P = 25.5e^{.022X} \tag{14-30}$$

The graph of (14-29) is the straight line in panel (b) of Figure 14-5, which becomes the curved exponential growth (14-30) in panel (a). The coefficient 25.5 (millions) is the P intercept—the fitted population in 1850 (when $X = 0$). The coefficient $b = .022 = 2.2\%$ is the fitted annual growth rate.

Although a fit like (14-30) may adequately summarize past growth, there are important reasons not to use it for forecasting. So many changes occur in the underlying process determining population growth (wars, recessions, the development and acceptance of birth control methods, etc.), that there is no single unchanging relationship that can be trusted for a long-term forecast. Rather than making this sort of projection, it may be more useful to look at the changes in the underlying growth process.

B—LOGS AS RELATIVE CHANGES

To generalize our experience with the exponential growth model, there are many models that consist of factors multiplied together, for example:

$$Y = b_0 X_1^{b_1} X_2^{b_2}$$

Such equations can be simplified by taking logs, which transform the multiplication into a more convenient sum that can be easily estimated with multiple regression.

Often, after the regression has been estimated, it is preferable to leave it in its log form instead of transforming it back into its original exponential form because logs have a very useful interpretation[1]: For small changes in any variable X,

$$\boxed{\text{Change in } \log X \simeq \textit{relative} \text{ change in } X \text{ itself.}} \qquad (14\text{-}31)$$

We can illustrate this equation for the variable P by using Figure 14-5b. Values of P are on the left axis, while corresponding values of $\log P$ are on the right axis. As P changes from 50 to 60, note that $\log P$ correspondingly changes from 3.9 to 4.1. Thus:

$$\text{Change in } \log P = 4.1 - 3.9 = .20$$

$$\text{Relative change in } P \text{ itself} = \frac{60 - 50}{50} = .20 \checkmark \qquad (14\text{-}31)\text{ confirmed}$$

To further illustrate how useful (14-31) is, we will give two more examples.

EXAMPLE **14-2**

In economics, a common model for the quantity supplied Q as a function of price P is

$$Q = AP^b \qquad (14\text{-}32)$$

When we take logs, we obtain a more convenient sum:

$$\log Q = \log A + b \log P \qquad (14\text{-}33)$$

Letting $\log Q \equiv Y$, $\log A \equiv a$, and $\log P \equiv X$, this equation takes the standard form

$$Y = a + bX$$

---

[1] The proof of (14-31) follows from a standard calculus formula:

$$\frac{d(\log X)}{dX} = \frac{1}{X}$$

equivalently,

$$d(\log X) = \frac{dX}{X}$$

That is, for infinitesimally small changes (differentials d),

$$\text{Change in } \log X = \text{Relative change in } X \qquad (14\text{-}31) \text{ proved}$$

The coefficients a and b can now be computed with a standard regression program. Suppose, for example, they turned out to be $a = 5.1$ and $b = 2.0$. In terms of (14-33), this means:

$$\log Q = 5.1 + 2.0 \log P \qquad (14\text{-}34)$$

If price P increases by 1%, how much change is there in Q, the quantity supplied?

Solution The *relative* change in P is given as 1% = .01. By (14-31), this is the change in $\log P$, the regressor. Because the regression coefficient is 2,

Change in response (i.e., change in $\log Q$)
$$= 2(.01) = .02 \quad \text{like (13-24)}$$

Since $\log Q$ thus increases by .02, (14-31) assures us that Q itself has a relative increase of .02. That is, quantity supplied increases by 2%.

In Example 14-2, a 1% increase in price resulted in a 2% increase in the quantity supplied. This value 2 is called the *elasticity* of supply (formally defined as the relative change in quantity ÷ relative change in price). In general:

$$\boxed{\begin{array}{l} \text{If } \log Q = a + b \log P, \\ \quad \text{then } b = \text{elasticity.} \end{array}} \qquad (14\text{-}35)$$

The higher the elasticity, the greater the quantity supplied as price increases.

EXAMPLE **14-3** Mexico's population P (in millions) has grown over time X (in years, since 1975) approximately according to

$$P = 60e^{.03X} \qquad\qquad (14\text{-}36)$$
$$\text{like (14-23)}$$

Taking logs, $\qquad\qquad \log P = 4.1 + .03X$

How much did the population grow in 1 year?

Solution The given change in the regressor X is 1 year. From (13-24), the change in the response (i.e., the change in $\log P$) is therefore .03. Accordingly, (14-31) assures us that the relative change in P is also .03 = 3%. That is, the annual *growth rate* in P is 3%.

In (14-36) the annual growth rate was just the coefficient of X. In general, this coefficient is called b, and so (14-24) is established.

To sum up: Log transformations are so useful that researchers often apply them routinely, unless they have good reason not to. Two important advantages of logs are:

1. Logs transform a multiplicative model into an additive model that can be estimated with a standard regression program.
2. Logs have a useful interpretation as relative changes. Further detail on logs is given in Appendix 14-4.

*PROBLEMS

14-15

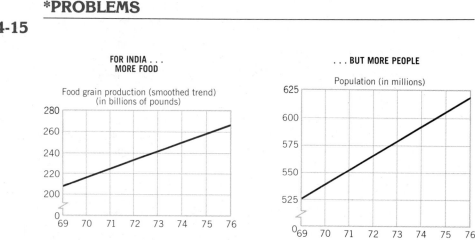

FOR INDIA . . . MORE FOOD

. . . BUT MORE PEOPLE

In 1977, the "gee whiz" graph above appeared in one of America's most popular newsweeklies. To make sure the readers got the point, a crowd of hungry people were shown on the background of the second graph.

a. Judging from a glance (as most readers do), how would you say India fared in 1976 compared to 1969 (in terms of food grain production per capita)?

b. To actually check your answer in part **a**, calculate the percentage change from 1969 to 1976 **(i)** in food grain production, **(ii)** in population. Which has increased more?

c. The graphs as they appear were obviously misleading. We can avoid this kind of deception by using an appropriate scale—the log scale that is specifically designed to show percentage changes.

Therefore on the log scale below (which is similar to the upper part of Figure 14-5b), plot both food grain production and population. Then at a glance, see which increased proportionately more.

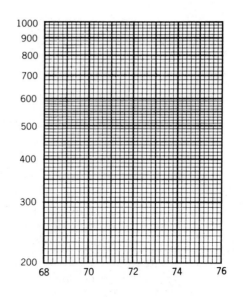

14-16 Suppose that a regression study of quantity supplied as a function of price yielded the following equation:

$$\log Q = 5.2 + 1.3 \log P$$

a. What is the elasticity of supply?

b. If price increased by 3%, by how much would the quantity supplied increase?

c. What price increase would be required to increase the quantity supplied by 10%?

14-17 In a 1964 study of how personal income was related to education, IQ, race and other variables, the following multiple regression was estimated from a sample of 1400 U.S. veterans (Griliches and Mason, 1973):

$$LINC = .046E + .0010 \text{ AFQT} + .17 \text{ WHITE} + \text{other variables}$$
$$(SE) \quad (.007) \quad (.0004) \qquad\qquad (.05)$$

where   LINC = log of the veteran's weekly income
            E = number of years of additional education (during and after military service)
     AFQT = the veteran's percentile rating on the Armed Forces Qualification Test Score, which was used as a rough measure of IQ
   WHITE = dummy variable for race, being 1 for whites and 0 otherwise

Other variables = age, amount of military service, amount of schooling before military service, father's education and occupational status, and degree of urbanization of his childhood home

Other things equal, it was estimated that back in 1964:

a. A veteran with 1 more year's additional education earned _____% more income.

b. A veteran who was white earned _____% more income than one who was not.

c. A veteran who rated 1 point higher on the AFQT (e.g., who was in the 51st percentile instead of the 50th percentile) earned _____% more income.

d. A non-white veteran who scored in the 80th percentile in the AFQT earned _____% more income than a white veteran who scored in the 50th percentile.

14-18   Assuming that the model in Problem 14-17 is specified correctly, by how much should you hedge your estimates in parts **a** and **b** in order to be 95% certain of being correct?

*14-5   Diagnosis by Residual Plots

In this chapter, we have suggested a wide variety of models besides ordinary multiple regression: dummy variables, polynomials, log transformations. How is one to decide which to use?

In the physical sciences, theory is often adequate to specify the appropriate model. For example, bacteria that are unrestricted by food or space limitations often grow exponentially (i.e., at a constant growth rate) for the same reasons that money grows exponentially when it is compounded at a given growth rate (i.e., interest rate).

Social scientists are less fortunate. With little theory to guide them, they often have to rely heavily on statistical analysis to choose an adequate model. For example, we have already discussed how the regressor X^2 in (14-21) can be tested to see whether it should be included, or dropped to simplify the equation. Now let's look at the other side of the coin: What are the warning signs that a model is too simple, and needs to be made more complex by including more regressors?

One way of diagnosing this kind of trouble is to examine the residuals $Y - \hat{Y}$. First, as a basis of comparison we show in Figure 14-6 what the residuals $Y - \hat{Y}$ ought to look like in a correctly specified model. All the assumptions of (12-3) are satisfied: The residuals are of constant variance, and independent. Because they contain no message or information, they are sometimes called *pure noise* or *white noise*. This means that the model has been correctly specified; that is, it has squeezed all the information from the data, spitting out absolutely useless residuals that tell us nothing.

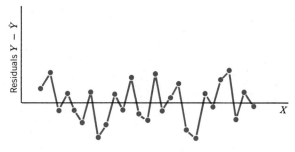

Figure 14-6
The residuals from a correctly specified model are statistically independent—pure noise.

On the other hand, Figures 14-7 to 14-10 show a few examples of simple regression in which the residuals do convey a message, telling us in each case how the model is misspecified, and how it should be corrected. For multiple regression, it is possible to plot the residuals against each of the regressors X in turn (or against $\hat{Y}$), to pick up an appropriate message.

Sometimes the data comes in the form of a *time series*—a sequence of observations over a period of time, such as the population growth of Figure 14-5. When we plot the residuals against time, we often detect *serial correlation*—that is, a tendency for each residual to be similar to the previous one, rather than independent white noise. For example, in Figure 14-11 we see that the last residual is positive, like the ones that preceeded it. And the next residual will likely be positive also.

When serial correlation like this is diagnosed, special techniques of time series analysis are required—especially if forecasts are to be made. For example, in Figure 14-11 we see how mistaken it would be to make a naive forecast using the fitted equation—that is, to forecast the graph

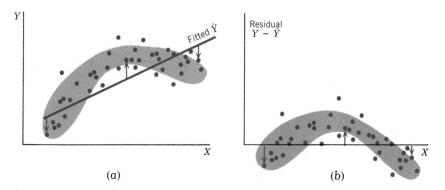

(a)   (b)

Figure 14-7
When residuals are bow-shaped, the model may require a term in X^2 to make it a parabola. (a) Original data, Y vs. X. (b) Corresponding residual plot $(Y - \hat{Y})$ plotted against X—with 3 typical residuals indicated by arrows.

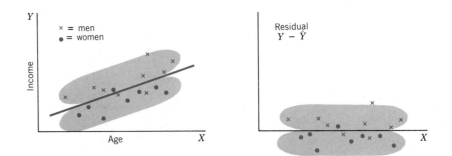

Figure 14-8
Residuals show that Y is related to sex. (At any age level X, men tend to have higher income; i.e., they tend to have positive residuals, and women negative residuals.) This indicates that a dummy sex variable should be included as another regressor.

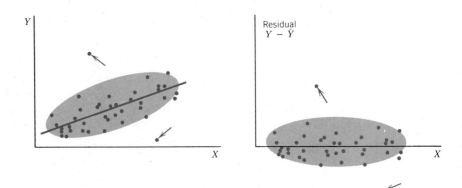

Figure 14-9
Outliers indicate data should be checked and/or robust regression used instead of ordinary least squares.

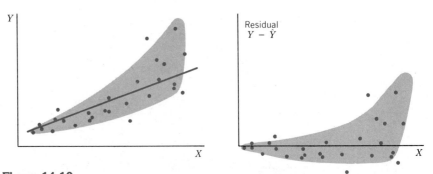

Figure 14-10
Residuals with a bow shape and increasing variability (i.e., the error increases as Y increases), indicate that a log transformation of Y is required; that is, instead of regressing Y on X, regress log Y on X.

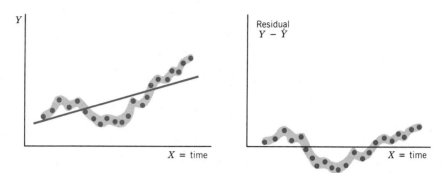

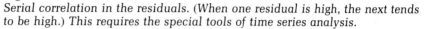

Figure 14-11
Serial correlation in the residuals. (When one residual is high, the next tends to be high.) This requires the special tools of time series analysis.

plummeting down to the fitted line. Because of the serial correlation, the graph is much more likely to change gradually, and special forecasting techniques have been developed to accomplish this.

*PROBLEMS

14-19 **a.** Let us reconsider the factory cost of producing hosiery, and the fitted regression $\hat{Y} = 3.0 + 2.0X$ that you graphed in Problem 12-10. To see whether the residuals show a pattern, circle the two points that have the largest negative residual (i.e., the two points furthest below the fitted line). Also circle the two points that have the largest positive residual.

b. A careful search of the production records showed that the two most negative residuals in part (a) occurred at $T = 1$ and 3 months, while the two most positive residuals occurred at $T = 33$ and 34. This suggests that time may be an influence, so it was included in the regression equation, in both linear and quadratic form. The resulting equation was:

$$\hat{Y} = -13.6 + 2.07X + 1.3T - .022T^2$$

On the diagram in Problem 12-10, graph $\hat{Y}$ as a function of X for each of the following 4 (equally spaced) time periods:

   i. $T = 1$ **ii.** $T = 17$ **iii.** $T = 33$ **iv.** $T = 49$

The four circled points in part **a** now have reduced residuals. Show them as arrows.

14-20 Continuing Problem 14-19, circle the correct choice in each pair of brackets: The p-value for H_0—that the population coefficients of T and T^2 are both zero—turned out to be less than .0001. This shows that there [*is, is not*] a discernible time dependence, and, consequently, it is [*almost cer-*

tain, *hardly possible*] that the sample is a random sample as first assumed in Problem 12-10.

CHAPTER 14 SUMMARY

14-1   A factor with two categories (such as male-female, or treatment-control) can be handled with one dummy variable. Similarly, a factor with c categories can be handled with $c - 1$ dummies.

14-2   When each factor is categorical and handled with dummy regressors, then OLS regression gives essentially the same results as ANOVA in Chapter 10. Moreover, interaction terms can be easily included by multiplying the regressors together.

14-3   Standard OLS computer programs can also handle regression models that are nonlinear in the variables if they are still linear in the coefficients b. For example, if $Y = b_0 + b_1X + b_2X^2$, the program can read the column of X^2 values just like any other column of numbers, and then use those values in calculating a standard regression.

*14-4   When a model is exponential and/or multiplicative, we transform it by taking logs so that standard OLS programs can then handle it.

   Another advantage of using logs is that they have a very useful interpretation: The change in log Y approximately equals the *relative* change in Y itself. Thus logs are particularly useful in estimating growth rates and supply (or demand) elasticities.

*14-5   An examination of residuals provides a diagnostic check on the model. When the model is inadequately specified, the residuals are not just pure noise; instead, they contain a message that can guide us to a better model.

REVIEW PROBLEMS

14-21   Multiple regression (or GLM, the General Linear Model) is very versatile and powerful. Choose the correct alternative in each pair of brackets:

   a.   It can handle categorical factors such as [income, race], by using [nonnegative variables, dummy variables].

   b.   It can handle ANOVA, which deals with [just categorical factors, just numerical factors]. In fact, it is particularly [suitable, unsuitable] to handle ANOVA when the number of observations in each cell varies.

   c.   It can handle polynomials, by using [dummy variables, X and X^2 and X^3 . . . as the regressors].

   d.   It can handle exponentials and multiplicative models by using [dummy variables, logarithms].

   e.   It [can, cannot] help to remove the bias of extraneous factors in observational studies, because each extraneous factor that can be [measured, conceived of] can be entered into the regression equation.

Then the coefficient of interest is interpreted as the change in Y that accompanies a [unit, significant] change in a regressor, if these extraneous factors were [insignificant, constant].

f. In observational studies, multiple regression [can, cannot] tell us exactly how X causes Y, because [some omitted extraneous factors may be causing X and Y to move together, the sign and size of the coefficient of X tells us the direction and size of the change that X produces in Y].

14-22 For each of the following models, outline the method you would use to estimate the parameters β_i:

a. $Y = \beta_0 + \beta_1 X + \beta_2 X^2 + \beta_3 X^3$

b. $Y = \beta_0 + \beta_1 T + \beta_2 \sin(\dfrac{2\pi T}{12})$

(linear growth over time T, plus a 12-month cycle)

c. $Y = \beta_0 (1 + \beta_1)^T$

d. $Y = \beta_0 \beta_1{}^T \beta_2{}^X$

e. $Y = \beta_0 + \beta_1 X + \beta_2 T + \beta_3 X^2 + \beta_4 T^2 + \beta_5 XT$

(fully quadratic model in 2 variables)

f. $Y = \beta_0 X_1{}^{\beta_1} X_2{}^{\beta_2}$ (Cobb-Douglas production function)

14-23 To see how movie prices were related to several factors, a study of urban theaters yielded the following regression (Lamson, 1970):

$$P = 5.8L - 8.2A - 7.7D + \cdots$$

where P = price of admission to the theater (adult evening ticket, in cents)

L = location dummy = 1 for suburbs, 0 for city center

A = age dummy = 1 if theater older than 10 years, 0 if newer

D = drive-in dummy = 1 if outdoor theater, 0 if indoor

a. What would you predict is the difference in the admission price if, other things being equal:

  i. the theater is in the suburbs instead of the city center,

  ii. the theater is old instead of young,

  iii. the theater is young and outdoors, instead of old and indoors.

b. At what sort of theater would you predict the admission price to be highest?

14-24 How is a nation's road system related to its wealth and population density? To answer this question, the following variables were deemed reasonably relevant and accessible for each of 113 countries in 1968 (Glover and Simon, 1975):

R = road mileage (number of miles of paved roads per square mile of land)

P = population per square mile

I = income per capita (U.S.$)

The logarithmic multiple regression was then computed to be

$$\log R = -6.43 + 1.17 \log P + 1.10 \log I$$
$$\text{(SE)} \qquad\qquad (.053) \qquad\quad (.070)$$

a. Does it make sense for both coefficients to be positive? Why?

b. What would be the expected difference in R for:

   i. a country that had 20% higher population density than another of the same per capita income,

   ii. a country that had 20% higher population density and 10% lower per capita income than another country?

14-25 A random sample of men and women in a large American university in 1969 gave the following annual salaries (in thousands of dollars, rounded):

Men	12,	11,	19,	16,	22
Women	9,	12,	8,	10,	16

Denote income by Y and denote sex by a dummy variable X, having $X = 0$ for men and $X = 1$ for women. Then

a. Graph Y against X.

b. Estimate by eye the regression line of Y against X. (*Hint:* Where will the line pass through the men's salaries? through the women's salaries?)

c. Estimate by least squares the regression line of Y against X. How well does your eyeball estimate in part **b** compare?

d. Construct a 95% confidence interval for the coefficient of X. Explain what it means in simple language.

e. Compare **d** with the solution in Problem 8-13.

f. Do you think that the answer to **d** is a measure of how much the university's sex discrimination affects women's salaries?

14-26 Now we can consider more precisely the regression equation for several hundred professors' salaries in Problem 13-12 by including some additional regressors (Katz, 1973):

$$\hat{S} = 230B + \cdots + 50T - 2400X + 1900P + \cdots$$

Standard error	(370)	(530)	(610)	
95% CI	( )	( )	( )	
t ratio	( )	( )	( )	
p-value	( )	( )	( )	

where S = the professor's annual salary (dollars)

T = 1 if the professor received a student evaluation score above the median; 0 otherwise

X = 1 if the professor is female; 0 otherwise

P = 1 if the professor has a Ph.D.; 0 otherwise

a. Fill in the blanks below the equation.

b. Answer True or False; if False, correct it.

    i. A professor with a Ph.D. earns annually $1900 more than one without a Ph.D.

    ii. Or, we might say that $1900 estimates the value (in terms of a professor's salary) of one more unit (in this case, a Ph.D.).

    iii. The average woman earns $2400 more than the average man.

c. Give an interpretation of the coefficient of T.

14-27 For the raw data of Problem 14-26, the mean salaries for male and female professors were $16,100 and $11,200, respectively. By referring to the coefficient of X there, answer True or False; if False, correct it:

After holding constant all other variables, women made $2400 less than men. Therefore, $2400 is a measure of the extent of sex discrimination, and $2500 (16,100 − 11,200 − 2400) is a measure of the salary differential due to other factors, for example, productivity and experience.

14-28 In Problem 14-26, the following additional independent variables were proposed:

a. Professor's faculty (arts, science, social science, business, or engineering).

b. Professor's rank (instructor, assistant professor, associate professor, or full professor).

c. Professor's marital status.

d. Professor's height.

In each case, state why or why not the variable would be a wise addition to the multiple regression, and if so, how it would be handled computationally (with dummy variables, or . . . ?).

*14-29 In Problem 14-26, teaching evaluation T appeared in a severely rounded form (which was explicitly stated earlier in Problem 13-12). What are the advantages and disadvantages of such rounding?

14-30 The manager of a delivery service decided on a critical examination of the costs of his fleet of 28 vans (maintenance and repairs). He therefore selected 12 vans of comparable size (1 ton load) to see if their records would give him some insight:

VEHICLE MAKE	AGE (IN YEARS)	COST (ANNUAL, IN $00)	MAKE	AGE	COST
F	0	12	G	1	18
F	2	17	G	3	22
F	1	20	G	4	22
F	1	14	G	2	14
G	2	15	G	3	18
G	4	22	G	3	16

a. His analysis led him to the following conclusions, which he has asked you to verify. (In each case, check both his arithmetic and his reasoning, and make constructive suggestions):

   i. "The average cost for the young vans (0 and 1 years old) is $1600; and for old vans (2,3,4 years old) is $1825. Since age makes little difference, I should keep the old vans rather than trade them in."

   ii. "The average repair cost for make F is $1575, and for make G is $1838. Clearly my next purchase should be F rather than G."

b. The data was run through a multiple regression program, with cars coded $D = 1$ for make F, and 0 for make G:

$$\text{COST} = 12.4 + 2.18 \text{ AGE} + 1.20D$$
$$\text{(SE)} \qquad\qquad (.89) \qquad\quad (2.28)$$

Interpret what this means to the manager. Since he hates computers but loves pictures, include graphs.

*c. Verify the equation in part b, using a computer.

14-31 In explaining how a nation's output Q is related to its input of capital K and labor L, the following *Cobb-Douglas* model is sometimes used:

$$Q = \beta_0 K^{\beta_1} L^{\beta_2}$$

If we take log Q, what does the model become? How can it be estimated?

14-32 Continuing Problem 14-31, suppose 25 observations yielded the following estimated equation:

$$\log Q = 1.40 + .70 \log K + .50 \log L$$

a. Estimate how much output increases if K increases by 10%, while L remains unchanged.

b. As wage rates have increased, capital has been substituted for labor. Specifically, suppose it has been found that a 12% increase in capital has been associated with a 3% reduction in labor. Would output increase or decrease as a consequence? By how much?

c. Suppose both capital and labor increase by 20%. Would output increase 20% too?

Chapter

Correlation

Q. What is the difference between capitalism and socialism?
A. Under capitalism, man exploits man. Under socialism, it's just the opposite.   ANONYMOUS

15-1 Simple Correlation

Simple regression analysis showed us *how* variables are linearly related; correlation analysis will show us the *degree* to which variables are linearly related. In regression analysis, a whole function is estimated (the regression equation); but correlation analysis yields a single number—an index designed to give an immediate picture of how closely two variables move together. Although correlation is a less powerful technique than regression, the two are so closely related that correlation often becomes a useful aid in interpreting regression. In fact, this is the major reason for studying it.

A—SAMPLE CORRELATION r

Recall how the regression coefficient of Y against X was calculated: We first expressed X and Y in deviation form (x and y), and then calculated

$$b = \frac{\Sigma xy}{\Sigma x^2} \tag{15-1}$$

$$\text{(11-5) repeated}$$

TABLE **15-1** **Math Score (*X*) and Corresponding Verbal Score (*Y*) of a Sample of Eight Students Entering College**

DATA		DEVIATION FORM		PRODUCTS		
		$x =$	$y =$			
X	Y	$X - \bar{X}$	$Y - \bar{Y}$	xy	x^2	y^2
80	65	20	15	300	400	225
50	60	-10	10	-100	100	100
36	35	-24	-15	360	576	225
58	39	-2	-11	22	4	121
72	48	12	-2	-24	144	4
60	44	0	-6	0	0	36
56	48	-4	-2	8	16	4
68	61	8	11	88	64	121
$\bar{X} = 60$	$\bar{Y} = 50$	$0\sqrt{}$	$0\sqrt{}$	$\sum xy =$ 654	$\sum x^2 =$ 1304	$\sum y^2 =$ 836

The correlation coefficient r uses the same quantities $\sum xy$ and $\sum x^2$, and uses $\sum y^2$ as well:

$$\boxed{\begin{array}{c} \textbf{Correlation of } X \textbf{ and } Y \\[6pt] r \equiv \dfrac{\sum xy}{\sqrt{\sum x^2}\,\sqrt{\sum y^2}} \end{array}} \qquad 1(15\text{-}2)$$

Note that in this formula, y appears symmetrically in exactly the same way as x. Thus the correlation r does not make a distinction between the response y and the regressor x, the way the regression coefficient b does. [In the denominator of the regression formula (15-1), the regressor x appears, but the response y does not.]

To illustrate, how are math and verbal scores related? A sample of the scores of 8 college students is given in the first two columns of Table 15-1. In the next two columns we calculate the deviations, and then the sums $\sum xy$, $\sum x^2$, and $\sum y^2$. If we wanted to see how Y can be predicted from X, we would calculate the regression coefficient:

$$b = \frac{\sum xy}{\sum x^2} = \frac{654}{1304} = .50 \qquad (15\text{-}3)$$

On the other hand, if we wanted to measure *how much* X and Y are related, we would calculate the correlation coefficient:

$$r = \frac{\sum xy}{\sqrt{\sum x^2}\,\sqrt{\sum y^2}} = \frac{654}{\sqrt{1304}\,\sqrt{836}} = .63 \qquad (15\text{-}4)$$

B—r MEASURES THE DEGREE OF RELATION

We have claimed that the correlation r measures how much X and Y are related. Now let's support that claim by analyzing what formula (15-2) really means. First, recall how we interpret a deviation from the mean:

$$x \equiv X - \overline{X} \qquad \text{(11-4) repeated}$$

The deviation x tells us how far an X value is from its mean $\overline{X}$. Similarly, the deviation y tells us how far a Y value is from its mean $\overline{Y}$. Therefore, when we plot an observed pair (x,y) in two dimensions, we see how far this observation is from the center of the data $(\overline{X},\overline{Y})$—as Figure 15-1 illustrates.

Suppose we multiply the x and y values for each student, and sum them to get Σxy. This gives us a good measure of how math and verbal scores tend to move together, as we can see in Figure 15-1: For any observation such as P_1 in the first or third quadrant, x and y agree in sign, so their product xy is positive.[1] Conversely, for any observation such as P_2 in the second or fourth quadrant, x and y disagree in sign, so their product xy is negative. If X and Y move together, most observations will fall in the first and third quadrants; consequently most products xy will be positive, as will their sum—a reflection of the positive relationship between X and Y. But if X and Y are negatively related (i.e., when one rises, the other falls), most observations will fall in the second and fourth quadrants, yielding a negative value for our Σxy index. We conclude that as an index of correlation, Σxy at least carries the right sign. Moreover, when there is no relationship between X and Y, with the observations distributed evenly over the four quadrants, positive and negative xy values will cancel, and Σxy will be 0.

To measure how X and Y vary together, Σxy suffers from just one defect: It depends on the units that X and Y are measured in. For example, suppose in Table 15-1 that X were marked on a different scale—specifically, a scale that is 10 times larger. (In other words, suppose the math score X is marked out of 1000 instead of 100.) Then every deviation x would be 10 times larger and so, therefore, would the whole sum Σxy.

We would like a measure of relation that is not so fickle—that is, one that remains "invariant to scale" even if we decide to use an X scale that is 10 times larger. How can we adjust Σxy to obtain such a measure? First, note that the quantity $\sqrt{\Sigma x^2}$ would also change by the same[2] factor 10. So if we divide Σxy by $\sqrt{\Sigma x^2}$, the factor of 10 cancels out, and we are left

---

[1]The point P_1 is given in the first line of Table 15-1. It represents a student with excellent scores, X = 80 and Y = 65—or in deviation form, x = 20 and y = 15. Note that the product xy is indeed positive (+ 300).

For the point P_2, x = − 10 and y = + 10. So now the product xy is negative (− 100).

[2]This is because each x^2 and consequently the whole sum Σx^2 would be 100 times larger. But when we take the square root, $\sqrt{\Sigma x^2}$ is reduced to being 10 times larger.

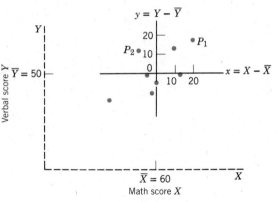

Figure 15-1
Scatter of math and verbal scores, from Table 15-1.

with an invariant measure. Of course, to protect against changes in the Y scale, we should divide by $\sqrt{\Sigma y^2}$ too. The result is

$$\frac{\Sigma xy}{\sqrt{\Sigma x^2}\,\sqrt{\Sigma y^2}}$$

Thus our search for an appropriate measure of how closely two variables are related has indeed led us to the correlation coefficient (15-2).

To get a further idea of the meaning of r, in Figure 15-2 we have plotted various scatters and their correlation coefficients. In panel (a), for example, we show a larger sample than the 8 points in Figure 15-1. Nevertheless, the outline of the scatter displays about the same degree of relation, so it is no surprise that r is about the same value .60.

In panel (b) of Figure 15-2 there is a perfect positive association, so that the product xy is always positive. Accordingly, r takes on its largest possible value, which turns out to be $+ 1$. Similarly, in panel (d), where there is a perfectly negative association, r takes on its most negative possible value, which turns out to be $- 1$. We therefore conclude:

$$\boxed{- 1.00 \leq r \leq 1.00} \tag{15-5}$$

Finally consider the symmetric scatters shown in panels (e) and (f) of Figure 15-2. The calculation of r in either case yields 0, because every positive product xy is offset by a corresponding negative product xy in the opposite quadrant. Yet these two scatters show quite different patterns: In (e) there is no relation between X and Y; in (f), however, there is a strong relation (knowledge of X will tell us a great deal about Y.) A zero value for r therefore does not necessarily mean "no relation." Instead, it means "no *linear* relation" (no straight-line relation). Thus:

$$\boxed{\textbf{\textit{r} is a measure of linear relation only}} \tag{15-6}$$

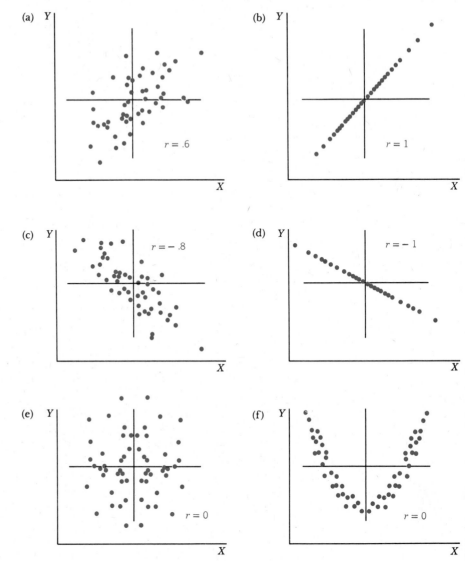

Figure 15-2
Various correlations illustrated: The vaguer is the scatter, the closer r is to 0.

C—POPULATION CORRELATION ρ

Once we have calculated the sample r, how can it be used to make inferences about the underlying population? For instance, in our example the population might be the math and verbal marks scored by *all* college entrants. This population might appear as in Figure 15-3, with millions of dots in the scatter, each representing another student. Let us assume the population is *bivariate normal*, which means, among other things, that the X scores are normally distributed, and so are the Y scores. Then

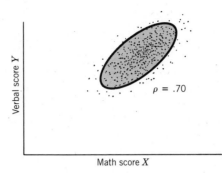

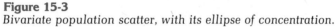

Figure 15-3
Bivariate population scatter, with its ellipse of concentration.

this scatter can be nicely represented by an ellipse that encloses most of the points (about 85%), called the *ellipse of concentration*.

If we were to calculate (15-2) using all the observations in the population, the result would be the *population correlation* coefficient ρ that we first encountered in Chapter 5. (This is proved in Appendix 15-1.

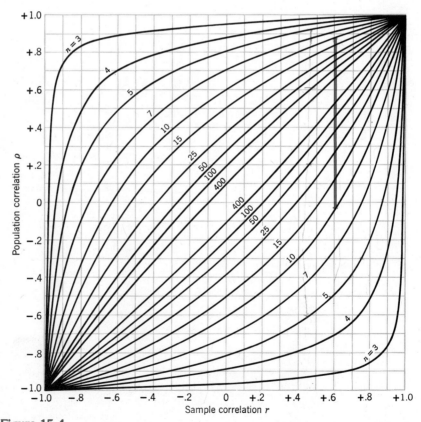

Figure 15-4
95% confidence bands for correlation ρ in a bivariate normal population, for various sample sizes n.

Notice that, as always, we use the Greek letter ρ—rho, the equivalent of r—to denote a population parameter.) Of course, as always in statistics, the population is unknown, and the problem is to infer ρ from an observed sample r. To do so, recall how we used P to make an inference about π in Figure 8-4. Similarly, in Figure 15-4 we can use r to make an inference about ρ.

For example, if a random sample of 10 students has r = .60, the 95% confidence interval for ρ is read vertically as $-.05 < \rho < .87$ as shown by the heavy blue line.

Because of space limitations, we will concentrate in the balance of this chapter on sample correlations. Each time a sample concept like r is introduced, it should be recognized that an equivalent population concept is similarly defined, and inferences may be made about it (assuming the population is bivariate normal).

When the population is not bivariate normal, however, such inferences may be risky or even impossible. For example, in Figure 11-2 the fertilizer levels X were not normally distributed. In fact, it is not proper to speak of them as having a probability distribution at all. Instead, the X values were selected at specified levels by the experimenter. In this case, it is not meaningful to even talk about a population correlation.

PROBLEMS

15-1 A random sample of 6 states gave the following figures for X = annual per capita cigarette consumption and Y = annual death rate per 100,000, from lung cancer (Fraumini, 1968):

STATE	X	Y
Delaware	3400	24
Indiana	2600	20
Iowa	2200	17
Montana	2400	19
New Jersey	2900	26
Washington	2100	20
Averages	2600	21
	$\bar{x}$	$\bar{y}$

   a. Calculate the sample correlation r. A correlation calculated from aggregated data such as this is called an *ecological* correlation.

   b. Find a 95% confidence interval for the population correlation ρ.

   c. At the 5% error level, test whether cigarette consumption and lung cancer are unrelated.

15-2 In Problem 15-1, suppose Y had been measured in annual deaths per *million*, from lung cancer. What difference would that make to:

   a. r?

   b. b?

    c. $t = b/SE$?

    d. p-value for H_o?

15-3 A random sample of baseball players was drawn, out of all the National League baseball players who came to bat at least 100 times in both the 1979 and 1980 seasons. Their batting averages were as follows:

PLAYER AND TEAM	1979	1980
Boroughs (Atlanta)	.220	.260
Cedeno (Houston)	.260	.310
Foote (Chicago)	.250	.240
Henderson (NY)	.310	.290
Scott (St. Louis)	.260	.250
Average	.260	.270

    a. Calculate r.

    b. Find the 95% confidence interval for ρ.

    c. Calculate the regression line you would use to predict the batting average in 1980 from 1979, and find a 95% confidence interval for β.

    d. Graph the estimated regression line and the scatter of 5 observations.

    e. At the 5% error level, can you reject:

        i. The null hypothesis $\beta = 0$?

        ii. The null hypothesis $\rho = 0$?

15-2 Correlation and Regression

A—RELATION OF REGRESSION SLOPE *b* AND CORRELATION *r*

As noted when we set out their formulas (15-1) and (15-2), b and r are very similar. In fact, it can easily be shown (in Appendix 15-2) that we can write b explicitly in terms of r as

$$b = r\frac{s_Y}{s_X} \tag{15-7}$$

Thus, for example, if either b or r is 0, the other will also be 0. Similarly, if either of the population parameters β or ρ is 0, the other will also be 0. Thus it is no surprise that in Problem 15-3e, the tests for $\beta = 0$ and for

$\rho = 0$ were shown to be equivalent ways of examining "no linear relation between X and Y."

B—EXPLAINED AND UNEXPLAINED VARIATION

The sample of math (X) and verbal (Y) scores are reproduced in Figure 15-5, along with the line regressing Y against X. Suppose we want to predict the verbal Y score of a given student—to be concrete, the student farthest to the right in Figure 15-5.

If the math score X were not known, the only available prediction would be to use the sample average $\overline{Y}$. Then the prediction error would be $Y - \overline{Y}$. In Figure 15-5, this appears as the large error (deviation) shown by the longest blue arrow.

However, if X is known we can do better: We predict Y to be $\hat{Y}$ on the regression line. Note how this reduces our error, since a large part of our deviation ($\hat{Y} - \overline{Y}$) is now explained. This leaves only a relatively small unexplained deviation ($Y - \hat{Y}$). The total deviation of Y is the sum:

$$(Y - \overline{Y}) = (\hat{Y} - \overline{Y}) + (Y - \hat{Y}) \qquad (15\text{-}8)$$
$$\text{total} \quad = \text{explained} + \text{unexplained}$$
$$\text{deviation} \quad \text{deviation} \quad \text{deviation}$$

Surprisingly, it can be proved that this same equality holds true when these deviations are squared and summed:

$$\Sigma(Y - \overline{Y})^2 = \Sigma(\hat{Y} - \overline{Y})^2 + \Sigma(Y - \hat{Y})^2 \qquad (15\text{-}9)$$
$$\text{total} \quad = \text{explained} + \text{unexplained}$$
$$\text{variation} \quad \text{variation} \quad \text{variation}$$

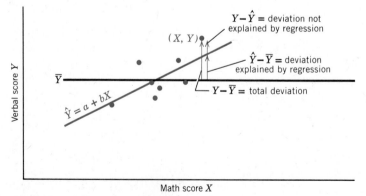

Figure 15-5
The value of regression in reducing variation in Y.

The explained variation is explained by the regressor; it is therefor no surprise that, as Appendix 15-2 shows, we can reexpress the middle term using b and x:

$$
\begin{array}{lcl}
\Sigma(Y - \overline{Y})^2 & = & b^2\Sigma x^2 \quad + \Sigma(Y - \hat{Y})^2 \\
\text{total} & = & \text{variation} + \text{unexplained} \\
\text{variation} & & \text{explained} \quad \text{variation} \\
& & \text{by } X \qquad \text{(residual)}
\end{array}
$$

(15-10)

like (10-14)

This breakdown or analysis of the total variation into its component parts is so important that it is illustrated in Figure 15-6. Again it is called *analysis of variance* (ANOVA), and displayed in an ANOVA table such as Table 15-2. In the last column, we test the null hypothesis $\beta = 0$ (i.e., Y has no linear relation to X). Just as in the standard ANOVA test in Chapter 10, the question is whether the ratio of explained variance to unexplained variance is large enough to reject H_0. Specifically, we form the ratio

$$
F = \frac{\text{variance explained by regression}}{\text{unexplained variance}}
$$

like (10-15)

$$
= \frac{b^2\Sigma x^2}{s^2}
$$

(15-11)

In panel (b) of Table 15-2 we give the computer printout for the math and verbal score example, which gives an F ratio of $328/84.67 = 3.87$. Scanning Table VI where d.f. $= 1$ and 6, we find this observed F value of 3.87 falls beyond $F_{.10} = 3.78$. Therefore,

$$
\text{p-value} < .10
$$

(15-12)

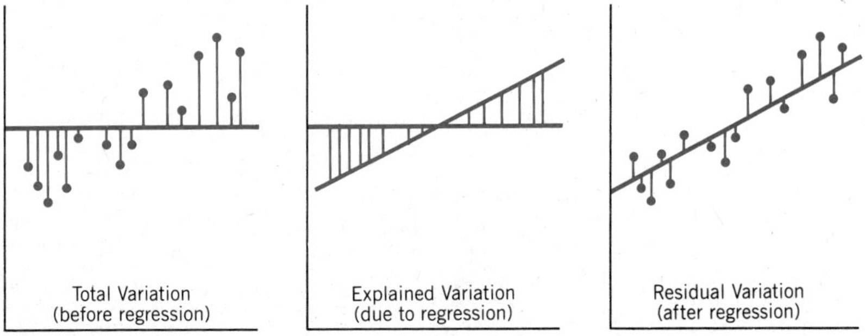

| Total Variation (before regression) | Explained Variation (due to regression) | Residual Variation (after regression) |

Figure 15-6
How regression reduces variation.

TABLE **15-2**   **ANOVA Table for Linear Regression (*a*) General**

SOURCE OF VARIATION	VARIATION (SUM OF SQUARES, SS)	d.f.	VARIANCE (MEAN SQUARE, MS)	F RATIO
Explained (by regression)	$\Sigma(\hat{Y} - \overline{Y})^2$ or $b^2\Sigma x^2$	1	$\dfrac{b^2\Sigma x^2}{1}$	$\dfrac{b^2\Sigma x^2}{s^2}$
Unexplained (residual)	$\Sigma(Y - \hat{Y})^2$	$n - 2$	$s^2 = \dfrac{\Sigma(Y - \hat{Y})^2}{n - 2}$	
Total	$\Sigma(Y - \overline{Y})^2$	$n - 1$		

(*b*) Computed for Math and Verbal Scores (in Table 15-1)

```
              ANALYSIS OF VARIANCE

          DUE TO      DF          SS      MS=SS/DF
          REGRESSION  1       328.00       328.00
          RESIDUAL    6       508.00        84.67
          TOTAL       7       836.00
```

(b)

Thus the F test (based on an ANOVA table) is just an alternative way of testing the null hypothesis that $\beta = 0$. The other method is to use $t = b/SE$ in (12-29)—which is equivalent,[3] and preferable if a confidence interval is also desired.

To sum up, there are three equivalent ways of testing the null hypothesis that X has no relation to Y: the ANOVA F test, the regression t test of $\beta = 0$, and the correlation test of $\rho = 0$. All three will now be illustrated in an example.

EXAMPLE **15-1**

Using the level $\alpha = 5\%$, test for a relationship between the math and verbal scores of Table 15-1 in the following 3 ways:

a. Using the calculations in the ANOVA Table 15-2, calculate the F test of the null hypothesis that $\beta = 0$.

---

[3]To show that the F and t tests are equivalent, we re-express (15-11) as

$$F = \frac{b^2}{s^2/\Sigma x^2} \tag{15-13}$$

According to (12-21), the denominator is just SE^2, so that

$$F = \left(\frac{b}{SE}\right)^2 = t^2 \tag{15-14}$$

H_0 is rejected whenever F is large, or equivalently, whenever t is very positive or negative—the appropriate test against a two-sided alternative hypothesis $\beta \neq 0$.

b. Test the same null hypothesis by alternatively using the t confidence interval.

c. Test the equivalent null hypothesis $\rho = 0$ using the confidence interval for ρ (based on our observed $r = .63$).

Solution

a. From the F-value calculated in Table 15-2, we have already concluded that

$$.05 < \text{p-value} < .10 \qquad\qquad \text{like (15-12)}$$

Since the p-value for H_o is more than 5%, H_o is too credible to be rejected.

b. We use the confidence interval

$$\beta = b \pm t_{.025} \frac{s}{\sqrt{\Sigma x^2}} \qquad \text{(12-23) repeated}$$

where $b = .50$ from (15-3)
$t_{.025} = 2.45$ from Appendix Table V
$s^2 = 84.7$ from Table 15-2b
$\Sigma x^2 = 1304$ from Table 15-1

Thus $\beta = .50 \pm 2.45 \dfrac{\sqrt{84.7}}{\sqrt{1304}}$

$$= .50 \pm 2.45 \,(.255)$$

$$= .50 \pm .62 \qquad\qquad (15\text{-}15)$$

Since $\beta = 0$ is included in the confidence interval, we cannot reject the null hypothesis at the 5% level.

c. In Figure 15-4, we must interpolate to find $n = 8$ and $r = .63$. This yields the approximate 95% confidence interval:

$$-.15 < \rho < +.90$$

Since $\rho = 0$ is included in the confidence interval, we cannot reject the null hypothesis at the 5% level. This agrees with the conclusions in **a** and **b**.

C—COEFFICIENT OF DETERMINATION, r^2

As proved in Appendix 15-2, the variations in Y in the ANOVA table can be related to r:

$$r^2 = \frac{\text{explained variation of } Y}{\text{total variation of } Y} \qquad (15\text{-}16)$$

This equation provides a clear intuitive interpretation of r^2. Note that this is the *square* of the correlation coefficient r, and is often called the *coefficient of determination. It is the proportion of the total variation in Y explained by fitting the regression.* Since the numerator cannot exceed the denominator, the maximum value of the right-hand side of (15-16) is 1; hence the limits on r are ± 1. These two limits were illustrated in Figure 15-2: in panel (b), $r = +1$ and all observations lie on a positively sloped straight line; in panel (d), $r = -1$ and all observations line on a negatively sloped straight line. In either case, a regression fit will explain 100% of the variation in Y.

At the other extreme, when $r = 0$, then the proportion of the variation of Y that is explained is $r^2 = 0$, and a regression line explains nothing. That is, when $r = 0$, then $b = 0$; again note that these are just two equivalent ways of formally stating "no observed linear relation between X and Y."

In Appendix 15-2, we finally show that r^2 is related to variances as well as variations:

$$s^2 \simeq (1 - r^2)\, s_Y^2 \qquad (15\text{-}17)$$

That is, the residual (unexplained) variance s^2 is just a fraction of the original total variance s_Y^2—with this fraction being the *coefficient of in-determination* $(1 - r^2)$. For example, if $r^2 = .40$,

$$s^2 = .60 s_Y^2 \qquad (15\text{-}18)$$

In other words, 60% of the total variance of Y is unexplained.

D—THE DIFFERENT ASSUMPTIONS OF CORRELATION AND REGRESSION

Both the regression and correlation models require that Y be a random variable. But the two models differ in the assumptions made about X. The regression model makes few assumptions about X, but the more restrictive correlation model of this chapter requires that X be a random variable, as well as Y. We therefore conclude that the regression model has wider application. It may be used for example to describe the fertilizer-yield problem in Chapter 11 where X was fixed at prespecified levels, or the verbal/math score problem in this chapter where math score X was a variable; however, the correlation model describes only the latter.

E—SPURIOUS CORRELATION

Even though simple correlation (or simple regression) may have established that two variables move together, no claim can be made that this necessarily indicates cause and effect. For example, the correlation of teachers' salaries and the consumption of liquor over a period of years turned out to be .90. This does not prove that teachers drink, nor does it prove that liquor sales increase teachers' salaries. Instead, both variables moved together, because both are influenced by a third extraneous variable—long-run growth in national income and population. (To establish whether or not there is a cause-and-effect relationship, extraneous factors like this would have to be kept constant, as in a randomized controlled study—or their effects allowed for, as in multiple regression.)

Such a correlation (or regression) that is attributable to an extraneous variable is often called a *spurious* or *nonsense* correlation. It might be more accurate to say that the correlation is real enough, but any naive inference of cause and effect is nonsense.

PROBLEMS

15-4 A random sample of 15 less-developed countries showed the following relation between population density X and economic growth rate Y (Simon, 1981):

COUNTRY	POPULATION DENSITY PER KM² (X)	PERCENT ANNUAL CHANGE IN PER CAPITA INCOME (Y)
A	27	3.3
B	32	0.8
C	118	1.4
D	270	5.4
E	10	1.4
.	.	.
.	.	.
.	.	.
Average	54	2.4
Variation	80,920 Σx^2	46.9
Variance	5780	3.35
St. dev.	76.0	1.83
Correlation, r		.54

a. Calculate the regression line of Y on X. Graph the regression line, along with the first 5 points.

b. Write out the ANOVA table for the regression of Y on X, using the following steps to obtain the variations column:

    i. Calculate the explained variation, using b found in part **a**, and $\Sigma x^2 = 80,920$ given in the table above.

    ii. Copy down the total variation $\Sigma(Y - \overline{Y})^2 = 46.9$ given in the table above.

    iii. Find the residual variation by subtraction.

    Carry through the ANOVA table as far as the p-value for H_0. Can you reject H_0 at the 5% error level?

c. Using the slope in part **a** and the residual variance in part **b**, calculate the 95% confidence interval for β. Can you reject H_0 at the 5% error level?

d. Find the 95% confidence interval for ρ. Can you reject H_0 at the 5% error level?

e. Do you get consistent answers in parts **b**, **c**, and **d** for the question "Are X and Y linearly related?"

f. From the ANOVA table in **b**, find the proportion of the variation that is explained by the regression. Does it agree with r^2?

    Also find the proportion left unexplained. Does it agree with $(1 - r^2)$?

15-5 A random sample of 50 U.S. women gave the following data on X = age (years) and Y = concentration of cholesterol in the blood (grams/liter):

	X	Y
	30	1.6
	60	2.5
	40	2.2
	20	1.4
	50	2.7
	.	.
	.	.
Average	41	2.1
Variation	10,600	11.9
Variance	216	.243
St. dev.	14.7	.493
Correlation	.693	

Repeat the same questions as in Problem 15-4.

15-3 The Two Regression Lines

A—REGRESSION OF Y AGAINST X

In Figure 15-7 we show a bivariate normal population of math and verbal scores, represented by the ellipse of concentration that outlines most of the possible observations. If we know a student's math score, say X_1, what is the best prediction of her verbal score Y? If we consider just the students who scored X_1 in math, they are represented by the vertical slice of dots above X_1. The question is: Among these likely dots (Y values), which is the best guess? The central one, of course, indicated by P_1.

Similarly, for any other given math score X, we could find the central Y value that is the best prediction. These best predictions, marked by the heavy dots, form the *population regression line* of Y against X (an exactly straight line, given that the distribution of X and Y is bivariate normal). The equation of this line is, of course:

$$Y = \alpha + \beta X \qquad\qquad (15\text{-}19)$$

This is the line that is estimated by the *sample* regression:

$$\hat{Y} = a + bX \qquad\qquad (15\text{-}20)$$

where, you recall,

$$\left.\begin{array}{c} b = \dfrac{\Sigma xy}{\Sigma x^2} = r\dfrac{s_Y}{s_X} \\[2mm] a = \overline{Y} - b\overline{X} \end{array}\right\} \qquad \begin{array}{l}(15\text{-}21)\\ \text{like (11-5)}\end{array}$$

B—"REGRESSION TOWARDS THE MEAN"

If we are given X_1 in Figure 15-7 and are asked to predict Y, it is important to fully understand why we would not select the point Q_1 on the major axis of the ellipse, even though Q_1 represents equivalent performance on

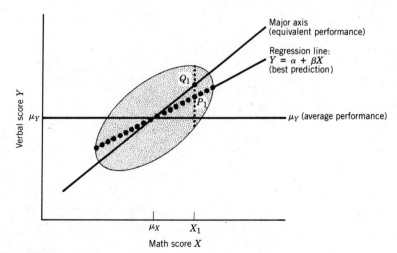

Figure 15-7
The population regression line of Y against X.

the two tests. Since the math score X_1 is far above average, an equivalent verbal score Q_1 seems too optimistic a prediction. Recall that there is a large random element involved in performance. There are a lot of students who will do well in one exam, but not so well in the other; in other words, the correlation is less than 1 for this population. Therefore, instead of predicting at Q_1, we are more moderate and predict at P_1—a compromise between equivalent performance at Q_1 and average performance at μ_Y.

Hence we have the origin of the word "regression." Whatever a student's score in math, there will be a tendency for her verbal score to regress toward the population average. It is evident from Figure 15-7 that this is equally true for a student with a very low math score; in this case, the predicted verbal score regresses upward toward the average.

C—THE REGRESSION OF X AGAINST Y

The bivariate distribution of Figure 15-7 is repeated in Figure 15-8, along with the regression line of Y on X.

Now let us turn around our question: If we know a student's *verbal* score Y_1, what is the likeliest prediction of his *math* score X? Can we still use the regression line $Y = \alpha + \beta X$? If we do, we obtain the estimate P_2, which is absurdly large—it even lies outside the scatter of likely values. Instead, we recognize that, with Y_1 given, the range of likely X values is the *horizontal* slice of dots, with the best prediction again being the midpoint P_1. As before, this is a compromise between equivalent performance Q_1 and average performance μ_X.

Similarly, for any other given verbal score Y, we could find the central X value that is the best prediction. These best predictions, marked by

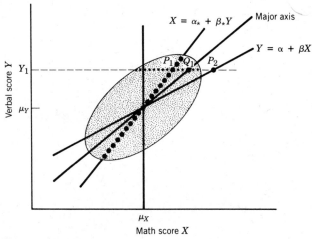

Figure 15-8
The two population regression lines.

heavy dots, again form a straight line, called the *population regression line* of X against Y:

$$X = \alpha_* + \beta_* Y \qquad (15\text{-}22)$$

This population line again may be estimated by the *sample regression line*:

$$\hat{X} = a_* + b_* Y \qquad (15\text{-}23)$$

where the estimates b_* and a_* are obtained from the usual regression formulas—interchanging X and Y, of course:

$$\left. \begin{array}{c} b_* = \dfrac{\Sigma yx}{\Sigma y^2} = r\dfrac{s_X}{s_Y} \\[2mm] a_* = \overline{X} - b_* \overline{Y} \end{array} \right\} \qquad \begin{array}{c} (15\text{-}24) \\ \text{like } (15\text{-}21) \end{array}$$

EXAMPLE **15-2**   Using the sample of math and verbal scores in Table 15-1 and their summary statistics:

  a. Calculate the regression of Y against X, and the regression of X against Y. Graph these two lines.

  b. For a student with a math score $X = 90$, what is the best prediction of the verbal score Y?

  c. For a student with a verbal score $Y = 10$, what is the best prediction of the math score X?

Solution   a. The appropriate calculations Σxy, Σx^2, and so on, have already been carried out in Table 15-1. So we can simply substitute into the appropriate formula:
  For the regression line of Y against X,

$$b = \frac{\Sigma xy}{\Sigma x^2} = \frac{654}{1304} = .50 \qquad (15\text{-}21) \text{ repeated}$$

$$a = \overline{Y} - b\overline{X} = 50 - .50(60) = 20$$

Thus

$$\hat{Y} = 20 + .50X \qquad (15\text{-}25)$$

For the regression line of X against Y,

$$b_* = \frac{\Sigma xy}{\Sigma y^2} = \frac{654}{836} = .78 \qquad (15\text{-}24) \text{ repeated}$$

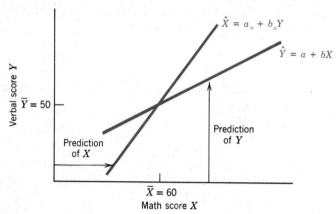

Figure 15-9
The two estimated regression lines for the verbal and math scores in Table 15-1.

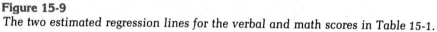

$$a_\star = \overline{X} - b_\star \overline{Y} = 60 - .78(50) = 21$$

Thus

$$\hat{X} = 21 + .78Y \qquad\qquad (15\text{-}26)$$

These two estimated regressions are graphed in Figure 15-9.

b. We substitute $X = 90$ into (15-25):

$$\hat{Y} = 20 + .50(90) = 65$$

c. We substitute $Y = 10$ into (15-26):

$$\hat{X} = 21 + .78(10) = 29$$

In both **b** and **c** note how the predicted grade regresses toward the average.

PROBLEMS

15-6   In a random sample of 200 pairs of father-son heights (X,Y), the summary statistics were computed as follows (in inches):

$$\overline{X} = 68 \qquad\qquad \overline{Y} = 69$$

$$\Sigma x^2 = 1920 \qquad\qquad \Sigma y^2 = 2040$$

$$\Sigma xy = 1010$$

$$n = 200$$

   a. Calculate and graph the regression line of:

      i. Y against X.

      ii. X against Y.

   b. Predict the height of a man drawn at random from the same population if:

      i. Nothing further is known.

      ii. His son's height is 73 inches.

      iii. His father's height is 64 inches.

   c. How much is:

      i. Y correlated to X?

      ii. X correlated to Y?

15-7 Suppose that a bivariate normal distribution of scores is perfectly symmetric in X and Y, with $\rho = .50$ and with an ellipse of concentration as follows:

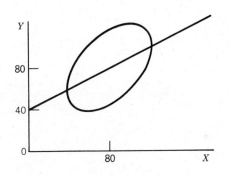

Answer true or false; if false, correct it.

   a. The regression of Y on X is the line shown, with equation

$$Y = 40 + .5X$$

   b. The variance of Y is 1/4 the variance of X.

   c. The proportion of the Y variation explained by X is only 1/4.

   d. Thus, the residual Y values (after fitting X) would have 3/4 the variation of the original Y values.

   e. For a student with a Y score of 70, the predicted X score is 60.

15-8 Let b and b_* be the sample regression slopes of Y against X, and X against Y for any given scatter of points. Answer True or False; if False, correct it:

   a. $b = r\dfrac{s_Y}{s_X}$

b. $b_* = r\dfrac{s_X}{s_Y}$

c. $bb_* = r^2$

d. $b_* = \dfrac{1}{b}$

15-9

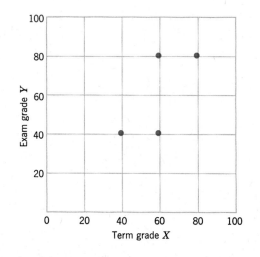

In the above graph of four students' marks, find geometrically (without doing any algebraic calculations):

a. The regression line of Y against X.

b. The regression line of X against Y.

c. The correlation r (*Hint*: Problem 15-8c).

d. The predicted Y for a student with $X = 70$.

e. The predicted X for a student with $Y = 70$.

15-4   Correlation in Multiple Regression

A—PARTIAL CORRELATION

In simple regression we have already seen that the coefficient b is closely related to the correlation r. Accordingly, the relation of Y to X can be expressed with either b or r.

In multiple regression the same general idea holds true: For each multiple regression coefficient there is an equivalent *partial correlation* coefficient.[4] For example, just as the multiple regression coefficient b_1 mea-

---

[4]Because of this equivalence, the multiple regression coefficient is sometimes called the *partial regression* coefficient. It also corresponds to the *partial* derivative in calculus (where all other variables are held constant).

sures how the response Y is related to the regressor X_1 (with the other regressors held constant), likewise the partial correlation coefficient r_1 is defined as the correlation between Y and X_1 (with the other regressors held constant). Thus the relation of Y to X_1 can be expressed with either b_1 or r_1, as we will illustrate in Example 15-3 below.

B—MULTIPLE CORRELATION R

Whereas the partial correlations measure how Y is related to each of the regressors one by one, the multiple correlation R measures how Y is related to *all* the regressors at once. To calculate R, we first use the multiple regression equation to calculate the fitted value $\hat{Y}$:

$$\hat{Y} = b_0 + b_1 X_1 + b_2 X_2 + \cdots$$

Then the multiple correlation R is defined as the ordinary, simple correlation between these fitted $\hat{Y}$ values and the observed Y values. That is,

$$\boxed{R \equiv r_{\hat{Y}Y}}$$

(15-27)

As a kind of simple correlation coefficient, this has the familiar algebraic properties already noted in describing simple correlation in Section 15-1 earlier. In particular, we note that when we square R, we get the *multiple coefficient of determination*:

$$\boxed{R^2 = \frac{\text{variation of } Y \text{ explained by all regressors}}{\text{total variation of } Y}}$$

(15-28)

like (15-16)

Since R^2 gives the proportion of the variation that is explained by all the regressors, it measures how well the multiple regression fits the data. As we add additional regressors to our model, we can see how much they add in explaining Y by noting how much they increase R^2.

C—CORRECTED R^2

Since the inclusion of even an irrelevant regressor will increase R^2 a little,[5] it usually is desirable to correct for this by reducing R^2 appropriately. If there are k regressors, we define the *corrected* R^2 as:

$$\overline{R}^2 \equiv \frac{(n-1)R^2 - k}{n - k - 1}$$

(15-29)

---

[5]By an irrelevant regressor (say X_4) we mean a regressor whose multiple regression coefficient β_4 is 0. Nevertheless, in a *sample* the estimated coefficient b_4 will be slightly nonzero— as some small part of the fluctuation in the irrelevant regressor will happen, by sheer coincidence, to coincide with the fluctuation in Y. Accordingly, that irrelevant regressor will appear to be explaining Y, and R^2 will accordingly increase a little.

Recall that R^2 gives the proportion of the total variation of Y that is explained. Correspondingly, $\overline{R}^2$ gives the proportion of the total *variance* of Y that is explained. (Since variance is variation/d.f., therefor $\overline{R}^2$ is more formally called R^2 *corrected for d.f.*)

Then, of course, $1 - \overline{R}^2$ gives the proportion of the total variance s_Y^2 that is left unexplained as the residual variance s^2; that is,

$$s^2 = (1 - \overline{R}^2)s_Y^2$$

(15-30)
like (15-17)

An example will illustrate these concepts.

EXAMPLE **15-3**

In a chemical plant (Brownlee, 1965), the percent of ammonia that was lost (Y) depended on the following factors:

X_1 = rate of operation
X_2 = temperature of the cooling water
X_3 = concentration of nitric acid

Based on $n = 21$ runs, the fitted regression equation was

$$\hat{Y} = -40.0 + .72X_1 + 1.29X_2 - .15X_3$$
Standard error (SE)   (.13)   (.37)   (.16)
$$R = .956$$

a. For each regressor:

   i. Calculate t and the p-value for H_o.

  ii. Is the regressor discernibly related to Y, at the level $\alpha = 5\%$?

 iii. Calculate the partial correlation with Y, from the formula:

$$\text{partial } r = \frac{b/SE}{\sqrt{(b/SE)^2 + (n-k-1)}}$$

(15-31)

b. Calculate $\overline{R}^2$.

c. If the variance of Y before regression was $s_Y^2 = 18.0$, what is the residual variance s^2?

Solution   **a.** We show in detail the calculations for the first regressor X_1.

   i. $$t = \frac{b}{SE} = \frac{.72}{.13} = 5.54$$ like (13-15)

To find the p-value, we use Table V with d.f. $= n - k - 1$ $= 21 - 3 - 1 = 17$. The observed $t = 5.54$ far exceeds the last tabled value of $t_{.0005} = 3.97$. Thus

$$\text{p-value} \ll .0005$$

ii. Since the p-value $< 5\%$, H_0 is easily rejected. Thus we conclude that ammonia loss Y is positively related to the rate of operation X_1 (discernible at the 5% level).

iii. $$\text{partial } r = \frac{.72/.13}{\sqrt{(.72/.13)^2 + (21 - 3 - 1)}} \qquad \text{like (15-31)}$$

$$= .80$$

When similar calculations for all three regressors are done, they are customarily recorded on the regression equation:

	$\hat{Y} = -40.0 +$	$.72X_1 +$	$1.29X_2 -$	$.15X_3$
(SE)		(.13)	(.37)	(.16)
t		5.54	3.49	.94
p-value		$\ll .0005$	$< .0025$	$< .25$
Partial r		.80	.65	.22
Discernible at 5%?		Yes	Yes	No

b. $$\overline{R}^2 = \frac{(n - 1)R^2 - k}{(n - k - 1)} \qquad\qquad \text{like (15-29)}$$

$$= \frac{(21 - 1)(.956)^2 - 3}{21 - 3 - 1} = .90$$

c. Since $\overline{R}^2 = .90$, therefore 90% of the variance in Y is explained by the three regressors, leaving 10% as the residual variance:

$$s^2 = (1 - \overline{R}^2)\, s_Y^2 \qquad\qquad \text{like (15-30)}$$
$$= (1 - .90)\, 18.0 = 1.80$$

In the rest of this chapter, we will have a chance to see how the multiple coefficient of determination illuminates further important issues in regression.

D—USING $\overline{R}^2$ IN STEPWISE REGRESSION

When there are a large number of regressors, a computer is sometimes programmed to introduce them one at a time, in a so-called *stepwise regression.* The order in which the regressors are introduced may be determined in several ways, two of the commonest being:

1. The statistician may specify a priori the order in which he wants the regressors to be introduced. For example, among many other regressors suppose there were 3 dummy variables to take care of 4 different geographical regions; then it would make sense to introduce these 3 dummy variables together in succession, without interruption from other regressors.
2. The statistician may want to let the data determine the order, using one of the many available computer programs (with names such as "backward elimination," "forward selection," "all possible subsets," etc.).

One of the commonest programs, called *forward selection,* introduces the statistically most important regressors first. It achieves this by having the computer choose the sequence that will make $\bar{R}^2$ climb as quickly as possible. Suppose, for example, that two regressors have already been introduced. In deciding which of the remaining regressors should be added in the next step, the computer tries each of them in turn, and selects the one that increases $\bar{R}^2$ the most.

Whichever method is used, the computer customarily prints the regression equation after each step (after each new regressor is introduced). Then at the end, the computer prints a summary consisting of the list of regressors in the order they were introduced and the corresponding value of $\bar{R}^2$ at each step. This is illustrated in Table 15-4, for some data on lung capacity (Lefcoe and Wonnacott, 1974). Since this table was produced by forward selection, it has screened the regressors in the rough order of their statistical importance—easily and automatically. Notice how the first regressors included make a big contribution in raising $\bar{R}^2$, while the later, less important regressors make less contribution.

TABLE **15-4** **Stepwise Regression of Lung Capacity for $n = 309$ Physicians**

STEP	NEW REGRESSOR ADDED	COEFF. OF DET. ($\bar{R}^2$)
1	age	.367
2	height	.459
3	cigarette smoking	.474
4	pipe and cigar smoking	.476
5	past cigarette smoking	.477

PROBLEMS

15-10 The regression of hay yield Y on rainfall X_1 and temperature X_2 was calculated in Problem 13-13 to be

$$\hat{Y} = 9.14 + 3.38X_1 + .0364X_2$$

Standard error	(.70)	(.090)
t ratio	( )	( )
Partial r	( )	( )
p-value	( )	( )

$$R = .803, n = 20$$

a. Fill in the blanks.

b. If the variance of Y is $s_Y^2 = 87$, what is the residual variance s^2 after regression?

15-11 a. In Table 12-1, values of $\hat{Y}$ and Y are given. Use them to calculate R from the definition (15-27).

b. Calculate r for the same yield-fertilizer example, using the definition (15-2). (In Table 11-2, calculate Σy^2. Then combine it with Σx^2 and Σxy already calculated there.) Does your answer agree with R in part (a)?

15-12 a. In the case of simple regression (with $k = 1$), show that the formula (15-31) reduces to

$$\text{Simple } r = \frac{t}{\sqrt{t^2 + (n-2)}} \tag{15-32}$$

b. For the 7 observations of yield and fertilizer in Problem 15-11, we found earlier that the regression coefficient had a t ratio of 5.2. Substitute this into the formula in **a** to calculate r. Does your answer agree with Problem 15-11?

c. Choose the correct term in the brackets:

When there is just one regressor, both the multiple correlation R and the partial correlation r_p [coincide with, are less than, are greater than] the simple correlation r.

In general, when there are several regressors, [R,r_p] measures how all of them together are related to Y, whereas [R,r_p] measures how each individual regressor is related to Y while all the other regressors are [ignored, held constant].

15-5 Multicollinearity

When two regressors are very closely related, it is hard to "untangle" their separate effects on Y. When one increases, the other increases *at the same time*. To which increase do we then attribute the increase in Y? It's very hard to say—and that is the problem of multicollinearity in a nutshell.

Since multicollinearity is such a common problem, however, it warrants a more detailed discussion—starting with an analogy in simple regression.

A—IN SIMPLE REGRESSION

In Figure 12-4, we showed how the slope b became unreliable if the X values were closely bunched—that is, if the regressor X had little variation. It will be instructive to consider the limiting case, where the X values are completely bunched at one single value $\overline{X}$, as in Figure 15-10. Then the slope b is not determined at all. There are any number of differently sloped lines passing through $(\overline{X},\overline{Y})$ that fit equally well: For each line in Figure 15-10, the sum of squared deviations is the same, since the deviations are measured vertically from $(\overline{X},\overline{Y})$. [This geometric fact has an algebraic counterpart. When all observed X values are the same—that is, when all deviations x are zero—then Σx^2 in the denominator of (11-15) is zero, and b is not defined.]

In conclusion, when the values of X show no variation, then the *relation* of Y to X can no longer be sensibly investigated. But if the issue is just *predicting* Y, this bunching of the X values does not matter *provided that* we confine our prediction to this same value $\overline{X}$—where we can predict that Y will be $\overline{Y}$.

B—IN MULTIPLE REGRESSION

Again consider the limiting case where the values of the regressors X_1 and X_2 are completely bunched up—on a line L in the three-dimensional Figure 15-11. Since L represents a linear relationship between X_1 and X_2, the regressors are said to suffer from *collinearity*. Since similar bunching can occur in higher dimensions when there are more than two regressors, it is more often called the problem of *multicollinearity*.

In Figure 15-11, multicollinearity means that all the observed points in our scatter lie in the vertical plane running up through L. You can think

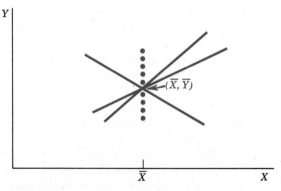

Figure 15-10
Degenerate regression: Because the X values are completely bunched at a point $\overline{X}$, *the regression slope b is not determined.*

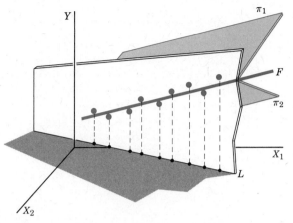

Figure 15-11
Multicollinearity: Because the X_1 and X_2 values are completely bunched on a line L, the regression slopes b_1 and b_2 are not determined.

of the three-dimensional space shown here as a room in a house: The observations are not scattered throughout this room, but instead lie embedded in the pane of glass shown standing vertically on the floor.

In explaining Y, multicollinearity makes us lose one dimension. In the earlier case of simple regression, our best fit for Y degenerated from a line to the point $(\overline{X}, \overline{Y})$. In this multiple regression case, our best fit for Y degenerates from a plane to the line F—the least squares fit through the points on the vertical pane of glass. In *predicting* Y no special problems will arise, *provided that* we confine our predictions to this same pane of glass—where we can predict Y from the regression line F on the glass.

But there is no way to determine the *relation* of Y to X_1 and X_2 separately. For example, any attempt to define the slope b_1 in the X_1 direction involves moving off that pane of glass; and we have no sample information whatsoever on what the world out there looks like. Or, to put it differently, if we try to explain Y with a plane—rather than a line F—we find that there are any number of planes running through F (e.g., π_1 and π_2) that fit our observations equally well: Since each passes through F, each yields an identical sum of squared deviations. [This is confirmed algebraically in the estimating equations (13-5) and (13-6): When X_1 is a linear function of X_2, it can be shown that these two equations cannot be solved uniquely for b_1 and b_2.]

Now let us be less extreme in our assumptions and consider the more realistic case where X_1 and X_2 are *almost* on a line (i.e., where all the observations in the room lie very close to a vertical pane of glass). In this case, a plane can be fitted to the observations, but the estimating procedure is very unstable; it becomes very sensitive to random errors, reflected in very large standard errors of the estimators b_1 and b_2. This is analogous to the argument in the simple regression case in Figure 12-4.

C—IN TERMS OF CORRELATION

The scatter of points (X_1, X_2) in Figure 15-11 that lie on the line L have a perfect correlation of 1.00. (Note how that pattern of dots is similar to Figure 15-2b, for example.) In general, the higher the correlation of the regressors, the worse the problem of multicollinearity—and the formulas for the standard errors reflect this. For example, b_1—the coefficient of X_1—has a standard error given by:

$$SE_1 = \frac{s}{\sqrt{\Sigma x_1^2} \sqrt{1 - R_1^2}}$$

(15-33)

like (12-21)

where

s^2 = the usual residual variance = $\Sigma(Y - \hat{Y})^2/(n-k-1)$
$x_1 = X_1 - \bar{X}_1$ = the regressor X_1 in its usual deviation form
R_1 = the multiple correlation of X_1 with all the other regressors

Thus we see how multicollinearity drives up the SE: The more X_1 is correlated with the other regressors (i.e., the larger is R_1), the greater is SE_1 and therefore the confidence interval for β_1. In other words, the more difficult it is to determine the relationship between Y and X_1.

D—IN DESIGNED EXPERIMENTS, MAKE THE REGRESSORS UNCORRELATED

In (15-33), note that SE_1 is at a minimum—as we would like it to be—when $R_1 = 0$—that is, when X_1 is completely uncorrelated with the other regressors. One of the first principles of experimental design, therefore, is to choose the levels of the X regressors to make them uncorrelated. One of the simplest ways to achieve this is with a square or rectangular array like the one on the "floor" of Figure 15-12. This design is a perfectly natural one: Each variable takes on a range of values independent of the other. Then there is no tendency for them to move together, and hence no problem of multicollinearity.

E—IN OBSERVATIONAL STUDIES, CHOOSE THE REGRESSORS WISELY

In observational studies, we usually have to take the values of the regressors as they come, and so the problem of multicollinearity seems incurable. There is a partial remedy, however: Choose the form of the regressors wisely.

For example, suppose that demand for a group of goods is being related to (1) prices and (2) income, with the overall price index being the first regressor. Suppose the second regressor is income measured in money

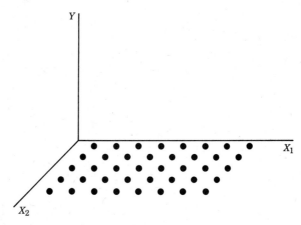

Figure 15-12
In a designed experiment, select an uncorrelated grid of X_1 and X_2 values to eliminate any problem of multicollinearity.

terms. Since this is real income multiplied by the same price index, the problem of multicollinearity may become a serious one. The solution is to use real income, rather than money income, as the second regressor. And in general, try to choose regressors that have as little inter-correlation as possible.

PROBLEMS

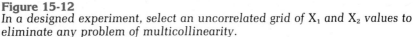

15-13 In the regression of Y on X_1 and X_2, match up the equivalent statements:

a. There is multicollinearity in the regressors.

b. Y has a nearly perfect linear relation to X_1 and X_2.

c. The multiple correlation of Y on X_1 and X_2 is nearly one.

d. The residual variance after regression is very small compared to the variance of Y without regression.

e. X_1 and X_2 have high correlation.

15-14 In Figure 14-4 and equation (14-21), the two regressors used were

$$X_1 = X \quad \text{and} \quad X_2 = X^2$$

a. Tabulate X_1 and X_2 for $X = 1,2,3,4,5$, and then calculate their correlation r.

b. Is multicollinearity therefore any problem?

c. The residual variance after regression was $s^2 = 15.7$. Using (15-33), calculate the standard error of b_1 and of b_2.

*d. Suppose the regressor X_2 had turned out to be statistically indiscernible, and it was decided to drop it.

   To predict Y, can you in fact just "drop off" the X_2 term from the multiple regression, or do you need to calculate the regression equation afresh, using just the regressor X_1?

15-15 Repeat Problem 15-14, if another statistician chose an alternative pair of regressors (called *orthogonal polynomials*):

$$X_1 = X - 3$$
$$X_2 = X^2 - 6X + 7$$

Assume the same values of X—that is, $X = 1,2,3,4,5$.

*15-16 In Problem 15-15, the two orthogonal polynomials were deliberately chosen (defined) to have zero correlation, and so are completely free of multicollinearity. They have a further advantage: the multiple regression equation is easy to compute—so easy, in fact, that it is feasible to do it by hand.

   Carry out this hand calculation on the data in Table 14-2 on page 406, using (13-5) and (13-6). Do you get the same answer as in (14-21)?

CHAPTER 15 SUMMARY

15-1 The correlation coefficient r measures how closely two variables are linearly related; its value is 0 if there is no relation, and $+1$ or -1 for a perfectly positive or negative relation.

15-2 Correspondingly, the coefficient of determination r^2 lies between 0 and 1 (0% and 100%). And r^2 gives the percent of the variation in one variable that is explained by regressing it on the other variable. There are other close connections between correlation and regression: The null hypothesis (that Y is unrelated to X), can equivalently be tested by using r or the regression slope b (or even the F test in the ANOVA table).

15-3 The regression line for predicting X from Y is quite different from the regression line for predicting Y from X. And each regression line is a compromise or "regression" from equivalent performance back towards the mean.

15-4 In multiple regression, to measure how closely Y is related to any particular regressor (while the other regressors are constant), we use the partial correlation. To measure how closely Y is related to the *whole set* of regressors, we use the multiple correlation R.

15-5 The problem of multicollinearity occurs when two (or more) regressors are highly correlated; that is, when one changes, there is a strong tendency for the other to change as well. Then the separate effects of the two regressors are difficult to sort out, and the two regression coefficients consequently have large standard errors.

REVIEW PROBLEMS

15-17 Suppose men always married women who were 4 inches shorter than themselves. What would be the correlation between husbands' and wives' heights?

15-18 **a.** Referring to the math and verbal scores of Table 15-1, suppose that only the students with math scores exceeding 65 were admitted to college. For this subsample of 3 students, calculate the correlation of X and Y.

b. For the other subsample of the 5 remaining students, calculate the correlation of X and Y.

c. Are these 2 correlations in the subsamples greater or less than the correlation in the whole sample? Do you think this will be generally true? Why?

15-19

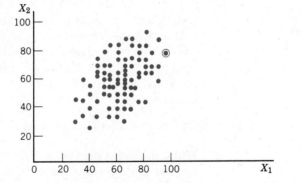

In an experimental program, 80 pilot trainees were drawn at random, and each given 2 trial landings. Their scores X_1 and X_2 were recorded in the above graph, and the following summary statistics were calculated:

$$\overline{X}_1 = 62, \qquad \overline{X}_2 = 61$$

$$\Sigma(X_1 - \overline{X}_1)^2 = 18,000 \qquad \Sigma(X_2 - \overline{X}_2)^2 = 21,000$$

$$\Sigma(X_1 - \overline{X}_1)(X_2 - \overline{X}_2) = 11,000$$

a. By comparison with Figure 15-2, guess the approximate correlation of X_1 and X_2. Then calculate it.

b. Draw in the line of equivalent performance, the line $X_2 = X_1$. For comparison, calculate and then graph the regression line of X_2 on X_1.

c. What would you predict would be a pilot's second score X_2, if his first score was $X_1 = 90$? Or $X_1 = 40$?

d. On the figure below we graph the distribution of X_1 and of X_2. The arrow indicates that the pilot who scored $X_1 = 95$ later scored $X_2 = 80$. (This is the pilot in the original graph who is circled on the extreme right.) Draw in similar arrows for all three pilots who scored

$X_1 = 90$. What is the mean of their three X_2 scores? How does it compare to the answer in **c**?

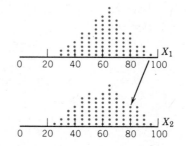

e. Repeat part (d) for all the pilots who scored $X_1 = 40$.

f. Answer True or False; if False, correct it.

The pilots who scored very well or very badly on the first test X_1 were closer to the average on the second test X_2.

One possible reason for this "regression toward the mean" is that a very high score probably represents some good luck as well as good skill. On another test, the good luck may not persist. So the second score will generally be not so good as the first, although still better than average.

15-20 Repeat Problem 15-19, interchanging X_1 and X_2 everywhere.

15-21 a. When the flight instructors in Problem 15-19 graded the pilots, they praised them if they did well, or criticized them if they did poorly. They noticed that the pilots who had been praised did worse the next time, while the pilots who had been criticized did better; so they concluded that criticism works better than praise. Comment.

b. The instructors therefore decided, for a fresh sample of 80 pilots, to criticize *every* pilot no matter how well or poorly the pilot did on the first test. They were disappointed to find that this uniform criticism made the second test scores no better, on average. Comment.

15-22 Answer True or False; if False, correct it:

a. $R^2 = \dfrac{\text{variation of } Y, \text{ explained by all regressors}}{\text{residual variation of } Y}$

b. Multicollinearity occurs when the response is highly correlated with the regressors.

c. Multicollinearity implies that the regression coefficients have large standard errors.

d. Some regressors therefore may be statistically indiscernible; they may be dropped from the model if the purpose is to predict Y for a combination of $(X_1, X_2, \ldots)$ that is similar to the given data (extrapolation instead of interpolation).

e. Suppose a regressor is dropped in part **d**. To obtain the appropriate prediction equation it is necessary to recalculate the regression equation with that regressor omitted.

f. Multicollinearity often is a problem in the social sciences, when the regressors have high correlation. On the other hand, in the experimental sciences, the values of the regressors often can be designed to avoid multicollinearity.

15-23 Suppose that all the firms in a certain industry recorded their profits P (after tax) in 1984 and again in 1985, as follows (hypothetical and simplified data):

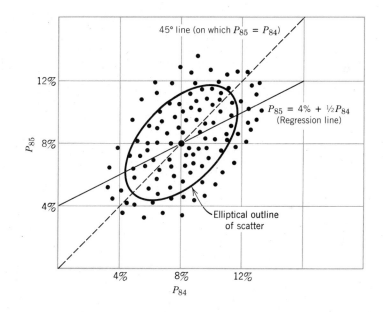

Answer True or False; if False, correct it:

a. The outstandingly prosperous firms in 1984, as well as the outstandingly poor firms in 1984, tended to become less outstanding in 1985.

b. This indicates, but does not prove, that some factor (perhaps a too progressive taxation policy, or a more conservative outlook by business executives, etc.) caused profits to be much less extreme in 1985 than in 1984.

c. This shows, among other things, how difficult it is to stay near the top.

d. The scatter is pretty well symmetrical about the 45° line where $P_{85} = P_{84}$. Therefore, for a firm with a 1984 profit of 11%, the best single prediction of its 1985 profit would be 11% too.

REVIEW PROBLEMS, CHAPTERS 11–15

15-24 Answer True or False; if False, correct it:

    a. In observational studies, simple regression can be very misleading about how X causes Y, because the simple regression coefficient picks up not only the effect of X, but all the extraneous variables too.

    b. Nevertheless, for *predicting* Y, simple regression can be very effective—especially if the correlation of X and Y is near 0.

    c. In observational studies, multiple regression can allow for extraneous factors by including them as regressors.

    d. In both observational studies and randomized experiments, multiple regression reduces s^2, and hence reduces the uncertainty in the regression coefficients and in predicting Y.

    e. One severe limitation of multiple regression is that it cannot include (1) factors that are categorical, or (2) nonlinearities such as polynomials and exponentials.

15-25 In 1970, a random sample of 50 American men (reconstructed from the U.S. Census Bureau's sample) aged 35 to 54 gave the following statistics on education X and income Y:

$$\overline{X} = 11.2 \text{ years} \qquad\qquad \overline{Y} = \$10,200$$

$$s_X = 3.5 \text{ years} \qquad\qquad s_Y = \$7,500$$

$$\text{Correlation } r = 0.43$$

    a. Calculate the estimated regression line of Y against X.

    b. Calculate the 95% confidence interval for the population slope.

    c. Estimate the income of a man who has 2 years of high school ($X = 10$). Include an interval allowance wide enough that you can be 90% sure of being right.

15-26 A group of 4 physicians hired a management consultant to see whether he could reduce the long waiting times of their patients. He randomly sampled 200 patients, and found their waiting time averaged 32 minutes, with a standard deviation of 15 minutes. To determine the factors that waiting time depended on, he ran a multiple regression:

$$\text{WAIT} = 22 + .09 \text{ DRLATE} - .24 \text{ PALATE} + 2.61 \text{ SHORT}$$
$$\text{(SE)} \qquad (.01) \qquad\qquad (.05) \qquad\qquad (.82) \qquad R^2 = .72$$

where WAIT = waiting time, in minutes
            DRLATE = the lateness of the doctors in arriving that morning (sum of their times, in minutes)
            PALATE = the lateness of the patient in arriving for his appointment (in minutes)

SHORT = 1 if the clinic was short staffed, and some of the appointments had to be rebooked; 0 if fully staffed with all 4 physicians

Answer True or False; if False, correct it:

a. Since the coefficient of SHORT is biggest, it is the most important factor in accounting for the variation in WAIT.

b. If two of the doctors were late that morning (by 20 minutes and 40 minutes), the expected increase in waiting time for a patient that day would be 2.1 minutes.

c. If the consultant had studied *all* the patients, he would have found, with 95% confidence, that:

    i. Their average waiting time would be somewhere between 2 and 62 minutes.

    ii. The regression coefficient of PALATE would be somewhere between $-.12$ and $-.36$.

d. Since patients who are late are likely to wait longer, the office staff is providing a strong incentive for patients to arrive on time.

e. If he included another factor in the multiple regression, R^2 would be larger, as would the corrected $\overline{R}^2$.

15-27 Continuing Problem 15-26:

a. For each of the three coefficients, calculate the 95% confidence interval, t ratio, and p-value. Also calculate the partial correlation of each regressor with the waiting time.

b. If a patient is drawn at random, predict how long he has to wait:

    i. If nothing else is known.

    ii. If he turns out to be 15 minutes late, on a day when the clinic was fully staffed but the four physicians were late by 10, 30, 0, and 60 minutes.

c. The patient in part **b** actually had to wait 22 minutes. Which of the predictions in **b** was closer, and by how much?

*d. Is the answer to part **c** typical of all 200 patients?

15-28 A certain drug (e.g., tobacco, alcohol, marijuana) is taken by a proportion of the American population. To investigate its effect on health (for example, mortality rate), suppose that certain people are to be studied for a five-year period, at the end of which each person's mortality will be recorded as follows:

$$M = 0 \text{ if he lives}$$

$$= 1 \text{ if he dies}$$

For each person, also let D represent his average monthly dose of the drug.

Criticize the scientific merit of the following four proposals. (If you like, criticize their ethical and political aspects too.) Which proposal do you think is scientifically soundest? Can you think of a better proposal of your own?

a. Draw a random sample of n persons. For each person, record the drug dose D that he chooses to take, and his mortality M after five years. From these n points, calculate the regression line of M against D, interpreting the coefficient of D as the effect of the drug.

b. Again, draw a random sample of n persons. For each person, record such characteristics as age, sex, grandparents' longevity, and so on, as well as drug dose D and mortality M after five years. Then calculate the multiple regression of M on all the other variables, interpreting the coefficient of D as the effect of the drug.

c. Once again, draw a random sample of n persons. Then construct a 95% confidence interval for the difference in mortality rates between drug users and nonusers, using (8-20):

$$\text{drug effect, } (\mu_1 - \mu_2) = (\overline{M}_1 - \overline{M}_2) \pm t_{.025} \, s_p \sqrt{(1/n_1) + (1/n_2)}$$

where n_1 and n_2 are the numbers of drug users and nonusers, respectively (so that $n_1 + n_2 = n$, the size of the random sample), and s_p^2 is the pooled sample variance.

d. Ask for volunteers who would be willing to use or not use the drug, as determined by the flip of a coin. The control group of volunteers is allowed no drug, while the treatment group is given a standard dose, over the 5-year period. Then a 95% confidence interval for the difference between drug users and nonusers would be the formula above.

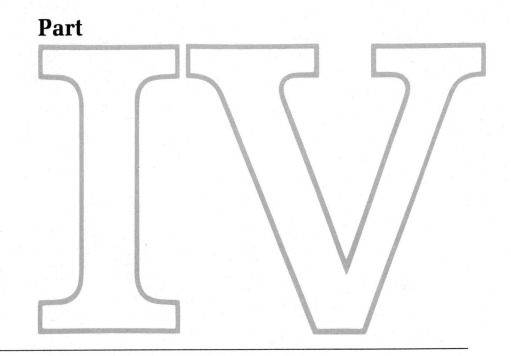

TOPICS IN CLASSICAL AND BAYESIAN INFERENCE

Nonparametric Statistics

(Requires Chapter 9)

In Section 7-3 we studied some robust estimates (such as the sample median or trimmed mean) that are more efficient than the sample mean when the population shape is very non-normal. In this chapter we will pursue this further by calculating confidence intervals and tests using an approach called *nonparametric statistics* (or *distribution-free* statistics, because they are free of the assumption that the population distribution is normal).

Because their historical development predates the computer, nonparametric statistics may not be quite as efficient as the more recent computer-oriented robust statistics (such as the trimmed or biweighted means in Chapter 7, or biweighted least squares in Chapter 11). However, nonparametric statistics offer some compensating advantages: They are easy to calculate, and easy to understand.

16-1 Sign Tests for Medians

Just as the sample mean $\overline{X}$ estimates the population mean μ, so the sample median $\overset{\vee}{X}$ estimates the population median ν (nu, pronounced "new," which follows μ in the Greek alphabet). In the next two sections, we will develop a nonparametric test and confidence interval for ν—analogous to the t test for μ developed in Chapter 8 and 9.

A—SINGLE SAMPLE

Suppose the median (ν) of family income in the U.S. South in 1971 was claimed to be \$6000. The sign test examines each observation to see whether

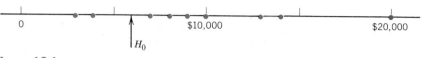

Figure 16-1
A sample of 9 incomes.

it is above (+) or below (−) this claimed value. For example, in the random sample of nine families shown as the dots in Figure 16-1, seven have an income above $6000, while only two have an income below. Does this evidence allow us to reject the claim (the null hypothesis) that

$$H_0: \quad \nu = \$6000?$$

If H_0 is true, half the population incomes lie above $6000. This does not mean that half the incomes *in a small sample* will necessarily be above $6000; but it does mean that if each sample observation is randomly drawn, the probability that it lies above $6000 is

$$H_0: \quad \pi = 1/2 \tag{16-1}$$

We recognize this as being just like the hypothesis that a coin is fair. To state it more explicitly, we have two events that are mathematically equivalent:

> **A "random observation will fall above the median"**
> **is equivalent to**
> **"A coin will turn up heads"**   (16-2)

If H_0 is true, the sample of $n = 9$ observations is just like tossing a coin 9 times. The total number of successes S (families above $6000) will have the binomial distribution, and H_0 may be rejected if S is too far away from its expected value to be reasonably explained by chance.

The alternative hypothesis, of course, is that the true median is higher than the $6000 claim. To judge between these two competing hypotheses, we calculate as usual the p-value for H_0—that is, the probability (assuming H_0 is true) that we would observe an S of 7 or more in 9 trials. (Recall that 7 is the number of "successes" we actually did get.) This is found in Appendix Table IIIc:

$$\begin{aligned} \text{p-value} &= \Pr(S \geq 7) \\ &= .090 = 9\% \end{aligned} \tag{16-3}$$

This means that if the population median really is $6000 (i.e., H_0 is true) there is only a 9% chance that the sample would be so lopsided. Therefore, H_0 does not have much credibility.

B—TWO PAIRED SAMPLES

With a little imagination, we can use the sign test for the difference in two paired samples [just as we used the earlier t test (8-24)].

EXAMPLE **16-1** Suppose a small sample of 8 men had their lung capacity measured before and after a certain treatment; the results are shown in Table 16-1. Use the binomial distribution to calculate the p-value for the null hypothesis that the treatment has no effect.

TABLE **16-1** **Lung Capacity of 8 Patients**

X (BEFORE)	Y (AFTER)	$D = Y - X$
750	850	+ 100
860	880	+ 20
950	930	− 20
830	860	+ 30
750	800	+ 50
680	740	+ 60
720	760	+ 40
810	800	− 10

Solution

In Table 16-1, the original matched pairs in the first two columns can be forgotten, once the differences (improvements) have been found in the last column. These differences D form a single sample to which we can apply the sign test. The null hypothesis (that the treatment provides no improvement) can be rephrased: In the population, the median difference is zero—that is,

$$H_0: \quad \nu = 0$$

or

$$\pi = \text{Pr (observing a positive } D) = 1/2$$

The question is: Are the 6 positive D's (6 "heads") observed in a sample of 8 observations (8 "tosses") consistent with H_0? The probability of this is found in Appendix Table IIIc:

$$\text{p-value} = \text{Pr}(S \geq 6)$$
$$= .145 = 14\% \tag{16-4}$$

This p-value is high enough that we cannot reject H_0 (that the treatment is ineffective) at the 5% level, or even the 10% level.

In applying the sign test, the occasional observation may yield $D = 0$ exactly, so it has no sign. This result is like a coin falling on its edge, and it is best just to discard the observation.

*C—EXTENSIONS

This analysis of medians may be extended in many ways. Whereas in Table 16-1 we were using paired samples (each before-and-after measurement was on the same individual), there is a similar test (Lindgren, 1976) called the *median test*, that can be applied if the two samples are independent. Moreover, for more than two samples, there is an analysis (Mosteller and Tukey, 1977) called *median polish*, which is like ANOVA using medians instead of means.

Even for regression there is a kind of median analysis. As we mentioned in Section 11-3, weighted least squares (with outliers given appropriately small weights) can be viewed as a move towards a "median-type fit."

PROBLEMS

16-1 When polarized light is passed through α-lactose sugar, its angle of rotation is 90°. An industrial chemist made the following 6 independent measurements of the angle of rotation of an unknown sugar:

a. What is the p-value for the hypothesis that the sugar is α-lactose?

b. What is the p-value for the hypothesis that the sugar is D-xylose (whose true angle of rotation is 92°)?

16-2 A random sample of annual incomes (thousands of dollars) of 10 brother-sister pairs was ordered according to the man's income as follows:

BROTHER'S INCOME (M)	SISTER'S INCOME (W)
9	14
14	10
16	8
16	14
18	13
19	16
22	12
23	40
25	13
78	24

Calculate the p-value for the following null hypotheses:

a. That the male median income is as low as 15.

b. That the female median income is as high as 15.

c. That on the whole, men earn no more than women.

16-3 Random Sample of Heights (Inches) of 8 Brother-Sister Pairs

BROTHER'S HEIGHT (M)	SISTER'S HEIGHT (W)
65	63
67	62
69	64
70	65
71	68
73	66
76	71
77	69

Calculate the p-value for the following null hypotheses:

a. That the male median height is as low as 66 inches.

b. That the female median height is as low as 63 inches.

c. That on the whole, men are no taller than women.

16-4

A random sample of 20 bolts produced by a new process showed the following shearing strengths:

$$10.4 \quad 9.6 \quad 10.6 \quad 11.0 \quad 9.9 \quad \quad 10.1 \quad 10.5 \quad 8.7 \quad 11.1 \quad 11.1$$

$$10.6 \quad 10.7 \quad 8.9 \quad 8.6 \quad 9.5 \quad \quad 8.4 \quad 11.5 \quad 11.3 \quad 10.7 \quad 9.1$$

a. Graph the data as 20 points on the x-axis.

b. The old process produced bolts with a median strength of 9.75. Calculate the p-value for the null hypothesis that the new process is no better.

*c. Calculate the p-value for the hypothesis that the lower quartile is 8.50.

16-2 Confidence Intervals for the Median

Recall the sample of 9 incomes from the U.S. South, shown again in Figure 16-2. When the X's are listed in order of size, they are called *order statistics* and given a *bracketed* subscript. Thus in Figure 16-2, $X_{(1)}$ is the smallest observation, $X_{(2)}$ the next, and so on. Then the median will be $X_{(5)}$, which leaves 4 below and 4 above. This sample median $\overset{\cdot}{X} = X_{(5)} = \$9000$ is a good point estimate of the unknown population median ν.

But how do we construct a confidence interval? That is, how far do we

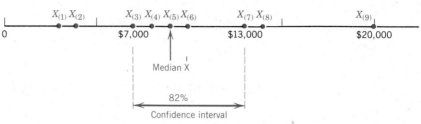

Figure 16-2
*Nine ordered observations of income, showing an 82% confidence interval for
the population median* ν.

go above and below this point estimate? One possible answer is to go two
observations on either side of the median, from $X_{(3)}$ to $X_{(7)}$:

$$X_{(3)} \le \nu \le X_{(7)} \tag{16-5}$$
$$7{,}000 \le \nu \le 13{,}000$$

But that leaves the question: How much confidence do we have that
this interval will cover the target ν? To answer this, panel (a) of Figure
16-3 illustrates the underlying population, from which the observations
are drawn. Note how half of this population is above ν, and half below.
(Also note that the population is not normal—the nonparametric analysis
of this chapter is perfectly valid for *any* population shape.) Panel (b) shows
a typical sample that yields a confidence interval covering ν. Panel (c)
shows a sample so lopsided that its confidence interval fails to cover ν.
How often will such a failure occur? Whenever the sample is this lopsided,
or worse. That is, whenever the number of observations above the median
is 7, or more. The chance of this is the same as 7 or more heads, in n =
9 tosses of a coin. From the binomial distribution in Table IIIc,

$$\Pr(S \ge 7) = .090 \approx 9\% \qquad \text{like (16-3)}$$

Of course, the confidence interval could go wrong equally often by being
lopsided in the other direction, with 7 or more observations to the left of
the median (like 7 or more tails). Thus the chance of going wrong one
way or another is 2(9%), and the confidence level is therefore:

$$100\% - 2(9\%) = 82\% \tag{16-6}$$

Of course, if we have a large sample size n, then we use the normal
approximation to the binomial. Moreover, with a large sample, we can
specify the confidence level in advance (at the customary 95%, for example). But in this case we have to go through the steps above in reverse
order, ending up with an estimated confidence interval. This is illustrated
in Problem 16-11, part **c**.

Finally, as always, a confidence interval like (16-5) can be used to test
a null hypothesis. We simply reject the null hypothesis if it falls outside
this confidence interval.

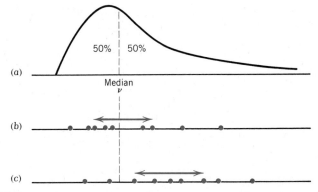

Figure 16-3
Confidence intervals are usually right, but occasionally wrong. (a) Underlying population, showing the target ν. (b) Typical sample, where the CI covers ν. (c) Unusually lopsided sample, where the CI fails to cover ν.

PROBLEMS

16-5 The industrial chemist who made the 6 measurements in Problem 16-1 thought the true angle would be somewhere between 89.9° and 93.4° (the smallest and largest measurements). How much confidence is there for this interval?

16-6 **a.** In Problem 16-2, construct a nonparametric confidence interval of the form $X_{(2)} \leq \nu \leq X_{(9)}$, for the median of (i) men's income, (ii) women's income, (iii) the difference between men's and women's income.

b. What is the confidence level for each of these intervals?

16-7 **a.** In Problem 16-3, construct a nonparametric confidence interval of the form $X_{(2)} \leq \nu \leq X_{(7)}$, for the median of (i) men's height, (ii) women's height, (iii) the difference between men's and women's height.

b. What is the confidence level for each of these intervals?

16-8 **a.** Referring to Problem 16-6, do you think that the population of men's incomes is distributed normally?

b. Do you think the nonparametric confidence interval would be narrower than the classical confidence interval?

c. To check your conjecture in **b,** go ahead and calculate the 98% confidence interval of the form (8-11).

16-9 **a.** Referring to Problem 16-7, do you think that the population of men's heights is distributed normally?

b. Do you think that the nonparametric confidence interval would be narrower than the classical confidence interval?

c. To check your conjecture in **b,** go ahead and calculate the 93% confidence interval of the form (8-11). [*Hint:* If you interpolate Table V, you will find $t_{.035} = 2.14$.]

16-10 Write a summary of what you learned from Problems 16-8 and 16-9.

16-11 A random sample of 25 working students yielded the following summer incomes (in \$100s, arranged in order):

$$4,5,5,5,6 \qquad 7,9,10,12,13 \qquad 14,14,15,17,18$$

$$18,19,23,25,27 \qquad 30,32,39,40,52$$

a. Estimate the population median income ν.

b. One possible confidence interval for the median is $7 \le \nu \le 27$. How much confidence would we have in an interval this wide? [*Hint:* Use the binomial approximation.]

c. Calculate the 95% confidence interval for ν. [*Hint:* Since you are now starting with the confidence *level* and working towards the confidence *interval* (rather than vice versa), you will have to reverse the steps in **b**.]

16-12 Using the data in Problem 16-4, calculate a 95% confidence interval for the median of the total production.

16-3 The Wilcoxon Rank Test

A—WHY USE RANKS?

The sign test for the median was extremely simple: It merely recorded whether or not each observation was above or below the hypothetical median ($+$ or $-$). On the other hand, the classical t test used much more information: In calculating the sample mean $\overline{X}$ it took into account the actual size of each observation. The *rank tests* we now consider are a compromise: In ranking numbers, we see which observation is first (or second, etc.), but we ignore *by how much* it is first.

As a compromise, rank tests are very robust, and therefore very popular. To see how they work, we shall study a typical and popular form, the Wilcoxon test[1] for two independent samples.

B—ILLUSTRATION

Suppose that independent random samples of annual income were taken from two different regions of the U.S. in 1980, and then ordered as in Table 16-2.

---

[1]The Wilcoxon test was discovered by Mann and Whitney in an equivalent form that is more complicated to understand and calculate; so we will stick with the Wilcoxon form.

TABLE **16-2** **Two Independent Samples of Income**

	SOUTH X_1	PACIFIC X_2
	6,000	11,000
	10,000	13,000
	15,000	14,000
	29,000	17,000
		20,000
		31,000
	Size $n_1 = 4$	$n_2 = 6$

Let us test the null hypothesis that the two underlying populations are identical. Suppose that the alternative hypothesis is that the South is poorer than the Pacific, so that a one-sided test is appropriate.

TABLE **16-3** **Combined Ranking Yields the W Statistic**

COMBINED ORDERED OBSERVATIONS		COMBINED RANKS	
X_1	X_2	R_1	R_2
6,000		1	
10,000		2	
	11,000		3
	13,000		4
	14,000		5
15,000		6	
	17,000		7
	20,000		8
29,000		9	
	31,000		10

$$W = 18$$

We first rank the combined X_1 and X_2 observations, as shown in Table 16-3. The actual income levels now are discarded in favor of this ranking. (Consequently, the test is not affected by skewness, or any other distributional peculiarity—in other words, it is a distribution-free test.) Then *Wilcoxon's Rank Sum W* is defined as the sum of all the ranks in the smaller sample; in this case:

$$W = 1 + 2 + 6 + 9 = 18 \tag{16-7}$$

Then the p-value for H_o can be found by looking up $W = 18$ in Appendix Table VIII:

$$\text{p-value} = .238 \simeq 24\% \tag{16-8}$$

This value is large, so the null hypothesis (of equal incomes) cannot be rejected.

C—GENERALIZATION

To keep the calculation of W as simple as possible, we arrange it so that we add up just a relatively few small numbers by following two conventions:

> **The W statistic is the rank sum of the *smaller* sample; and we start ranking at the end where this smaller sample is concentrated.** (16-9)

For example, in Table 16-3, the South (X_1) was the smaller sample; and it was concentrated at the low-income end, so that is where we started ranking. (Note that adding up the ranks in the other sample would mean more work: more numbers, and bigger numbers).

Appendix Table VIII gives the corresponding one-sided p-value, if both sample sizes ≤ 6. For larger sample sizes, W is approximately normal, with:

$$E(W) = \frac{1}{2} n_1(n_1 + n_2 + 1)$$

$$SE = \sqrt{\frac{1}{12} n_1 n_2 (n_1 + n_2 + 1)}$$

(16-10)

where n_1 is the size of the smaller sample (whose ranks sum to W). An example will illustrate these formulas:

EXAMPLE 16-2

Although the formulas (16-10) are most useful for large samples, they often give a good approximation for small samples too. To illustrate, use (16-10) instead of Table VIII to calculate the p-value for H_o in Table 16-3.

Solution

The smaller sample has $n_1 = 4$ observations, while $n_2 = 6$. Thus (16-10) becomes

$$E(W) = \frac{4}{2} (4 + 6 + 1) = 22$$

$$SE = \sqrt{\frac{4(6)}{12} (4 + 6 + 1)} = 4.69$$

In Table 16-3, the rank sum of the smaller sample has already been observed to be $W = 18$. Since p-value is the probability of getting a value as extreme as the one we've observed, the p-value is therefor $Pr(W \leqslant 18)$. To calculate this, we first must standardize the critical value $W = 18$, using the mean and standard error:

$$Z = \frac{W - E(W)}{SE} \qquad\qquad (16\text{-}11) \\ \text{like } (9\text{-}11)$$

$$= \frac{18 - 22}{4.69} = -.85$$

$$Pr(W \le 18) = Pr(Z \le -.85)$$

$$= .198$$

$$p\text{-value} = 20\%$$

This is a good approximation to the correct value of 24% found in (16-8). With continuity correction ($W \le 18.5$), the approximation is even better (p-value = 23%).

We emphasize that the approximately normal distribution of the test statistic W must not be confused with the population of the X values, which may be very non-normal. In fact the W test, like all nonparametric tests, is specifically designed for very non-normal populations.

D—TIES

Observations that are tied with one another should be given the same rank—their average rank. Unless most of the observations are ties, we can then continue as usual.

For example, suppose the samples of income were the ones ranked in Table 16-4. The first tie occurs at the 3rd and 4th incomes, and so each is given the average rank, 3½. Continuing on, the next tie occurs at the 7th, 8th, and 9th incomes, where each is given the average rank, 8. Once this ranking is complete, we proceed exactly as before.

TABLE 16-4 **_W_ Statistic, When Ties Occur**

COMBINED ORDERED OBSERVATIONS, X		COMBINED RANKS, R	
SOUTH X_1	PACIFIC X_2	R_1	R_2
6		1	
10		2	
11 tie	11	3.5	3.5
	13		5
	14		6
15,15 tie	15	8,8	8
	18		10
	20		11
		$W = 22.5$	

PROBLEMS

16-13 A random sample of 6 men's heights and an independent random sample of 6 women's heights were observed, and ordered as follows:

MEN'S HEIGHTS	WOMEN'S HEIGHTS
65	62
67	63
69	64
70	66
73	68
75	71

Calculate the p-value for the null hypothesis H_0 that on the whole, men are no taller than women.

16-14 Two makes of cars were randomly sampled, to determine the mileage (in thousands) until the brakes required relining. Calculate the p-value for the null hypothesis that make A is no better than B.

MAKE A	MAKE B
30	22
41	26
48	32
49	39
61	

16-15 Recalculate the p-value in Problem 16-14, using the normal approximation, just to see how well it works.

16-16 Sometimes data is not collected in numerical form; instead, it is only ordered from best to worst, or highest to lowest. For example, to compare the wine from vineyards C and M, 8 bottles were selected at random from each, and labeled $C_1 \ldots C_8$ and $M_1 \ldots M_8$. Then in a blind test a wine-tasting expert listed them in order of preference. The results were as follows (listed from highest to lowest preference):

$$M_7 \; M_5 \; M_3 \; C_2 \; M_2 \; M_6 \; C_8 \; M_1 \; C_5 \; M_4 \; C_3 \; C_4 \; M_8 \; C_1 \; C_6 \; C_7$$

A rank test is ideally suited to test the null hypothesis that the two vineyards produce wine of equal quality. What is the p-value for this hypothesis?

16-4 Rank Tests in General

Rank tests are so very efficient and understandable that it is worthwhile extending them to cover other cases besides the Wilcoxon test. In fact, we can develop several rank tests at once from the following general strategy:

> If outliers (or some other form of non-normality) seem to be a problem, rank the data. Then just perform the usual test (such as t or F) on the *ranks,* instead of the original data. (16-12)

One of the most straightforward examples of this strategy is to extend the Wilcoxon test to a *rank test for k independent samples.* An example will illustrate:

EXAMPLE **16-3**

Suppose we have three regions to compare, instead of just the two given earlier. Table 16-5 gives independent random samples (of annual income in the U.S. in 1980) from the three different regions.

TABLE **16-5** **Three Independent Samples of Income (compare to Table 16-2)**

SOUTH X_1	PACIFIC X_2	NORTHEAST X_3
6,000	11,000	7,000
10,000	13,000	14,000
15,000	17,000	18,000
29,000	131,000	25,000

Carry out the rank test in two steps:

a. Rank the 12 observations.

b. Do the usual one-way ANOVA F test on the ranks to get the p-value for H_o.

Solution **a.** The ranks are found just as in the Wilcoxon test of Table 16-3:

Table of ranks

R_1	R_2	R_3
1	4	2
3	5	6
7	8	9
11	12	10
$\overline{R} = 5.50$	7.25	6.75

b. Using the average ranks calculated above, we obtain the following ANOVA table:

SOURCE	SS	df	MS	F
Between regions	6.50	2	3.15	.21
Within regions	136.5	9	15.2	(p > .25)
Total	143.0 $\checkmark$	11 $\checkmark$		

Since $p > .25$, H_o is not rejected.

Using the same two steps—transform the data into ranks, and then apply a standard test—we can similarly carry out a nonparametric test in a wide variety of situations. Such an approach is much easier than working through one rank test after another in exacting detail. And there are only two costs:

1. Using (16-12) gives only approximate p-values, not exact ones. (But we *already* had to be satisfied with approximate p-values for large samples, as we saw at the end of Example 16-2. In any case, approximate p-values are perfectly satisfactory in practice.)
2. Although (16-12) is often perfectly straightforward, it sometimes has to be given a slight twist. (For example, see Problems 16-19 and 16-20, or Conover, 1980.)

PROBLEMS

In each of the following problems, carry out the two steps of (16-12):

a. Transform to ranks.

b. Carry out the classical test.

16-17 (Wilcoxon rank test for two independent samples)

a. For the data in Table 16-3, verify that the ranks are as follows:

ORIGINAL DATA, ORDERED		RANKS	
6,000	11,000	1	3
10,000	13,000	2	4
15,000	14,000	6	5
29,000	17,000	9	7
	20,000		8
	31,000		10
		$\overline{R} = 4.5$	6.2

b. Do the usual two-sample t test (8-20) on the ranks (i.e., test the difference between the two means 4.5 and 6.2) to get the p-value for H_o.

16-18 (Kruskal-Wallis rank test for k independent samples, or one-factor ANOVA)

a. For the data in Table 10-1, verify that the ranks are as follows:

ORIGINAL DATA, ORDERED			RANKS		
46	52	49	1	9	3.5
47	54	50	2	11.5	5.5
49	55	51	3.5	13	7.5
50	58	51	5.5	14	7.5
53	61	54	10	15	11.5

b. Do the usual one-factor ANOVA on the ranks to get the p-value for H_o.

16-19 (Wilcoxon rank test for two paired samples)

a. For the data in Table 16-1, let us temporarily ignore the + and − signs of the differences, and rank them from smallest to largest. Then let us put back the + and − signs on the ranks. Thus verify that we obtain the following "signed ranks":

ORIGINAL DATA, ORDERED BY SIZE OF DIFFERENCES			SIGNED RANKS OF DIFFERENCES
X	Y	D	
810	800	− 10	− 1
950	930	− 20	− 2.5
860	880	+ 20	+ 2.5
830	860	+ 30	+ 4
720	760	+ 40	+ 5
750	800	+ 50	+ 6
680	740	+ 60	+ 7
750	850	+ 100	+ 8

b. Treating the final column like the differences calculated in the usual paired t test (8-24), calculate the p-value for H_o.

c. Is your answer in part **b** the same as the p-value found earlier in (16-4)? Why?

16-20 (Friedman rank test for k matched samples, or two-factor ANOVA)

a. Consider again the data in Table 10-6. For each operator, let us rank the 3 machines. (This is *not* the usual combined ranking that we did in Problem 16-18, for example. Instead, we begin by ranking the 3 machines for the first operator, temporarily ignoring the others. Then we start afresh, ranking the 3 machines for the second operator. And so on.) Verify that we obtain the following ranks:

ORIGINAL DATA (UNORDERED			RANKED IN EACH ROW		
53	61	51	2	3	1
47	55	51	1	3	2
46	52	49	1	3	2
50	58	54	1	3	2
49	54	50	1	3	2

b. Carry out the usual two-factor ANOVA on these ranks. There will be absolutely no differences in rows (since all row averages will be 2.0). But the differences in columns (machines) will be interesting, and the F ratio will give the desired p-value for H_o.

16-21 In the preceding four problems, we have illustrated a good variety of rank tests, one for each cell of the following table:

	2 SAMPLES	k SAMPLES
Independent		
Matched		

In each cell, write in the name of the nonparametric rank test (e.g., Friedman two-factor ANOVA).

16-22 In Problem 16-20, use the Friedman test to find the p-value for the null hypothesis of no differences between *operators* (rows).

16-23 (Spearman rank correlation) Treat the data in Problem 15-1 with the general rank test (16-12). Specifically:

a. Verify that the X and Y ranks are as follows:

ORIGINAL DATA		RANKED IN EACH COLUMN	
X	Y	R_X	R_Y
3400	24	6	5
2600	20	4	3.5
2200	17	2	1
2400	19	3	2
2900	26	5	6
2100	20	1	3.5

b. Calculate the rank correlation coefficient, and from Figure 15-4 determine whether it is discernibly different from 0 at the 5% level (2-sided).

16-5 Runs Test for Independence

One of the crucial assumptions we have used throughout this text (much more important than the normality assumption) is the assumption that our sampling is random. Now we shall develop a test of whether or not this assumption is valid.

By definition, a random sample consists of observations that are drawn *independently* from a *common* population. Thus, if the observations are graphed in the time order in which they were sampled, the graph should look somewhat like panel (a) in Figure 16-4. On the other hand, if the observations are correlated, they will display some "tracking," as in panel (b). Or, if the observations come from two different populations, they may

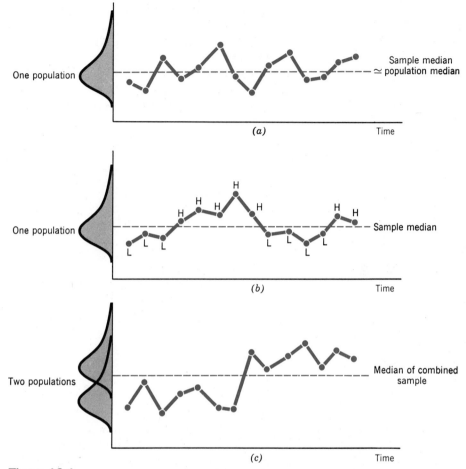

Figure 16-4
(a) *Independent observations from one population.* (b) *Serially correlated observations from one population.* (c) *Observations from two populations.*

appear to be displaced, as in panel (c). How can we quantify these differences that are obvious to the eye, and find some numerical measure to test the null hypothesis of randomness? We note that when H_o is true in panel (a), the path of the observations crosses the median line quite frequently; but when H_o is not true, this happens much less frequently. This is the basis for the runs test.

For example, in panel (b), we mark observations H (for high) or L (for low), depending on whether they fall above or below the sample median. With slashes indicating the crossovers, this sequence is

$$LLL/HHHHH/LLLL/HH \qquad (16\text{-}13)$$

A run is defined as an unbroken sequence of H or L values. In this case the number of runs $R = 4$.

Let us suppose in general that there are n observations. (If the sample size is odd, the median line will pass through the median observation, which should be counted neither L nor H. With this observation discarded, n then refers to the even number of observations remaining.) When H_o is true, the number of runs R has a sampling distribution that is approximately normal, with:

$$\boxed{\begin{aligned} E(R) &\simeq \frac{n}{2} + 1 \\[2mm] SE &\simeq \frac{\sqrt{n-1}}{2} \end{aligned}} \qquad (16\text{-}14)$$

For example, in Figure 16-4b, $n = 14$, so that (16-14) yields:

$$E(R) \simeq \frac{14}{2} + 1 = 8$$

$$SE \simeq \frac{\sqrt{14-1}}{2} = 1.80$$

In Figure 16-4b, the observed number of runs is $R = 4$. What p-value for H_o does this give? Since p-value is the probability of getting a value as extreme as the one we've observed, the p-value is therefor $Pr(R \leqslant 4)$. To calculate this, we first must standardize the critical value $R = 4$, using the mean and standard error:

$$Z = \frac{R - E(R)}{SE} \qquad (16\text{-}15)$$
like (16-11)

$$= \frac{4 - 8}{1.80} = -2.22$$

$$\begin{aligned} Pr(R \leq 4) &= Pr(Z \leq -2.22) \\ &= .013 \\ \text{p-value} &= 1\% \end{aligned}$$

This provides strong evidence that the sample is not random.

PROBLEMS

16-24

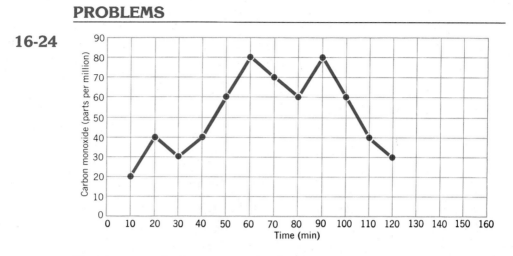

The above graph shows a sample of 12 air pollution readings, taken every 10 minutes over a period of 2 hours. To what extent can we claim that these are statistically independent observations from a fixed (rather than a drifting) population? Answer by calculating the p-value for the null hypothesis.

16-25 Repeat Problem 16-24 for the following sample of 25 observations (read across):

67, 63, 58, 79, 62,   55, 56, 50, 57, 55,   43, 47, 23, 31, 38,

49, 33, 43, 34, 42,   51, 66, 54, 46, 55

CHAPTER 16 SUMMARY

16-1 Nonparametric statistics were the earliest form of robust statistics, and are still popular. While they may not attain quite the high efficiency or generality of the robust statistics developed recently for the computer—for example, those we have described in Chapters 7 and 11—nonparametric statistics are nevertheless familiar in the literature and very easy to calculate.

The simplest nonparametric test is the sign test for the median.

16-2 Using the same logic as the sign test, a confidence interval for the population median can be constructed. We simply count off an appropriate number of observations on either side of the sample median.

16-3 Rank tests use more information than sign tests, and are usually more efficient. The simplest one is the Wilcoxon test that ranks the observations in two independent samples.

16-4 There is a very simple way to carry out a nonparametric test in many situations where the classical test seems unsatisfactory because of outliers (or some other form of non-normality): Just transform all measurements to ranks, and then perform the classical test on the ranks (instead of on the original measurements).

16-5 Independence of successive observations—a basic assumption of random sampling—can be tested by counting runs (that is, unbroken sequences of values above or below the median).

REVIEW PROBLEMS

In all problems, calculate *nonparametric* statistics unless stated otherwise.

16-26 Make the correct choice in each square bracket:

a. In using ranks, the Wilcoxon test treats outliers in a compromise fashion—a compromise between $[\dot{X}, \overline{X}]$, which uses the numerical value of each outlier no matter how large, and $[\dot{X}, \overline{X}]$, which does not change even if the outlier is moved much farther out.

b. To see whether a series of hourly temperature readings has serial correlation, the [sign test, runs test, W test] can be used. If the p-value for H_o is relatively [high, low], it indicates there is indeed serial correlation.

16-27 A firm wished to test two different programs designed to improve the effectiveness of its sales staff. It therefore took a random sample of 8 salesmen who had been working in each program. The improvement that each salesman showed during the course of his program was recorded on the following graph:

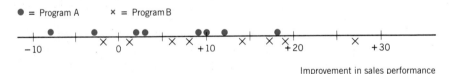

Improvement in sales performance

Calculate the p-value for each of the following null hypotheses:

a. Program A has a median improvement of zero.

b. Program B has a median improvement of zero.

c. The two programs are equally effective.

16-28 In view of its better performance in Problem 16-27, program B was being considered for widespread adoption. Calculate an appropriate confidence interval for the median improvement it would achieve in the population of the firm's entire sales staff.

16-29 Two samples of children were randomly selected to test two art education programs, *A* and *B*. At the end, each child's best painting was judged by an independent artist. In terms of creativity, the ranking was as follows:

RANK OF CHILD	1	2	3	4	5	6	7	8	9	10	11	12	13	14
Art Program	B	A	B	B	B	A	B	B	A	B	A	A	A	A

State H_o in words, and calculate its p-value.

16-30 A random sample of 10 pairs of twins was used in a certain study. In each pair, one twin was chosen at random for an enriched education program (E), and the other twin was given the standard program (S). At the end of the year, the performance scores were as follows:

TWIN'S SURNAME	JONES	ABLE	MISAK	BAKER	GOOD	LEE	BERK	WONG	RAKOS	SCOTT
S Group	57	91	68	75	82	47	63	72	67	68
E Group	64	93	72	72	91	52	79	81	77	80

Is the enriched program really effective? To answer this, calculate:

a. An appropriate confidence interval (including a definition of the parameter).

b. An appropriate p-value.

16-31 A company recorded a sequence of 17 weekly sales (in thousands of dollars, for the first 4 months of 1985) as follows:

33, 26, 28, 24   28, 34, 28, 23   27, 29, 31, 32, 35   30, 32, 26, 31

a. Graph, and then calculate the p-value for the hypothesis of independent observations.

b. Can you validly test the hypothesis that the median meets the target value of 33? (Include a brief reason.) If so, calculate the p-value.

*16-32 Penicillin treatment was administered to 55 patients randomly selected from 107, leaving 52 as controls. The frequencies of various outcomes were as follows (same data as Problem 8-16, but coded differently):

IMPROVEMENT AFTER 6 MONTHS	TREATED PATIENTS f	CONTROL PATIENTS f
Considerable improvement	28	4
Moderate or slight improvement	10	13
No material change	2	3
Moderate or slight deterioration	5	12
Considerable deterioration	6	6
Death	4	14
Total	55	52

a. State H_o in words.

b. Calculate the p-value for H_o.

Chi-Square Tests

(Requires Chapter 9)

The age of chivalry is gone; that of sophisters, economists, and calculators has succeeded.   EDMUND BURKE

Chi square (χ^2) provides a simple test based on the difference between observed and expected frequencies. Because it is so easy to understand and calculate, it is a very popular form of hypothesis testing. And it makes so few assumptions about the underlying population that it is commonly classified as a *nonparametric test*.

17-1 χ^2 Tests for Multinomials: Goodness of Fit

A—EXAMPLE

Let us test the null hypothesis that births in Sweden occur equally often throughout the year. Suppose the only available data is a random sample of 88 births, grouped into seasons of differing length. The observed frequencies O are given in Table 17-1.

How well does the data fit the null hypothesis (of no difference in the birth rate between seasons)? The notion of goodness of fit is developed in the following four steps (calculated in the last four columns of Table 17-1):

1. First consider the implications of the null hypothesis. H_o means that every birth is apt to occur in any season with a probability proportional

TABLE **17-1**   Observed Frequencies of 88 Births Classified by Season, and Subsequent χ^2 Calculations

GIVEN DATA		4-STEP χ^2 CALCULATIONS			
		(1)	**(2)**	**(3)**	**(4)**
SEASON (CELL)	**OBSERVED FREQUENCY** O	**PROBABILITY (if H_0 true)** π	**EXPECTED FREQUENCY** $E = n\pi$	**DEVIATION** $(O - E)$	**DEVIATION SQUARED AND WEIGHTED** $(O - E)^2/E$
Spring Apr–June	27	91/365 = .25	.25(88) = 22.0	27 − 22 = + 5.0	$5^2/22$ = 1.44
Summer July–Aug	20	.17	15.0	+ 5.0	1.67
Fall Sept–Oct	8	.167	14.7	− 6.7	3.05
Winter Nov–Mar	33	.413	36.3	− 3.3	.30
$n = 88$		1.00 ✓	88 ✓	0 ✓	$\chi^2 \simeq 6.16$ p-value $\simeq$ .10

to the *length* of that season. For example, spring is defined to have 3 months, or 91 days; thus the probability of a birth occurring in the spring is $\pi = 91/365 = .25$. Similarly, all the other probabilities π are calculated in column (1).

2. Now calculate what the expected frequency in each season would be if the null hypothesis were true. For example, for spring we found $\pi = .25 = 25\%$, so that the expected frequency would be 25% of 88 = 22 births. Similarly, all the other frequencies[1] E are calculated in column (2):

$$E = n\pi \qquad (17\text{-}1)$$

3. The question now is: "By how much does the observed frequency deviate from the expected frequency?" For example, in spring this deviation is $27 - 22 = 5$. Similarly, all the other deviations $(O - E)$ are set out in column (3).

4. To get some idea of the collective size of these deviations, it is pointless to add them up, since their sum must always be zero—a problem we have met before. Once again, we solve it by squaring each deviation $(O - E)^2$. Then to show its relative importance, we compare each squared deviation with the expected frequency E in its cell, $(O - E)^2/E$, as shown in column (4). Finally, we sum the contributions from all cells, and obtain 6.16.

---

[1] Formula (17-1) is just a restatement of formula (4-13) for the mean of a binomial distribution. We should have $E \geq 1$ in each cell (more conservative authors say $E \geq 5$). If this condition is not met, then χ^2 should be used with considerable reservation. Or, cells can be redefined more broadly until this condition is met.

Note that we had a similar condition for the binomial: As suggested following (8-27), use of the normal approximation required at least 5 successes and 5 failures.

This overall measure of deviation is called chi-square:

$$\text{chi-square, } \chi^2 \equiv \Sigma \frac{(O - E)^2}{E} \qquad (17\text{-}2)$$

A large value of χ^2 indicates a large deviation from H_o, and consequently little credibility for H_o. To determine just how small this credibility is, the tail area or p-value can be determined from the χ^2 distribution in Appendix Table VII (just like the tail area was found from the t distribution in Table V).

Note that the four cell frequencies are not independent: Since $O_1 + O_2 + O_3 + O_4 = n$, any one of them may be expressed in terms of the others. For example, $O_4 = n - (O_1 + O_2 + O_3)$; thus, the last cell is determined by the previous three, and does not provide fresh information. Therefore, in this case χ^2 has only 3 pieces of information (degrees of freedom, d.f.). And in general, with k cells,

$$\text{d.f. } = k - 1 \qquad (17\text{-}3)$$

Since d.f. = 3, we scan along the third row of Appendix Table VII. We find that the observed χ^2 value of 6.16 is about the same as $\chi^2_{.10} = 6.25$. Thus

$$\text{p-value} \simeq .10 \qquad (17\text{-}4)$$

At the customary 5% level, therefore, the null hypothesis (of the same birth rate in all seasons) cannot be rejected.

This example has shown how the χ^2 test can be applied to data counted off in several cells (such as seasons). This χ^2 test is sometimes called the *multinomial* test to indicate that it is an extension of the binomial, where data was counted off in two cells (such as male and female, or Democrat and Republican.)

B—LIMITATIONS OF HYPOTHESIS TESTS

Continuing the example of Swedish births above, suppose that we gather a very large sample; in fact, *all* the births in Sweden in 1935 are shown in Table 17-2 (Cramer, 1946). With a little stretch of the imagination, this may be considered a random sample from an infinite conceptual population. χ^2 is calculated to be 128, which exceeds the last value in Table VII by so much that we conclude that the p-value $\ll$ .001. At any reasonable level, H_0 is rejected.

At this point, we see that such a hypothesis test is asking the wrong question. To see why, recall H_0: births occur with probability proportional to the length of the season. Even before *any* data was gathered, we knew that H_0 could not be *exactly* the truth—a good approximation, perhaps, but not exactly true. So when the data calls for rejection of H_0, the only sensible reaction is "But we could have told you so, *even before we collected the data.* All we have to do to reject *any* specific null hypothesis is collect a large enough sample."

TABLE **17-2** **Observed Frequencies of 88,273 Swedish Births Classified by Season and Subsequent χ^2 calculations**

	GIVEN DATA		4-STEP χ^2 CALCULATIONS			
		(1)	(2)	(3)	(4)	
	OBSERVED FREQUENCY	PROBABILITY (IF H_0 TRUE)	EXPECTED FREQUENCY			
SEASON	O	π	$E = n\pi$	$(O - E)$	$(O - E)^2/E$	
Spring Apr–June	23,385	.24932	22,008	1,377	86.16	
Summer July–Aug	14,978	.16986	14,994	− 16	.02	
Fall Sept–Oct	14,106	.16712	14,752	− 646	28.29	
Winter Nov–Mar	35,804	.41370	36,519	− 715	14.00	
	88,273	1.0000 ✓	88,273 ✓	0 ✓	$\chi^2 = 128$ p-value $\ll$ .001	

C—ALTERNATIVE TO χ^2: CONFIDENCE INTERVALS

With the χ^2 test (or for that matter, any hypothesis test) becoming increasingly irrelevant as sample size grows larger and larger, what then *is* relevant? Confidence intervals are sometimes helpful, and indeed become increasingly precise and useful as sample size grows. Let us see how they could illuminate the data in Table 17-2.

Let us continue to regard the 88,273 observed births as a very large sample from an infinite conceptual population. Now consider P, the sample proportion of births observed in the first season, spring:

$$P = \frac{23,385}{88,273} = .265 \qquad (17\text{-}5)$$

Next, use this to construct a 95% confidence interval for the corresponding population proportion (probability of spring births). From (8-27):

$$\pi = .265 \pm 1.96 \sqrt{\frac{(.265)(.735)}{88,273}}$$

$$= .265 \pm .0029 \qquad (17\text{-}6)$$

Now compare this with the probability of births in the spring if H_0 was true. Since spring was defined to have 91 of the 365 days of the year, we already found in Table 17-2 that

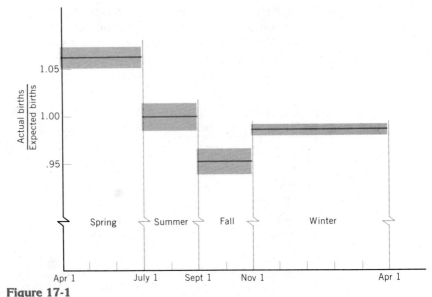

Figure 17-1
Ratio of actual births to expected births (if H_0 is true). Shading shows 95% confidence intervals.

$$\text{null } \pi = \frac{91}{365} = .249 \tag{17-7}$$

Consider finally the ratio:

$$\frac{\pi}{\text{null } \pi} = \frac{(.265 \pm .0029)}{.249}$$

That is,

$$\frac{\text{Actual births}}{\text{Expected births if } H_0 \text{ true}} = 1.06 \pm .01 \tag{17-8}$$

This is easy to interpret: In spring, births were about 6% above an "equal rate" pattern (more precisely, somewhere between 5% and 7% above, with 95% confidence).

As with any other confidence interval, we can immediately use (17-8) to test H_0. [If the null hypothesis is true, then $\pi/\text{null } \pi = 1.00$. Since this value does not fall within the confidence interval (17-8), H_0 can be rejected.] While this confirms our earlier χ^2 hypothesis test,[2] we have already seen that such hypothesis testing is inadequate in these circum-

---

[2]Because the hypothesis test here is about *spring* births, it is not quite identical to our earlier hypothesis test, which considered births in *all* seasons. This subtle difference in fact is what the χ^2 test is explicitly designed to pick up (Wonnacott, 1977).

stances. Far more important is the information (17-8) provides about the unexpectedly heavy frequency of spring births—information that was not available from our earlier hypothesis test.

The shaded bands in Figure 17-1 illustrate (17-8) and similar confidence intervals for the other seasons.

PROBLEMS

17-1

PERIOD OF DAY	NUMBER OF ACCIDENTS
8–10 A.M.	31
10–12 A.M.	30
1–3 P.M.	41
3–5 P.M.	58

This table classifies last month's accidents in a large steel plant into 4 equal time periods. Regarding it as a random sample,

a. Find the p-value for H_0 (that accidents are equally likely to occur at any time of day).

b. Can you reject H_0 at the 5% error level?

c. Analyze in a more graphical way, as in Figure 17-1.

17-2 Parents of blood type AB will produce children of three different types: AA, AB, and BB. If the hypothesis of Mendelian inheritance is true, these three types will be born 25%, 50%, and 25% of the time in the long run. The following data gives the blood types of the 284 children born of 100 AB couples. What p-value does it yield for the Mendelian hypothesis?

BLOOD TYPE	NUMBER OF CHILDREN
AA	65
AB	152
BB	67

17-3 Throw a fair die 30 times (or simulate it with random digits in Table I).

a. Use χ^2 to calculate the p-value for H_0 (that it is a fair die).

b. Do you reject H_0 at the 5% level?

c. If each student in a large class carries out this test, approximately what proportion will reject H_0?

17-4 Repeat Problem 17-3 for an unfair die. (Since you do not have an unfair die available, use the table of random digits to simulate a die that is biased toward aces; for example, let the digit 0 as well as 1 represent the ace, so that the ace has twice the probability of any other face.)

*17-5 Is there a better test than χ^2 for the die in Problem 17-4 that is suspected of being biased toward aces? If so, use it to recalculate Problem 17-4.

*17-6   For another useful interpretation of d.f., prove that the expectation of χ^2 in (17-2) is $(k - 1)$. [*Hint:*   Use (5-17) and (4-13).]

17-2 χ^2 Tests for Independence: Contingency Tables

A—EXAMPLE

Contingency means dependence, so a contingency table is simply a table that displays how one characteristic depends on another. For example, Table 17-3 shows how income is observed to depend on region in a sample of 400 U.S. families in 1971. To test the null hypothesis of no dependence in the underlying population, χ^2 again may be used.

TABLE **17-3**   **Observed Frequencies for 400 Families Classified by Region and Income, 1971**

j REGION	i	1	2	3	4	TOTAL FREQUENCY	$P_j =$ RELATIVE FREQUENCY
		0–5	5–10	10–15	15 –		
1 South		28	42	30	24	124	.31
2 North		44	78	78	76	276	.69
Total Frequency		72	120	108	100	400	
P_i = Relative Freq.		.18	.30	.27	.25		

In Table 17-3 let π_{ij} denote the underlying bivariate probability distribution; for example, π_{41} is the probability that a family earns above \$15 thousand, and is in the South. Let π_i and π_j similarly denote the marginal probability distributions. Then the null hypothesis of statistical independence may be stated precisely

$$H_0: \quad \pi_{ij} = \pi_i \pi_j$$

(17-9)
like (5-6)

To test how well the data fits this hypothesis, we set out the 4 steps in Table 17-4, analogous to the χ^2 calculations in Table 17-1:

1.   First, work out the implications of H_0. The best estimate of π_i and π_j are the marginal relative frequencies P_i and P_j calculated in the last row and column of Table 17-3. Substituting them into (17-9) yields the estimated probabilities P_{ij} for each cell as set out at the top of Table 17-4.[3]

---

[3]This step is very much like the fitting of each cell in 2-way ANOVA, except that here a probability is fitted by multiplying two component probabilities, whereas in ANOVA a *numerical* response is fitted by the *addition* of two component effects, as in (10-17). This analogy can be pushed further by taking logarithms, obtaining a *log-linear* model (Wonnacott, 1984 *IM*).

TABLE **17-4** **4-Step χ^2 Calculations**

1. Assuming independence, estimate the bivariate probabilities $P_{ij} = P_iP_j$

				P_j
.056	.093	.084	.077	.31
.124	.207	.186	.173	.69
P_i				
.18	.30	.27	.25	

2. Calculate the expected frequencies $E = nP_{ij}$

22.3	37.2	33.5	31.0
49.7	82.8	74.5	69.0

3. Calculate the deviations $(O - E)$

5.7	4.8	-3.5	-7.0
-5.7	-4.8	3.5	7.0

4. Calculate $(O - E)^2/E$

1.45	.62	.36	1.58
.65	.28	.16	.71

sum, $\chi^2 = 5.81$

2. Calculate the expected frequencies $E = nP_{ij}$.
3. Calculate the deviations $(O - E)$.
4. Square and weight by the expected frequencies; in other words, calculate $(O - E)^2/E$. Then sum to get an overall measure of discrepancy:

$$\text{chi-square for independence, } \chi^2 \equiv \Sigma\Sigma\frac{(O - E)^2}{E} \qquad \text{(17-10)} \\ \text{like (17-2)}$$

We have written the Σ sign twice in (17-10) to indicate that we sum over the whole table (c columns and r rows). Then the degrees of freedom for this test are[4]:

$$\text{d.f.} = (c - 1)(r - 1) \qquad \text{(17-11)}$$

---

[4]The d.f. may be calculated from a general principle that is useful for many applications:

$$\text{d.f.} = (\# \text{ cells}) - 1 - (\# \text{ estimated parameters}) \qquad \text{(17-12)}$$

To apply this, we have to know the number of estimated parameters (in this case, estimated probabilities). Consider first the c estimated column probabilities P_i. Once the first $(c - 1)$ are estimated, the last one is strictly determined, since $\Sigma P_i = 1$. Thus, there are only $(c - 1)$ independently estimated column probabilities, and by the same argument, only $(r - 1)$ row probabilities. Thus, from (17-12):

$$\text{d.f.} = cr - 1 - [(c - 1) + (r - 1)] \\ = (c - 1)(r - 1) \qquad \text{(17-11) proved}$$

Thus, for Table 17-4,

$$\text{d.f.} = (4 - 1)(2 - 1) = 3$$

Finally, the value $\chi^2 = 5.81$ calculated at the end of Table 17-4 can be converted to a p-value. Since d.f. $= 3$, we scan along the third row of Table VII. We find that the observed χ^2 value of 5.83 lies between $\chi^2_{.25} = 4.11$ and $\chi^2_{.10} = 6.25$. Thus

$$.10 < \text{p-value} < .25 \tag{17-13}$$

This p-value is too high to reject H_0 at the customary 5% level (or even at the 10% level). That is, at this level χ^2 fails to establish any dependence of income on region.

B—ALTERNATIVE TO χ^2: A CONFIDENCE INTERVAL ONCE AGAIN

Since the χ^2 test does not exploit the numerical nature of income, it misses the essential question: *How much* do incomes differ between regions?

This question can be answered by reworking the data in Table 17-3. Since income is numerical, we can calculate the mean income in the North ($\overline{X}_1$), and the South ($\overline{X}_2$). Then, as detailed in Problem 17-10, we can find the 95% confidence interval for the difference in the population mean incomes:

$$(\mu_1 - \mu_2) = (\overline{X}_1 - \overline{X}_2) \pm t_{.025}\, s_p \sqrt{\frac{1}{n_1} + \frac{1}{n_2}} \qquad \text{like (8-20)}$$

$$= (10.87 - 9.52) \pm 1.96\,(7.03) \sqrt{\frac{1}{276} + \frac{1}{124}}$$

$$= 1.35 \pm 1.11 \text{ thousand dollars} \tag{17-14}$$

That is, the North has a mean income that is $1,350 \pm \$1,110$ higher than the South.

The secondary question of testing H_0 (no difference between regions) can be answered immediately: At the 5% level, H_0 now can be rejected, since 0 does not lie in the confidence interval (17-14). That is, there is a discernible difference between the two regions. This is a much stronger conclusion than we obtained from the χ^2 test following (17-13), where we failed to find a discernible difference between the two regions. The 2-sample t proved to be a more powerful test, because it takes into account the *numerical* nature of income, which the χ^2 test ignores.

Of course, if more than two regions were to be compared, we would use ANOVA instead of the two-sample t. But the conclusion would be generally the same: Any such test that fully exploits the numerical nature of the data will be more powerful than χ^2. Thus we conclude,

> **Numerical** variables should be analyzed with a tool (such as the 2-sample t, ANOVA, or regression) that exploits their numerical nature. The χ^2 test is appropriate if all the variables are categorical. (17-15)

PROBLEMS

17-7 In a study of how occupation is related to education, the following random sample of 500 employed men was drawn in the U.S. in 1980. (This sample has been reconstructed from the 1981 Statistical Abstract of the U.S.)

	OCCUPATION			
EDUCATION	WHITE COLLAR	BLUE COLLAR	SER-VICES	FARM WORK
4 or more years of high school	194	146	27	10
Less than 4 years of high school	18	79	18	8

a. State H_0 in words.

b. Calculate χ^2 and the p-value for H_0.

c. For each occupation, graph the estimated proportion of better-educated workers, with its surrounding confidence interval.

17-8 The research division of a large chemical company tried each of three termite repellants on 200 wooden stakes driven into random locations. Two years later, the number of infected stakes was as follows:

	TREATMENT		
INFECTED?	T_1	T_2	T_3
Yes	26	48	18
No	174	152	182

Note that this experiment was designed to have exactly the same number of stakes (200) for each treatment. Thus the relative frequency for each treatment is 200/600 = 1/3, and does not really estimate any underlying population proportion. So the sampling differs from the simple random sample of Table 17-3.

Nevertheless, the standard χ^2 test still remains valid. When used this way, it is often called a χ^2 test with *fixed marginal totals*, or a χ^2 test of *homogeneity* (Are the various treatments homogeneous, i.e., similar in terms of infection rate?)

a. Calculate χ^2 and the p-value.

b. Are the treatments discernibly different at the 5% level?

17-9 A large corporation wishes to test whether each of its divisions has equally satisfactory quality control over its output. Suppose output of each division, along with the number of units rejected and returned by dealers, is as follows:

	DIVISION		
	A	B	C
Output	600	400	1000
Rejects	52	60	88

Assuming this output can be regarded as a random sample, calculate the p-value for H_0: no difference in divisions. (*Hint*: Since χ^2 requires "pigeonholing" into mutually exclusive cells, the data will first have to be retabulated.)

17-10 Verify (17-14). (In Table 17-3, approximate the incomes from 0 to 5 by the cell midpoint 2.5. Continue for the other cells. In the last cell, which is open-ended and has no midpoint, use 17.5 as a rough approximation to the average income within the cell.)

17-11 In a survey for its advertisers, a newspaper chain randomly sampled 100 readers of each of its 3 major newspapers, with the following results:

	NEWSPAPER		
SOCIAL CLASS	A	B	C
Poor	31	11	12
Lower Middle Class	49	59	51
Middle Class	18	26	31
Rich	2	4	6

a. State H_0 in words.

b. Calculate χ^2 and the p-value for H_0. Is the difference between newspapers discernible at the .001 level?

*17-12 **a.** Analyze Problem 17-11 in a way that exploits the ordered nature of social class: Since the four social classes are ordered from poor to rich, a reasonable strategy is to number them 1, 2, 3, 4 (call it a *social class score* if you like). With the 300 people all having their social class transformed into a numerical score, it is now possible to calculate the mean score for newspaper A, and compare it to the mean scores for newspapers B and C using ANOVA.

b. Did you find the difference between newspapers discernible at the .001 level? Is this ANOVA test better than χ^2 in this particular case?

CHAPTER 17 SUMMARY

17-1 χ^2 is a hypothesis test based on the difference between observed values and expected values (expected under the null hypothesis). In the simplest case, χ^2 can be applied to data sorted into several cells, according to a single factor such as season.

17-2 χ^2 can also be applied to data sorted according to two factors, to test their independence of each other. χ^2 tests are designed for purely categorical variables such as sex or nationality (while earlier procedures such as regression remain appropriate for numerical variables such as income or productivity).

REVIEW PROBLEMS

17-13 A random sample of 1250 university degrees earned in 1976 gave the following breakdown:

	DEGREE		
SEX	BACHELORS	MASTERS	DOCTORATE
Male	501	162	27
Female	409	143	8

a. State the null hypothesis in words.

b. Calculate the χ^2 p-value for H_0.

c. For a more graphic alternative to χ^2, calculate the proportion of women in each of the 3 different degree categories. Include confidence intervals, and a graph.

17-14 According to the Mendelian genetic model, a certain garden pea plant should produce offspring that have white, pink, and red flowers, in the long-run proportions 25%, 50%, 25%. A sample of 1000 such offspring was colored as follows:

white, 21%; pink, 52%; red, 27%

a. Find the p-value for the Mendelian hypothesis.

b. Can you reject the Mendelian hypothesis at the 5% level?

17-15 In a study of how the burden of poverty varies among U.S. regions, a random sample of 1000 families in each region of the U.S. in 1976 yielded the following information on poverty (defined in 1976 as an income below $5000 for a family of 4 people, roughly).

Calculate the p-value for the null hypothesis that poverty is equally prevalent in all regions.

Incidence of Poverty by Region

	NORTHEAST	NORTH-CENTRAL	SOUTH	WEST
Poor	57	59	121	83
Not poor	943	941	879	917

17-16   A multinational construction firm had a labor force of widely varied ages:

	AGE GROUP		
	18–25	25–40	OVER 40
Number of workers	21,000	34,000	25,000

To test whether its accidents have been dependent on the age of the labor force, a random sample of 200 workers in each age group was carefully observed for a year.

	AGE GROUP		
	18–25	25–40	OVER 40
Number of workers in the sample of 200 who had one or more accidents	27	16	20

Calculate the p-value for the null hypothesis that accident rate is independent of age.

Chapter

*Maximum Likelihood Estimation

(Requires Chapter 7)

Doubt is the beginning, not the end, of wisdom.   ANONYMOUS

The stars on Chapters 18, 19 and 20 indicate that they are a little more challenging. Readers who work their way through this material, however, will be amply rewarded: They will discover the philosophical foundations of modern statistics at the same time as they find practical solutions to important problems.

18-1 Introduction

In this chapter we will study a very powerful method for deriving estimates, called maximum likelihood estimation (MLE). To put it into perspective, we begin by reviewing some earlier estimation techniques.

A—METHOD OF MOMENTS ESTIMATION (MME)

The first principle of estimation that we used, back in Chapter 6, was the most obvious and intuitive one: estimate a population mean μ with the sample mean $\overline{X}$; estimate a population variance σ^2 with the sample variance s^2; and so on. This principle of estimating a population moment by

499

using the equivalent sample moment is called *method of moments esti-mation* (MME). Often it works very well; but now we consider a case where it does not.

EXAMPLE **18-1**

In World War II, the allies used sampling theory to measure German production very effectively (Wallis and Roberts, 1962). The Germans gave their rockets serial numbers from 1 to N. The problem for the allies was to estimate the size of the German arsenal; that is, how large was N?

To estimate this, captured German rockets were viewed as provid-ing a random sample of serial numbers. To illustrate, suppose a total of 10 captured weapons had serial numbers as follows:

$$77, 30, 05, 39, 28, 10, 27, 12, 73, 49$$

The following sequence of questions will lead us to an estimate of N:

a. What is the sample mean $\overline{X}$?

b. For the population of serial numbers 1, 2, 3, . . . , N, what is the mean μ in terms of N?

c. Using the method of moments approach, set $\mu = \overline{X}$. Then solve for N.

d. Can you think of a better estimate?

Solution

a. $\overline{X} = \dfrac{77 + 30 + \ldots + 49}{10} = \dfrac{350}{10} = 35$

b. The mean of the serial numbers 1,2,3,4,5, for example, is the middle number 3; formally, this is just (5 + 1)/2, the number midway between the last and the first. Similarly, for any set of serial numbers 1,2, . . . , N, the mean is

$$\mu = \frac{N + 1}{2}$$

c. $$\mu = \frac{N + 1}{2} = \overline{X}$$

Solving for N, the MME is:

$$N = 2\overline{X} - 1 \tag{18-1}$$
$$= 2(35) - 1 = 69$$

d. When we look closely at the data, we see that N = 69 can be easily improved. The largest observation is 77, so that N must be at least this big:

$$N \geq 77$$

So N = 77 itself would clearly be a better estimate than the MME of 69.

B—MAXIMUM LIKELIHOOD ESTIMATION (MLE)

Example 18-1 illustrated what can go wrong with MME. A better way is needed to generate estimators in new and complex situations, and this is where statistics as a science stood about 1910. To fill this need, Maximum Likelihood Estimation (MLE) was developed by Sir Ronald Fisher (who also developed randomized control in experiments).

We will see how MLE provides a superior estimate in several situations, including Example 18-1 above. But first we will show that in some other more familiar situations, MLE provides the same estimate as the method of moments. For example, the MLE of a population proportion π is the sample proportion P. Thus MLE provides further justification for these familiar estimates.

18-2 MLE for Some Familiar Cases

A—EXAMPLE OF HOW MLE WORKS

Suppose that a shipment of radios is sampled for quality, and that 3 out of 5 are found defective. What should we estimate is the proportion π of defectives in the whole shipment (population)? Temporarily, try to forget the common sense method of estimating the population proportion π with the sample proportion $P = 3/5 = .60$. Instead, let us investigate an alternative method: consider a whole range of possible π that we might choose, and then try to pick out the one that best explains the sample.

For example, is $\pi = .1$ a plausible value for the population? If $\pi = .1$, then the probability of S = 3 defectives out of a sample of n = 5 observations would be given by the binomial formula (4-8):

$$\binom{n}{S} \pi^S (1 - \pi)^{n-S} = \binom{5}{3} .1^3 .9^2 \simeq .008 \qquad (18-2)$$

In other words, if $\pi = .1$, there are only about 8 chances in a thousand of getting the actual sample that we observed.

Similarly, if $\pi = .2$, we would find from the binomial formula—or even better, from Table IIIb—that there are about 50 chances in a thousand of getting the sample that we observed. In fact, it seems quite natural to try out all values of π, in each case finding out how likely it is for such a population π to generate the sample that we actually observed. We simply read across Table IIIb, and so obtain the graph in Figure 18-1. In this situation, where the sample values $n = 5$ and $S = 3$ are fixed, and the only variable is the hypothetical value of π, we call the result the *likelihood function* $L(\pi)$:

$$L(\pi) = \binom{5}{3} \pi^3 (1 - \pi)^2 \qquad (18\text{-}3)$$

The *maximum likelihood estimate* (MLE) is simply the value of π that maximizes this likelihood function. From Figure 18-1 we see that this turns out to be $\pi = .60$—which coincides nicely with the common sense estimate $P = 3/5$.

In general, we similarly define:

> **The MLE is the hypothetical population value that maximizes the likelihood of the observed sample.** $\qquad (18\text{-}4)$

In other words, the MLE is the hypothetical population value that is more likely than any other to generate the sample we actually observed.

To show these same issues geometrically, in Figure 18-2 we graph the binomial probabilities as a function of both S and π. In Chapter 4, we thought of π fixed and S variable. For example, the dotted distribution in

From Table IIIb, where $n = 5$ and $S = 3$:

π	(0)	.1	.2	.3	.4	.5	.6	.7	.8	.9	(1.0)
$L(\pi)$	0	.008	.051	.132	.230	.312	.346	.309	.205	.073	0

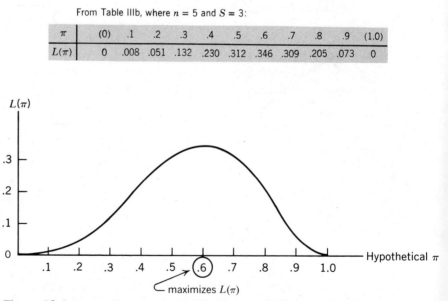

Figure 18-1
The likelihood function $L(\pi)$ copied down from Table IIIb and then graphed. It gives the likelihood that various hypothetical population proportions would yield the observed sample of 3 defectives in 5 observations.

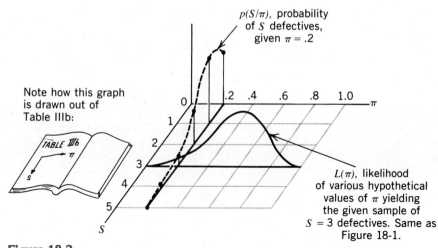

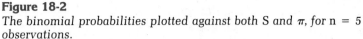

Figure 18-2
The binomial probabilities plotted against both S and π, for n = 5 observations.

the S direction shows the probability of getting various numbers of defectives, if the population proportion were π = .2. But in this chapter, we regard S—the observed sample result—as fixed, while the population π is thought of as taking on a whole set of hypothetical values. For example, the solid curve in the π direction shows the likelihood that various possible population proportions π would yield S = 3 defectives. Slices in this π direction are called likelihood functions (whereas slices in the S direction are called probability distributions).

B—MLE FOR A PROPORTION π IN GENERAL

We show in Appendix 18-2 that our result above was no accident, and that the maximum likelihood estimate of the binomial π is *always* the sample proportion P:

$$\boxed{\text{MLE of } \pi = P} \tag{18-5}$$

In Chapter 1, we appealed to common sense (the method of moments) in using the sample proportion to estimate the population proportion. Now we can add the more rigorous justification of maximum likelihood: a population with π = P is more likely than any other to generate the observed sample.

C—MLE FOR THE MEAN OF A NORMAL POPULATION

Suppose we have drawn a random sample of 3 observations, say, X_1, X_2, X_3, from a parent population that is normal, with mean μ and variance

σ^2. Our problem is to find the MLE of the unknown μ. Deriving the MLE with calculus would involve trying out all of the hypothetical values of μ; we illustrate this geometrically by trying out 2 typical values, μ_* and μ_o, as shown in the two panels of Figure 18-3.

First, in panel (a) we consider the likelihood of observing the sample X_1, X_2, X_3, if the population mean is μ_*. The probability (strictly speaking, the probability density) of observing X_1 is the arrow above X_1, which is quite large. So too is the probability of observing X_2. However, the probability of X_3 is very small because it is so distant from μ_*. Because they are randomly drawn, these sample values $X_1, X_2,$ and X_3 are independent. Therefore, we can multiply their probabilities to get their joint probability—that is, the likelihood of this sample. Because of the very small probability of X_3, this likelihood is quite small.

On the other hand, a population with mean μ_o as in panel (b) is more likely to generate the sample values. Since the X values are collectively closer to μ_o, the likelihood—derived by multiplying their probabilities together—is greater. Indeed, very little additional shift in μ_o is apparently required to maximize the likelihood of the sample. It seems that the MLE of μ might be centered right among the observations X_1, X_2, X_3—that is,

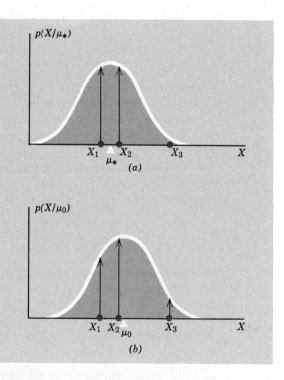

Figure 18-3
Likelihood for the mean μ of a normal population, based on a sample of three observations (X_1, X_2, X_3). (a) Small likelihood $L(\mu_*)$, the product of the three probabilities. (b) Larger likelihood $L(\mu_o)$.

might be just the sample mean $\overline{X}$. In fact, this is proved in Appendix 18-2:

$$\boxed{\begin{array}{c} \text{For a normal population,} \\[6pt] \text{MLE of } \mu = \overline{X} \end{array}}$$ (18-6)

D—MLE FOR NORMAL REGRESSION

Let us again assume the population shape is normal—this time, in the regression model:

$$E(Y) = \alpha + \beta X \qquad \text{(12-1) repeated}$$

Then our argument will be similar to the argument given above. Specifically, the MLE estimation of α and β requires selecting those hypothetical population values of α and β that are more likely than any others to generate the sample that we observed. For example, suppose we have observed the sample of three points shown in both panels of Figure 18-4.

First, let us try out the line shown in panel (a)—at first glance, a pretty bad fit for the three observed points. Temporarily, suppose that this were the true regression line; then the distributions of Y would be centered around it, as shown. The likelihood that they would generate the three observations in the sample is the product of the three probabilities (arrows above Y_1, Y_2, and Y_3). This likelihood seems relatively small, mostly because of the minuscule probability of Y_1. Our intuition that this is a bad estimate is confirmed; such a hypothetical population is not very likely to generate the given sample.

In panel (b) we show the same given sample, but now another hypothetical line. This line seems more likely to give rise to the sample we observed. Since the Y values are collectively closer to the hypothetical regression line, their probability is consequently greater.

The MLE technique speculates on possible populations (regression lines). How likely is each to give rise to the sample that we observed? Geometrically, our problem is to try them all out, by moving the population through all its possible values—that is, *by moving the regression line and its surrounding distributions through all possible positions*. Different positions correspond to different trial values for α and β. In each case, the likelihood of observing Y_1, Y_2, Y_3, would be evaluated. For our MLE, we choose that hypothetical population which maximizes this likelihood. It is evident that little further adjustment is required in panel (b) to arrive at the MLE. This procedure intuitively seems to result in a good fit; moreover, since it seems similar to the least squares fit, it is no surprise that we are able to prove in Appendix 18-2, that the two coincide:

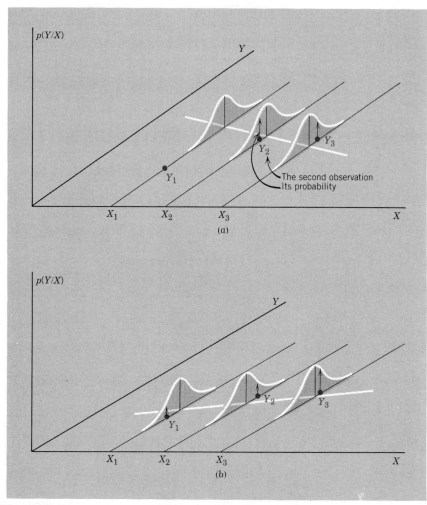

Figure 18-4
*Likelihood for regression. (a) Small likelihood, the product of three
probabilities. (b) Larger likelihood as the trial line comes closer to the three
observed points.*

| In a normal regression model, |
| MLE is identical to least squares. |

(18-7)

This establishes an important theoretical justification of least squares: It
is the estimate that follows from applying maximum likelihood to a regression model with normal distribution shape.

PROBLEMS

18-1 Suppose that a quality controller for an auto firm takes a random sample of n cars from the "population" coming off the assembly line, in order to estimate the proportion π of all cars with defective paint jobs. Graph the likelihood function, and show the MLE of π for each of the following cases:

   a. $n = 8$, with 2 defectives.

   b. $n = 8$, with 4 defectives.

   c. $n = 2$, with 1 defective.

   d. $n = 2$, with 0 defective.

18-2 In Problem 18-1, for which case is the likelihood function a parabola? For which cases is the MLE the sample proportion P?

18-3 MLE for the Uniform Distribution

So far, we have seen how MLE justifies some familiar estimates like P, $\overline{X}$, and least squares in regression. It is time to see how well it works in a less familiar case—the problem of estimating the number of German rockets in Example 18-1.

Recall that we are trying to estimate the population size N, with a sample of 10 serial numbers reproduced in blue in panel (a) of Figure 18-5. We speculate on various possible values of N, in each case calculating the likelihood of the given sample of 10 observations. In panel (a), for example, we show one possible value, $N_* = 120$. The rectangular or uniform distribution shows the probability of observing any specific serial number. (With 120 numbers, all equally probable, the probability of observing any one is 1/120.) The probability of each of the 10 blue observations is shown as usual by an arrow. Then the likelihood or joint probability of all 10 observations is the product[1] of these 10 probability arrows—namely, $(1/120)^{10}$. This likelihood $L(N_*)$ is then graphed as the small value in panel (c).

In panel (b), we consider another possible value of N—this time, $N_0 = 80$. The uniform distribution in this panel shows the probability of observing any of the now 80 equiprobable serial numbers. Because of this smaller range of serial numbers, there is now a greater probability of each; therefore, the arrows in panel (b) are larger than in panel (a). Thus their product—the likelihood of getting the sample we observed—is greater than in panel (a). This larger likelihood $L(N_0)$ is graphed as the higher value in panel (c).

---

[1] In multiplying the 10 individual probabilities, we have assumed independence—that is, random sampling with replacement. Random sampling without replacement would still give approximately the same likelihood.

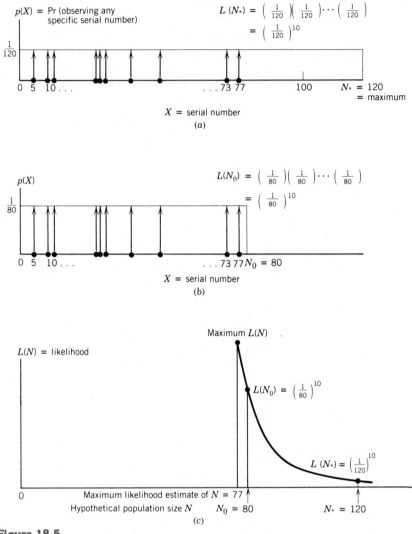

Figure 18-5
Likelihood for the uniform distribution. (a) Small likelihood, the product of 10 small probabilities. (b) Larger likelihood as the trial population N gets smaller. (c) The likelihood function plotted for these 2 values of N, and all others as well.

If we reduce N even further, the 10 arrows would of course get even larger, as would their product—the likelihood $L(N)$. In other words, up to a point, smaller and smaller values of N result in larger and larger values of the likelihood function in panel (c).

However, for any value of N less than 77, $L(N)$ suddenly falls to zero, as shown in panel (c). It's easy to see why. If the entire population of rockets had serial numbers ranging, say, from 1 to 75 (i.e., if $N = 75$), it would be impossible to get the rocket with serial number 77 that we actually observed. That is, the likelihood of this would be zero. (Of course, this is also true of any other value of N below 77.)

The maximum likelihood estimate is that value of N which is more likely than any other to generate the sample we observed—in other words, the value of N where the likelihood function $L(N)$ reaches a maximum. In panel (c) we see that this is $N = 77$, the largest value (serial number) that we observed.

> **The MLE of the largest value in a uniform distribution (of serial numbers) is the largest value observed in the sample.**   (18-8)

The MLE of 77 is obviously superior to the Method of Moments Estimate (MME) of 71, calculated in Example 18-1. Whereas the MLE of 77 was more likely than any other to generate the sample we observed, in this particular case it was *impossible* for the MME of 71 to generate the sample we observed.

Although the MLE of 77 is in some sense the best single estimate (*point estimate*), in this case it is an extreme estimate: It's the lowest possible estimate of N that can be sensibly made. It would be much more informative to give the readers *the whole likelihood function* in panel (c), which shows that N must be at least 77, and there is relatively little likelihood that N is above 90 or 100.

PROBLEMS

18-3 **a.** For $N = 90$, what is $L(N)$ in Figure 18-5c? (Just write it out; don't calculate it.) What does that number mean?

   b. What is the maximum value of the likelihood function in Figure 18-5c? (Again, just write it out; don't calculate it.)

18-4 Suppose a sample of $N = 10$ serial numbers of German rockets were as follows:

$$396, 576, 454, 519, 906, 964, 612, 26, 363, 162$$

What is the MME of total production? And the MLE?

18-5 The sample in Problem 18-4 was in fact a *simulated* sample drawn from the population 1,2,3, . . ., 999. (We started at the beginning of Table I, drawing triples of digits.) The value of such a simulation is that since N is actually known ($N = 999$), we can judge how good any given estimate is.

   a. In Problem 18-4, was MME or MLE closer to the target N?

   b. Let everyone in the class do the same sort of simulation, and have the instructor graph the Monte Carlo distribution of the MME and MLE. Which is closer to the target on the whole? In what sense?

18-6 Continuing Problem 18-5:

a. Is the MME unbiased? Prove it.

b. Is the MLE unbiased? Prove it.

c. It can be shown that the bias of the MLE estimate can be removed if we blow it up by the factor $(n + 1)/n$. Calculate this unbiased estimate of N, for the sample in Problem 18-4.

d. It can be proved that the unbiased version of MLE in part (c) is very efficient:

$$\text{Efficiency of MLE relative to MME} = \frac{n + 2}{3}$$

(Note that they are equally efficient for a sample size $n = 1$, because they coincide then: The single observation—doubled—gives both the MLE and the MME.)

Now suppose in Problem 18-4, rather than the unbiased version of MLE based on $n = 10$ observations, the MME was used instead. To be equally accurate, how much larger would the sample have to be?

Repeat for $n = 1000$ observations.

18-4   MLE in General

A—DEFINITION OF MLE FOR ANY POPULATION PARAMETER

Suppose a sample $(X_1, X_2, \ldots, X_n)$ is drawn from a population with probability function $p(X/\theta)$, where θ is the unknown population parameter we wish to estimate. If the sample is random (VSRS), then the X_i are independent, each with the probability function $p(X_i/\theta)$. Hence the joint probability function for the whole sample is obtained by multiplying:

$$p(X_1, X_2, \ldots, X_n/\theta) = p(X_1/\theta)\, p(X_2/\theta) \ldots p(X_n/\theta) \qquad (18\text{-}9)$$

But we regard the observed sample as fixed, and ask "Which of all the hypothetical values of θ maximizes this probability?" This is emphasized by renaming (18-9) the likelihood function $L(\theta)$, and the MLE is the hypothetical value of θ that maximizes this function.

B—THE ADVANTAGES OF MLE

MLE is a method of estimation with strong intuitive appeal: Since the MLE is the population value that is most likely to generate the observed sample, it is the population value that best "matches" the sample. In familiar situations it produces the familiar estimates, such as P for π, and $\overline{X}$ for μ. Its advantages, however, go far beyond these intuitively appealing ones, and include the following:

1. Where MLE is not the same as MME (method of moments estimation), then MLE is generally superior (as we saw in estimating N for the uniform distribution in Section 18-3).
2. MLE is very straightforward: Just write out the likelihood function, and maximize it. Even in very complex cases where a formula cannot be derived, powerful computers can nevertheless calculate a good numerical approximation.
3. As well as providing the MLE estimate, the whole likelihood function itself is useful to show the *range* of plausible values for the parameter (as we noted in estimating N at the end of Section 18-3, and will pursue in the next chapter).
4. MLE has many of the attractive properties described in Chapter 7. Specifically, under broad conditions, an MLE estimate has the following large-sample (asymptotic) properties. It is:

   i. *Unbiased.*

   ii. *Efficient,* with smaller variance than any other unbiased estimator. (However, this is only true if the parent population does have the specific shape—normal, uniform, or whatever—that is assumed in deriving the MLE. Thus, MLE may lack robustness in dealing with a population of unknown shape.)

   iii. *Normally distributed* in its sampling distribution, with easily computed mean and variance. (That is, if we did a Monte Carlo study of how the MLE estimate varies from sample to sample, it would vary normally.) Thus, confidence intervals and tests are easy to carry out.

For example, we already have seen that the three asymptotic properties are all true for $\overline{X}$, the MLE of μ in a normal population. They are equally true for other MLE estimates too, in large samples.

However, in a small sample, MLE (like all the other estimators we have encountered so far) can often be improved upon, as we shall see in the next chapter.

C—CALCULATING THE MLE IN PRACTICE

If we take logs in (18-9), we reduce the product to a simpler sum:

$$\text{log likelihood} = \log p(X_1/\theta) + \log p(X_2/\theta) + \cdots$$

Since whatever maximizes likelihood must simultaneously maximize log likelihood, we can restate the MLE in more practical terms:

> MLE is that value of θ that maximizes
>
> log likelihood, $\mathscr{L}(\theta) \equiv \Sigma \log p(X_i/\theta)$ (18-10)

An example will illustrate how log likelihood simplifies finding the MLE, in sampling from any kind of population distribution.

EXAMPLE 18-2

The Poisson distribution has probability function

$$p(X/\theta) = \frac{e^{-\theta}\theta^X}{X!}$$ (18-11)

where θ is the unknown parameter.

From a sample of n observations $X_1, X_2, \ldots, X_n$, calculate the MLE of θ. Find its specific numerical value if the sample turns out to be: 15, 8, 13.

Solution

We first calculate the log of the given probability distribution:

$$\log p(X/\theta) = \log \left[\frac{e^{-\theta}\theta^X}{X!} \right]$$

$$= -\theta + X \log \theta - \log X!$$

Thus,

$$\mathscr{L}(\theta) = \Sigma \log p(X_i/\theta)$$ (18-10) repeated

$$= \sum_{i=1}^{n} (-\theta + X_i \log \theta - \log X_i!)$$

$$= -n\theta + (\Sigma X_i) \log \theta - \Sigma \log X_i!$$

Setting the derivative equal to zero:

$$\frac{d\mathscr{L}(\theta)}{d\theta} = -n + \frac{\Sigma X_i}{\theta} = 0$$

$$\theta = \frac{\Sigma X_i}{n}$$ (18-12)

$$= \frac{15 + 8 + 13}{3} = 12$$

PROBLEMS

18-7 **a.** The MLE of the regression slope β (when populations are normal) is just the OLS estimate $b = \Sigma xy/\Sigma x^2$. Does it satisfy the three asymptotic properties given on page 511?

 b. Is the MLE estimate of N in (18-8) unbiased? Asymptotically unbiased? Is this consistent with the first asymptotic property given on page 511? (Hint: See Problem 18-6c.)

18-8 The waiting time X until the next telephone call arrives at a switchboard has a geometric distribution:

$$p(X) = \theta\, e^{-\theta X}, \qquad 0 \leq X$$

Find the MLE of θ:

 a. For a sample of n observations $(X_1, X_2 \ldots X_n)$.

 b. For the following sample of 5 observations (in minutes):

$$1.2,\ 7.5,\ 1.8,\ 3.7,\ 1.1.$$

18-9 A sample of n observations $(X_1, X_2 \ldots X_n)$ are taken from a normal population with known mean μ but unknown standard deviation σ. Derive the MLE of σ.

CHAPTER 18 SUMMARY

18-1 While simple methods of estimation are often adequate, more complex situations require a systematic and powerful method of deriving estimates—which maximum likelihood estimation (MLE) provides.

18-2 In many cases, MLE yields the same estimate as the one we have been using in earlier chapters. (For example, the MLE of the population proportion π is the sample proportion P.) This reassures us that MLE is a reasonable procedure. Even more important, the long list of valuable characteristics of MLE (given in Section 18-4) justifies these earlier estimates.

18-3 Estimating the largest number in a population of serial numbers is an example of a more complex situation when the naive method of moments fails to give a good estimate. Fortunately, MLE gives an excellent estimate—the largest serial number observed in the sample.

18-4 MLE has so many valuable characteristics that it is by far the most popular method of deriving estimates—especially in complicated situations, where computers can calculate good approximations.

REVIEW PROBLEMS

18-10

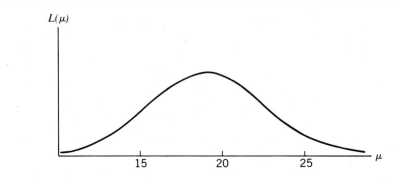

    a. Given the likelihood function above, what is the MLE of μ?

    b. The given likelihood function in fact is that of a random sample of 3 observations (X = 17, 22, and 18) from a normal population with σ = 6 and unknown mean μ. What principle does your answer in **a** therefore illustrate?

    *c. Using a computer and the formula (4-26) for the normal distribution, verify that the graph above really does represent the likelihood function of part **b.**

18-11 In each of the following cases, calculate the MME and MLE, and state which is better (if they are different):

    a. An arbitrator of a teachers' salary dispute sampled the annual salaries of 25 teachers randomly chosen from a population of 12,000 high school teachers in the state, and found $\overline{X}$ = \$21,000 and s = \$11,000. Assuming the population of salaries is normally distributed, estimate the mean salary in the state.

    b. As delegates arrived at a convention, they were given serial tags numbered 1, 2, 3, ... A brief walk in the corridor showed 5 tags, numbered 37, 16, 44, 43, 22. Assuming this is a random sample, estimate the total number of delegates.

    c. The number of customer complaints at a large department store were carefully recorded for 10 randomly chosen days as follows:

$$22, 21, 14, 32, 15, 19, 21, 27, 16, 27$$

Assuming the daily number of complaints for all 310 working days in the year is normally distributed, estimate the total number of complaints for the year.

*Bayesian Inference

(Requires Chapter 8)

> *Life is the art of drawing sufficient conclusions from insufficient premises.*
> SAMUEL BUTLER

In several places already (especially Chapters 9 and 13), it has become clear that prior belief about a parameter may play a key role in its estimation. Bayesian theory is the means of formally taking such prior information into account.

19-1 Posterior Distributions

A—EXAMPLE

The Bayesian theory introduced in Section 3-6 showed how to combine prior information with sample data, to produce posterior probabilities. Since this theory provides the basis for the next two chapters, we will review it with an example:

EXAMPLE **19-1** Over the past year, the Radex Corporation has made 200 shipments of radios to their retailers, and has earned a notorious reputation for bad quality control: 44% of the radios were defective in the first 128 shipments (the "bad" shipments). After a shakeup in production, the defective rate in the remaining 72 shipments was reduced to 15% (the "good" shipments).

Now suppose you have just been appointed purchaser for a large department store. Your first job is to make a decision on a shipment of radios your store has received from Radex that has been lying in the warehouse for 2 weeks. You want to know if it's one of the bad shipments—because if it is, you want to return it.

a. Without any further information, what is the chance it is a bad shipment? A good shipment?

b. To test the shipment, you draw a radio at random from it. If this sample radio turns out to be defective, now what is the chance it is a bad shipment? A good shipment?

Solution **a.** The solution is given in Figure 19-1, which is just like Figure 3-8. Some brief notation will be helpful: Let

$$B = \text{bad shipment} \qquad DEF = \text{defective radio}$$
$$G = \text{good shipment} \qquad OK = \text{satisfactory radio.}$$

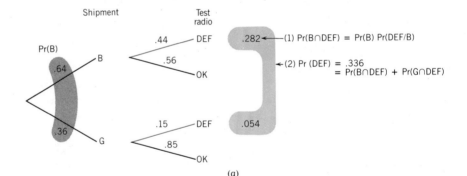

(a)

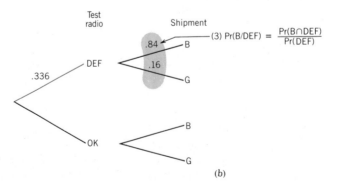

(b)

Figure 19-1
Bayes Theorem. (a) Initial tree. (b) Reverse tree.

There were 200 shipments of radios, of which 128 were bad (B) and 72 were good (G). Therefore,

$$\Pr(B) = \frac{128}{200} = .64 \qquad\qquad (19\text{-}1)$$

$$\Pr(G) = \frac{72}{200} = .36$$

These *prior* probabilities are shown in panel (a), shaded in gray.

b. In addition to these prior probabilities, we were also told that in the bad shipments, 44% of the radios were defective—that is,

$$\Pr(DEF/B) = .44$$

Similarly, in the good shipments, only 15% were defective, that is,

$$\Pr(DEF/G) = .15$$

This information is shown in the second set of branches in panel (a). Now we can calculate the posterior probabilities in 3 steps:

(1) The probability that a test radio will come from a bad shipment and will also be defective is

$$\Pr(B\cap DEF) = (.64)\,(.44) = .282 \qquad (19\text{-}2)$$

$$= \Pr(B)\cdot\Pr(DEF/B)$$

This calculation, and other similar ones, are shown as we move to the right in panel (a).

(2) What is the probability that a test radio will be defective? There are two ways this can happen: We can get a defective radio out of a bad shipment (with probability .282), or out of a good shipment (with probability .054). Thus,

$$\Pr(DEF) = .282 + .054 = .336$$

$$= \Pr(B\cap DEF) + \Pr(G\cap DEF)$$

This probability is shaded in color in panel (a) and is then entered on the first branch of panel (b).

(3) If we have a radio that has tested out defective, it is far more likely to have come from a bad shipment (.282) than from a good one (.054). Formally,

$$Pr(B/DEF) = \frac{.282}{.336} = .84 \qquad (19\text{-}3)$$

$$= \frac{Pr(B \cap DEF)}{Pr(DEF)}$$

The probability it has come from a good shipment is therefore only $1.00 - .84 = .16$. These *posterior* probabilities are shaded in color in panel (b).

These answers seem intuitively correct: Initially the probability that the shipment is bad is 64%; it rises to 84% if the tested radio is defective.

Alternatively, this whole problem could have been illustrated with a rectangular sample space instead of a pair of trees, as Figure 19-2 shows.

B—GENERALIZATION

In Figure 19-3 we show how the analysis of Figure 19-1 can be applied generally: We let θ_1 and θ_2 represent any two states—that is, characteristics of the population—that we are trying to discern (such as the quality of the shipments B and G in our earlier example); and we let X_1 and X_2 represent the data—that is, the test or sample results (such as the DEF or

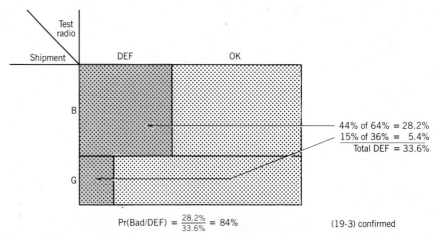

$$Pr(Bad/DEF) = \frac{28.2\%}{33.6\%} = 84\% \qquad (19\text{-}3) \text{ confirmed}$$

Figure 19-2
Alternative to tree reversal: Bayes Theorem represented as a rectangular sample space. Each of the hundreds of possible shipments is shown as a dot, and those that produced a defective test radio are shaded.

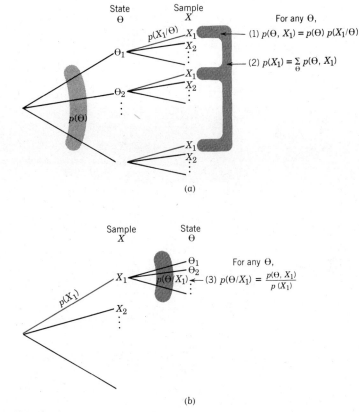

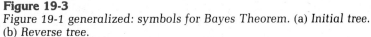

Figure 19-3
Figure 19-1 generalized: symbols for Bayes Theorem. (a) Initial tree.
(b) Reverse tree.

OK of our sample radio). Moreover, to generalize this approach, we can easily visualize the more complicated diagram that results if we consider a larger number of possible states θ_1, θ_2, θ_3, . . . , and a larger number of sample results X_1, X_2, X_3, ...

The last equation (3) in Figure 19-3 showing the posterior probabilities is of particular interest:

$$p(\theta/X_1) = \frac{p(\theta, X_1)}{p(X_1)} \tag{19-4}$$

Substituting the top right-hand equation (1) in Figure 19-3 yields:

$$\boxed{p(\theta/X_1) = \frac{1}{p(X_1)} \, p(\theta)p(X_1/\theta)} \tag{19-5}$$

On the right side, let us look at the three factors one by one. When X_1 is fixed, $p(X_1)$ is a fixed constant too. Next, $p(\theta)$ is the prior distribution,

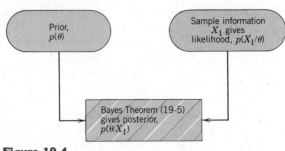

Figure 19-4
The logic of Bayes Theorem (compare to Figure 3-11).

incorporating all our prior knowledge about θ. Finally consider the last term $p(X_1/\theta)$; when X_1 is fixed while θ varies, it is called the likelihood function [as in (18-3)].

If we let $\propto$ denote *equals except for a constant, or is proportional to*, then we can write (19-5) in words as:

$$\boxed{\text{posterior} \propto \text{prior} \times \text{likelihood}} \qquad (19\text{-}6)$$

This is *Bayes Theorem*, and is illustrated in Figure 19-4.

PROBLEMS

19-1   A factory has three machines (θ_1, θ_2, and θ_3) making bolts. The newer the machine, the larger and more accurate it is, according to the following table:

MACHINE	PROPORTION OF TOTAL OUTPUT PRODUCED BY THIS MACHINE	RATE OF DEFECTIVE BOLTS
θ_1 (Oldest)	10%	5%
θ_2	40%	2%
θ_3 (Newest)	50%	1%
	100% $\checkmark$	

Thus, for example, θ_3 produces half of the factory's output; and of all the bolts θ_3 produces, 1% are defective.

a.   Suppose that a bolt is selected at random; *before* it is examined, what is the chance that it was produced by machine θ_1? by θ_2? by θ_3?

b.   Suppose that the bolt is examined and found to be defective; *after* this examination, what is the chance that it was produced by machine θ_1? by θ_2? by θ_3?

19-2   Suppose that you are in charge of the nationwide leasing of a specific car model. Your service agent in a certain city has not been perfectly reliable:

He has shortcut his servicing in the past about 1/10 of the time. Whenever such shortcutting occurs, the probability that an individual will cancel his lease increases from .2 to .5.

    a. If an individual has canceled his lease, what is the probability that he received shortcut servicing?

    b. Suppose that the service agent is even more unreliable, shortcutting half the time. What is your answer to **a** in this case?

19-3 Your firm has just purchased a 13 million dollar piece of machinery, and your engineers have found it to be substandard. You know that the manufacturer that produced it substitutes cheaper components 25% of the time. You further know that such a substitution doubles the probability that the machine will be substandard from .1 to .2.

    If your machine has cheaper components, you could win a lawsuit against the manufacturer. What is the chance of this?

19-2 The Population Proportion π

A—EXAMPLE

Now we shall apply the general principle (19-6) to an example in which there are many, many states of nature θ, each one being a proportion π.

EXAMPLE **19-2**

Let us consider a variation of the Radex problem in Example 19-1. Suppose now that the shipments of radios are not just good or bad, but have a wide variety of quality levels. Some have a high proportion π of defective radios, some have a small proportion. For all the possible levels of π, the record of shipments from Radex (i.e., the prior distribution of π) is given in the first 3 columns of Table 19-1.

    a. Graph the prior distribution of π.

    b. Now suppose that you take a random sample of $n = 5$ radios out of the shipment you have received, in order to get sample evidence on π; and further suppose that 3 of these 5 turn out to be defective. Using (19-6), calculate the posterior distribution of π, and graph it.

    c. Suppose that your department store will regard this shipment of radios as acceptable only if π is less than 25%. What is the probability of this:

        i. Before the sample?

        ii. After the sample?

Solution   **a.** The prior distribution of π is given in the third column of Table 19-1, and graphed in Figure 19-5.

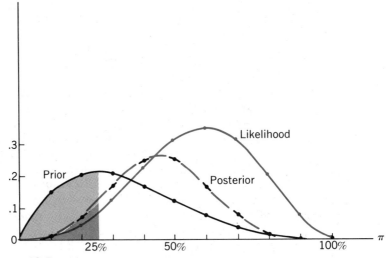

Figure 19-5
The prior distribution and likelihood function (based on sample information) are multiplied together to give the compromise posterior distribution of π.

b. To calculate the posterior distribution using (19-6), we will first need the likelihood function—that is, the likelihood of getting the 3 defectives that we observed in our sample of 5. This, of course, is given by the binomial formula, for a fixed $S = 3$, and various values of π. The easiest way to obtain it is to read Table IIIb horizontally along the row where $n = 5$ and $S = 3$ (obtaining the same likelihood function as in Figure 18-1). We record it in column (4) of Table 19-1, and graph it in Figure 19-5.

Now, following (19-6), we multiply the likelihood function in Table 19-1 by the prior distribution. This gives us the posterior distribution in column (5), except that it does not sum to 1. To achieve this, we divide all the probabilities in this column by .161 [just as we divided by $p(X_1)$ in the fundamental Bayesian theorem (19-5)]. The result is the posterior distribution in the last column of Table 19-1, which we also graph in Figure 19-5.

c. In Figure 19-5 and Table 19-1 we see that $\pi < 25\%$ means a π value of 0, 10%, or 20%.

i. From column (3) of Table 19-1, the prior probability of less than 25% defectives is:

$$Pr(\pi < 25\%) = .01 + .15 + .20 = .36$$

ii. From the last column of Table 19-1, the posterior probability is:

$$Pr(\pi < 25\%/S = 3) = 0 + .01 + .06 = .07$$

TABLE **19-1**

Calculating the Posterior Distribution of a Population Proportion π

(a)			(b)		
GIVEN PRIOR DISTRIBUTION FOR π			CALCULATIONS TO OBTAIN POSTERIOR DISTRIBUTION		
(1)	(2)	(3)	(4)	(5)	(6)
PROPORTION (PROBABILITY) DEFECTIVE π	NUMBER OF SHIPMENTS	RELATIVE NUMBER OF SHIPMENTS	LIKELIHOOD OF π (FROM TABLE IIIb, GIVEN SAMPLE $n = 5$, $S = 3$)	PRIOR TIMES LIKELIHOOD (3) × (4)	DIVIDING BY .161, YIELDS POSTERIOR
0%	2	.01	0	0	0
10%	30	.15	.008	.001	.01
20%	40	.20	.051	.010	.06
30%	42	.21	.132	.027	.17
40%	34	.17	.230	.039	.24
50%	26	.13	.313	.041	.25
60%	16	.08	.346	.028	.18
70%	8	.04	.309	.012	.08
80%	2	.01	.205	.002	.01
90%	0	0	.073	0	0
100%	0	0	0	0	0
	200	1.00 ✓		.161	1.00 ✓

Thus the sample, with its large proportion of defectives, has lowered the probability that this is an acceptable shipment from .36 to .07.

This example displays several features that will be found to be generally true:

1. In Figure 19-5 we see that the posterior distribution is a compromise, peaking between the prior distribution and the likelihood function.
2. If we had multiplied either the prior or the likelihood by some convenient constant, it merely would have changed the second-last column of Table 19-1 by the same constant. But the adjusted or "normed" values in the last column would be exactly as before. Accordingly:

> **Multiplying the prior or likelihood by a convenient constant does not affect the posterior.**   (19-7)

3. The problem as stated had discrete values for π (0, .1, .2, ...). However, this was just a convenient way of tabulating a variable π that really is continuous. So we have sketched all the graphs in Figure 19-5 as continuous.[1]

---

*[1] Is it really legitimate to shift the argument back and forth between discrete and continuous models? For our purposes, yes. The essential difference between a discrete probability function and the analogous continuous probability density is simply a constant multiplier (the constant being the cell width, as shown in (4-14)). And constant multipliers do not really matter, as stated in (19-7), above.

B—GENERALIZATION

To generalize Example 19-2, let us consider first the possible prior distributions for π. There is a whole family of distributions, called the β distributions, that serve as a convenient approximation for many of the priors that we might encounter in practice; and their formula is very simple:

$$\beta \text{ distribution for the prior, } p(\pi) \propto \pi^a(1 - \pi)^b \qquad (19\text{-}8)$$

This formula was deliberately chosen to be of the same form as the likelihood function (19-10) below, and is called the *conjugate* prior. This similarity will prove very convenient in the subsequent calculation of the posterior distribution.

In (19-8), the parameters a and b may be any numbers, although positive small integers are most common. For example, you can verify that the prior in column (3) of Table 19-1 is approximately the β distribution with $a = 1$ and $b = 3$:

$$p(\pi) \propto \pi(1 - \pi)^3 \qquad (19\text{-}9)$$

Next, to generalize the likelihood function, consider a sample of n observations that results in S "successes" and F "failures" (where $F = n - S$, of course). The likelihood function then is given by the general binomial formula:

$$p(S/\pi) = \binom{n}{S} \pi^S(1 - \pi)^F \qquad (19\text{-}10)$$
$$\text{like} (4\text{-}8)$$

Since the sample has already been observed and only π is viewed as variable, we can write this more briefly as:

$$\text{likelihood } L(\pi) \propto \pi^S(1 - \pi)^F \qquad (19\text{-}11)$$

When the prior (19-8) is multiplied by the likelihood (19-11), we obtain the posterior:

$$p(\pi/S) \propto \pi^a(1 - \pi)^b \pi^S(1 - \pi)^F$$

$$\propto \pi^{a+S}(1 - \pi)^{b+F} \qquad (19\text{-}12)$$

The logic of this posterior is so simple and so useful that it is worth reviewing briefly:

$$\left.\begin{array}{ll} \textit{If } \text{prior} & \propto \pi^a(1 - \pi)^b \\ \textit{and } \text{likelihood} \propto \pi^S(1 - \pi)^F \\ \textit{Then } \text{posterior} \propto \pi^{a+S}(1 - \pi)^{b+F} \end{array}\right\}$$ (19-13)

Thus we see the real advantage of using a conjugate prior (19-8), of the same form as the likelihood (19-11). In (19-13) it greatly simplifies the derivation of the posterior distribution. In fact, all three distributions—prior, likelihood, and posterior—now have the same β function form. Furthermore, many such β functions are already tabulated, by reading *across* Table IIIb (ignoring the constant binomial coefficient, which does not really matter). This greatly reduces the computations, as the next example illustrates.

EXAMPLE **19-3**

Referring back to Example 19-2, suppose Radex has much improved its quality control. In fact, the prior distribution has so improved that it now may be approximated with a β function with $a = 0$, and $b = 4$:

$$p(\pi) \propto \pi^0(1 - \pi)^4$$

Suppose, however, that the sample turns out the same way (3 defectives out of 5). Graph the prior, the likelihood, and the posterior.

Solution

As already remarked in (19-7), we may ignore constants such as binomial coefficients, which do not depend on π.

First we extract each distribution from the appropriate row of Table IIIb. For example, the likelihood function is $L(\pi) \propto \pi^3(1 - \pi)^2$, which we find in Table IIIb—under $n = 5$ and $S = 3$. The prior distribution is $p(\pi) \propto \pi^0(1 - \pi)^4$, which we can also find in Table IIIb—under $n = 4$ and $S = 0$. Finally, the posterior distribution is given by (19-13) as:

$$p(\pi/S) \propto \pi^3(1 - \pi)^6$$ (19-14)

Again, we can find this in Table IIIb—under $S = 3$ and $n = 3 + 6 = 9$. These three distributions are copied down for easy reference:

π	0	.1	.2	.3	.4	.5	.6	.7	.8	.9	1.0
$L(\pi)$	0	.008	.051	.133	.230	.313	.346	.309	.205	.073	0
$p(\pi)$	1.00	.656	.410	.240	.130	.063	.026	.008	.002	.000	0
$p(\pi/S)$	0	.045	.176	.267	.251	.164	.074	.021	.003	.000	0

The graphs of all three distributions are given in Figure 19-6. Note again that the posterior is a compromise, peaking between the prior and the likelihood.

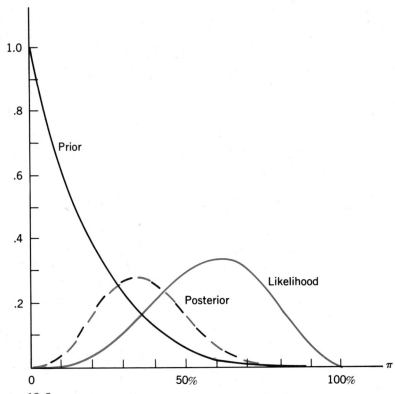

Figure 19-6
Different prior, hence different posterior, than in Figure 19-5.

C—THE PRIOR AS A QUASI-SAMPLE

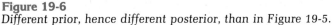

The posterior formula (19-13) is very convenient not only for computations, but also to illustrate advantages of the Bayesian approach—as an example will illustrate:

EXAMPLE **19-4**   In Example 19-3, suppose now our prior view is that all possible values of π are equally likely. That is, the prior distribution is completely flat, and thus might be called the *prior of pure ignorance*. It can be represented by a special β function, with $a = b = 0$:

$$p(\pi) \propto \pi^0(1 - \pi)^0 \tag{19-15}$$

Let us also suppose that a larger sample of $n = 9$ observations was taken, and had $S = 3$ defective radios.

a. What is the posterior distribution? How does it compare to the likelihood?

b. In terms of whether or not the shipment should be returned, how would the overall information available to the decision-maker here differ from the information available in Example 19-3?

Solution **a.**
$$\text{prior} \propto \pi^0(1 - \pi)^0$$
$$\text{likelihood} \propto \pi^3(1 - \pi)^6$$
$$\text{Thus} \quad \text{posterior} \propto \pi^3(1 - \pi)^6 \qquad \begin{array}{l}(19\text{-}16)\\ \text{like}(19\text{-}13)\end{array}$$

Since the posterior distribution summarizes all the information—the prior and the data—it is the relevant distribution for making decisions. But in this case the posterior is exactly the same as the likelihood function, so that any decision should be based on the likelihood function alone. This seems intuitively correct: Because of the prior of pure ignorance, it is appropriate to consider only the sample information (contained in the likelihood function).

b. The posterior distribution here in (19-16) is exactly the same as in the previous situation (19-14). Consequently, the decision-maker would be in exactly the same position—absolutely no difference.

In Example 19-4, we combined a kind of informationless prior with a relatively large sample of 9 observations. On the other hand, in Example 19-3 we started with information in the prior, and a sample of 4 fewer observations. But in both cases, we got the same posterior distribution; and therefore the same decision about the shipment in the warehouse. So the prior in Example 19-3 may be thought of as providing exactly the same information as an extra 4 observations.

In terms of its impact on decision-making, we can therefore think of the prior distribution as equivalent to an additional sample—and so we call it a *quasi-sample*. In general, the prior distribution in (19-13) can be considered a quasi-sample of a successes and b failures, to be combined with the actual sample of S successes and F failures.

The concept of the quasi-sample makes the point of this chapter very clear: Ignoring prior information is like throwing away sample data.

D—CONFIDENCE INTERVALS

A Bayesian 95% confidence interval for π can be constructed by finding the interval that contains 95% of the posterior distribution. As one would expect, it takes account of not just the observed sample, but also the quasi-sample (in the prior). In other words, the number of successes is essentially $(S + a)$ rather than S, while the number of failures is essentially $F + b$ rather than F.

More precisely, set

$$\left. \begin{array}{l} S^* \equiv S + a + 1 \\ F^* \equiv F + b + 1 \end{array} \right\} \tag{19-17}$$

The extra 1 observed success and failure is a mathematical quirk, which, if you like, you can think of as a "bonus" for using such an ingenious technique. Then we naturally continue defining:

$$\left. \begin{array}{l} n^* \equiv S^* + F^* \\ P^* \equiv \dfrac{S^*}{n^*} \end{array} \right\} \tag{19-18}$$

In these terms, the Bayesian confidence interval is easily stated (and proved, in Appendix 19-2). It is just like the classical confidence interval, using P^* and n^* instead of P and n:

Bayesian 95% confidence interval:

$$\pi = P^* \pm 1.96 \sqrt{\dfrac{P^* (1 - P^*)}{n^*}} \tag{19-19} \text{ like (8-27)}$$

Like the classical confidence interval, this formula requires S^* and F^* to be each at least 5.

EXAMPLE **19-5**   In a sample taken from a shipment of radios, 6 were defective and 14 okay.

a. Calculate the 95% classical confidence interval for the proportion of defectives in the shipment.

b. Suppose the record of past shipments showed relatively few defectives, and could be approximated by the prior distribution

$$p(\pi) \propto \pi^1 (1 - \pi)^9$$

Calculate the 95% Bayesian confidence interval. How is it different from the classical?

Solution   **a.** The sample proportion is $P = 6/(6 + 14) = .30$. Hence the 95% confidence interval is:

$$\pi = P \pm 1.96 \sqrt{\frac{P(1-P)}{n}} \qquad \text{(8-27) repeated}$$

$$= .30 \pm 1.96 \sqrt{\frac{.30\,(.70)}{20}}$$

$$= .30 \pm .20$$

b. To get the Bayesian 95% confidence interval, we first calculate:

$$\left.\begin{array}{l} S^* = S + a + 1 = 6 + 1 + 1 = 8 \\ F^* = F + b + 1 = 14 + 9 + 1 = 24 \end{array}\right\} \quad \text{(19-17) repeated}$$

Then

$$\left.\begin{array}{l} n^* = S^* + F^* = 8 + 24 = 32 \\ P^* = \dfrac{S^*}{n^*} = \dfrac{8}{32} = .25 \end{array}\right\} \quad \text{(19-18) repeated}$$

Hence the 95% confidence interval is:

$$\pi = P^* \pm 1.96 \sqrt{\frac{P^*(1-P^*)}{n^*}} \qquad \text{(19-19) repeated}$$

$$= .25 \pm 1.96 \sqrt{\frac{.25(.75)}{32}}$$

$$= .25 \pm .15$$

Because it exploits prior information, the Bayesian interval is centered better (at a compromise) and is more precise than the classical interval.

PROBLEMS

19-4 To decide on whether to include expensive blood freezing equipment, a proposed blood bank needed to estimate the proportion π of its blood donations that would be unsatisfactory (due to hepatitis traces or other reasons). The records of similar well established clinics showed their proportions π tended to be quite small, with a distribution roughly given by

$$p(\pi) \propto \pi (1 - \pi)^4$$

a. Graph this distribution.

b. In a random sample of 5 prospective donors, 3 provided unsatisfactory blood. Show graphically how this evidence changes the range of plausible values of π.

c. Calculate the Bayesian 95% confidence interval for π. Show it on the graph in part **b.** Does it seem to cut about $2\frac{1}{2}$% from each tail of the posterior distribution?

19-5   In Problem 19-4, suppose a larger sample of 50 prospective donors was taken, with 30 providing unsatisfactory blood.

a. How would you expect the Bayesian 95% confidence interval to be different from:

    i. The former answer in Problem 19-4?

    ii. The classical 95% confidence interval?

b. Calculate the Bayesian and classical confidence intervals to confirm your expectations in **a.**

19-6   The manager of a newly opened franchise for auto brake relining wanted to estimate what proportion π of his customers would also need their steering aligned. If π was high enough, he planned to offer alignment as an additional service.

a. He randomly sampled 40 cars, and inspection showed 8 of them did indeed need alignment. What is the 95% confidence interval for π?

b. Fortunately, the experience of 86 other franchised shops was available to the manager on request. After eliminating the 11 shops that clearly serviced a different clientele from his own, the remaining 75 shops showed values of π ranging from .15 to .63. In fact, a relative frequency histogram of these values of π showed it was closely approximated by the function $280\pi^3(1 - \pi)^7$.

    What is the Bayesian 95% confidence interval for π?

19-3   The Mean μ in a Normal Model

A—EXAMPLE

Once more, we shall apply the general principle (19-6), this time to estimate the mean μ of a normal population.

EXAMPLE **19-6**   Steelco sells thousands of shipments of steel beams. Within each shipment, the breaking strengths of the beams are distributed normally around a mean μ, with standard deviation $\sigma = 17.3$. But μ changes from shipment to shipment because of poor quality control. In fact, suppose that when all shipment means μ are recorded in a bar graph, the result is Figure 19-7. That is, the distribution of μ is

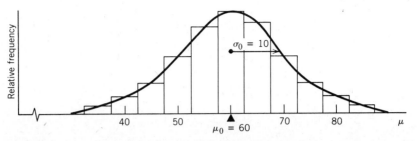

Figure 19-7
The distribution of shipment means is approximately normal.

approximately normal, with mean $\mu_o = 60$ and standard deviation $\sigma_o = 10$. This is our prior information.

Now suppose we have received one specific shipment of beams from Steelco, and have to estimate its mean strength μ. (If μ is too low, we should return the shipment.) To estimate μ, we take a random sample of 12 beams from the shipment, and $\overline{X}$ turns out to be 70.

a. Sketch the likelihood function.

b. Sketch the posterior distribution obtained by multiplying the prior distribution of Figure 19-7 by the likelihood function.

c. Suppose that the shipment must be regarded as unsatisfactory if μ is less than 62.5. What would you say is the probability of this, as estimated from the sketched distributions:

    i. Before the sample was taken?
    ii. After the sample?

Solution   **a.** What is the likelihood of getting $\overline{X} = 70$? The distribution of $\overline{X}$ is normal, with standard error $\sigma/\sqrt{n} = 17.3/\sqrt{12} = 5.0$. Thus its equation is[2]:

$$p(\overline{X}/\mu) \propto e^{-\frac{1}{2}\left(\frac{\overline{X}-\mu}{5}\right)^2} \qquad \text{(19-20)}$$
$$\text{like (4-26)}$$

where $\overline{X} = 70$, and μ is regarded as the variable. As usual, to emphasize its dependence on μ, we rename (19-20) the likelihood function:

$$L(\mu) \propto e^{-\frac{1}{2}\left(\frac{\mu-70}{5}\right)^2} \qquad \text{(19-21)}$$

---

[2]For simplicity, we have ignored the constant multiplier $1/\sqrt{2\pi}\,\sigma$ in the normal distribution, so we can plot it with a maximum value of 1. In view of (19-7), this will not essentially affect the posterior.

We recognize this likelihood as a normal curve, centered at 70 with a standard deviation of 5. We roughly sketch its graph in Figure 19-8b. [To do so, we recall from Figure 4-10 that a normal curve contains about ⅔ (68%) of its area within one standard deviation (in this case, 5 units) of its center, and 95% of its area within two standard deviations.]

b. The prior distribution in Figure 19-7 is similarly graphed in Figure 19-8a.

For the posterior distribution, we must multiply the prior times the likelihood. We do this for several values[3] of μ, to obtain several points on the posterior distribution. From them, we sketch the graph of the posterior distribution shown in Figure 19-8c.

c. i. Before the sample is taken, the probability that μ is below 62.5 must be estimated from the only available information—the prior. This probability, shown as the shaded area in Figure 19-8a, is approximately 60%.

ii. After the sample is taken, the best available information is the posterior. The desired probability, shown as the shaded area in Figure 19-8c, is approximately 10%.

In this example, the posterior seems to be distributed normally, which is no surprise, since it is the product of a normal prior and normal likelihood. (This is just like our previous example: the posterior distribution for π was a β-function, just like the prior and likelihood.) Again, we note that the posterior peaks between the prior and the likelihood, and that it is closer to the likelihood, the curve with the least variance. That is, the more concentrated curve, which accordingly provides the more precise and reliable information, has the greater influence in determining the posterior.

B—GENERALIZATION

The features of Figure 19-8 are generally true, and may be expressed with convenient formulas. We calculate the posterior distribution of μ from two pieces of information:

1. The *actual* sample centered at $\overline{X}$, which gives us the likelihood function.
2. The prior distribution, which is *equivalent* to additional sample information, called the quasi-sample.

---

[3]For example, consider $\mu = 65$ shown in Figure 19-8. We read off the prior probability (.9) and likelihood (.6) from the first two graphs. Their product, $.9 \times .6 = .54$, is the desired posterior in the third graph.

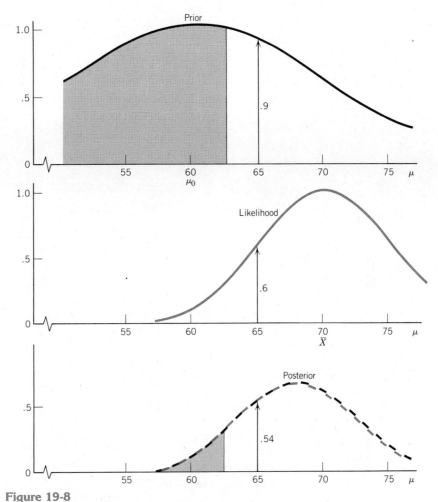

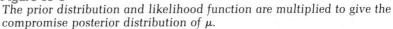

Figure 19-8
*The prior distribution and likelihood function are multiplied to give the
compromise posterior distribution of μ.*

The concept of the quasi-sample has already been encountered (in estimating π in Section 19-2). Of course, the prior distribution is not really
a sample. But its effect on the posterior distribution is exactly equivalent
to the effect of an additional sample with a certain number (n_0, say) of
observations. And the quasi-sample will provide a nice intuitive way to
understand and remember the formulas. (The proofs are given in Appendix 19-3.)

The quasi-sample would, of course, be centered on the prior mean μ_0.
But how many quasi-observations n_0 should we say it contains? That is,
how heavily should we weight the quasi-sample when we combine it with
the actual sample? The smaller the prior variance σ_0^2 (relative to σ^2), the

more certainty or information is contained in the prior; consequently the larger n_0 will be[4]:

$$n_0 = \frac{\sigma^2}{\sigma_0^2}$$

(19-22)

When the n_0 quasi-observations centered at the prior mean μ_0 are combined with the n actual observations centered at $\overline{X}$, the overall posterior mean is the weighted average:

$$\text{Posterior mean} = \frac{n_0\mu_0 + n\overline{X}}{n_0 + n}$$

(19-23)
like (2-7)

Since the total number of observations is $n_0 + n$, the variance of the posterior distribution is[5]:

$$\text{Posterior standard error} = \frac{\sigma}{\sqrt{n_0 + n}}$$

(19-24)
like (6-7)

Finally, when the prior is normal and the likelihood is normal (by the Normal Approximation Rule), then the posterior can be shown to be normal too. An example will illustrate these formulas.

EXAMPLE **19-7**   Let us rework Example 19-6 with formulas instead of graphs.

    a. Express the value of the prior distribution in terms of quasi-observations.

    b. Calculate the posterior distribution.

    c. Calculate the probability that μ is below 62.5:

        i. Before the sample is taken.

       ii. After the sample has been taken.

Solution   **a.**

$$n_0 = \frac{\sigma^2}{\sigma_0^2}$$

(19-22) repeated

---

[4]Of course, σ^2—the variance of the population from which we are sampling—is typically not known, so in practice we use the sample s^2 to estimate it. But we use σ^2 here to make the argument more transparent.

[5]Why doesn't σ_0^2 appear in this formula? Implicitly it does, because (19-22) defined n_0 in terms of σ_0^2.

$$= \frac{17.3^2}{10^2} = \frac{300}{100} = 3$$

That is, the prior distribution is equivalent to 3 observations.

b. The posterior is normal, with:

$$\text{Posterior mean} = \frac{n_0\mu_0 + n\overline{X}}{n_0 + n} \qquad \text{(19-23) repeated}$$

$$= \frac{3(60) + 12(70)}{3 + 12} = 68 \qquad \text{(19-25)}$$

$$\text{Posterior standard error} = \frac{\sigma}{\sqrt{n_0 + n}} \qquad \text{(19-24) repeated}$$

$$= \frac{17.3}{\sqrt{3 + 12}} = 4.5 \qquad \text{(19-26)}$$

c. **i.** Before the sample, we use the prior distribution. The standardized Z value is, therefore:

$$Z = \frac{\mu - \mu_0}{\sigma_0} = \frac{62.5 - 60}{10} = .25$$

Thus

$$\Pr(\mu < 62.5) = \Pr(Z < .25) = 1 - .401 = .599$$
$$= 60\%$$

ii. After the sample, we use the posterior distribution calculated in **b**. Then:

$$Z = \frac{62.5 - 68}{4.5} = -1.22$$

Thus

$$\Pr(\mu < 62.5) = \Pr(Z < -1.22) = .111$$
$$= 11\%$$

Note that the precise answers calculated from formulas in Example 19-7 correspond very nicely to the rough graphical answers that we found before in Example 19-6. Since the formulas are easier and more accurate, from now on we shall use them instead of graphs.

C—CONFIDENCE INTERVALS

Since the posterior distribution of μ is normal, 95% of its probability lies within ± 1.96 standard deviations from its mean. This gives us:

> **Bayesian 95% confidence interval for a mean, assuming a normal prior:**
>
> $$\mu = \left(\frac{n_0 \mu_0 + n\overline{X}}{n_0 + n} \right) \pm 1.96 \frac{\sigma}{\sqrt{n_0 + n}}$$
>
> (19-27) like (8-8)

where, as always, $n_0 = \sigma^2 / \sigma_0^2$. An example will illustrate:

EXAMPLE **19-8**

For Example 19-7, calculate:

a. The classical 95% confidence interval.
b. The Bayesian 95% confidence interval.

Solution

a.
$$\mu = \overline{X} \pm 1.96 \frac{\sigma}{\sqrt{n}} \qquad \text{like (8-9)}$$

$$= 70 \pm 1.96 \frac{17.3}{\sqrt{12}} = 70 \pm 10$$

b. Substitute (19-25) and (19-26) into (19-27):

$$\mu = 68 \pm 1.96(4.5) = 68 \pm 9 \qquad (19\text{-}28)$$

Because it exploits prior information, the Bayesian interval is centered better (at a compromise between $\overline{X} = 70$ and $\mu_0 = 60$), and is more precise than the classical interval.

PROBLEMS

19-7 Continuing Example 19-6, suppose the sample size is $n = 48$ instead of $n = 12$, but everything else is the same, namely: a normal prior with $\mu_0 = 60$, $\sigma_0 = 10$; and the n observations fluctuate with $\sigma = 17.3$, and have a mean $\overline{X} = 70$.

a. Graph the likelihood function, as well as the old prior.
b. Where does the likelihood function peak? What is this number called?
c. Graph the posterior also. Does the posterior peak between the prior and likelihood? Which is the posterior closer to? How much closer?
d. Calculate the classical and Bayesian 95% confidence intervals for μ, and comment on their difference.

19-8   Repeat Problem 19-7, if the sample size is $n = 2$.

19-9   From your experience in Problems 19-7 and 19-8, answer True or False; if False, correct it:

     The Bayesian confidence interval is very similar to the classical when n is small. It is only when n grows large that the Bayesian confidence interval shows its real superiority to the classical.

19-10   Suppose that it is essential to estimate the length θ of a beetle that accidentally gets caught in a delicate piece of machinery. A measurement X is possible, using a crude device that is subject to considerable error: X is distributed normally about the true value θ, with a standard deviation $\sigma = 5$ mm. Suppose a sample of 6 independent observations yields an average $\overline{X} = 20$ mm.

     Some information also is available from the local agricultural station. They tell us that this species of beetle has length that is distributed normally about a mean of 25 mm with a standard deviation of 2.5 mm.

     a.   Find the Bayesian 95% confidence interval for the beetle's length θ.

     b.   What is the probability that θ is at least 26 mm.?

*19-11   The value of wheat falls as its moisture content increases, as measured by the electrical conductivity of the wheat. Suppose the measured moisture value is distributed normally about the true value, with an error that has a standard deviation of only one-fifth of a percentage point.

     Suppose that the loads of wheat that are brought to a grain elevator during a certain week have moisture contents m varying from 14% to 16%, roughly; specifically, the values of m are distributed normally about a mean of 15%, with a standard deviation of half a percentage point.

     If one such load has a measured value of 13.8%, its true value m may be slightly different, of course. Specifically,

     a.   Find the Bayesian 95% confidence interval for m.

     b.   What is the probability that m is less than 14%?

19-4   The Slope β in Normal Regression

Our treatment of the simple regression slope β will be very similar to the population mean μ in the previous section. We assume, for example, that the prior distribution of β is normal, with mean β_0 and standard deviation σ_0.

A sample of n observations (X,Y) gives further information about β, summarized by the estimate $b = \Sigma xy / \Sigma x^2$. To measure the reliability of b, we need σ (the standard deviation of the Y values about the line), and also σ_X (the standard deviation of the X values[6]). In fact, their ratio (σ/σ_X)

---

[6] Of course, both σ and σ_X are typically unknown, and have to be estimated with s and s_X (just as σ was estimated with s in footnote 4 earlier).

plays the same role as σ formerly played for μ. Specifically, the amount of information in the prior is described by the size n_0 of the quasi-sample:

$$n_0 = \frac{(\sigma/\sigma_X)^2}{\sigma_0^2}$$

(19-29)
like (19-22)

When the n_0 quasi-observations centered at β_0 are combined with the n actual observations in the sample (from which b was calculated), the result is the weighted average (as proved in Appendix 19-4):

$$\text{Mean of the posterior distribution of } \beta = \frac{n_0\beta_0 + nb}{n_0 + n}$$

(19-30)
like (19-23)

Since the total number of observations is $n_0 + n$, and noting again that σ/σ_X plays the former role of σ,

$$\text{SE of the posterior distribution of } \beta = \frac{\sigma/\sigma_X}{\sqrt{n + n_0}}$$

(19-31)
like (19-24)
and (12-6)

When the prior and likelihood are normal, the posterior can also be shown to be normal. Then any desired probability can be calculated from the normal table, and a 95% confidence interval can be determined:

Bayesian 95% confidence interval for a regression slope:

$$\beta = \left(\frac{n_0\beta_0 + nb}{n_0 + n}\right) \pm 1.96 \frac{\sigma/\sigma_X}{\sqrt{n_0 + n}}$$

(19-32)

An example will illustrate:

EXAMPLE **19-9**   Suppose that the prior distribution for a regression slope β is normal with a central value of 5.0 and a standard deviation of .40. In addition, suppose that a sample of 8 observations of (X,Y) are taken, and yield the following statistics:

$$\Sigma xy = 2100, \qquad \Sigma x^2 = 350, \qquad \text{residual } s = 12.7$$

To be more realistic than in our earlier Examples, let us now suppose σ is unknown, so that s will replace σ where necessary, and correspondingly, $t_{.025}$ will replace 1.96. Then find:

a. The classical 95% confidence interval.

b. The Bayesian 95% confidence interval. How does it compare?

Solution   **a.**
$$b = \frac{\Sigma xy}{\Sigma x^2} = \frac{2100}{350} = 6.0$$

Also, d.f. $= n - 2 = 8 - 2 = 6$, so that $t_{.025} = 2.45$. Thus the 95% classical confidence interval is

$$\beta = b \pm t_{.025} \frac{s}{\sqrt{\Sigma x^2}}$$

$$= 6.0 \pm 2.45 \frac{12.7}{\sqrt{350}}$$

$$= 6.0 \pm 1.7$$

b. In standard notation, the prior information is:

$$\beta_0 = 5.0$$
$$\sigma_0 = .40$$

In the sample, we estimate σ with $s = 12.7$, and σ_X with $s_X = \sqrt{\Sigma x^2/(n-1)} = \sqrt{350/7} = 7.1$. Hence the quasi-sample size (reflecting the strength of the prior) is:

$$n_0 = \frac{(\sigma/\sigma_X)^2}{\sigma_0^2} \qquad \text{(19-29) repeated}$$

$$\simeq \frac{(12.7/7.1)^2}{.50^2} \simeq 13$$

Because of the replacement throughout of σ by s, we need to replace 1.96 with $t_{.025}$ in the confidence interval. Since the total number of observations $n_0 + n = 13 + 8 = 21$, we have d.f. $= 21 - 2 = 19$, and hence $t_{.025} = 2.09$. Thus the Bayesian 95% confidence interval is

$$\beta = \frac{n_0\beta_0 + nb}{n_0 + n} \pm t_{.025} \frac{s/s_X}{\sqrt{n_0 + n}} \qquad \text{like (19-32)}$$

$$= \frac{13(5.0) + 8(6.0)}{13 + 8} \pm 2.09 \frac{12.7/7.1}{\sqrt{13 + 8}}$$

$$= 5.4 \pm 0.8$$

Because it exploits prior information, the Bayesian interval is centered better (at a compromise), and is more precise than the classical interval.

Bayesian techniques can easily be applied to multiple regression, too (see, for example, Press, 1972). The theory is basically the same, but the calculations are best left to a computer, of course.

PROBLEM

19-12 A consulting economist was given a prior distribution for a regression slope β (a marginal propensity to save) that was distributed normally about a mean of .20 with a standard deviation of .08. Calculate the Bayesian 95% confidence intervals for β as he successively gathers more data:

a. $n = 5$, $\Sigma xy = 5.0$, $\Sigma x^2 = 25$, $s = .36$

b. $n = 10$, $\Sigma xy = 8.0$, $\Sigma x^2 = 50$, $s = .28$

c. $n = 20$, $\Sigma xy = 15.0$, $\Sigma x^2 = 100$, $s = .31$

19-5 Bayesian Shrinkage Estimates

A—PRIOR DISTRIBUTIONS: OBJECTIVE, SUBJECTIVE, AND NEUTRAL

In the examples so far in this chapter, the prior distribution has been based on empirical evidence as solid as the sample evidence on which the likelihood function is based. When we get such an empirical or *objective* prior, we are fortunate indeed.

Often, however, the prior distribution is not so clear-cut and may be based on the statistician's personal judgment. In principle, such a *personal* or *subjective* prior can be developed quite easily (as we discussed in Section 3-7). In practice, however, it requires considerable skill to develop a personal prior that is sufficiently free of prejudice.

If there seems to be no prior knowledge at all, even this can be formulated as an *informationless* or *neutral* prior. We shall now show how this will allow certain parts of the data (that classical statistics does not fully exploit) to help estimate the unknown parameter. The ingenious result is called a *Bayesian shrinkage estimate*, and we will see how nicely it resolves some of the dilemmas of classical estimation.

B—THE DILEMMA OF CLASSICAL ESTIMATION

As a concrete example, in Table 19-2 we repeat the data of Table 10-3. Suppose we need an estimate of μ_2, the long-run mean output of the second machine. What do we use?

The obvious estimate is $\overline{X}_2 = 56$, the *sample* mean of that machine. But what if H_0 is true (i.e., there is no difference in machines, so that all 15 observations come from the same population)? Then a better estimate would be the grand mean $\overline{\overline{X}} = 52$, based on all the data in Table 19-2.

TABLE **19-2** **Production of 3 Machines (repeat of Table 10-3)**

	MACHINE 1	MACHINE 2	MACHINE 3
	50	48	57
	42	57	59
	53	65	48
	45	59	46
	55	51	45
	$\overline{X}_1 = 49$	$\overline{X}_2 = 56$	$\overline{X}_3 = 51$

Grand mean $\overline{\overline{X}} = 52$

$$F = \frac{65}{39.0} = 1.67$$

The problem in a nutshell is this: *We don't know which of these two estimates to choose because we don't know whether H_0 or H_1 is true:* If H_0 is true (i.e., if there is no difference in machines), the better estimate is $\overline{\overline{X}} = 52$; but if H_1 is true (and the machines do differ), then the better estimate is $\overline{X}_2 = 56$. The classical way out of this dilemma is to use the F test at some customary level, say 5%. In our example in Table 19-2, F was calculated to be 1.67. Since this falls short of the critical $F_{.05} = 3.89$, we cannot reject H_0; so we use the H_0 estimate of 52.

But imagine that Table 19-2 represented only one experiment in a whole sequence of possible experiments; and suppose the F values ranged continuously from 1.67 on up. Classical accept-or-reject theory would require us to use $\overline{\overline{X}}$ whenever F was less than $F_{.05} = 3.89$ (no matter how slightly) and $\overline{X}_2$ whenever F was above 3.89 (no matter how slightly). That is, our estimate could only take on two values ($\overline{\overline{X}}$ or $\overline{X}_2$), and would jump from one to the other because of a trivial change in the data (F changing, for example, from 3.88 to 3.90). This is clearly unsatisfactory. Instead, we should develop an estimate that changes *continuously* between $\overline{\overline{X}}$ and $\overline{X}_2$, depending on the strength of the sample message we get from F.

C—BAYESIAN SOLUTION IN GENERAL

To the extent that the F statistic is large, H_1 is more credible; to the extent that F is small, H_0 is more credible. The Bayesian solution (derived in Wonnacott, 1981) makes this idea precise:

Bayesian Shrinkage Estimate (for $F \geq 1$):

Give H_0 a weight $\dfrac{1}{F}$

Give H_1 the remaining weight $= 1 - \dfrac{1}{F}$ (19-33)

For $F < 1$, we would run into trouble if we tried to use (19-33): H_0 would get a weight of more than 1, and H_1 would get a negative weight. Therefor to the above Bayesian estimate we must add the following guideline:

$$\text{For } F < 1: \\ \left.\begin{array}{l} \text{Give } H_0 \text{ a weight } = 1 \\ \text{Give } H_1 \text{ a weight } = 0 \end{array}\right\} \qquad (19\text{-}34)$$

To know when the Bayesian shrinkage estimate is appropriate, we must be clear on its basic assumption. It assumes the prior is vaguely distributed around the null hypothesis H_0; that is, values on either side of H_0 roughly seem equally likely. (For a more careful statement of assumptions and derivations, see Box and Tiao, 1973.)

Now let us see how the shrinkage formulas work in specific cases:

D—BAYESIAN SHRINKAGE ESTIMATES FOR ANOVA

For ANOVA, we saw that the null hypothesis H_0 used the grand mean $\overline{\overline{X}}$, and the alternative hypothesis H_1 used the specific machine mean $\overline{X}_2$. If we denote the Bayesian estimate by BE, then (19-33) becomes:

> **Bayesian estimate of the typical (second) population mean in ANOVA:**
>
> $$\text{BE}(\mu_2) = \left(\frac{1}{F}\right)\overline{\overline{X}} + \left(1 - \frac{1}{F}\right)\overline{X}_2 \qquad (19\text{-}35)$$

An example will show how the formula works, and how it can be interpreted graphically:

EXAMPLE **19-10**

Recall that in Table 19-2, from a sample of $n = 5$ observations on each machine, the three machine means were 49, 56, and 51. The grand mean was $\overline{\overline{X}} = 52$, and the F statistic was 1.67.

a. Calculate the Bayesian estimate of the second machine mean, and also the first and third.

b. On a graph, contrast the Bayesian and classical estimates.

Solution

a. The Bayesian estimate is a compromise between $\overline{\overline{X}} = 52$ and $\overline{X}_2 = 56$, with the weights determined by $F = 1.67$. When these numbers are substituted into (19-35), we obtain

$$\text{BE}(\mu_2) = \frac{1}{1.67}\overline{\overline{X}} + \left(1 - \frac{1}{1.67}\right)\overline{X}_2$$

$$= .60(52) + .40(56) = 53.6 \qquad (19\text{-}36)$$

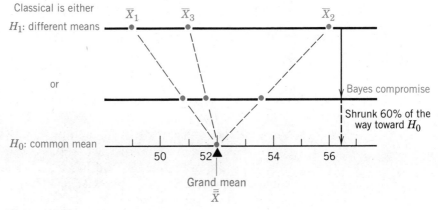

Figure 19-9

The Bayesian estimates are H_1 *estimates shrunk toward* H_0. *(In our example, with F = 1.67, there is a 60% shrinkage.)*

The Bayesian estimate is thus a .60 − .40 compromise. For the other machines, the compromise is still .60 − .40, because the F statistic is still 1.67. Therefore,

$$BE(\mu_1) = .60(52) + .40(49) = 50.8$$
$$BE(\mu_3) = .60(52) + .40(51) = 51.6$$

b. As Figure 19-9 shows, Bayesian estimates are shrunk toward the grand mean $\overline{X}$. So they are sometimes called *shrinkage* estimates.

E—BAYESIAN SHRINKAGE ESTIMATES FOR REGRESSION

We can also apply the basic Bayesian solution (19-33) to regression, provided once more our prior expectation is that the parameter is somewhere in the neighborhood of the null hypothesis.

Let us consider the simplest case, where the null hypothesis is $\beta = 0$. In other words, let us suppose the slope β is judged about as likely to be negative as positive, before we collect the sample. [This assumption that the prior distribution is centered at zero produces an estimate shrunk towards 0 in (19-37) below. If the prior distribution is centered at some other point, as it often is in regression, then the estimate should be shrunk toward that point, of course.]

When we substitute the null hypothesis ($\beta = 0$) and the alternative hypothesis (the classical estimate b) into (19-33), we obtain the compromise Bayesian estimate of the slope β:

$$BE(\beta) = \left(\frac{1}{F}\right) 0 + \left(1 - \frac{1}{F}\right) b$$

$$= \left(1 - \frac{1}{F}\right) b \tag{19-37}$$

where F is the customary F ratio for testing the null hypothesis:

$$F = t^2 = \left(\frac{b}{SE}\right)^2$$

(19-38)
(15-14) repeated

EXAMPLE **19-11**

The Statistics 200 course at a large university has many sections, each taught by a different instructor. Every year, the students are all given a common final exam, and in each section the average mark is calculated—the students' performance Y. In turn, the students of that section fill out an evaluation form on how good they thought their instructor was—the instructor's popularity X. How is Y related to X?

It's not clear whether this is a positive or negative relationship. (If instructors who are judged good get this rating because they teach really well, then their students will do well and Y will be positively related to X. On the other hand, if instructors who are judged good get this rating primarily because they demand very little work, then their students will do badly and Y will be negatively related to X.) Thus the prior judgment is that β is about as likely to be positive as negative, so the Bayes compromise with 0 given in (19-37) is okay.

Of the hundreds of sections evaluated over the years, a random sample of 10 was selected, and plotted in Figure 19-10. The fitted least squares line showed a quite negative slope ($b = -2.4$)—due perhaps to the large degree of scatter, and subsequent high standard error (SE = 1.94).

a. What is the classical p-value for H_0 ($\beta = 0$, i.e., no relation of Y to X)?

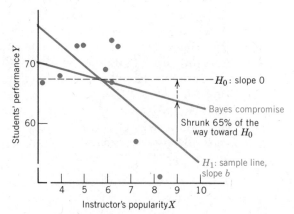

Figure 19-10
The Bayesian line is the least squares line shrunk toward H_0. (In our example, with F = 1.53, the shrinkage is 65%.)

b. What is the Bayesian shrinkage estimate of the slope β? Compare it graphically to the OLS slope.

Solution   **a.** From (19-38),

$$F = \left(\frac{b}{SE}\right)^2 = \left(\frac{-2.4}{1.94}\right)^2 = 1.53$$

We refer to Appendix Table VI, with F having d.f. $= 1$ and $n - 2 = 8$. The observed value $F = 1.53$ is close to $F_{.25} = 1.54$, so that we conclude:

$$\text{p-value} \simeq .25$$

b. Because H_0 had substantial credibility in part **a**, we expect the Bayes compromise slope to be shrunk substantially toward 0. Formula (19-37) gives it explicitly:

$$\text{BE}(\beta) = \frac{1}{1.53}(0) + \left(1 - \frac{1}{1.53}\right)(-2.4)$$

$$= .65(0) + .35(-2.4)$$

$$= -.8 \tag{19-39}$$

As shown in Figure 19-10, this means that the classical slope b is shrunk 65% of the way toward 0, from -2.4 to $-.8$.

F—SOME OTHER APPROACHES

There are ways other than Bayesian shrinkage to modify classical estimates. They range from the very theoretical to the very practical, and are known by such names as *James-Stein estimation*, *ridge regression*, *cross validation*, and *minimum MSE* (Wonnacott, 1984 IM). All these modifications share a common goal of minimizing estimation error, and all produce roughly the same answer—the Bayesian shrinkage estimate given in (19-33). Since each of these philosophically different approaches yields much the same estimator, each tends to confirm the validity of the others, and in particular, the validity of Bayesian shrinkage. -

As well as providing an estimate more valid than the classical, Bayesian shrinkage also gives a more precise confidence interval around it (as described in Appendix 19-5).

PROBLEMS

19-13   Recall that in Table 10-1, a sample of $n = 5$ observations from each machine gave 3 machine means: 49, 56, and 51. The grand mean was 52 and the F statistic was 8.3.

    a. Calculate the Bayesian shrinkage estimates for the population means of all 3 machines.

    b. On a graph, contrast the Bayesian shrinkage and classical estimates.

    c. Comparing the graph in **b** to the graph in Figure 19-8

        i. Which graph shows more shrinkage toward the null hypothesis value $\overline{\overline{X}}$?

        ii. In which case was there more evidence that H_0 was true, as given by the classical test?

19-14 An experiment to compare three treatments was conducted. Eighteen rats were assigned at random, 6 to each of the 3 groups. Their lengths of survival (in weeks) after being injected with a carcinogen were as follows:

	T_1	T_2	T_3
	13	3	10
	9	6	7
	16	6	8
	7	5	12
	11	7	14
	10	9	9
Mean $\overline{X}$	11	6	10
$\Sigma(X - \overline{X})^2$	55	20	34

On a graph, contrast the classical and Bayesian shrinkage estimates of the mean survival time for the 3 treatments.

19-15 A random sample of 7 women showed a positive slope of blood cholesterol level on age. If the least squares slope was .024 with a standard error of .010, what is the Bayesian shrinkage slope?

19-16 The GMAT exam was given to a random sample of 5 men and an independent random sample of 5 women graduating from an East Coast university. The percentage scores obtained were:

$$\text{Men:} \quad 73 \quad 61 \quad 60 \quad 70 \quad 76$$
$$\text{Women:} \quad 71 \quad 85 \quad 72 \quad 93 \quad 84$$

For the difference between the male and female scores (average over the whole class), calculate the classical and Bayesian shrinkage estimates.

19-17 Suppose that a year after having taken the exam in Problem 19-16, the 5 men were again tested. The second test scores (along with the corresponding first test scores) were as follows:

$$\text{First score:} \quad 73 \quad 61 \quad 60 \quad 70 \quad 76$$
$$\text{Second score:} \quad 79 \quad 67 \quad 68 \quad 75 \quad 81$$

For the improvement in score (average for the whole graduating class of men), calculate the classical and Bayesian shrinkage estimates. What assumptions are you making for the latter?

19-6   Classical and Bayesian Estimates Compared

A classical estimate such as $\overline{X}$ or a least squares slope b is entirely based on the sample. The Bayesian estimate is a more subtle one that sees all shades of gray—a compromise between the sample and the prior distribution. If the prior is not known, but seems to be vaguely centered around the null hypothesis H_0, then a neutral prior will allow the F ratio of the sample to determine the degree of compromise (that is, the degree of shrinking toward H_0).

As the sample becomes larger and more reliable, there is a greater and greater relative weight placed on it, so that:

> **For large samples, the Bayesian estimate is practically the same as the classical estimate (based on the sample alone).**   (19-40)

In other words, if the sample is large, there is no need to derive the Bayesian estimate; the classical estimate will be similar, and easier. Thus the Bayesian estimate is primarily a *small-sample modification*.

For example, it would be foolish to ignore prior information in trying to guess an election if a sample contained only 15 voters. Yet it would be unnecessary to bother with prior information if the sample contained 1500 voters (as does the typical Gallup poll).

CHAPTER 19 SUMMARY

19-1   Bayesian statistics provide a whole new way of looking at estimation, as a combination of sample information *and* prior information. It is based on posterior probabilities, which are easily obtained from tree reversal (Bayes Theorem).

19-2   For a proportion π, the prior information can be regarded as a quasi-sample of a successes and b failures. They can then be combined naturally with the actual S successes and F failures in the sample.

19-3   For a mean μ, the prior information can be regarded as a quasi-sample of n_0 observations centered at μ_0 (the mean of the normal prior distribution). This can then be combined naturally with the n actual observations in the sample, to produce a compromise estimate between μ_0 and $\overline{X}$.

19-4   For a regression slope β, the prior information can similarly be regarded as a quasi-sample of n_0 observations. The same is true in multiple regression, where a computer is customarily used.

19-5   When the prior is vaguely distributed around the null hypothesis H_0, then the Bayesian estimate turns out to be the classical estimate shrunk toward H_0. The degree of shrinking is greatest when the sample F statistic is least (and consequently the p-value for H_0—the credibility of H_0—is greatest).

19-6   In large samples, the Bayesian estimate is practically the same as the classical estimate based on the sample alone. Thus the Bayesian modification is primarily for small samples.

REVIEW PROBLEMS

19-18   A shipment of natural sponges is to be sampled to determine three characteristics:

1.   The proportion π that are torn.
2.   The mean weight μ.
3.   The slope β of the graph of absorbency Y as a linear function of weight X.

Suppose, on the basis of past shipments, that the following priors are deemed appropriate:

1. $p(\pi) \propto \pi^3 (1 - \pi)^{12}$
2. μ is normal, with mean 150 and standard deviation 20.
3. β is normal, with mean 4.0 and standard deviation .50.

A sample of $n = 20$ sponges yielded the following statistics:

1.   Sample proportion that are torn, $P = 10\%$.
2.   For weight, $\overline{X} = 140$, $\Sigma(X - \overline{X})^2 = 22,800$.
3.   For regression, $\Sigma(X - \overline{X})^2 = 22,800$, $\Sigma(X - \overline{X})(Y - \overline{Y}) = 114,000$, and residual $s = 53.4$.

Graph the posterior distribution and 95% confidence interval for

a. π

b. μ

c. β

19-19   An engineer was asked to estimate the mean breaking strength of a new shipment of cables. A large number of past shipments had varied considerably in their mean strength μ, having an overall mean of 3400 and standard deviation 100.

   The engineer decided that this past record was so variable that further evidence was required, so he sampled the new shipment, obtaining the following 5 breaking strengths:

$$3600, \quad 3300, \quad 3500, \quad 3700, \quad 3900$$

a.   What is the best estimate of the new shipment mean μ? What assumptions is it based upon?

b.   Construct an interval around your estimate, wide enough so as to be 95% certain to contain the actual μ.

19-20 A statistician was asked to summarize what she knew about the proportion π of state senators who would favor higher gasoline taxes.

  a. Before any sampling was done, her guess for π ranged generally on the low side—in fact, her personal prior distribution was

$$p(\pi) \propto \pi (1 - \pi)^3.$$

  Graph this distribution.

  b. A random sample of 6 senators turned out to have 5 in favor of higher gasoline taxes. Show graphically how this evidence changes the statistician's guess for π.

  c. Construct the Bayesian 95% confidence interval for π. Show it on the graph in **b.** Does it seem to take in about 95% of the posterior distribution?

19-21 In testing out a new variety of hybrid corn, a small pilot experiment on $n = 100$ randomly selected plots found the average yield to be $\overline{X} = 115$ and $s = 21$.

  a. Construct a 95% confidence interval for the mean μ.

  b. Dr. C said that he would bet 95¢ to a nickel that the confidence interval in **a** covers the true mean μ.

  Dr. B doubted that she would give such high odds. She had discussed the new variety with an expert who thought it would be at least as productive as the standard variety, whose record was well established: $\mu = 120$. Therefore, if she had to bet on the confidence interval, her odds would be in the neighborhood of 80¢ to 20¢.

  Settle the dispute as well as you can.

  c. Suppose the expert in **b** roughly quantified her knowledge by giving μ a normal prior distribution centered at 121, with a standard deviation of 3. What Bayesian confidence interval for μ would this give?

  d. Suppose that instead of talking to the expert, they had gathered more data: Suppose 50 more plots had an average $\overline{X} = 121$ and $s = 20$.

  Using all 150 observations, what is the best estimate of μ? What is the 95% confidence interval? (*Hint:* For all 150 observations, an elaborate calculation would show that $s \approx 21$ still.)

  e. How do the estimates in **c** and **d** compare?

*19-22 **a.** Suppose that, in an earlier quality control study, 100 steel bars had been selected at random from a firm's output, with each subjected to strain until it broke:

Breaking strength, θ	200	210	220	230	240	250
Relative frequency	.10	.30	.20	.20	.10	.10

Suppose that this is the only available prior distribution of θ. Graph it.

b. Now suppose that a strain gauge becomes available that gives a crude measurement X of any bar's breaking strength θ (without breaking it). By "crude" we mean that the measurement error ranges from -20 to $+20$, so that X has the following distribution:

X	$\theta - 20$	$\theta - 10$	θ	$\theta + 10$	$\theta + 20$
$p(X/\theta)$	.2	.2	.3	.2	.1

Suppose that the measurement of a bar purchased at random turns out to be $X = 240$. What is the (posterior) distribution of θ, now that this estimate is available? Graph the likelihood and posterior distributions. Where does the posterior distribution lie relative to the prior distribution and the likelihood?

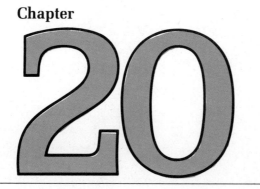

*Bayesian Decision Theory

(Requires Chapter 19)

In the face of uncertainty, the wise decision is the one we can live with serenely—no matter how unlucky it may turn out to be.
—ANONYMOUS

In the last chapter we saw how Bayesian theory could be used in estimation. In this chapter we will see how useful it can be in making decisions.

20-1 Maximizing Gain (or Minimizing Loss)

In Example 19-6, we considered a shipment of beams. On the basis of the sample and prior information, we found a 10% probability of its being unsatisfactory ($\mu < 62.5$). Should it be returned or not? To make this decision, we also need to know the costs of a wrong decision. For example, if it's a good shipment but we err by deciding it is a bad one—and therefore return it—then we unnecessarily incur costly delays. On the other hand, if it is a bad shipment, but we erroneously decide it is a good one—and keep it—then the cost may be huge: A bridge using this inferior steel may collapse as a consequence.

Thus in this chapter we seek to develop a formal way of making the best decision by taking such losses into account. We begin with the simplest possible example.

EXAMPLE **20-1**

John Nelson runs the refreshment concession at a football stadium, selling drinks and umbrellas. He is paid a flat fee of $100 a game, from which he must deduct his losses; these, in turn, depend on how badly he matches his merchandise with the weather. Suppose that he has just three possible options (actions, a):

$$a_1 = \text{sell only drinks}$$

$$a_2 = \text{sell some drinks, some umbrellas}$$

$$a_3 = \text{sell only umbrellas}$$

If he sells only drinks (a_1) and it rains, his loss is $70. If it shines, however, he just loses $10.

    Similarly, if he chooses action a_2 or a_3, there will be other losses, and we can show them all in the following loss table:

TABLE **20-1**

Loss Function $l(a,\theta)$

STATE θ	ACTION		
	a_1	a_2	a_3
θ_1 (Rain)	70	40	20
θ_2 (Shine)	10	40	60

Suppose further that the probability distribution (long-run relative frequency) of the weather is as follows:

TABLE **20-2**

Probability Distribution of θ

STATE θ	$p(\theta)$
θ_1 (Rain)	.40
θ_2 (Shine)	.60

If he wants to minimize long-run losses, what is the best action for him to take? (As always, try to work this out before reading on; it will make the discussion much easier.)

Solution

If he chooses a_1, what would his loss be, on the average? Let us call it $L(a_1)$. To calculate it, we simply weight each possible loss with its relative frequency, and obtain

$$L(a_1) = 70(.40) + 10(.60) = 34 \qquad (20\text{-}1)$$

like (4-4)

TABLE **20-3**

Calculation of the best action a, using the prior distribution $p(\theta)$

$p(\theta)$	STATE θ	ACTION		
		a_1	a_2	a_3
.40	θ_1 (rain)	70	40	20
.60	θ_2 (shine)	10	40	60
	Average loss $L(a)$	34	40	44

minimum

Similarly, for each of his other actions, we again calculate his average loss:

$$L(a_2) = 40(.40) + 40(.60) = 40 \qquad (20\text{-}2)$$

$$L(a_3) = 20(.40) + 60(.60) = 44 \qquad (20\text{-}3)$$

All these calculations are summarized in Table 20-3. We see that the average loss $L(a)$ is a minimum at a_1. Thus the best action is to sell drinks only.

Of course, this problem can be generalized to any number of states θ or actions a. For any action a that is taken, and any state θ that occurs, there is a corresponding loss[1] $l(a,\theta)$. Our decision is called a *Bayesian* decision if—as in the example above—we choose the action that minimizes average (expected) loss:

$$\boxed{\text{Choose } a \text{ to minimize } L(a) \equiv \sum_{\theta} l(a,\theta)p(\theta)} \qquad (20\text{-}4)$$

The probabilities $p(\theta)$, of course, should represent the best possible intelligence on the subject. For example, if the salesman cannot predict the weather, he will have to use the prior probabilities (.40 and .60) set out in Table 20-2; but if he can predict the weather by using a barometer, for example, then of course he should use the posterior probabilities that result from exploiting this sample information. An example will illustrate:

---

[1]The formulation of the problem in terms of losses is perfectly general, since gains may be represented simply as negative losses. We choose to use loss rather than gain, however, since loss is more natural than gain in the context of this chapter.

The losses we consider are losses of money. A more general theory can be developed based on losses of perceived value called *utility* (Wonnacott, 1984 *IM*).

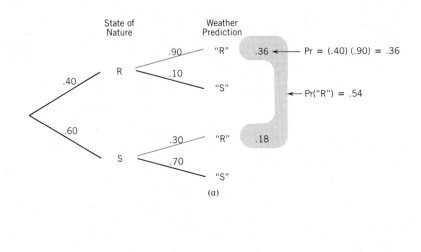

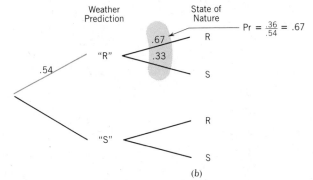

Figure 20-1

Calculation of the posterior probabilities by Bayes Theorem. (a) Original tree. (b) Reverse tree.

EXAMPLE **20-2**

Suppose John Nelson in Example 20-1 has kept records of the local TV weather report: On rainy days it correctly predicts "rain" 90% of the time; on shiny days, it correctly predicts "shine" 70% of the time. (Note that the TV predictions are indicated with quotation marks.)

If the report predicts "rain," which of the three possible actions (a_1, a_2, or a_3) should he choose now?

Solution

We first find the relevant posterior probabilities in Figure 20-1 (in exactly the same way that we found posterior probabilities in Figure 19-1). The posterior probabilities of rain and shine (given the "rain" prediction) are circled in Figure 20-1b and are copied down as the first column of Table 20-4. Using these best possible probabilities, the analysis then proceeds just as it did in Table 20-3.

TABLE **20-4**

Calculation of the Best Action a, Using the Posterior Distribution, $p(\theta/X_1)$

$p(\theta/X_1)$	θ	ACTION a_1	a_2	a_3
.67	θ_1	70	40	20
.33	θ_2	10	40	60
Average loss $L(a)$		50	40	33
				↑ minimum

We conclude that his best action, once he has the forecast of "rain" (i.e., sample information), is to sell umbrellas only (action a_3).

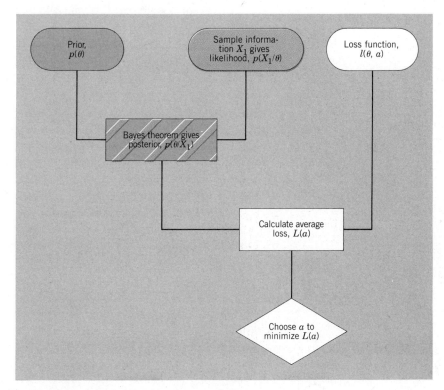

Figure 20-2
The logic of Bayesian decisions. The three given components—prior information, sample information, and the loss function—produce the best action a.

The logic of calculating the best decision in general—*Bayesian decision theory*—is summarized in Figure 20-2. It combines a new element (the loss function) with the two elements (the prior and the sample) already studied in Chapter 19. It produces the decision that has the least possible long-run (average) loss.

20-1   PROBLEMS

Suppose that John Nelson who is running the refreshment concession faces the following losses:

STATE θ	ACTION		
	a_1	a_2	a_3
θ_1 (Rain)	50	30	5
θ_2 (Shine)	15	30	35

a. If it rains 20% of the time and shines the remaining 80%, what is the best action?

b. Suppose that the local weather prediction has the following record of accuracy: of all rainy days, 70% are predicted correctly to be "rain"; of all shiny days, 90% are predicted correctly to be "shine." What is the best action if the prediction is "rain"? If it is "shine"?

c. Is this summary True or False? If False, correct it: If John must decide before the weather report, he should choose the compromise a_2. However, if he can wait for the weather report, then he should choose a_3 if the report is "rain," or a_1 if the report is "shine." But this solution is obvious, without going to all the trouble of learning about Bayesian decisions.

d. How much is the weather report worth on average? This is called the Expected Value of Sample Information, EVSI. (*Hint:* What is the expected loss when "rain" is predicted? When "shine" is predicted? Overall? Then compare to the loss without prediction.)

20-2 A farmer has to decide whether to sell his corn as feed for animals (action A) or for human consumption (action H). His losses depend on the corn's water content (determined by the mill after the farmer's decision has been made), according to the following loss table:

STATE θ	ACTION a	
	A	H
Dry	30	10
Wet	20	40

a. If his only additional information is that, through long past experience, his corn has been classified as dry 1/3 of the time, what should
his decision be?

b. Suppose that he has developed a rough-and-ready means of determining whether it is wet or dry—a method that is correct 3/4 of the
time, regardless of the state of nature. What should his decision be
if this method indicates his corn is "dry"? If it indicates "wet"?

c. As in Problem 20-1, find the EVSI (Expected Value of Sample
Information).

20-3 a. Suppose that the steel bars in a large shipment have the following
distribution of breaking strengths:

Breaking strength θ	200	210	220	230	240	250
Relative frequency	.10	.30	.20	.20	.10	.10

You have just purchased a bar at random, and want an estimate a of
its breaking strength θ. You consult three statisticians, and receive
three proposed estimates:

$$a_1 = \text{the mode of the distribution}$$

$$a_2 = \text{the median}$$

$$a_3 = \text{the mean}$$

Calculate each of these estimates.

b. Suppose that the loss due to estimation error is proportional to that
error:

$$l(a,\theta) = |a - \theta|$$

For the 6 possible values of θ and the 3 proposed values of a, tabulate
$l(a,\theta)$. Then decide which estimate a is best (gives the minimum
expected loss).

c. Repeat **b** if losses for large errors are especially heavy, now being
proportional to the *square* of the error:

$$l(a, \theta) = (a - \theta)^2$$

d. Repeat **b** if the losses for large errors are not a bit more than for small
errors. Thus we can take the loss to be a constant value (say, 1 unit)
no matter how large the error—and zero, of course, if there is no error
at all:

$$l(a,\theta) = 0 \text{ if } a = \theta$$
$$= 1 \text{ otherwise}$$

20-4   (Generalization of Problem 20-3)

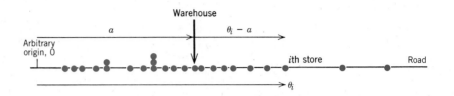

A warehouse is to be built to service many retail stores, all of the same size, and strung out at various positions θ_i along a single main road. In order to minimize total transportation cost, guess where the warehouse should be located (at the mean, median, mode, or midrange of the distribution of θ) in each of the following cases:

a. If the transportation cost is zero for the stores right at the warehouse, and if it is a constant value (irrespective of distance) for the other stores.

b. If the transportation cost for each store is strictly proportional to its distance from the warehouse. Thus, we wish to minimize:

$$\Sigma \, |\theta_i - a|$$

c. If there are not only transportation costs but also inventory costs that increase sharply with distance from the warehouse. Specifically, suppose that the cost is proportional to the square of the distance. Thus, we wish to minimize:

$$\Sigma \, (\theta_i - a)^2$$

d. If the only cost involved is the time delay in reaching the farthest store. (This might occur, for example, if all stores had to receive a display-room sample of the new model before a crucial advertising campaign could start.) Thus we wish to minimize the maximum distance.

20-5 Suppose that there were thousands of stores in Problem 20-4, and that the diagram showing their distribution now becomes:

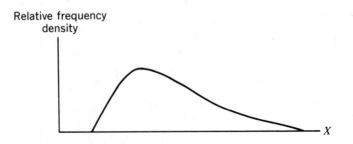

On this graph, roughly indicate the 4 solutions that you came up with for Problem 20-4, **a** through **d**.

*20-6 Prove your answers in Problem 20-4.

20-2 Point Estimation as a Decision

A—THE BEST ESTIMATE DEPENDS ON THE LOSS FUNCTION

Recall that in Chapter 19 we combined the prior and sample information to create the posterior distribution. Then we could use the posterior distribution to construct a 95% confidence interval for the parameter θ.

Now we will use the posterior distribution for another purpose—to make a decision on a single *point* estimate of θ. We will see that the loss function vitally affects what the point estimate should be. Although we learned in Chapter 7 that means and medians are often good point estimates for general scientific purposes, we will see that for *specific* decisions, we can often do better. An example will illustrate:

EXAMPLE **20-3** Professor Williams is to teach Economics 350 next quarter, and in ordering the texts, she faces a dilemma: if she orders too many, the excess must be returned—at a cost of $2 each for handling and postage; if she orders too few, there is an even bigger cost—each student who is missing a text must buy a temporary substitute costing $8 each.

Although Williams doesn't personally pay any of these costs, she would like to find the solution that would involve the least cost to everybody concerned. So she gathers whatever prior information she can: She finds that student enrollment in this course over the past 20 quarters has been quite stable, fluctuating between 40 and 80 with the following frequency distribution (grouped into cells of width 10).

ENROLLMENT θ	RELATIVE FREQUENCY $p(\theta)$
40	10%
50	20%
60	40%
70	20%
80	10%
Total	100%

If she has to order the texts without any further information, then she will have to make a decision based on the above frequency distribution; it's the only information she has. Assuming the bookstore has a regulation that books must be ordered in multiples of 10, how many texts should she order?

TABLE **20-5**

		ACTION a (ESTIMATED ENROLLMENT—I.E., TEXTS ORDERED)				
$p(\theta)$	STATE θ (ACTUAL ENROLLMENT)	$a_1(40)$	$a_2(50)$	$a_3(60)$	$a_4(70)$	$a_5(80)$
.10	40	0	$20	$40	$60	$80
.20	50	$80	0	$20	$40	$60
.40	60	$160	$80	0	$20	$40
.20	70	$240	$160	$80	0	$20
.10	80	$320	$240	$160	$80	0
	Average loss $L(a)$	$160	$90	$40	$30	$40

Loss function $l(a,\theta)$ yields average loss $L(a)$

$\uparrow$ minimum

Solution

This problem is very similar to Example 20-1, with the possible actions $a_1, a_2 \ldots$ now being the number of texts to be ordered ($a_1 = 40$, $a_2 = 50$, and so on). For each of these actions, we first tabulate the loss function in Table 20-5.

This loss function reflects the costs of misestimation: When the estimated true enrollments are equal ($a = \theta$), note that there is a 0 loss. From there, the costs of misestimation in each column rise: $80 for each 10 books if actual enrollment turns out to be higher than estimated, and $20 for each 10 books, if it turns out to be lower.

Now we proceed exactly as in Example 20-1. If, for example, the estimate a_3 were chosen, we can calculate the expected loss:

$$L(a_3) = \$40(.10) + \$20(.20) + \cdots + \$160(.10) = \$40 \quad (20\text{-}5)$$

which is then recorded in the bottom margin. We similarly calculate the loss for each of the other actions. We then see that the minimum loss occurs at:

$$a_4: \text{ order 70 texts} \quad (20\text{-}6)$$

Example 20-3 illustrated an important point. We did not select the mean (60) of the $p(\theta)$ distribution (even though it was the median and mode too). The reason is that relatively heavy losses occurred on one side. (The loss function was asymmetrical, which we confirm with a glance at the numbers in, say, column a_3.) This pulled our best estimate off center.

> **The best (Bayesian) point estimate of θ depends upon the loss function as well as the distribution of θ.** (20-7)

With the pressure created by an asymmetrical loss function now clear, we restrict ourselves in the rest of this section to symmetric loss functions.

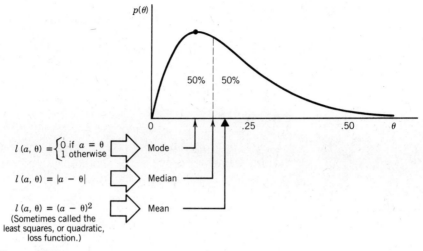

Figure 20-3
Different symmetric loss functions l (a,θ) lead to different central values of θ.

B—SYMMETRIC LOSS FUNCTIONS

With a symmetric loss function, the best estimate often does turn out to be one of the familiar centers of the θ distribution—the mean, median, or mode. Figure 20-3 gives three symmetric loss functions, and shows that each leads the decision-maker seeking the best point estimate to a different center of the distribution. It is worthwhile to explain the reason why (leaving the proofs for Problems 20-3 to 20-6).

The 0-1 criterion means that "a miss is as good as a mile." Hitting the target exactly is all that counts. To maximize this chance, we go to the point where the distribution has the highest probability—the mode.

The least squares criterion listed last in Figure 20-3 is exactly the same criterion already used in least squares estimation. And you may recall from Section 11-3 that the least squares criterion always produces the mean.

The absolute value criterion is between the other two. It is therefore no surprise that the best estimate it produces is the median—the central measure that lies between the mode and the mean.

EXAMPLE **20-4**   Suppose we are asked to estimate the proportion π of defective radios in a shipment, and we incur a cost of $50 for every 1% we err in our estimate. For example, if we guessed 8%, and π turned out to be 13%, the error would cost us 5 × $50 = $250.

Suppose Figure 20-3 shows the posterior distribution of π (which, we suppose, has been derived from prior and sample information in the usual way.) What would be your best estimate of π, roughly?

Solution   Since the loss function is essentially the absolute-value function, the estimate that minimizes cost is the *median* of the posterior distribution in Figure 20-3. This appears to be about $\pi = .15$.

PROBLEMS

20-7   A newsstand operator was in a dilemma about how many copies of the *Sunday Times* to order, since demand fluctuated from week to week. In fact, to help make an informed decision, over the past 100 weeks he kept records of sales:

SALES	RELATIVE FREQUENCY
20	10%
25	30%
30	20%
35	10%
40	10%
45	10%
50	5%
55	5%

Since the profit on each sale was $.40, he was tempted to order 55 papers every Sunday to be sure he never ran out. Every copy left unsold cost him an equal amount, however—$.40 in handling costs.

To maximize his net profit, how many should he order? (Assume it has to be a multiple of 5.)

20-8   In Problem 20-7, suppose the *Times* increased the profit to $1.20 (by lowering the wholesale price by $.80, so that the retail price and sales potential didn't change). And suppose handling cost remains at $.40.

a.   Now what is the best number to order?

b.   How much would this increase the average weekly retail sales?

20-9   Suppose that the steel beams manufactured by a new process have breaking strengths that are distributed normally about a mean of 220 with a standard deviation of 15. A beam is sampled at random to estimate its particular breaking strength θ, and it is measured with strain gauges. Five such independent measurements turned out to be 230, 245, 235, 255, 235. Past experience shows that such measurements have a standard deviation of 10, and are unbiased and normally distributed. What would you estimate is the breaking strength θ of this beam:

a.   If the cost of estimation error is proportional to the error. For example, an estimate that is 4 times as far off is 4 times as costly.

b. If far-off estimates are more serious. For example, an estimate that is 4 times as far off is $4^2 = 16$ times as costly.

c. If far-off estimates are no more serious than a slightly-off estimate, so that we can say "a miss is as good as a mile."

20-3 Hypothesis Testing as a Decision

In the last section, we saw that in treating estimation as a form of decision, we were able to derive optimal estimators for many common situations. In this section, we likewise shall treat hypothesis testing as a form of decision, and so derive optimal tests that are free of the arbitrary features of classical hypothesis testing introduced in Section 9-3.

A—FINDING THE BEST WORKING HYPOTHESIS

Since hypothesis testing is just a form of decision, with its own special notation and calculations, we begin with an example whose form is again very similar to our very first decision in Example 20-1.

EXAMPLE **20-5**

Suppose there are two species of beetle that might infest a forest. The relatively harmless species, S_0, would do only \$10,000 damage; the more harmful species, S_1, would do \$100,000 damage. A perfectly harmless and effective insecticide spray is available for \$30,000.

Now suppose that an infestation of beetles is sighted, but that unfortunately there is no information about which species it may be. Should the forest ranger decide to spray or not, if the only further information is that past records show:

a. Species S_0 is three times as frequent as S_1?

b. Species S_0 is nine times as frequent as S_1?

Solution

a. It is convenient to summarize the given information in Table 20-6, where the losses (in \$000) are set out in the main body of the table. But first, on the left we list the probabilities for the two species (.75 and .25, in the given 3 to 1 ratio). These probabilities as usual are used as weights to calculate the average loss for each of the two actions. The smaller loss occurs for action a_1, (spray), so this is the best action.

TABLE **20-6**   **Calculation of the best action a if the only available information is the prior distribution $p(\theta)$**

| | | ACTION | |
| | | a_0 (DON'T SPRAY) | a_1 (SPRAY) |
$p(\theta)$	STATE θ		
.75	S_0 (Harmless)	10	30
.25	S_1 (Harmful)	100	30
	Average loss $L(a)$	32.5	30
			↑ minimum

b. For the prior probabilities of .9 and .1, a similar calculation in the bottom row would give average losses of \$19,000 for a_0 and \$30,000 for a_1. Thus the best action in this case is a_0 (don't spray). It is now so unlikely that the beetle is harmful that it is worth taking the risk.

The decision in this example would have been trivial if it were possible to identify the species for sure: If it were the relatively harmful species S_1, we would merely scan the S_1 row, and come up with the smallest loss of 30 for the action a_1 (spray). Similarly, if it were species S_0, the best action would be a_0 (don't spray).

It is interesting to note that the best action is still a_0 when S_0, although uncertain, is sufficiently *likely*—for example, 9 times as likely as S_1, as in question **b**. That is, if S_0 is sufficiently likely, we act *just as if* it were certain. We accept S_0 as a *working hypothesis*; even though we know it may not be true, it is better than the alternative hypothesis S_1. This exemplifies what hypothesis testing is all about— a search for the best working hypothesis, rather than absolute truth.

B—GENERALIZATION

For a perfectly general solution, we need to make only a few changes to Example 20-5, mostly notational:

1. We will suppose that some sample data X has also been observed so that the posterior probabilities $p(\theta/X)$ are available instead of just $p(\theta)$.

2. Instead of S_0 and S_1, we will call the two possible states θ_0 and θ_1, or H_0 and H_1 (null and alternative hypotheses). Then the possible actions are "accept H_0" and "accept H_1."

3. Since the loss function $l(a,\theta)$ takes on so few values, we can express it briefly as l_{00}, l_{01}, and so on in Table 20-7.

In terms of the losses, we introduce a natural concept, *regret*. The regret (r_0) if H_0 is true is defined as the *extra* cost incurred if we make the wrong decision (that is, if we accept H_1 when H_0 is true). This is read off the first line of Table 20-7:

$$r_0 \equiv l_{10} - l_{00} \tag{20-9}$$

Similarly, the regret if H_1 is true (r_1) is the extra cost if we erroneously accept H_0. This is read off the second line of Table 20-7:

$$r_1 \equiv l_{01} - l_{11} \tag{20-10}$$

Note that Table 20-7 has the same headings as Table 9-1 and so we are solving the same hypothesis testing problem. But our procedure now is rational rather than arbitrary: We calculate the average loss, and minimize it. The final result (proved in Appendix 20-3) is very easy to state:

> For optimal decisions, accept H_0 if, and only if,
>
> $$\frac{p(X/\theta_1)}{p(X/\theta_0)} < \frac{r_0 p(\theta_0)}{r_1 p(\theta_1)}$$

$$\tag{20-11}$$

This criterion certainly is reasonable. If θ_0 is a sufficiently plausible explanation of the data—that is, if $p(X/\theta_0)$ is sufficiently greater than $p(X/\theta_1)$—then the ratio on the left-hand side of (20-11) will be small enough to satisfy this inequality. Thus H_0 will be accepted, as it should be.

To illustrate further, consider the very simple case in which the regrets r_i are equal, and the prior probabilities $p(\theta_i)$ also are equal. Then the right-hand side of (20-11) becomes 1. Consequently, H_0 is accepted if $p(X/\theta_0)$ is greater than $p(X/\theta_1)$. Otherwise, the alternative H_1 is accepted. In simplest terms: In this special case, we accept whichever hypothesis is more likely to generate the observed X, as shown in panel (a) of Figure 20-4.

TABLE **20-7** **The Symbols for Bayesian Hypothesis Testing, when posterior $p(\theta/X)$ is available**

		ACTION	
$p(\theta/X)$	STATE θ	a_0 (ACCEPT H_0)	a_1 (ACCEPT H_1)
$p(\theta_0/X)$	H_0	l_{00}	l_{10}
$p(\theta_1/X)$	H_1	l_{01}	l_{11}

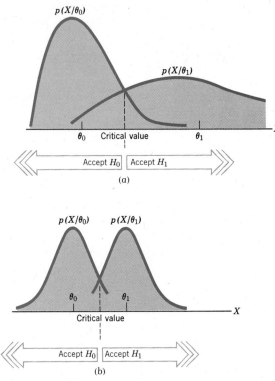

Figure 20-4
Hypothesis testing, using the Bayesian criterion (20-11) in the special case when $r_0 = r_1$ and $p(\theta_0) = p(\theta_1)$. (a) For any $p(X/\theta_i)$. (b) For $p(X/\theta_1)$ having the same symmetric shape as $p(X/\theta_0)$. Compare to Figure 9-6.

In panel (b), we make the further assumption that $p(X/\theta_0)$ and $p(X/\theta_1)$ have the same symmetric and unimodal shape. Then (20-11) reduces to the very reasonable criterion:

> Accept H_0 whenever X is observed closer to θ_0 than to θ_1 (20-12)

C—USING A SAMPLE MEAN $\overline{X}$ TO TEST THE MEAN μ OF A NORMAL POPULATION

As an example of how useful the Bayesian criterion (20-11) is, let us see what it yields in a very familiar case: Suppose a normal population has an unknown mean, and so we are testing H_0: $\mu = \mu_0$ against H_1: $\mu = \mu_1$ (to simplify the algebra, we call the larger mean μ_1, so that $\mu_1 > \mu_0$).

Suppose also that, from this unknown population, we have some data— a whole *sample* of n observations ($X_1, X_2, \ldots, X_n$). Then the Bayesian criterion for accepting H_0 reduces to a very reasonable form (proved in Appendix 20-3):

Accept H_0 if, and only if,

$$\overline{X} < \frac{\mu_1 + \mu_0}{2} + \frac{\sigma^2/n}{\mu_1 - \mu_0} \log \frac{r_0\, p(\mu_0)}{r_1\, p(\mu_1)} \qquad (20\text{-}13)$$

Here log is the natural log (base e). If a single observation is taken, then this criterion is still appropriate, of course, with $n = 1$.

Criterion (20-13) tells us two important things:

1. When the population is normal, the sample mean $\overline{X}$—not the sample median, or something else again—is the appropriate sample statistic to use for μ. (A point first mentioned following equation (7-6), and again in Section 18-2 when we proved $\overline{X}$ was the MLE).
2. In the simplest case where the regrets r_i are equal, and the prior probabilities $p(\mu_i)$ also are equal, then the log in (20-13) is zero, and we have the simple result: Accept H_0 whenever

$$\overline{X} < \frac{\mu_1 + \mu_0}{2}$$

Since $(\mu_1 + \mu_0)/2$ is the half-way point between μ_0 and μ_1, this is just the same sensible criterion for accepting H_0 that we already found in (20-12): Accept H_0 whenever $\overline{X}$ is observed closer to μ_0 than to μ_1.

In more complicated cases, (20-13) tells us to move the critical point up or down, depending on the regrets and prior probabilities. An example will illustrate:

EXAMPLE **20-6**

To continue Example 20-5, part **a**, suppose that in each species, beetle length X is distributed normally with a standard deviation $\sigma = 8$ mm. But the two species can be distinguished somewhat by their different mean lengths: $\mu_0 = 24$ mm, while $\mu_1 = 30$ mm.

If a sample of $n = 4$ beetles is taken, how small must $\overline{X}$ be in order to accept H_0? In particular, if $\overline{X}$ turns out to be 26 mm, should we choose H_0 or H_1?

Solution

We first calculate the regrets from Table 20-6:

$$r_0 = 30 - 10 = 20 \qquad \text{like (20-9)}$$

$$r_1 = 100 - 30 = 70$$

In Table 20-6, we also find the prior probabilities, $p(\mu_0) = .75$ and $p(\mu_1) = .25$. When all these values are substituted into (20-13), we obtain:

$$\overline{X} < \frac{30 + 24}{2} + \frac{8^2/4}{30 - 24} \log \left[\frac{20}{70} \frac{.75}{.25} \right]$$

$$\overline{X} < 27 + 2.67 \log [.86]$$

$$\overline{X} < 26.6 \tag{20-14}$$

Thus, the critical point below which we accept H_0 is 26.6. In particular, for $\overline{X} = 26$, we should accept H_0—that is, not spray.

Recall that following Example 20-5, we said "If H_0 is sufficiently likely, we act just as if it were certain." This is a nice way to interpret our answer in this example, too: Because $\overline{X} = 26$ is close enough to $\mu_0 = 24$, we can now view H_0 as likely enough for us to accept it as the working hypothesis.

PROBLEMS

20-10   As in Example 20-5, suppose that there was a threat from either a relatively harmless (S_0) or a harmful (S_1) species of beetle. Their relative probabilities and the losses associated with various actions are as follows:

		ACTION	
PROB	SPECIES	DON'T SPRAY (ACCEPT H_0)	SPRAY (ACCEPT H_1)
.4	S_0	1	3
.6	S_1	10	2

Should H_0 be accepted or not?

20-11   Continuing Problem 20-10, suppose that one beetle in the threatening swarm has been captured, and its length X measured. Also suppose that a biologist provided the following distribution of lengths for each species:

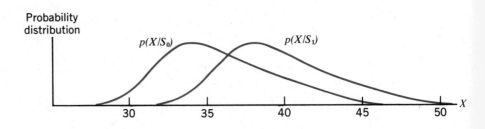

Now should H_0 be accepted or not, if X turns out to be:

a. 30   **c.** 33   **e.** 38

b. 50   **d.** 35   **f.** 40

20-12 Continuing Problem 20-10 in a different way, suppose that the distribution of beetle length $p(X/S_i)$ is normal with $\sigma = 2$, and with $\mu = 35$ for species S_0, or $\mu = 40$ for species S_1.

 a. Roughly graph these distributions.

 b. From the graph and (20-11), determine roughly the critical value for X (below which we should accept H_0).

 c. Using (20-13) with $n = 1$, determine exactly the critical point in **b**.

 d. If a sample of 5 beetles is available rather than just one, what is the critical value for $\overline{X}$ (below which we should accept H_0)?

20-13 Repeat Problems 20-10 to 20-12, assuming that the table of Problem 20-10 is replaced with:

i.

		ACTION	
		ACCEPT	ACCEPT
PROB	SPECIES	H_0	H_1
.25	S_0	5	10
.75	S_1	20	10

ii.

		ACTION	
		ACCEPT	ACCEPT
PROB	SPECIES	H_0	H_1
.75	S_0	0	200
.25	S_1	700	100

*20-14 Suppose a neurologist has to classify his most serious patients as requiring exploratory brain surgery or not. He makes the decision on the basis of a combined score X from several tests. Past experience has shown that X is normally distributed with $\sigma = 8$, and mean $\theta_0 = 100$ for those who do not require surgery, and $\theta_1 = 120$ for those who do.

The losses (regrets) of a wrong classification are obvious: An unnecessary operation means resources are wasted and the patient himself may be hurt. Yet the other loss may be worse: If a patient requiring surgery does not get it, the time lost until clear symptoms appear may be crucial. Suppose that this second loss is considered roughly five times as serious as the first. From past autopsies, it has been found that of the people taking the test, 60% actually needed the operation, while 40% did not.

 a. What should be the critical score above which the person is classified as requiring the operation?

 b. Then what is α? (Probability of type I error.)

 c. What is β? (Probability of type II error.)

20-15 Continuing Problem 20-14, suppose instead of the optimal test, a classical test is used, arbitrarily setting $\alpha = 5\%$.

 a. What is the critical score?

 b. Then what is β?

c. By how much has the average loss increased by using this less-than-optimal method?

20-4 Classical and Bayesian Statistics Compared

A—GENERAL

Now that we have finished the technical details, it is important to stand back for perspective, and see how the Bayesian methods of Chapters 19 and 20 fit in with the rest of the book. Let us begin with a brief review.

If we want to summarize sample data, classical statistics are appropriate (as in Chapters 1 to 18). For example, $\overline{X}$ and b, along with the confidence intervals constructed around them, provide an appropriate summary of the sample evidence.

If prior information is available, it can be incorporated with the data to give the improved posterior distribution and confidence interval (as in Chapter 19).

Finally, if a *decision* has to be reached, then the loss function is also essential (Chapter 20). We cannot possibly make a decision that minimizes losses (or equivalently, maximizes benefits) if we don't take into account what the losses are.

Of the three components—the data, prior distribution, and loss function—the data is usually the most solid and objective; it properly deserves the first 18 chapters of this text. The remaining two components are often subjective. Since decisions are made by humans, however, subjectivity will always be with us; and Bayesian analysis is a good way of dealing with it explicitly, rather than trying to sweep it under the rug.

B—SUBJECTIVE PRIORS AND LOSSES

When prior information is objective—as in Examples 20-1 or 20-3—there is of course no problem. Often an objective prior is not available, however, and a subjective prior must be used instead. Subjective priors are often not as difficult to reasonably define as one might expect. (Recall Section 3-7 where we showed how a subjective prior might be calibrated.) Nonetheless, specification of subjective priors remains one of the most controversial aspects of Bayesian analysis—because people have a very human tendency to exaggerate how good their guesses are (they tend to give subjective prior distributions too little variance). Accordingly, it is important in any Bayesian analysis to give the reader not only the final posterior estimate, but also the two components—the prior and the sample data. If the reader has a different prior, he will then be able to substitute it and come up with his own posterior. In fact, some authors give a range of several reasonable priors themselves in order to show the range of reasonable posteriors.

If there is no suitable prior, neither objective nor subjective, we can always use *informationless* or *neutral* priors that lead to shrinkage estimates. As we saw in Section 19-5, they are still better than falling back on classical statistics.

Losses are often subjective, too. How, for example, can we compare the loss of a limb with the loss of a million dollars? These questions are terribly difficult, but they *must* be answered by someone, somehow, in decisions such as how much to compensate workers for industrial accidents. Fortunately, statistical decision theory has developed a tool called *utility theory* for comparing things that at first glance seem incomparable (Raiffa, 1968, or Wonnacott, 1984 IM). And surprisingly, economists are now developing methods of at least roughly approximating the most difficult loss of all to pin down: the loss of a human life (Bailey, 1980). Such developments greatly extend the applicability of this chapter on decisions.

Having dealt with the controversies of Bayesian decision theory in general, let us next see what they mean for the specific issues addressed in Chapters 19 and 20.

C—ESTIMATION

We have often emphasized the inadequacy of using a single point estimate to describe a population parameter. As early as Chapter 8, for example, we saw that an interval estimate was far more informative. Now we have seen that a posterior (Bayesian) interval estimate is even better because it incorporates prior information as well as the data. Of course, the most informative and detailed way of all to describe a population parameter is to show the whole posterior distribution. As computers take over the burden of computing and graphing, this is becoming a more and more viable alternative.

We must remember that as the sample becomes large enough (and hence swamps the prior) such posterior information reduces to the classical equivalent. For example, the posterior confidence interval reduces to the classical confidence interval; and the whole posterior distribution reduces to the classical likelihood function. (This is yet another reason to study classical statistics for 18 chapters.)

D—ESTIMATION AS A DECISION

While a population parameter is generally best described by a posterior confidence interval (or even better, by the whole posterior distribution) for specific decisions one often requires a specific point estimate—determined by the loss function. Ideally, this loss function should be applied to the posterior distribution. But in the face of more limited information, it may be applied to the prior (if no data exists), or to the classical likelihood function (if no prior exists, or the sample data is so plentiful that it swamps the prior). In short, the loss function should be applied to the best available distribution that describes the population parameter.

This allows us to put the classical technique of maximum likelihood estimation (MLE) into perspective. By using the likelihood function alone, MLE ignores any prior information; and by selecting the point where the likelihood function peaks (the mode), MLE implicitly uses the arbitrary 0-1 loss function. Accordingly, MLE should be viewed as a useful summary of sample data, rather than a guide for decision-making.

E—HYPOTHESIS TESTING AS A DECISION

The trouble with classical hypothesis testing, as first noted in Section 9-3, is that it arbitrarily sets the level α, usually at 5%. As Problem 20-15 showed, when we compare this sort of classical test to the Bayesian test that minimizes average loss, the cost of this arbitrariness can be very high.

PROBLEMS

20-16   Answer True or False; if False, correct it:

a.   Bayesian estimation combines sample data with a prior distribution and a loss function, whereas classical estimation uses only the data.

b.   A Bayesian test uses the prior and loss functions to determine the critical point. The classical test instead uses an arbitrary level of α, usually $\alpha = 5\%$. The cost of this arbitrariness, however, is small.

c.   When an objective prior distribution and loss function are not available, subjective ones can be used instead, based on the individual decision-maker's best judgment. This means that different decision-makers would reach different decisions from the same data—an intolerable state of affairs.

   By contrast, classical testing at the 5% level is entirely data based, so that all decision-makers who used the same data would arrive at the same decision.

d.   If the prior distribution is flat, then the posterior distribution coincides with the likelihood function.

   If, in addition, the loss function is quadratic, then the Bayes estimate coincides with the MLE.

20-17   Assume that the loss function is quadratic, that the prior distribution of μ is normal, and that the observations X_i are drawn from a normal population with mean μ and variance 150. Then, in each case, calculate the best Bayesian estimate of μ. Also calculate the MLE estimate for comparison.

	PRIOR PARAMETERS		SAMPLE DATA		ESTIMATE OF μ	
	μ_0	σ_0^2	$\overline{X}$	n	BAYESIAN	MLE
a.	100	25	200	6		
	100	250	200	6		
	100	2500	200	6		
b.	100	25	200	6		
	100	25	200	60		
	100	25	200	600		

20-18 Based on your answers in Problem 20-17, answer True or False; if False, correct it:

As the prior information becomes more and more vague (as σ_0^2 decreases), the Bayesian estimate of μ gets closer and closer to the MLE estimate; this also occurs as the sample size gets bigger and bigger. In both cases, it is because the likelihood function overwhelms the prior distribution.

20-19 Suppose a revolutionary phenomenon has been proposed by the famous Professor Brainstorm, and yet if the truth were known, it really doesn't exist. In formal language, the null hypothesis is actually true.

a. If 100 researchers investigated this phenomenon, about how many would obtain "statistically significant" results so that they could claim the Brainstorm effect really existed (at the 5% level)?

b. Now suppose the *Journal of Statistically Significant Results* (JSSR) used "statistical significance at the 5% level" as its sole criterion for publishing. (An exaggeration, of course, but we will use it to illustrate some real problems.) If all 100 researchers submitted their papers to this journal, which ones would get published? (Would the researchers who didn't get statistically significant results even submit their papers?) For anyone who looks at the published evidence, what would he conclude about the Brainstorm effect?

CHAPTER 20 SUMMARY

20-1 The best (Bayesian) decision is the one that minimizes average loss. It incorporates all three components of a decision: (1) the current data (as in classical statistics, Chapters 1 to 18), (2) the prior information, combined with the data to give the posterior distribution (as in Bayesian estimation, Chapter 19), and finally (3), the loss function (the new component).

20-2 For a specific decision, often a specific point estimate is needed (as opposed to the interval estimate, which is used for more general purposes of information). Various symmetric loss functions lead to various centers of the posterior distribution as the optimal point estimate—the mean, median, or mode. An asymmetric loss function, which penalizes an error in one direction more heavily than in the other, will naturally move the optimal point estimate off center.

20-3 To determine which hypothesis is least costly to take as our working hypothesis, the Bayesian criterion states: If $p(X/\theta_0)$ is sufficiently higher than $p(X/\theta_1)$ ("sufficiently" being defined in terms of priors and losses), then we accept H_0.

20-4 When objective evidence is unavailable, the decision-maker's own subjective knowledge and values can be used for the prior distribution and loss function. This makes the personal nature of decision-making explicitly clear—and open to analysis and improvement.

As sample size grows and the data swamps the prior, Bayesian estimation approaches classical estimation. This is why 18 chapters were devoted to classical estimation, and why Bayesian statistics are called primarily a *small-sample technique*.

REVIEW PROBLEMS

20-20 When 1000 aircraft radar were tested in flight, the test scores were classified according to whether careful examination of the radar later proved it to be okay or defective (suggested by Malcolm, 1983). This produced the following two histograms—approximately normal, with $\sigma = 15$, and means of 300 and 250—for okay and defective radars, respectively:

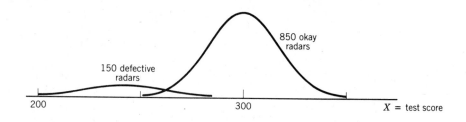

If the cost of a false alarm is one-half the cost of a missed fault, find the best cutoff test score, below which a radar set should be pronounced "faulty." [*Hint:* When μ_0 is above μ_1, then the inequality in (20-13) has to be reversed, of course.]

20-21 A TV manufacturer has a choice of 3 kinds of components to fill a crucial position in his TV sets; ordinary, at $2.00 each; reliable, at $2.55 each; or very reliable, at $3.00 each. The only difference the better kind makes is that it is less likely to break if the TV set is badly jarred in shipping; in fact, past shipping records show that the 3 kinds of components break 4%, 2%, and 1% of the time, respectively. If such a break costs $30 in labor and parts to repair, what kind of component should be used?

20-22 An economist constructed a 95% confidence interval for a marginal propensity to save, based on the best available data and prior (all normally distributed):

$$.02 < \beta < .10$$

This estimate proved too much for his client to digest. She insisted instead on a single number. What is the best point estimate of β if, on careful probing, she felt it would be equally serious to overestimate as to underestimate?

20-23 To screen applicants for a new position of loan officer, the manager of a bank required them to pass a lie detector (polygraph) test. The polygraph operator explained that the data is summarized into a normally distributed score X. For subjects who are lying, X has a mean of 70 and standard deviation of 10. For subjects who are telling the truth, X has a mean of 40 and a standard deviation of 10.

The manager estimated from past experience that 90% of the applicants give honest responses, while 10% lie. At the same time, she judged that the error of failing an honest candidate was about as serious as the error of passing a dishonest candidate. In order to minimize overall losses, what should be the passing value of X?

20-24 Repeat Problem 20-23, with one slight change: For subjects who are lying, X now has mean 45 and standard deviation 5. (Note that *both* these moments are different from those recorded by subjects who are telling the truth. A rough graphical solution will be easiest.)

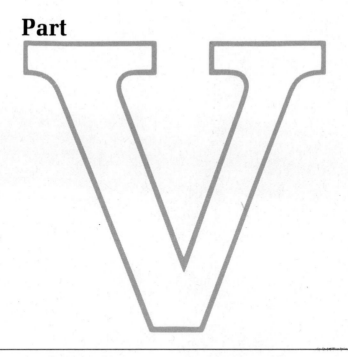

Part

V

SPECIAL TOPICS
FOR BUSINESS AND
ECONOMICS

Decision Trees

(Requires[1] Chapter 3)

The approach employs systematic analysis, with some number-pushing, which helps the decision-maker clarify in his own mind which course of action he should choose . . . It is . . . designed for normally intelligent people who want to think hard and systematically about some important real problems.

RAIFFA AND KEENEY, 1977

Decision trees, like probability trees in Chapter 3, are simply a graphic way to organize the calculations that will lead us to the best possible decisions in the face of uncertainty. We will see, for example, how a firm can weigh its risks in order to increase its profits, or how an investment portfolio might be selected to satisfy the twin goals of growth and security. As well as better business decisions that involve money, we will also see how to make better personal decisions that involve more subjective benefits.

21-1 The Basic Tree

A decision tree is like a probability tree in Chapter 3, with one important additional component: Decisions have to be made at various branchings of the tree.

---

[1]Although Chapter 21 covers many of the same issues as Chapters 19 and 20, it is more intuitive and can be read independently.

A—EXAMPLE

Suppose the owners of BEA, a very large oil exploration company, hold a lease that must be either (1) sold now, (2) held for a year and then sold, or (3) exercised now. If they decide to exercise it, the cost of drilling will be 200K ($200,000) and will lead to one of the consequences (well types) listed in Table 21-1. (For example, reading along the first row, there is a 50% chance that the well will turn out to be dry, with a zero payoff.)

If they decide instead to sell their lease now, they can get 125K for it. Finally, if they decide to hold it for a year in a gamble that oil prices will rise, they face a risk: If this gamble fails—and they estimate that the chance of this is 90%—then they will take a slight loss; they will have to sell their lease for 110K. On the other hand, if this gamble succeeds and oil prices rise, they will be able to sell their lease for 440K.

Jack Moor, their vice president, has read somewhere that statistics will help make "wise decisions in the face of uncertainty," so he consults us to help determine the best strategy. What should we advise?

B—SETTING UP THE TREE

To outline the dilemma facing the BEA managers, we set up a tree, starting with their first decision: Should they sell now, sell in a year, or drill? This *decision fork* is shown as the square on the left in Figure 21-1, with the three lines coming from it representing the three possible alternatives that can be chosen.

Now we must work out the consequences of each of these three alternatives. The consequence of the first—the upper branch of the tree—is immediately seen to be 125K and we show this value with the "price tag" marked 125K.

The second branch of the tree is more complicated. If they sell their lease a year from now, the price will be determined by chance (or at least, by something completely outside their control—the price of oil). This *chance fork* or *gamble* is shown as a circle to distinguish it from the *decision fork* on the left, which is shown as a square:

Decision forks are shown as squares, chance forks as circles.

(21-1)

TABLE **21-1**

Possible Outcomes of Drilling[a]

WELL TYPE	PROBABILITY	PAYOFF
Dry	.50	0
Wet	.40	400K
Gusher	.10	1500K

[a]It is assumed that drilling involves practically no time delay. And in the last column, the symbol K means $1000.

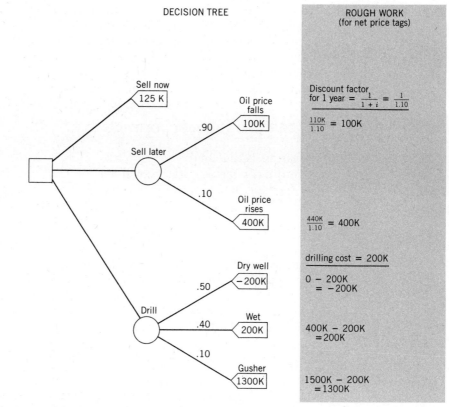

Figure 21-1
Setting up the decision tree.

The two possible outcomes branching from this "sell later" chance fork are marked with their probabilities, exactly as in the trees of Chapter 3.

How do we calculate the price tags for these two possible outcomes? Consider the first outcome, if the oil price falls. The sale would yield 110K next year. But at the prevailing interest rate $i = 10\%$, this 110K next year is worth only 100K this year. (The reason these two amounts have the same value is that you can take 100K this year, and by investing at the prevailing 10% interest rate, transform it into 110K next year.) Thus this discounted *present value* of 100K is entered in the price tag. [We must do this discounting in order to make a fair comparison with the earlier price tag of 125K (from selling the lease now). In other words, we must get all price tags into comparable terms (present values).]

Of course, we must similarly discount the value that we place on the next price tag, since the 440K we get in this case is also delayed a year— and hence is only worth 400K now. These and other calculations are recorded briefly in the right-hand margin of Figure 21-1.

Turning to the lowest branch of the tree, we encounter a circle that represents an even more complicated chance fork, representing the uncertainties of drilling. Again, the price tags are calculated in the margin: In each of these cases, a drilling cost of 200K has to be subtracted, in order to get the net return from each of these outcomes.

This completes our tree. The decision forks (squares) and chance forks (circles) have been laid out in sequence in Figure 21-1 so we can clearly follow them through to their consequences.

C—DERIVING THE BEST SOLUTION

So far, in setting up the decision tree, we have worked from left to right—recording on the way all the available information (appropriately adjusted in the right-hand column). Turning now to *solving* the decision tree in Figure 21-2, we will work *backwards* from right to left, with each step marked in color. These steps will be of two types: *averaging* each chance fork, and *pruning back* each decision fork:

  i. *Averaging each chance fork.* With the first alternative (sell now) having a price tag of 125K, for purposes of comparison how do we find a single price tag for the next alternative (sell later)? We take the *average* or *expected value* (EV), by weighting each outcome by its probability:

$$\text{EV of selling later} = .90(100K) + .10(400K) \qquad (21\text{-}2)$$
$$= 130K$$

This 130K is recorded in the "sell later" circle in blue.

Similarly, the expected value for the final alternative (drilling) is calculated to be

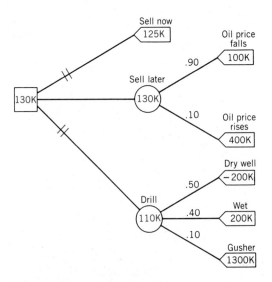

Figure 21-2
Solving the decision tree. We work backwards, from right to left, averaging the chance forks (circles) and pruning the decision forks (squares). This simplifies the tree until it displays the best choice.

$$\text{EV of drilling} = .50\,(-200K) + .40(200K) + .10(1300K) \quad (21\text{-}3)$$
$$= 110K$$

This 110K is similarly recorded in the "drill" circle in blue.[2]

It is worth briefly reviewing why this calculation of expected value can be applied in this case. Since BEA is a very large exploration company, we can imagine it gambling on a hundred such wells. Then:

About 50 would be dry, @ $-200K$ each   = $50(-200K)$
About 40 would be wet, @ $200K$ each   = $40(200K)$
About 10 would be gushers, @ $1300K$ each = $10(1300K)$

The average gain (expected value) per well is thus

$$\text{EV} = \frac{\text{total gain}}{100} = \frac{50(-200) + 40(200K) + 10(1300K)}{100}$$
$$= 110K$$

This is what justifies the calculation of expected value in (21-3).

ii. *Pruning back the decision fork.* The last step is to compare the three expected values—125K, 130K, and 110K. We simply slash off the inferior ones, which leaves the best policy (sell later); then we record its expected value (130K) in the decision square. This completes the decision, and clearly displays how it was derived.

D—DECISION-MAKING IN GENERAL

In Figure 21-3, we set out a more complicated tree. (Ignore the color for now.) Let us continue to suppose the decision maker can easily afford to take risks (i.e., is *risk neutral* because of sufficient capital to finance gambles over and over, so that averaging out is relevant. For decision makers who are risk averse, see Section 21-3.) Then we can determine the best

---

[2]You may have noticed that there is an easier way of calculating this expected value of 110K. The procedure we used was to:

1. Subtract out the 200K drilling costs (in the right hand column) in order to get the *net* price tags ($-200K$, $200K$, $1300K$).
2. Average these.

A simpler computation is this:

1. Initially ignoring the 200K drilling costs, average the *gross* price tags (0, 400K, 1500K); this will yield 310K.
2. We can then subtract the 200K drilling cost to arrive at the same 110K result.

Equation (2-17) confirms that this is always true: We can subtract before *or* after averaging. We prefer to subtract before because the price tags shown in Figure 21-2 then give an accurate picture of the *range of risk*; and we shall see later in Section 21-3 that this may be very important.

decision by working our way backwards through this tree, applying the following general rules:

1. Whenever we encounter a chance circle, we calculate the *average* (expected value) of this gamble, and record it in the circle. Having obtained this summary figure, we can then ignore the potentially confusing brush (all the branchings of the tree) to the right of this circle.
2. Whenever we encounter a decision square, we *prune back* (slash off) all inferior policies, leaving the best one recorded in the square. Once again, we can then ignore the brush to the right of this square.

The complete solution is shown in color in Figure 21-3. The question is: How would you explain this to a vice-president who has never seen a decision tree before? You might describe your finding in this way:

Option C is the best, with an expected value of 130K, *provided* you play it right—like this:

1. Initially *choose option C*, which is a gamble. If its first outcome occurs (as it may, with a 60% chance), then you'll have to make another choice—between options G and H.
2. *Choose G,* which involves another gamble—this is a high risk one, but with an expected value of 143K.
3. When option C is initially chosen in step 1 above, suppose its *second,* rather than its first outcome occurs. That's good news: Your expected value has now risen to 146K; but you face another gamble. If *its* first outcome occurs, you're home free—with 150K. If its second outcome occurs, you must make another choice between options I and J. *Choose J,* which leaves you with 130K.

This analysis can be applied to a wide range of business problems. Notice that we haven't had to specify in any way in Figure 21-3 what sort of a business this is, or what sort of problems it faces. (We've only had to specify price tags and probabilities.) It might, for example, describe decision-making by an oil company. In fact, notice that the top branching (option A) in Figure 21-3 is just the decision BEA faced in Figure 21-2. Thus you can view Figure 21-3 as the problem BEA would have faced if it had introduced this form of decision-making one month earlier, say, before it had bought the exploration lease and when it had to decide between this lease and two other ways of spending its money (options B and C). In Figure 21-3, we now see that the purchase of the lease was a mistake; if we had introduced this analysis a month earlier, and chosen option C instead of the purchase of the lease (option A), we could have increased the company's expected earnings by 144K − 130K = 14K.

Alternatively, Figure 21-3 could describe an entirely different firm facing a sequence of risky decisions. For example, it could represent a retail firm deciding between three different levels of inventory, where the chance forks represent the uncertain sales that are expected to absorb this inventory.

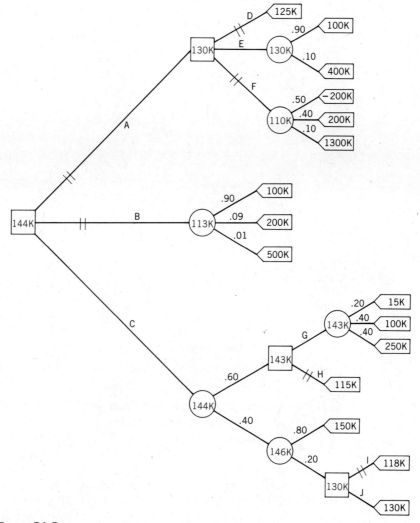

Figure 21-3
A more complicated tree.

PROBLEMS

In all the Problems in this section and the next, unless otherwise stated assume that decision makers are wealthy enough to be risk neutral, that is, they can afford to go for the long-run average.

21-1 Karen Becker faced some major decisions on whether to expand her small riverside café. After careful thought, she found her decision tree reduced to the following (in thousands of dollars):

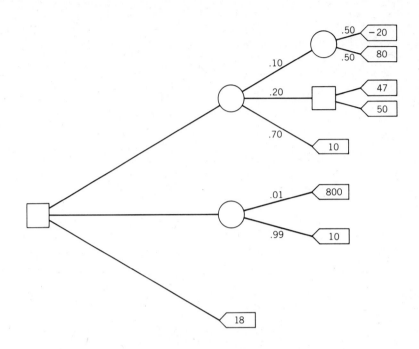

    a. By averaging out and folding back, show her best strategy.

    b. Suppose she is relatively poor and risk averse. Might that change her decision? Why?

21-2 A large oil corporation developed a new, but very expensive, process for extracting oil from its shale fields in Colorado, and had to decide whether to go ahead with such a high-stakes game. The rough guess of the planning staff was that it would cost 200 million just to set up the plant; there was also a 60% chance the project would fail and they would lose their 200 million. Yet there was a 30% chance that it would work and produce a 100 million profit (above their 200 million investment) and a 10% chance that it would work spectacularly and produce a whopping 1000 million profit.

    Assuming all these figures are present values (allowing for interest), should they go ahead with the plant?

21-3 In the 1960s, experimental work on the seeding of hurricanes at sea with silver oxide indicated it would probably reduce the force of the hurricane. Because of high risks, initially the seeding was entirely limited to experimental work on hurricanes that spent their life at sea. Before applying this research to hurricanes that threatened the coast, the U.S. government had to face the possibility that seeding might sometimes make hurricanes worse rather than better, and so commissioned a formal decision analysis (Howard, 1972). The following table outlines the estimates of the various chances and losses involved:

Probabilities and Losses of Various Wind Changes Occurring in the 12 Hours Before Hurricane Landfall

INTERVAL OF CHANGES IN MAXIMUM SUSTAINED WIND	APPROXIMATE MIDPOINT OF INTERVAL	CONSEQUENT DAMAGE ($1,000,000)	PROBABILITY THAT WIND CHANGE WILL BE WITHIN INTERVAL	
			IF SEEDED	IF NOT SEEDED
Increase of 25% or more	+32%	336	.038	.054
Increase of 10% to 25%	+16%	191	.143	.206
Less than 10% change	0%	100	.392	.480
Reduction of 10% to 25%	−16%	47	.255	.206
Reduction of 25% or more	−34%	16	.172	.054

If the government was interested in minimizing long-run costs, would seeding be worthwhile? Work out the answer under two different assumptions:

a. The only cost of seeding was the cost of the operation—$250,000 for silver oxide and airplanes.

b. There was an additional cost of "government responsibility." If they seeded and the storm became worse, whether or not it was the fault of the seeding, they might be sued for damages, or suffer politically from a storm of criticism. Suppose this cost was considered roughly equivalent to an extra $168 million for the highest increase in wind (25% or more), an extra $57 million for a moderate increase in wind (10% to 25%), and an extra $5 million for very little change in wind (−10% to +10%).

21-4 Jack Campbell has just arrived at UCLA for a year's sabbatical leave from the University of Edinburgh. He needs a car but cannot decide which of the following three to choose:

i. He could lease a new Lasalle for 12 months at $300 per month, which includes insurance and all repairs.

ii. He could buy a new Renown for $6500, pay $500 insurance, and be free of repair bills because of the car's excellent warranty. At the end of 12 months, he expects he could sell it back to the dealer for $4000. Or better yet, there is a 30% chance his sister will be coming over to California, and would be happy to buy it from him for $4500.

iii. He could buy a 3-year-old Q-car from "Honest Ed" for $2500. The record of these cars is excellent—repair free, in fact—except for transmission trouble: Every year, 20% of them develop transmission trouble that is impossible to fix—it renders the car worthless. Furthermore, he would then have to lease a car for the remaining months at $400 per month. And his insurance of $500 would not be refundable at all.

> Even if the car survived the twelve months, it would not command much of a resale price—about $800 on the used car market, or $1200 from his sister if she came over.

Jack doesn't mind taking a few risks, nor does he mind the extra wheeling and dealing required to deal with Honest Ed, his sister, or whomever. He heard you are taking a course in decision analysis, and wants your advice. Work out two answers:

a. Ignoring interest.

b. Assuming interest is 10% per year.

21-2   Testing to Revise Probabilities: Bayes Theorem

A—EXAMPLE, WITH TEST INFORMATION ADDED

To return to the oil lease example in Figure 21-2, note again the drilling branch at the bottom of this diagram showing the high cost if a well turns out to be dry: The loss is 200K in drilling costs, with nothing in return. It would be ideal if we could wait to make the drilling decision *until after we knew* whether or not the well would be dry, and so avoid this waste.

   In practice, we cannot know for sure without actually drilling the hole. However, suppose we can get a good idea from a preliminary seismic analysis that will determine the geological structure of the area in which we are drilling. There are 3 kinds of structure: no structure (N)—bad news; open structure (O)—better news; and closed structure (C)—most encouraging of all. Table 21-2 shows why N is bad news: From past records, geologists have found that most (70%) of the dry wells have been in this N structure, while most (60%) of the gushers have been in the C structure.

   The cost of the seismic analysis to determine the geological structure is about 10K. Is it worth it? If so, how should the result be used?

B—SOLUTION: REVISED PROBABILITIES

Does Table 21-2 really give us useful information? It tells us, once we know the kind of well, the probabilities of various geological structures.

TABLE **21-2**   **Conditional Probabilities of Various Geological Structures for Each Type of Well**

GIVEN THIS TYPE OF WELL...	... THESE ARE THE PROBABILITIES OF VARIOUS KINDS OF GEOLOGICAL STRUCTURE.		
	N	O	C
Dry	.70	.20	.10
Wet	.30	.40	.30
Gusher	.10	.30	.60

But we want to know something quite different: Given (seismic) knowledge of the geological structure, what are the probabilities of various kinds
of wells? In other words, we want information in the reverse order. Fortunately, this can be provided by the Bayesian analysis developed in
Section 3-6 (Figure 3-9, for example). Figure 21-4 shows how it works in
this case: Starting from the left in panel (a), we display the best information
we have so far on the various kinds of well—the prior probabilities (.50,
.40, .10, taken from Figure 21-2) that apply if we drill "blind," without
first doing a seismic analysis. Moving to the right, we set out the three
"branchings" (taken from the three rows in Table 21-2) that show the
probability of each geological structure, once the type of well is known.
Outlined on the extreme right is our calculation that there is a .48 probability of a N structure; this is reproduced in panel (b) as the first branch.
Then the three branchings to the right give the desired *posterior proba-*

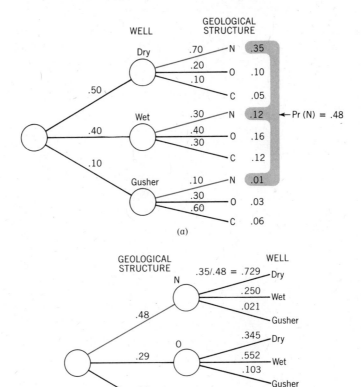

Figure 21-4
*Bayes Theorem to revise probabilities. (a) Initial tree. (b) Reverse tree. These
are not decision trees; they are Bayesian probability trees, which provide better
probabilities to plug into the decision tree in Figure 21-5.*

bilities of various kinds of well, once the geological structure is known through our seismic analysis.

Panel (b) now provides exactly the information we need to describe our new "fourth option" (first doing a seismic analysis to find out the structure, and only then deciding on whether or not to drill). When it is plugged into Figure 21-2, we get the complete decision tree shown in Figure 21-5.

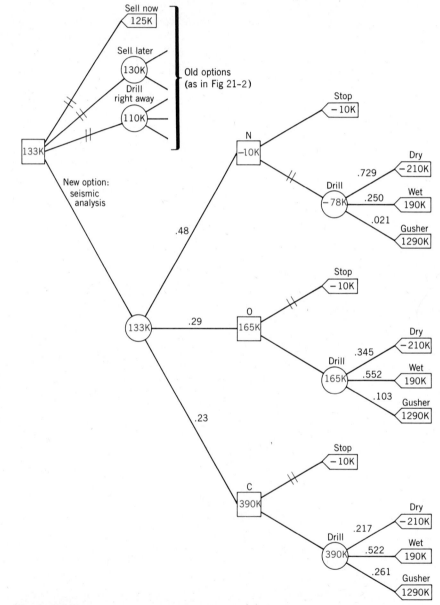

Figure 21-5
Drill decision (Figure 21-2) revised by seismic test results (from Figure 21-4b).

Note that the right-hand price tags (for dry, wet, and gusher wells) are the same as before, except of course that now we must deduct the 10K cost of the seismic analysis. This 10K cost also appears in the 3 price tags marked "stop" that allow us to stop and not drill once we know the structure.

When we average out and prune back (shown in blue in Figure 21-5) we find that this new "fourth option" is worth 133K, and is now the best decision: We should begin with the seismic analysis of structure, and only then decide whether or not to drill.

C—WHAT IS THE SEISMIC ANALYSIS WORTH?

When we compare the new value of the tree (133K) with the old value (130K in Figure 21-2), we can calculate the improvement in the expected value of the enterprise due to the seismic analysis (or, in general, due to whatever sample information we may choose to collect):

$$\text{Expected Value of the Sample Information,}$$
$$\text{EVSI} = 133K - 130K \qquad (21\text{-}4)$$
$$= 3K$$

Thus we see how Bayesian theory can improve business decisions.

PROBLEMS

21-5 In Problem 21-2, suppose a small pilot plant was studied as a possible way to get valuable information. From past experience, it was guessed that with a cost of <u>40 million</u>, a viable pilot plant could be built that would reasonably reflect the operation of the main plant itself, as follows: if the main plant works (or works spectacularly), the pilot plant has an 80% chance of working, and a 20% chance of failing; if the main plant fails, the pilot plant is sure to fail too, being smaller and cheaper.

a. Should they build the pilot plant?

b. A year later, having already invested 20 million in the pilot plant, the price of oil drops so that their profits would be only 30% of their formerly estimated value. Now what should they do?

21-6 The Arvida Flour Mill was debating whether or not to change its packaging from a bag to a box. The increased cost would be 0.6¢ each, reducing the profit of each item from 5.3¢ down to 4.7¢. For this extra cost, however, Arvida would get a more convenient package, which, it was hoped, might improve sales: They guessed there was a 60% chance that sales would increase by 20%. (Otherwise, sales would remain unchanged.)

The planning horizon for Arvida was 5 years; beyond that, technology might change to make both boxes and bags obsolete. During that 5-year period, Arvida's projected sales were 15 million bags.

    a. Is it worthwhile changing to a box?

    b. In the past, Arvida has sometimes conducted a consumer survey to predict whether its sales would increase or not. This survey cost $10,000, and had a good record: Whether sales increased or not, there was a 75% chance of a correct prediction.

        An intensive survey by a market research firm would cost $30,000, and increase the chances of a correct prediction from 75% to 90%. Should Arvida hire them? Or go with its own $10,000 survey? Or forget them both?

21-7 Replace Table 21-2 with the following "half-price" test that costs only 5K:

Conditional Probabilities of Various Geological Structures for Each Type of Well

GIVEN THIS TYPE OF WELL . . .	. . . THESE ARE THE PROBABILITIES OF VARIOUS KINDS OF GEOLOGICAL STRUCTURE		
	N	O	C
Dry	.50	.30	.20
Wet	.40	.30	.30
Gusher	.20	.20	.60

    a. Would you say this seismic test is better or worse than in Table 21-2?

    b. Rework the decision tree in Figure 21-5, including as an option this half-price seismic test.

21-3 Utility Theory to Handle Risk Aversion

A—THE LIMITATIONS OF EXPECTED MONETARY VALUE (EMV)

On the basis of Figure 21-5, Jack Moor recommends to BEA that it should proceed with the fourth option, worth 133K: Do a seismic analysis and then (unless the structure is N) drill for oil. But now suppose that Moor himself, rather than BEA, owns this lease. As an *individual* faced with exactly these same numbers, would he make the same decision?

His answer is *no way*—and he gives good reasons. True, if everything goes right he becomes a very rich man. (He ends up with a gusher netting 1290K.) But is it worth the risk? If everything goes wrong, he's wiped out. (He ends up having to pay for a seismic analysis and drilling a dry well that puts him 210K in debt.) He therefore opts for selling the lease and thus becomes just comfortably rich, for certain.

Clearly, this is a reasonable decision. Why then does he advise BEA to make an entirely different decision, and take a risk on drilling? The answer is that BEA can essentially ignore risk because, as a very large company, it is drilling hundreds of wells (or, equivalently, taking hundreds of other major decisions). True, some of these wells will be dry, but some will be gushers. As we saw in Figure 21-5, the average or expected value that BEA will earn on each well will be 133K, which is better than what it would get from selling.

In conclusion, it is appropriate for BEA to use the money price tags and the resulting calculations of *expected monetary value* (EMV) in Figure 21-5. But these calculations are inappropriate for an individual who can only "spin the wheel once" and has a normal aversion to putting all his money (plus a lot he has to borrow) on the table. For such an individual, averaging these purely monetary price tags must be inappropriate, because it points to the wrong decision. This is our problem, then, in the rest of this chapter: to find a better set of price tags that reflect an individual's aversion to risk.

B—ALL DOLLARS DON'T HAVE THE SAME UTILITY

To confirm that individuals are typically averse to risk, and that EMV calculations are therefore inappropriate, consider the following question:

EXAMPLE **21-1**

Suppose a generous friend offers you a once-in-a-lifetime gamble:

Option 1. On the flip of a fair coin, you get $1,500,000 if heads, $0 if tails (21-5)

Alternatively, if you prefer, you can choose:

Option 2. $500,000 for certain (21-6)

Which option would you choose?

Solution

Of course there is no "right" or "wrong" answer. Different people have different individual values that lead them to different but perfectly valid answers. Yet most people would prefer option 2, even though its expected monetary value is less:

$$\text{EMV of option 2} = \$500,000 \ (1.0) = \$500,000 \quad (21\text{-}7)$$

whereas

$$\text{EMV of option 1} = \$1,500,000 \ (.50) + 0 \ (.50) = \$750,000 \quad (21\text{-}8)$$

The reason is simple: Most people value their first half a million more than their second or third. (Technically, we say that they have a declining marginal utility of money.) With the first $500,000 they can buy the important things—a home, a couple of cars, a good pension, and so on. With another $500,000, or even another $1,000,000, they can only buy less important things—a second

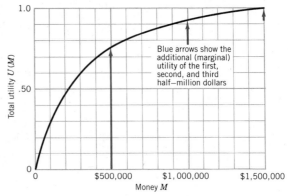

Figure 21-6
Jack Moor's utility of money.

> house, several more cars, and so on. Why should they risk losing
> that first $500,000 in order to get just another $1,000,000?

Thus we conclude that everyone has a subjective valuation of money,
called the "utility of money," and this depends on what money will buy
(the first or second home, etc.). In Figure 21-6, we graph the utility of
money for a typically risk-averse individual, say Jack Moor. The utility
of the first half-million dollars is high, and is shown as the first vertical
arrow. The additional (i.e., marginal) utility of the second or third half-
million dollars is much less.[3] (To see just how a utility curve like this is
derived, and why it is legitimate to then use it in decision analysis, see
Raiffa, 1968, or Wonnacott, 1984 IM.)

C—DECISIONS BASED ON EXPECTED UTILITY

A wise decision is one that maximizes a person's happiness or satisfaction
as measured by his utility function, rather than mere money. Thus, to
solve a decision tree, the decision-maker must first convert each payoff
on the right from dollars to utility—by reading utility values vertically
off a graph reflecting his personal values, such as Fig. 21-6. Then he
proceeds as before, averaging out and folding back, to find the decision
with highest expected utility. An example will illustrate:

EXAMPLE **21-2**   Suppose the decision-maker is no longer the BEA Corporation, but
instead is Jack Moor personally, with his utility function given in

---

[3]Although we show just a limited range, and arbitrarily scale the U axis from 0 to 1 over
this range, the utility curve continues to slowly rise on the right since human wants are
never completely satisfied.

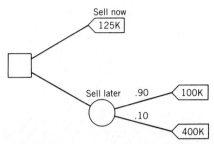

Figure 21-7
The decision tree set out in terms of money (reduced version of Figure 21-1).

Figure 21-6. Also suppose the drilling option in Figure 21-1 is not available, so that the decision tree is the reduced version set out in Figure 21-7. Find Jack's optimal solution, that maximizes his expected utility.

Solution   The money value of each outcome shown in Figure 21-7 is converted (using Figure 21-6) to utilities in Figure 21-8. These utilities, shown in special brackets, are then averaged out and folded back, to finally show the option with highest expected utility: the first option (sell now).

In this example, it is interesting to compare the best decision ("sell now") with the earlier one in Figure 21-2 ("sell later"). The difference reflects the different values of the decision-makers. Specifically, in Jack Moor we now have a decision-maker who is risk averse, since he no longer has the large capital behind him that would enable him to go for long-run expected monetary value. Thus, he finds the risk-free choice ("sell now") more attractive.

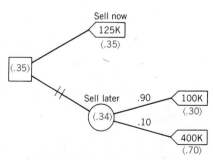

Figure 21-8
The decision tree converted to bracketed utilities, and solved.

PROBLEMS

21-8   Two forms of treatment are available for lung cancer—surgery or radiation—and they produce different survival patterns as roughly shown below (McNeil and others, 1978):

X = NUMBER OF YEARS OF ADDITIONAL LIFE	RELATIVE FREQUENCY, $p(X)$	
	IF SURGERY	IF RADIATION
0	24%	6%
2	31%	45%
5	20%	28%
10	5%	5%
15	5%	5%
20	5%	5%
25	10%	6%

a. If patients want to maximize their life expectancy, which form of
therapy should they choose? (This is the criterion traditionally recommended by doctors.)

b. A sample of 14 patients was asked a series of questions that uncovered
their utility of various lengths of life. The typical patient had the
following utility curve:

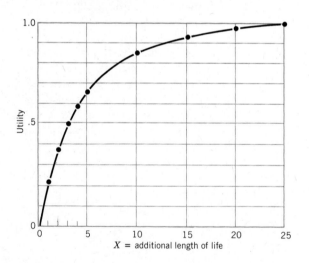

Now what is the best treatment for such a patient?

c. If you were a doctor, would you recommend surgery or radiation?
Why?

d. Do you agree with the following conclusion of the authors (McNeil
and others, 1978)? If not, why not?

"These results emphasize the importance of choosing therapies not only on the basis of objective measures of survival but also on the basis of patient attitudes."

21-9 Suppose a fire insurance policy costs $115 per year, and compensates the homeowner against the following losses. (Frequency is given as fires per 100,000 homes, annually. Data is hypothetical.)

DAMAGE (IN $000)	AVERAGE DOLLAR LOSS	FREQUENCY
Small fire (0–5)	$ 1,000	2100
Medium fire (5–20)	10,000	300
Major fire (20–60)	30,000	20
Beyond repair (60)	60,000	5

a. The owners of a mortgaged home always have to buy fire insurance. (The value of the house must be protected, or lenders will not provide mortgage money.) For someone who has just paid off the mortgage, however (and is thus free of all such restrictions imposed by lenders), is this insurance wise to keep?

b. Of all the income the insurance company receives from premiums, what proportion actually is paid out for fire losses? (The rest goes to administration costs and profit.)

c. Suppose the insurance company has a "$5000 deductible" option— they only pay damage after the first $5000. What percentage of their former losses would they have to pay out? If they passed that percentage on to the customer, what would the premium be instead of $115?

21-10 Suppose an insurance company insures against the $50 loss of a contact lens, in exchange for a premium of $12 per year. Its experience every year has been that 10% of its customers make one claim, and another 3% make two claims. (After that, the policy is cancelled.) Would you buy this insurance? Why or why not?

CHAPTER 21 SUMMARY

21-1 Decision problems can be graphed as a tree, with circles showing the chance forks, and squares showing the decision forks. For big companies that can average out their risk, the best decision can be found by using the criterion of expected monetary value (EMV). To calculate this, the circles (chance forks) are averaged out, and the squares (decision forks) are pruned back, to display the best choice.

21-2 Sample information, using Bayes Theorem, can reduce uncertainty and so improve the value of the enterprise.

21-3 For decisions that involve big risks relative to financial capacity, payoffs must be expressed in terms of utility, rather than money. In fact, utility is a versatile enough tool to allow us to compare *all* kinds of payoffs, not just money payoffs.

REVIEW PROBLEMS

21-11   Hunter Creations wholesales a singing toy for $12. Although it was a best seller last year, the sales manager is worried about its future, since 6% of total sales were returned for a broken spring in the windup mechanism. The cost of making good the guarantee was $15—a little more than the wholesale price of the replacement toy. As well, there were an unknown number of customers who didn't bother to claim the guarantee, but who were nonetheless very dissatisfied. It is estimated that each such dissatisfied customer costs the company about $30 on average, in terms of lost future sales (not only of this toy, but also of the seventeen others that Hunter sells).

The sales manager therefore suggests that a better spring be used. Production reports that an unbreakable alloy spring is available, but costs $1.40 instead of the present $0.15. They feel that the extra cost is not justified.

Since the projected sales are substantial—300,000 toys of this model—this is an important decision. The sales manager hires you as a consultant to advise her.

a.   In order to make a recommendation, what further information do you need?

b.   Suppose the sales manager, on the basis of similar past experience, guessed that the proportion of dissatisfied customers (customers with broken springs but not claiming the guarantee) could be as high as 10%. When pressed further, she roughly guessed that this proportion π might be 0%, 5%, or 10%, with probabilities of .40, .40, and .20, respectively. On the basis of this information, what would you recommend?

21-12   In Problem 21-11, a sample survey of 100 customers was proposed to estimate what π really is. Although it would cost $1000, it would give some objective evidence. In fact, a consulting statistician was able to calculate just how the sample proportion P would be related to π:

IF THE TRUE PROPORTION π WERE ...	... THEN THESE ARE THE PROBABILITIES OF VARIOUS RANGES OF THE SAMPLE PROPORTION P		
	0–2%	3–7%	8% or more
0	1.00	0	0
5%	.13	.74	.13
10%	.01	.19	.80

We find in the last cell, for example, that if the true proportion is 10%, it is likely that the sample proportion will be near 10% too. Specifically, the chances are .80 that P will be 8% or more.

a. Is this sample survey worth the cost? If so, how should the company act on the evidence it will provide?

b. If you have studied Chapter 6, verify the table given above.

21-13 a. John Makeham's dream home has turned into a nightmare. Nine months ago, Regal Lumber contracted to supply all the lumber, precut and delivered for $40,000. However, because the price of lumber has increased since by 25%, Regal is now trying to back out. They offer to sell him the lumber for $50,000 or return his deposit of $5000, whichever he prefers.

If he now buys the lumber elsewhere, it will cost him $55,000 to have it precut to the specifications already built into the foundation. His lawyer advises him that, as soon as Regal has returned his $5000 deposit, he can then sue the company for the $15,000 difference he paid in excess of the contracted price. The chances of his winning the case, including his court costs, are estimated at 80%. If he loses, he will have to pay all the court costs of $10,000. His lawyer recommends that before he definitely decides on this risky option, he should consult a legal expert who, for a $4000 fee, could do extensive research and guarantee to let him know for sure whether or not he will win in court.

Since he has invested so much already in this house and lot, he feels he has to finish building it one way or another. Find his best course of action, assuming he is wealthy enough that he can easily absorb financial setbacks and therefore use expected monetary value as his criterion.

b. To what value would the legal expert have to reduce his fee in order for it to be a worthwhile buy? (This is sometimes called the Expected Value of Perfect Information, EVPI.)

c. Find his best course of action, if he is risk-averse. In fact, suppose large costs are particularly worrisome, as indicated by his "disutility curve":

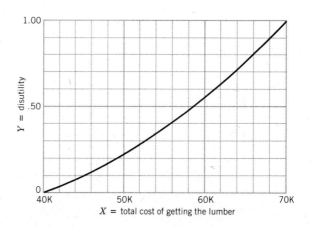

21-14   Suppose a very large company has an oil lease they could sell now for 250K, or else drill. Fortunately, to help them decide, they have not only a seismic test available, but also records of how the test and the subsequent drilling have turned out in hundreds of similar situations:

Bivariate Frequency Distribution for $n = 500$ Drillings

	GEOLOGICAL STRUCTURE		
TYPE OF WELL	N	O	C
Dry	170	80	30
Wet	30	40	30
Gusher	20	40	60

The test costs 20K and the drilling costs 300K. The expected returns are 2000K for a gusher, 500K for a wet well, and nothing of course for a dry well. What should they do?

21-15   Having developed a new quick-open can that keeps coffee fresher, a large food company now has to decide whether or not to market it. The marketing manager guesses that sales might roughly change 2%, 1%, 0, or −1%—with each of these possibilities being equally likely. After change-over costs, this would mean profit changes of 75, 30, −15, −60 thousand dollars, respectively.

a.   Is it worthwhile introducing the new can?

b.   What would it be worth to be able to predict the market perfectly; that is, what is the Expected Value of Perfect Information (EVPI)?

c.   A 6-month trial is suggested to get the reaction to the new can in Middletown—a small city, typical of the whole U.S. market, where the company has often tested new products. In fact, from past tests, the research staff guessed that the Middletown reaction will be closely related to U.S. sales, as outlined in the following table of probabilities:

POSSIBLE CHANGE IN U.S. SALES	REACTION IN MIDDLETOWN			
	FAVORABLE	NEUTRAL	UNFAVORABLE	TOTAL PROBABILITY
+2%	.8	.2	0	1.00
+1%	.6	.2	.2	1.00
0	.3	.4	.3	1.00
−1%	.1	.4	.5	1.00

     i.   If the trial costs 40 thousand dollars, is it worth it?

     ii.   If the trial costs 10 thousand dollars, is it worth it? If so, how should its results be used?

     iii.   At what price does the trial become too expensive?

Chapter

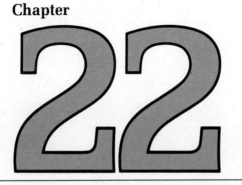

Index Numbers

(Requires Nothing Previous)

The time has come, the Walrus said,
To speak of many things.
Of shoes, and ships, and sealing wax,
Of cabbages—and kings.

LEWIS CARROLL

A *price index* is a single figure that shows how a whole set of prices has changed. For example, if we are asked what has happened to prices over the last 12 months, it is far simpler to reply that the overall price index has risen by 5%, say, rather than that the price of eggs is up 20%, the price of TVs is down 10%, and so on.

Similarly, if we wish to compare the quantity or output of industrial goods in the United States and Canada, it is convenient to state that the *industrial output index* of Canada is 8% that of the United States, rather than state that Canada produces, say, 9% of the U.S. output of autos, 2% of the U.S. output of aircraft parts, and so on.

22-1 Price Indexes

A—PRICE RELATIVES

Let us develop a price index for food, assuming the national diet consists of just the three items listed in Table 22-1—steak, pepper, and bread—at

TABLE **22-1**   **Price Relatives for a Hypothetical Diet**

| | GIVEN PRICES | | RATIO | PRICE RELATIVE |
| | 1980 | 1985 | $\dfrac{P_t}{P_o}$ | $\dfrac{P_t}{P_o}(100)$ |
ITEM	P_o	P_t		
Steak (per pound)	$2.20	$3.00	1.36	136
Pepper (per ounce)	$2.00	$2.00	1.00	100
Bread (per pound)	$ .50	$ .60	1.20	120

prices initially denoted by P_o, and t years later by P_t. The price of steak, for example, has increased by the factor $3.00/2.20 = 1.36$. Customarily it is multiplied by 100 to get rid of the decimal point, and then it is called the *price relative*:

$$\text{Price relative} \equiv \frac{P_t}{P_o}(100) \tag{22-1}$$

How can we summarize the overall price trend with a single number? A natural suggestion is to average the price relatives for all items:

$$\text{Simple average} = \frac{136 + 100 + 120}{3} = 119$$

The problem with the simple average, of course, is that it gives pepper as much weight as bread or steak. We therefore must look for an index that gives a heavier weight to important items.

B—THE LASPEYRES AND PAASCHE INDEXES

A price index is intended to measure the overall increase in prices, so let us just see how much the average person's physical diet—the quantities called Q_o in Table 22-2—has increased in cost. The numbers along the first line of this table show that in 1980, this diet included 50 pounds of steak at a cost of $2.20 per pound, for a total of $110. Similarly, the expenditures on pepper and bread totaled $4 and $40, so that the total cost was $154. In general terms:

$$\text{Initial cost} = \Sigma P_o Q_o \tag{22-2}$$

What would this same diet cost later? We merely use the later prices P_t instead of P_o:

$$\text{Later cost} = \Sigma P_t Q_o \tag{22-3}$$

TABLE **22-2** **From Prices and Quantities, the Laspeyres or Paasche Price Index Can Be Calculated**

	GIVEN DATA				LASPEYRES PRICE INDEX		PAASCHE PRICE INDEX	
					COST OF 1980 BASKET		COST OF 1985 BASKET	
	PRICES		QUANTITIES					
	1980	1985	1980	1985	IN 1980	IN 1985	IN 1980	IN 1985
ITEM	P_o	P_t	Q_o	Q_t	P_oQ_o	P_tQ_o	P_oQ_t	P_oQ_t
Steak (lb)	$2.20	$3.00	50	40	$110	$150	$ 88	$120
Pepper (oz)	$2.00	$2.00	2	3	$ 4	$ 4	$ 6	$ 6
Bread (lb)	$ .50	$ .60	80	100	$ 40	$ 48	$ 50	$ 60
			↑		$154	$202	$144	$186
			1980 basket of goods					

$$\text{Index} = \frac{\$202}{\$154}(100) \qquad \text{Index} = \frac{\$186}{\$144}(100)$$
$$= 131 \qquad\qquad\qquad = 129$$

The price index is calculated by comparing the costs in the two years, for this *fixed basket of goods*:

$$\boxed{\textbf{Laspeyres price index} \equiv \frac{\Sigma P_tQ_o}{\Sigma P_oQ_o}(100)} \qquad (22\text{-}4)$$

$$= \frac{\$202}{\$154}(100) = 131$$

Note that the cost of pepper does influence the index very little indeed, since it was only $4 in a budget totalling well over $100.

The question naturally arises: Why do we use as weights the *initial* basket of goods; that is, why do we use weights Q_o in (22-4)? We could construct an equally valid index using the *later* basket of goods (given by Q_t). This gives us a slightly different answer:

$$\boxed{\textbf{Paasche price index} \equiv \frac{\Sigma P_tQ_t}{\Sigma P_oQ_t}(100)} \qquad (22\text{-}5)$$

$$= \frac{\$186}{\$144}(100) = 129$$

Although there may be no theoretical reason for preferring one index to the other, there is a practical reason for using the Laspeyres index when indexing is done year after year: The Laspeyres index uses the same base weights Q_o in all calculations, whereas the Paasche index has to use different weights Q_t every year.

Clearly, in applying the Laspeyres index, the selection of the base year becomes a crucial issue. For example, the index would become almost meaningless if the initial base period were a wartime year in which steak was rationed, and very little was consumed. In this case, steak would essentially disappear from the calculation of price indexes for the later peacetime period when it was an important item in consumption. Hence, great care must be taken that the base year is a reasonably typical one, free of disasters or other unusual events that distort consumption patterns.

C—FISHER'S IDEAL INDEX: THE GEOMETRIC MEAN

Recall that the ordinary mean (arithmetic mean) was defined by first adding all n items. We now define the *geometric mean* by first *multiplying* all n items—and then taking the nth root:

$$\text{Geometric mean} \equiv \sqrt[n]{X_1 X_2 \ldots X_n}$$

This is the mean that is appropriate in averaging ratios or factors of increase. A small digression will illustrate. Suppose the population of a suburb increased by a factor of 2 in the first decade, and by a factor of 8 in the second, so that the overall increase was 16 times:

	POPULATION	INCREASE PER DECADE
1950	1,000	
1960	2,000	2 times
1970	16,000	8 times
Increase overall		16 times

Does the arithmetic mean

$$\frac{2 + 8}{2} = 5$$

describe the typical increase per decade? The answer is no: If a fivefold increase occurred over two successive decades, the overall increase would be twenty-fivefold, not sixteen.

Instead, let us try the geometric mean:

$$\sqrt{(2)(8)} = 4$$

This is just right: If a fourfold increase occurred over two successive decades, the overall increase would be the sixteenfold that actually did occur.

Now, consider the Laspeyres and Paasche indexes. Since they measure ratios, the geometric mean is the appropriate way to obtain their average too:

$$\boxed{\text{Fisher's ideal index} \equiv \sqrt{\text{(Laspeyres index) (Paasche index)}}} \qquad \text{(22-6)}$$

$$= \sqrt{(131)(129)} = 130$$

22-2 Further Indexes

A—QUANTITY INDEXES

A price index measures the increase in the price or "cost of living". Similarly, we could measure the increase in the *quantity* or "*standard of living*". For example, back in Table 22-2, notice that steak purchases went down while bread purchases went up. Overall, has the standard of living gone up or down? That depends on the relative importance of steak and bread in the consumer budget, which depends on the price of the two. That is, prices should be used as weights in calculating a quantity index, just as quantities were used as weights in calculating the price index. We thus obtain (using the sums already calculated in Table 22-2):

$$\text{Laspeyres quantity index} \equiv \frac{\Sigma Q_t P_o}{\Sigma Q_o P_o}(100) \qquad \text{(22-7)}$$

$$= \frac{144}{154}(100) = 94$$

$$\text{Paasche quantity index} \equiv \frac{\Sigma Q_t P_t}{\Sigma Q_o P_t}(100) \qquad \text{(22-8)}$$

$$= \frac{186}{202}(100) = 92$$

$$\text{Fisher's ideal index} \equiv \sqrt{\text{(Laspeyres index) (Paasche index)}} \qquad \text{(22-9)}$$
$$= \sqrt{(94)(92)} = 93 \qquad \text{like (22-6)}$$

B—THE TOTAL COST INDEX

So far, we have defined both price and quantity indexes. It finally is possible to measure their combined effect, by seeing how much *total cost* increased (due to changes in *both* price and quantity):

$$\boxed{\text{Total cost index} \equiv \frac{\Sigma P_t Q_t}{\Sigma P_o Q_o}(100)} \qquad \text{(22-10)}$$

$$= \frac{186}{154}(100) = 121$$

Now, if our indexes of price, quantity, and total cost are good indexes, they should satisfy the basic relation called the *factor-reversal test*:

$$(\text{price index}) (\text{quantity index}) = \text{total cost index} \qquad (22\text{-}11)$$

For example, if the price index increases twofold and the quantity index increases threefold, then the total cost index should increase sixfold; if not, there is something peculiar about the price and quantity indexes we are using. Unfortunately, (22-11) often is *not* satisfied, as a simple calculation will show[1]:

$$
\left.
\begin{array}{lll}
\text{Laspeyres indexes} & (131)(94) \overset{?}{=} 121 \\
& \text{No: } 123 \;\; > 121 \\[4pt]
\text{Paasche indexes} & (129)(92) \overset{?}{=} 121 \\
& \text{No: } 119 \;\; < 121 \\[4pt]
\text{Fisher's ideal indexes } & (130)(93) \overset{?}{=} 121 \\
& \text{Yes: } \;\; 121 \overset{\checkmark}{=} 121
\end{array}
\right\} \qquad (22\text{-}12)
$$

It easily is proved (in Problem 22-5) that Fisher's ideal index *always* satisfies this factor reversal test. This is one of the properties that justifies the name "ideal."

C—FURTHER ADVANTAGES OF FISHER'S IDEAL INDEX

We note in (21-12) that the Laspeyres indexes of price and quantity both are larger than the corresponding ideal indexes; thus their product is larger than it should be (i.e., larger than the cost index). This is not an isolated example; the Laspeyres indexes are too large in nearly all other cases as well, and it is interesting to see why. We shall give the reason in the case of the Laspeyres price index (leaving the quantity index as an exercise).

The key to the argument is the numerator, given in (22-4) as $\Sigma P_t Q_o$. We see that current prices are weighted by the old quantities. Thus, whatever goods currently have relatively high prices P_t and hence are not bought very much, are nonetheless weighted with the old higher quantities Q_o, making the product $P_t Q_o$ too high. Hence the Laspeyres index overstates.

In a similar way, we may argue that the Paasche index understates. Consequently the ideal index, being the geometric mean between the Laspeyres and Paasche indexes, is preferred.

---

[1] To be mathematically correct, we ought to express indexes with their decimal point included of course (i.e., without multiplying by the customary 100). For example, in (22-12), for the first equation we actually should write:

$$(1.31)(.94) \overset{?}{=} = 1.21$$

PROBLEMS

22-1

	PRICES		QUANTITIES (PER PERSON)	
ITEM	1970	1980	1970	1980
Flour (pound)	$ .25	$ .30	100	80
Eggs (dozen)	1.00	1.25	50	40
Milk (pint)	.30	.35	100	100
Potatoes (pound)	.05	.05	60	100

From the above hypothetical data, and using 1970 as the base year:

a. Calculate the three price indexes for 1980—Laspeyres, Paasche, and Fisher. Order them according to size. Is it the same order as in equation (22-12)?

b. Calculate the same three *quantity* indexes. Order them also according to size. Is it the same order as in equation (22-12)?

c. Calculate the total cost index, and verify which of the three indexes passes the factor-reversal test.

d. What difference to the various indexes would it make if all quantities were measured, not on a per person basis, but rather as total national consumption of the good? (That is, all figures in the last two columns would be 200 million times higher.)

22-2 **a.** For the data in Problem 22-1, calculate the price relatives for each good, and pick out the smallest and largest. Does the Laspeyres index lie between them? Does the Paasche index? The Fisher index?

b. Someone claims that the Laspeyres price index is just a weighted average of price relatives—the weight for each item being just the amount spent on it initially (P_0Q_0). Is this claim true for the data in Problem 22-1? Is it true in general?

22-3

	PRICES				QUANTITIES (PER PERSON)			
ITEM	1940	1950	1960	1970	1940	1950	1960	1970
Steak (pound)	$1.00	$1.20	$1.40	$1.20	40	50	70	50
Bread (pound)	.10	.20	.40	.20	80	100	110	100

From the above hypothetical data, answer the following questions:

a. Using 1940 as the base period, fill in the following table:

TYPE OF PRICE INDEX	1950	1960	1970
Laspeyres			
Paasche			
Fisher			

b. Which of the three indexes is simplest to calculate and measure?

c. Fill in the following table:

TYPE OF PRICE INDEX	1960 INDEX, USING 1950 BASE	1970 INDEX, USING 1960 BASE	PRODUCT OF THESE 2 INDEXES
Laspeyres			
Paasche			
Fisher			

Note that in 1970, both prices and quantities reverted to what they were in 1950; that is, during the 1960s, everything that happened in the 1950s was reversed. It therefore would be appropriate if the product of the two indexes, which measures the total change over the two decades, was 1.00 exactly. This *time reversal test* is passed by which indexes?

22-4 Explain why:

a. The Paasche price index usually is an understatement.

b. The Paasche quantity index usually is an understatement.

*22-5 **a.** Prove that the Fisher indexes in general pass the time reversal test stated in Problem 22-3c.

b. Prove that the Fisher indexes in general pass the factor reversal test stated in (22-11).

22-3 Indexes in Practice

A—PUBLISHED INDEXES

The U.S. Government publishes a wide variety of price indexes. The one most commonly called the *"cost of living"* is the *Consumer Price Index* (CPI) published monthly by the Bureau of Labor Statistics. A Laspeyres-type index, it traces the living costs of U.S. urban consumers. It includes food, rent and home ownership, energy, clothing, transportation, and medical care—each component having its own index published as well. The CPI is further broken down by region and even by city, making it altogether a very, very detailed report of consumer price changes. To complement

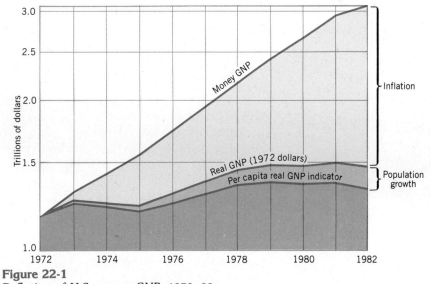

Figure 22-1
Deflation of U.S. money GNP, 1972–82.

the CPI, for primary commodities the *Producer Price Index* (PPI) is published, also in great detail.

For everything counted in the Gross National Product (GNP), the broadest possible index is required—called *the GNP implicit price deflator*. To see how it is used, Figure 22-1 shows GNP growth. The top curve is money GNP—the actual money paid on the nation's GNP every year. This reflects not only an increase in the physical quantity of goods and services produced but also an increase in their prices. Accordingly, to obtain the real (physical) GNP we must divide money GNP with the GNP implicit price deflator.

If we wish to show the growth in *per capita* real income (roughly, the improvement in real income[2] of the average American), we must further divide by an index of population growth. This leaves us with the lower curve in Figure 22-1—a sort of *quantity* or *"standard of living"* index.

B—WHAT'S REALLY HAPPENING TO THE COST OF LIVING?

There are several reasons why changes in the consumer price index (CPI), especially small changes of 1% or 2%, cannot be taken too seriously:

1. Since only some of the items are sampled, and only in some of the stores, there is substantial sampling error—especially in the more

---

[2]Although GNP and national income are not identical concepts, they move together closely enough so that percentage changes in GNP can be used to roughly approximate percentage changes in income.

finely divided components (such as the CPI for one city) where the sample size is small.

2. Actual transaction prices are sometimes difficult to determine, especially on items such as cars where the amount of discount from the list price varies with the skill of the buyer and many other factors.

3. As previously noted, the Laspeyres' index used by the government tends to overstate price increases because it doesn't take into account the "partial escape from inflation" open to consumers—namely, their ability to substitute goods that have had a relatively low price increase for goods that have had a relatively high price increase.

4. A serious difficulty in calculating any index occurs because technology changes the basket of goods in ways that make comparisons of prices difficult. For example, suppose we want to compare prices in 1985 with prices in 1975. How do you deal with video cassette recorders, which are an important purchase in 1985, but weren't even available in 1975—at any price? Since the basket of available goods is thus changing, the base period cannot be too far distant in the past. Any extended index series generally involves a number of points in time at which base weights have been changed in recognition of the introduction of new products. In fact, there is some merit in changing the base every year, forming what is called a *chain index*.

5. A related problem of technology is that, even when data does exist, it often is not comparable. Televisions provide a good example. Suppose that we want to compare prices in 1977 with prices in 1947. Although TVs existed in 1947, they were not at all comparable with the 1977 models. To observe that their price has doubled may be of little consequence; if the increased size of screen, the addition of color, and so on have made a TV now "twice" as good, then isn't this twice the product for twice the price? Have prices changed at all? Unless quality adjustments are accounted for explicitly, such index calculations may not indicate the increase in the cost of living, but rather the increase in the cost of *better* living.

The difficulty with quality judgments is that it is almost impossible to make them. Is the present TV set twice as good, or three times as good as the first model? Autos are another example. At one extreme there are some who are skeptical that the auto is really any better today than it was 30 years ago. On the other hand, we might look to specific improvements such as seat belts or disc brakes, with a view to making a quality adjustment for each. The question is: how much? One answer might be to look at each of the quality improvements that were first introduced as options (before they became standard equipment) and find out the extra price charged for each. But even this is not satisfactory. From either the producer's or consumer's point of view, the value of a car does not necessarily increase by the sum of a set of options that become standard equipment. For the producer, the cost of these optional items decreases when they become standard equipment, since they no longer require the special care and attention of custom items. For the consumer, there may be some items that he simply would not buy if they were options—hence he values them at less than their option price. (Indeed there may even be items—such as automatic transmission—that some buyers would prefer not to have, even if they were *free* options.)

It may be concluded, therefore, that adjustments for changes in product quality should be made; yet in practice they become extraordinarily difficult. Our example of auto improvements (initially priced as options) is perhaps the most conducive to cold calculation; yet even here, any estimate becomes very difficult to defend.

These various practical complications have led many observers to conclude that, in a world of rapidly changing products, small changes in the cost of living index on the order of, say, 1% or 2% may not mean much.

PROBLEMS

22-6 In a number of U.S. industries, wages now are tied to the CPI; that is, indexing in wage contracts ensures that wage rates automatically adjust for any increase in the CPI.

a. How are these justified?

b. Do you see any problem arising if such indexing becomes widespread for all (or most) wages? Can you suggest some solutions?

c. Do you think it would be wise to index the salaries of politicians and civil servants?

22-7 **a.** Would you prefer to spend an (after-tax) annual income of $10,000 by purchasing *entirely* out of the 1980 Sears catalogue, or *entirely* out of the 1940 Sears catalogue (with its lower prices, but more restricted selection of goods, which also are less up-to-date; and of course there will be many goods—like TVs—in the 1980 catalogue that just are not available in the 1940 catalogue). Assume that not only all goods but also all services (such as medical care and transatlantic trips) are included. In other words, would you prefer to spend your entire $10,000 on 1980 goods and services at 1980 prices, or on 1940 goods and services at 1940 prices?

b. Let us sharpen the answer to **a**: how much money do you think you would need to spend in the 1980 catalogue in order to equal the satisfaction you would get by spending $10,000 in the 1940 catalogue? This ratio of expenditure (1980 to 1940) might be called the "subjective" price index.

c. During this period, the U.S. Consumer Price Index actually rose from 42 to 246 (base = 1967). How does this compare with your answer in **b?** For your personal values, therefore, has the CPI overstated or understated price increases?

d. Repeat **a** to **c** for an income of $500,000 instead of $10,000.

22-8 By reading off the appropriate values in Figure 22-1, roughly calculate how much higher was:

    a. The GNP deflator in 1982, compared to 1972.

    b. The population size in 1982, compared to 1972.

CHAPTER 22 SUMMARY

22-1 A price index measures the change in price of a *fixed* basket of goods and services, and so is a measure of inflation. The Laspeyres index uses the *initial* basket, while the Paasche index uses the *present* basket. A subtle difference is produced, which the Fisher ideal index resolves by taking the geometric mean of these two.

22-2 A quantity index measures the change in the amount of physical goods and services consumed, and so is a measure of economic growth. It is calculated by interchanging price with quantity in the formula for the price index.

22-3 The U.S. Government publishes many indexes in great detail. Three of the most commonly quoted are the *Consumer Price Index* (CPI), the *Producer Price Index* (PPI), and the *GNP implicit price deflator*.

    In constructing an index, there are many practical difficulties that limit its reliability: sampling fluctuations have to be allowed for; list prices are discounted to an unknown extent; and changes in product quality and availability make it very difficult, if not impossible, to define a "fixed" basket of goods.

REVIEW PROBLEMS

22-9 The following table gives the prices and quantities of a basket of goods chosen to be the basis for the price and quantity consumer indexes (hypothetical data):

	1960		1970	
GOODS	Q_o	P_o	Q_t	P_t
Beans (16 oz. can)	40	$ .24	40	$ .28
Milk (quarts)	150	$ .20	100	$ .30
Gasoline (gal.)	200	$ .35	250	$ .40
Shoes (pair)	3.5	$ 10.00	4.0	$ 12.00
TV (B and W set)	.05	$200.00	.10	$150.00

    a. Calculate the 3 price indexes (Laspeyres, Paasche, Fisher).

    b. Calculate the 3 quantity indexes.

    c. Calculate the total cost index.

    d. Confirm that the "factor reversal" test (22-11) is true only for the Fisher index.

22-10 In each blank, write in one (*or more*) of "Laspeyres, Paasche, or Fisher."

    a. The _____ price index is the price that the old basket of goods would cost today, compared to what it cost then.

    b. The _____ price index is the price that today's basket of goods costs, compared to what it would have cost then.

    c. The _____ index is usually higher than the _____ index, while the _____ index is invariably between them.

    d. The _____ price index is invariably between the lowest and the highest price relative.

    e. The _____ index passes the time reversal test and the factor reversal test.

    f. The U.S. CPI is a kind of _____ price index.

22-11 Choose the correct alternative in each bracket:

    a. A [price, quantity] index may be called a "standard of living" index, while a [price, quantity] index may be called a "cost of living" index.

    b. The U.S. CPI grew from 133 in 1973 to 171 in 1976, and then to 217 in 1979. Thus the higher inflation occurred during the three year period [1973–6, 1976–9].

Chapter

Sampling Designs

(Requires Chapter 8)

There are many in this old world of ours who hold that things break about even for all of us. I have observed for example that we all get the same amount of ice. The rich get it in the summertime and the poor get it in the winter. BAT MASTERSON

The simple random sampling of Chapters 6 and 8 is conceptually and computationally the most straightforward method of sampling a population. However, we can get estimates that have smaller variance (are more efficient) by *combining* random sampling with some special design features.

23-1 Stratified Sampling

A—SIMPLIFIED EXAMPLE

Let us consider first an exaggerated example, where the benefits of stratified sampling are clearest. Suppose that a large population is known to be 60% urban, 40% rural, and that the problem is to estimate average income. Suppose that the unknown income levels are $10,000 for each urban worker and $5,000 for each rural worker. Accordingly, the population earns a mean income:

$$\mu = (10{,}000)(.6) + (5{,}000)(.4) = 8000 \qquad (23\text{-}1)$$

like (4-4)

614

TABLE **23-1** Simple Random versus Stratified Sampling: an Oversimplified Example

	POPULATION		VARIOUS SAMPLES				
			STRATIFIED SAMPLES		TYPICAL SIMPLE RANDOM SAMPLES		
STRATUM	PROPORTION (1)	MEAN INCOME (IN $1000 UNITS) (2)	(i) PROPORTIONAL (3)	(ii) TO BE WEIGHTED (4)	(5)	(6)	(7)
Urban	.60	10	10	10	10	5	5
			10		5	5	10
			10		10	5	10
					10	10	5
Rural	.40	5	5	5	10	5	10
			5				
Total	1.00	$10(.6) + 5(.4)$ $\mu = 8$					
		Sample Means	$8\surd$	$10(.6) + 5(.4)$ $\overline{X} = 8\surd$	9	6	8

How well would a simple random sample estimate μ? Three possible random samples of size 5 are displayed in the last three columns of Table 23-1. Because the samples are completely random, the number of urban and rural workers fluctuates; this in turn causes the sample mean $\overline{X}$ to fluctuate. Of course, occasionally we will be lucky, as in the very last column, where there are two "5 thousand" observations, that is, two rural workers. Then in this sample, the proportion of rural workers happens to be exactly the same as in the population, and $\overline{X}$ is dead on target.

But if we have prior knowledge of the proportion of urban and rural workers, why depend on luck? Surely it would be wise to design the sampling proportions to be exactly equal to the population proportions. This proportional sampling is shown in column (3); note that the result in this greatly oversimplified case is an estimate that is dead on target. In fact, this same favorable result is guaranteed with a sample of just one observation from each stratum, shown in column (4)—provided, of course, that instead of computing the ordinary mean, each observation is weighted according to the size of its stratum. This weighted or *stratified* sampling yields:

$$\overline{X}_s = 10,000(.6) + 5,000(.4) = 8,000 \qquad (23\text{-}2)$$

which, of course, exactly coincides with μ in (23-1). This is the perfect estimator, dead on, with a sample of only two (the number of strata).

We emphasize that the reason we have been able to do so well is because the weights w_1 and w_2 (i.e., population proportions in each stratum) were *known*. For example, a source like the U.S. population census could be drawn upon for information on urban/rural proportions.

B—GENERAL CASE

Table 23-1 is contrived, of course, because each stratum is perfectly homogeneous. In practice, each stratum will have some variability, as we outline in Table 23-2. The mean and variance within the first stratum are denoted by μ_1 and σ_1^2, and so on. The most general stratified sample consists of drawing a sample of size n_1 from the first stratum, n_2 from the second, and so on, so that the total sample size $n = \Sigma n_i$.

Now the overall population mean is the weighted mean of the strata means μ_i:

$$\mu = w_1\mu_1 + w_2\mu_2 + \cdots \qquad \text{(23-3)}$$
$$\text{like (23-1)}$$

where the weight w_i is the proportion of the population in the ith stratum. To estimate μ, it is natural to estimate each stratum mean μ_i with the sample mean $\overline{X}_i$ in that stratum. This produces the stratified sampling mean $\overline{X}_s$:

$$\boxed{\overline{X}_s \equiv w_1\overline{X}_1 + w_2\overline{X}_2 + \cdots} \qquad \text{(23-4)}$$
$$\text{like (23-3)}$$

In words: We weight the *sample* means with the *correct* and *known* population weights. Using the rules for linear combinations in Chapter 5, we can show (in Problem 23-6) that $\overline{X}_s$ is an unbiased estimator:

$$\boxed{E(\overline{X}_s) = \mu} \qquad \text{(23-5)}$$
$$\text{like (6-5)}$$

TABLE **23-2** **Generalization of Table 23-1**

	POPULATION			STRATIFIED SAMPLE	
STRATUM	PROPORTION OF WHOLE POPULATION (WEIGHT)	MEAN	VARIANCE	SAMPLE SIZE	SAMPLE MEAN
1	w_1	μ_1	σ_1^2	n_1	$\overline{X}_1 = \dfrac{\Sigma X_{1j}}{n_1}$
2	w_2	μ_2	σ_2^2	n_2	
$\vdots$					
i	w_i	μ_i	σ_i^2	n_i	$\overline{X}_i = \dfrac{\Sigma X_{ij}}{n_i}$
$\vdots$					
r	w_r	μ_r	σ_r^2	n_r	
	$\Sigma w_i = 1$	$\mu = \Sigma w_i\mu_i$		$n = \Sigma n_i$	$\overline{X}_s = \Sigma w_i\overline{X}_i$

Furthermore:

$$\operatorname{var} \overline{X}_s = w_1{}^2 \operatorname{var} \overline{X}_1 + w_2{}^2 \operatorname{var} \overline{X}_2 + \cdots \qquad \text{like (5-22)}$$

$$= w_1{}^2 \frac{\sigma_1{}^2}{n_1} + w_2{}^2 \frac{\sigma_2{}^2}{n_2} + \cdots \qquad (23\text{-}6)$$

In practice, in each stratum the variance $\sigma_i{}^2$ is unknown and has to be estimated with the *sample* variance s_i^2, so that

$$\operatorname{var} \overline{X}_s \simeq w_1{}^2 \frac{s_1^2}{n_1} + w_2^2 \frac{s_2^2}{n_2} + \cdots \qquad (23\text{-}7)$$

This leaves the problem of how the n_i are specified; that is, how many observations should be drawn from each stratum? Three common methods are given next.

C—PRESPECIFIED PROPORTIONAL STRATIFICATION

The most obvious way to choose n_i is to make the sample proportion exactly equal to the population proportion, as illustrated earlier in column (3) of Table 23-1:

Proportional sampling requires setting the n_i so that

$$n_i = w_i\, n \qquad (23\text{-}8)$$

D—POST STRATIFICATION, AFTER SAMPLING

Suppose, as always, that we know the population weights w_i. For example, we might know that 70% of the population of a city is white and 30% black. In taking a house-to-house sample of 1000, we would like to use these population weights in (23-8) to fix our sample at 700 whites and 300 blacks. But we cannot. There is no way of knowing before we knock on a door whether the respondent will be a white or a black. So we have to knock on 1000 doors (i.e., take a simple random sample of 1000), and take whatever black and white proportion we happen to get. Suppose our sample turns out to include 680 whites and 320 blacks. We use the 680 white observations to calculate the mean $(\overline{X}_1)$ and variance $(s_1{}^2)$ for the whites; and similarly use the 320 black observations to calculate $\overline{X}_2$ and $s_2{}^2$ for the blacks. Then we use these sample means to estimate the population mean μ in (23-4) (noting, of course, that in this formula we use the *true* population weights .70 and .30, rather than the less accurate sample weights .68 and .32). This provides an estimate that is about as good as the prespecified proportional sampling in (23-8).

An example will show how effective this stratification can be:

EXAMPLE **23-1**

When a shipment traveled via several railroads (or airlines, or other carriers), the freight charges were all recorded on a *waybill*. The total charge was customarily collected by the originating railroad, and later divided appropriately. Before computers, this division required a calculation that was complex and time-consuming; at that time, the complete annual accounting of thousands of waybills became so expensive in time and money that a sampling scheme was tried—and found very successful (Neter, 1976).

Here is how it worked: Suppose, for example, that Railroad A took a simple random sample of 1000 of its 200,000 waybills. The amount X that it owed Railroad B on each waybill was calculated, and all 1000 of these observed debts to B were summarized with $\overline{X} = \$13.90$ and $s = \$18.71$.

Since stratifying would improve accuracy, the population and sample were sorted according to the total charge T on the waybill, with the following results:

STRATUM DEFINED BY THE TOTAL CHARGE *T* ON THE WAYBILL	KNOWN NUMBER OF WAYBILLS IN EACH STRATUM OF THE POPULATION	NUMBER IN EACH STRATUM OF THE SAMPLE, n_i	AVERAGE AMOUNT THAT B IS OWED, $\overline{X}_i$	STANDARD DEVIATION s_i
$0 < T \leq \$25$	104,000	540	$ 4.20	$ 4.50
$\$25 < T \leq \$100$	68,000	330	$18.70	$12.10
$\$100 < T$	28,000	130	$42.00	$31.60
Totals	200,000	1000		

Estimate the total amount owed B, including a 95% confidence interval:

a. Using the simple random sample.

b. Using stratification after sampling.

Solution   **a.**

$$\text{var } \overline{X} = \frac{\sigma^2}{n} \approx \frac{s^2}{n} \qquad \text{(6-6) repeated}$$

$$= \frac{18.71^2}{1000} = .35 \qquad \text{(23-9)}$$

Then the 95% confidence interval for μ is:

$$\mu = \overline{X} \pm 1.96 \text{ SE} \qquad \text{like (8-8)}$$

$$= 13.90 \pm 1.96 \sqrt{.35} \qquad \text{(23-10)}$$

$$= 13.90 \pm 1.16$$

Finally, the *total* amount owed B is just the population size times the mean μ:

$$\text{Total} = 200{,}000 \, (13.90 \pm 1.16)$$
$$= \$2{,}780{,}000 \pm \$232{,}000 \tag{23-11}$$

b. We first need the population proportions w_i found from the second column of the given table:

$$w_1 = \frac{104{,}000}{200{,}000} = .52$$

Similarly, $w_2 = .34$ and $w_3 = .14$.

Then we can calculate the stratified sample mean:

$$\overline{X}_s = w_1\overline{X}_1 + w_2\overline{X}_2 + \cdots \tag{23-4 repeated}$$

$$= .52 \, (4.20) + .34 \, (18.70) + .14 \, (42.00)$$

$$= 14.42$$

Similarly, we can calculate its variance:

$$\text{var } \overline{X}_s \simeq w_1{}^2 \frac{s_1{}^2}{n_1} + w_2{}^2 \frac{s_2{}^2}{n_2} + \cdots \tag{23-7 repeated}$$

$$= .52^2 \frac{(4.50)^2}{540} + .34^2 \frac{(12.10)^2}{330} + .14 \frac{(31.60)^2}{130}$$

$$= .20$$

We note with satisfaction that this variance is about half the variance of $\overline{X}$ in (23-9). This illustrates the value of stratification.

To find the 95% confidence interval for μ, we follow the same steps as in **a**:

$$\mu = 14.42 \pm 1.96 \sqrt{.20} \qquad \text{like (23-10)}$$

$$= 14.42 \pm .88$$

$$\text{Total} = 200{,}000 \, (14.42 \pm .88)$$

$$= \$2{,}884{,}000 \pm \$176{,}000 \tag{23-12}$$

That is, Railroad A would pay $2,884,000 to Railroad B, and both railroads could be assured this figure comes reasonably close to what a full and expensive accounting would disclose.

E—OPTIMAL STRATIFICATION

If we can allocate the sample sizes n_i in advance, can we improve upon proportional stratification? In the proportional approach (23-8), recall that we take more observations in stratum i if its size w_i is large—that is, if this stratum is a large part of the population. Now consider two other reasons for taking more observations in stratum i:

i. If σ_i is large—that is, if this stratum is highly varied, and hence requires a large sample to estimate $\overline{X}_i$ accurately. (Or, to restate: The smaller is σ_i, the smaller is the necessary sample. The limiting case is a stratum with very little variance, when only 1 observation is required—as shown already in column (4) of Table 23-1.)

ii. If the cost c_i of sampling each observation in this stratum is relatively low. That is, we should take more observations where they cost less. (We assume a simple model, where each observation within stratum i costs the same amount, c_i)

If we have a fixed budget (an allowable cost C_o let us say), how can we get the most accurate estimate of μ? All three factors above (w_i, σ_i, and c_i) can be taken into account in a mathematical analysis that finally arrives at the appropriate formula:

> Optimal sampling requires setting the sample size in stratum i at:
>
> $$n_i = k \frac{w_i \, \sigma_i}{\sqrt{c_i}}$$

(23-13)

where the constant of proportionality k can be determined from the budget C_o and turns out to be:

$$k = \frac{C_o}{\Sigma \, w_i \sigma_i \sqrt{c_i}}$$

(23-14)

Generally, the variances σ_i^2 and the costs c_i are not known beforehand, so it is advisable to proceed in two stages:

1. After part of the sample has been taken on a simple random basis, the s_i observed so far may be used to estimate σ_i, and the costs incurred in the sampling so far may be used to estimate c_i.
2. These estimates are then used (along with prior information about w_i) in (23-13) to estimate the best n_i—that is, the best design for the rest of the sampling.

PROBLEMS

23-1 In the same kind of sampling scheme as Example 23-1, airline A took a simple random sample of 5000 of the 100,000 tickets purchased from it. (On each ticket, the passenger was carried by both airline A and airline B.) The amount X that airline A owed B on each ticket was calculated, and all 5000 were summarized with $\overline{X}$ = \$53.20 and s = \$108.00.

Stratification according to the total cost T of the ticket gave further detail:

STRATUM DEFINED BY THE TOTAL COST T OF THE TICKET	KNOWN NUMBER OF TICKETS IN EACH STRATUM OF THE POPULATION	NUMBER IN EACH STRATUM OF THE SAMPLE, n_i	AVERAGE AMOUNT THAT B IS OWED, $\overline{X}_i$	STANDARD DEVIATION s_i
$0 < T \leq \$100$	50,000	2470	\$ 16	\$ 32
$\$100 < T \leq \$250$	20,000	1020	\$ 29	\$ 51
$\$250 < T \leq \$500$	20,000	990	\$ 67	\$ 86
$\$500 < T$	10,000	520	\$251	\$202
Totals	100,000	5000		

Calculate the total amount owed B, including a 95% confidence interval:

a. Using the simple random sample.

b. Using stratification after sampling.

23-2 In Problem 23-1, suppose stratification *before* sampling had been planned, so that the sample sizes n_i *perfectly* reflected the population sizes:

STRATUM	POPULATION FREQUENCY	PRESPECIFIED SAMPLE FREQUENCY, n_i	$\overline{X}_i$	s_i
$0 < T \$100$	50,000	2500	\$ 15	\$ 35
$\$100 < T \leq \$250$	20,000	1000	\$ 29	\$ 62
$\$250 < T \leq \$500$	20,000	1000	\$ 63	\$ 75
$\$500 < T$	10,000	500	\$246	\$164
Totals	100,000	5000		

Now calculate the total amount owed B, including a 95% confidence interval.

23-3 A sample survey of TV watching habits was commissioned by a local TV station for its advertisers. Of the 1000 adults who were randomly selected to be mailed a questionnaire, 800 failed to respond. Rather than pursue all 800, a 10% subsample of 80 was randomly selected for a costly followup that persisted until all 80 finally responded.[1]

This can be regarded as a stratified sample, and here are the results for X = the number of hours of TV watched in the previous 7 days.

---

[1]In practice, there are always a few who *never* respond, and require special allowance. For simplicity, however, we ignore this problem here.

STRATUM	$\overline{X}_i$	s_i	SAMPLE SIZE n_i	APPROXIMATE POPULATION PROPORTION w_i
Early responders	36	14	200	.20
Reluctant responders	11	8	80	.80

   a. Calculate an unbiased estimate of the population mean μ.

   b. If the survey team had decided to ignore the nonrespondents, what would their estimate have been? How useful would that answer be?

23-4 Referring to Problem 23-1, suppose the large tickets were more expensive to process—the cost for a ticket in stratum 1, 2, 3, and 4 averaged out to be $.25, $.36, $.64, and $1.00, respectively.

   a. What was the cost of the stratified sampling in Problem 23-1?

   b. For the same cost, what would the sample sizes be if optimal stratified sampling were used?

   c. Then how much would the optimal design reduce the uncertainty (SE) of the estimate?

   d. The three important tools we have studied in this case are:

     i. *Random sampling* (simple random sampling).

     ii. *Stratified* random sampling.

     iii. *Optimal* stratified random sampling.

   Briefly summarize the value of each successive level of sophistication.

23-5 Repeat Problem 23-4 for the data in Example 23-1. Assume the cost of sampling a waybill in stratum 1, 2, and 3 averaged out to be $.25, $.36, and $.64, respectively.

*23-6 Prove formulas (23-5) and (23-6).

23-2 Other Sampling Designs

A—MULTISTAGE AND CLUSTER SAMPLING

When a population is expensive to sample because of high transportation costs, multistage sampling may provide a way to cut costs, as we illustrated already in Section 1-1. For a more detailed example, consider the three-stage sampling used by the U.S. Bureau of the Census in their Current Population Survey:

   1. The U.S. is divided up roughly into 2000 areas, called primary sampling units (PSU), each one being a city or small group of contiguous counties. At the first stage, about 500 PSUs are randomly selected.

   2. Each of these PSUs is divided up very roughly into 100 Enumeration Districts (ED). About 5 of these EDs are then randomly selected.

3. Each of these EDs is divided up very roughly into 100 Ultimate Sampling Units (USU). About 5 of these USUs are then randomly selected.

Each USU consists of roughly 10 people in a cluster of 3 or 4 households, and all of them are selected for the survey. When *all* of the people in the ultimate unit or cluster are selected like this, it is called *cluster sampling*. So the Current Population Survey may be called a multistage cluster sample or, more specifically, a *three-stage cluster sample*.

It would be interesting to calculate the chances that *you* (if you are an adult American) will get selected in the next Current Population Survey. In Table 23-3 we see, for example, that about 500 out of the 2000 PSUs are chosen; then in each of these PSUs, about 5 out of the 100 EDs are chosen; and so on. Thus the chances you will be chosen are $125,000/200,000,000 \simeq .006$.

Clearly, multistage cluster sampling provides great savings in transportation costs by concentrating the sampling in a few clusters in a few areas. Yet this same clustering produces less accuracy: To the extent that the people in a chosen cluster are "homogeneous"—have similar incomes, for example—only the first observation provides completely fresh information, while later observations just tend to repeat it. (But this cost is often small compared to the benefit, so that cluster sampling is very widely used.)

B—SYSTEMATIC SAMPLING

As an example, we may wish to draw a sample of 100 from a deck of 5000 IBM cards. One straightforward way is to be *systematic*: pull every 50th card. This method usually has about the same precision as simple random sampling, so those formulas may be used.

The only drawback of systematic sampling is that its precision may badly deteriorate if systematic fluctuations (*periodicities*) occur in the data (as would be the case, for example, if you were to sample the density of traffic at 5 p.m. every Friday). If this is suspected as a possible problem, one could break the sample down into several subsamples that can be compared. To return to our example of IBM cards, instead of one systematic sample of 100 cards, one could select 5 samples of 20 cards each

TABLE **23-3** **The Three Stages of Sampling Give the Rough Probability that You Will Be Selected for Next Month's Current Population Survey**

At each stage, sampling selects ...	PSUs		EDs		USUs		People		
Number sampled	500	×	5	×	5	×	10	=	125,000
Number available →	2000		100		100		10		200,000,000

(with each of these 5 samples beginning at a *randomly* chosen point in the deck).

C—SEQUENTIAL SAMPLING

In sequential sampling, the sample size n is no longer fixed, but instead is revised as we accumulate sample information. For example, suppose we are testing whether a new drug is an improvement on the old drug, using a series of matched pairs (e.g., twins, one taking the new drug and the other taking the old). If, in every one of the first 20 pairs sampled, the new drug is better than the old, then a decision can be made immediately. A predetermined sample size like $n = 100$ is just not necessary.

This procedure already has been encountered in the footnote following Table 9-1. The idea is to take a small sample, and if the result is sufficiently clear (as in the case above), the job is done; but if the sample is ambiguous, a second, and perhaps even further stage of sampling is required until the ambiguity is resolved. This procedure is justified if the expected advantage of the smaller sample size outweighs the disadvantage of having to resume sampling several times.

PROBLEMS

23-7   Choose the correct alternative in each bracket:

  a. In the last stage, multistage sampling often observes everyone in the final sampling unit—or, as it's called by the Current Population Survey, the [USU, PSU]. Then it is called [cluster, systematic] sampling.

  b. Cluster sampling—and multistage sampling—work best when each cluster is relatively [heterogeneous, homogeneous].

  c. When testing an hypothesis H_o against an alternative H_1, [systematic, sequential] sampling allows you to exploit the early results to determine whether it is necessary to continue sampling. Thus the sample size n will be very short when the early results are [overwhelming, indecisive], and longer otherwise. So n must be regarded as a [random variable, fixed parameter].

   The advantage of this kind of sampling is that the expected sample size will be (less, greater) than the sample size of the comparable simple random sample.

  d. Systematic sampling [works best, may be disastrous] when there are unsuspected periodicities in the data. For example, to determine the average outgoing quality of automobiles from the Ford Ypsilanti plant, sampling each Friday afternoon would be [more biased, more fairly representative] than simple random sampling.

CHAPTER 23 SUMMARY

23-1   When a population consists of several relatively *homogeneous* strata, then stratified sampling is much more efficient than simple random sampling.

23-2   When a population consists of relatively *heterogeneous* strata, then multistage cluster sampling can greatly cut down sampling costs with little corresponding loss in information.

There are many other convenient and ingenious variations of sampling as well—such as systematic sampling and sequential sampling.

REVIEW PROBLEMS

23-8   In each bracket, select the correct choice:

a.   Stratified sampling is relatively effective when most of the variation occurs [between, within] strata. Similarly, cluster sampling is relatively effective when most of the variation occurs [between, within] clusters. Yet at some stage, [the first design, each design] requires random sampling as part of its procedure.

b.   Stratified and multistage cluster sampling designs can be used together, often very effectively. Whether used together or alone, each design is intended to provide more information per budgeted dollar than [simple random sampling, optimal stratified sampling].

c.   Stratified sampling yields an estimator of μ that is [biased, unbiased] and has [smaller, larger] variance than the simple random sampling estimate $\overline{X}$.

d.   The crucial prior knowledge required in stratified sampling is to know [the variance σ_i^2, the relative size w_i] of each stratum.

23-9   In order to investigate student political attitudes, suppose you wish to take a sample of students from your university. How would you take:

a.   A simple random sample?

b.   A stratified sample?

c.   A multistage cluster sample?

23-10   In order to estimate the total inventory of its product that is being held, a tire company conducts a stratified sample of its dealers, with these dealers being stratified according to their inventory held over from the previous year. For a total sample size of $n = 1000$, the following data were obtained:

STRATUM (LAST YEAR'S INVENTORY)	POPULATION		STRATIFIED SAMPLE (CURRENT INVENTORY)		
	NUMBER OF DEALERS	PROPORTION w_i	PRESPECIFIED n_i	$\overline{Y}_i$	s_i^2
0–99	4,000	.20	200	105	1600
100–199	10,000	.50	500	180	2500
200–299	5,400	.27	270	270	2500
300–	600	.03	30	390	5600
Totals	20,000	1.00	1000		

a. Is the design a *proportional* stratified sample?

b. From this stratified sample, estimate the mean inventory μ, and hence the total inventory. Include a 95% confidence interval.

*23-11 Compare the stratified sampling in Problem 23-10 with the regression analysis in Problem 12-17. Then answer True or False; if False, correct it:

a. Both methods involve the use of prior information (about last year's inventory X) to improve the estimate of this year's inventory Y.

b. Proportional stratified sampling holds two advantages over regression:

i. It can be applied to either a numerical stratification (e.g., size of inventory), or a categorical stratification (e.g., male/female); but regression can only be used for a numerical stratification.

ii. It does not require the assumption that the relation between X and Y is linear (which is required in regression analysis).

*23-12 Ten days before the 1980 Presidential election, a Gallup poll of 600 men showed 51% in favor of Reagan; and of 600 women, 42% in favor of Reagan (ignoring third-party candidates, as usual). Construct an appropriate 95% confidence interval for:

a. The difference between men and women in the whole population.

b. The proportion of all adults (men *and* women) who were in favor of Reagan, assuming:

i. That men and women were equally numerous.

ii. That women outnumbered men by about 52% to 48%—the 1978 voting figures. (Hint: The variance in each stratum is estimated from (6-20), with P substituted for π.)

*23-13   Referring to an earlier ingenious sampling design for sample surveys discussed in Problem 3-48, answer True or False; if False, correct it:

    a.   Randomized response is an ingenious way to obtain reliable statistics for a population, while protecting the individual's confidentiality.

    b.   This protection is absolutely airtight—there is no way the interviewer, or the courts, or a computer bandit, or *anyone*, can know about any given individual.

Chapter

Time Series

(Requires Chapter 15)

*My interest is in the future because I am going to spend the rest
of my life there.*   C. F. KETTERING

There are two major categories of statistical information: cross section and
time series. To illustrate, econometricians estimating how U.S. consumer
expenditure is related to national income ("the consumption function")
sometimes use a detailed breakdown of the consumption of individuals
at various income levels at one point in time (cross section); at other times,
they examine how total consumption is related to national income over
a number of time periods (time series); and sometimes they use a combination of the two. In this chapter, we shall use some familiar techniques
(especially regression) and develop some new methods to analyze time
series. Although our examples will often use annual or quarterly data, the
techniques are equally applicable to monthly data, weekly data, and so
on.

24-1   Two Special Characteristics of a Time Series

Figure 24-1 shows a time series with a *trend*, a simple regression of wheat
price on time. In earlier chapters, any such regression fit came with a
warning: There are important characteristics of a time series that keep us
from using this sort of simple trend line to forecast. We begin by considering two.

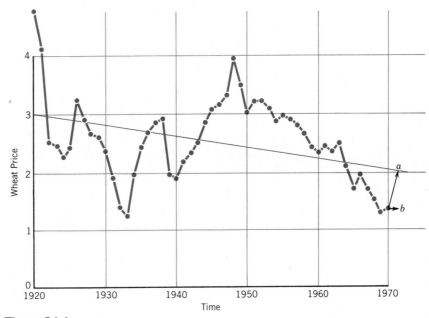

Figure 24-1
Serial correlation in a time series: When previous observations are low, the next tends to be low. (The price of wheat relative to the consumer price index, i.e., adjusted for inflation. From Simon, 1982).

A—TIME SERIES HAVE SERIAL CORRELATION

One special pattern of time series is serial correlation—when each observation is statistically dependent on the previous ones. In Figure 24-1, for example, after years of low prices, next year's price is more likely to be low again.

This is more difficult to analyze than the customary regression models we have dealt with so far, where observations were independent. For example, suppose we have reached 1970 in Figure 24-1, and wish to forecast next year's price. It would be foolish indeed to use the fitted regression line (i.e., use arrow *a* to predict) because this would imply a sudden upward leap in price. It would be much better to ignore the fitted regression and simply draw in a freehand prediction such as arrow *b*, which assumes that next year's price will be the same as this year's. In this chapter, we will see how we can do even better than either of these naive alternatives.

Since serial correlation can cause such difficulties, it is important to have a simple and routine test for whether or not it is present. This is provided by the Durbin-Watson test, which is customarily computed at the end of every regression, and which we return to later.

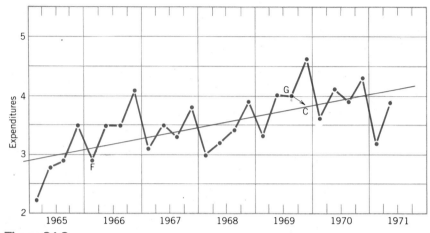

Figure 24-2

Seasonal variation in a time series—fourth quarter is always high (U.S. new plant and equipment expenditures, billions of dollars, for manufacturing durable goods).

B—TIME SERIES OFTEN HAVE A SEASONAL PATTERN TOO

Another feature that occurs in many time series is *seasonal variation* (or monthly or other periodic variation). The quarterly data of Figure 24-2, for example, displays a seasonal pattern so strong that we can discern its main features at a glance: The fourth quarter of every year tends to be high, followed by a big drop in the next quarter.[1]

To see why it is essential to take account of seasonal variation, we might ask: "Was observation F in the first quarter of 1966 disappointing?" (This might be important, for example, in evaluating a new government policy such as tax advantages to stimulate expenditures.) At first glance, it seems that F was indeed disappointing, since it was a big drop. But then we notice that it was not as big a drop as in the other first quarters. So we conclude that F reflects relatively good performance.

Other examples abound: In evaluating whether or not a government has been successful in increasing employment, the observation that employment has increased in the spring of 1980 would not be convincing evidence if, as part of a seasonal pattern, employment customarily rises every spring anyhow.

Thus when a series has seasonal variation, we must keep this in mind in evaluating a single observation—or in forecasting. For example, if we have reached point G in 1969, it would be a mistake to forecast the next (fourth quarter) observation on the fitted regression line, that is, to follow arrow C down. This would ignore the large increase that has occurred in all the recent fourth quarters.

---

[1]Do wheat prices in Figure 24-1 have a seasonal pattern? There is no way to know because the *annual* data in that figure provides no information on the different seasons.

24-2 Decomposition and Forecasting Using Regression

In this section, we will use the familiar technique of multiple regression to illustrate two major objectives of time series analysis:

1. To *understand* the time series better, by *decomposing* it. That is, we estimate each of the patterns that make up the time series in order to get a picture of their relative importance. (The estimate of the seasonal pattern is also valuable for its own sake: We can use it for seasonal adjustment.)
2. To *forecast* future values from the observed past values of the time series.

If a time series followed only one of the patterns we have described above (trend, seasonal, or serial correlation), there would be few problems. In practice, however, it is typically a mixture of all three that is very difficult to unscramble. Consider again, for example, the quarterly data on capital expenditures shown in Figure 24-2. What combination of these three patterns can be perceived in this time series? We've already noted that there obviously appears to be some trend and some seasonal pattern. There is also some serial correlation, but precisely *how much* of each of these components is a mystery. Let's now solve that mystery.

A—TREND

Trend is often the most important element in a time series. It may be linear, as illustrated in Figure 24-2, growing at a *constant amount* over time. Or it may follow any of the nonlinear patterns discussed in Chapter 14; for example, it may be exponential, growing at a *constant rate* over time (in which case it would be linear on a log scale).

We will show how regression can be used to estimate trend in a way that will also capture the seasonal pattern described next.

B—SEASONAL PATTERN

There may be seasonal fluctuation in a time series for many reasons: A holiday such as Christmas results in completely different purchasing patterns. Or the seasons may affect economic activity; in the summer, agricultural production is high, while the sale of ski equipment is low. Or a seasonal pattern may result from tax laws; at the end of the year, people may buy or sell stocks simply to cash in their gains or losses for tax reasons.

To illustrate how regression can handle trend and seasonal patterns together, we take a really spectacular example—jewelry sales, as shown in Figure 24-3a. While there seems to be a slight upward trend (which we assume is linear for simplicity) the clearest pattern is a seasonal one, marked by the sharp rise in sales every fourth quarter because of Christmas.

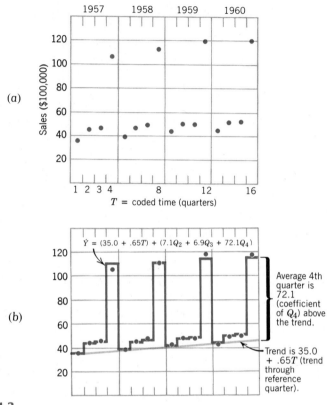

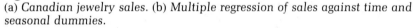

Figure 24-3
(a) *Canadian jewelry sales.* (b) *Multiple regression of sales against time and seasonal dummies.*

These fourth quarter observations can be handled very easily with a dummy variable:

$$Q_4 = 1 \text{ if fourth quarter} \atop = 0 \text{ otherwise} \Big\}$$

$$\text{(24-1)}$$
$$\text{like (14-2)}$$

Similarly, we can define dummy variables Q_2 and Q_3 for the second and third quarters. (The first quarter is left as the reference quarter, and so does not need a dummy variable. Thus, as always, the number of dummies will be 1 less than the number of categories.) These dummy variables are tabulated in Table 24-1 along with the rest of the data ready to input to the computer. The computed least squares fit[2] turned out to be:

---

[2]For time series, ordinary least squares (OLS) can be improved upon. A technique called *generalized* least squares (GLS) allows for serial correlation and so is more efficient (as we explain in Section 24-6).

TABLE **24-1** **Canadian Department Store Jewelry Sales and Seasonal Dummies**

SALES, Y ($100,000's)	TIME, T (QUARTER YEARS)		Q_4	Q_3	Q_2
36	1957	1	0	0	0
44		2	0	0	1
45		3	0	1	0
106		4	1	0	0
38	1958	5	0	0	0
46		6	0	0	1
47		7	0	1	0
112		8	1	0	0
42	1959	9	0	0	0
49		10	0	0	1
48		11	0	1	0
118		12	1	0	0
42	1960	13	0	0	0
50		14	0	0	1
51		15	0	1	0
118		16	1	0	0

$$\hat{Y} = \underbrace{35.0 + .65T}_{\text{trend}} + \underbrace{7.1Q_2 + 6.9Q_3 + 72.1Q_4}_{\text{seasonal pattern}} \qquad (24\text{-}2)$$

This is graphed as the jagged set of blocks in Figure 24-3b. Notice that we have used a technique that ensures the seasonal pattern is exactly the same every year; for example, each fourth quarter is the same amount (72.1) above the reference quarter (the first quarter).

Now we can decompose the original time series in the left-hand (unshaded) side of Figure 24-4. Panels (a) and (b) show the fitted trend and seasonal components, in sum comprising the fitted regression (in Figure 24-3). The difference between the fitted regression and the observed dots (in Figure 24-3) is the residual series, reproduced in panel (c) of Figure 24-4. Thus the total jewelry sales series in panel (d) has been broken down into its three components in panels (a), (b), and (c).

With the trend and seasonal components captured by regression, this leaves the residual yet to be analyzed.

C—SERIALLY CORRELATED RESIDUAL

We have seen that, for any time period t, the difference between the least squares fit $\hat{Y}_t$ and the actual observed value Y_t is the *residual* term $\hat{e}_t$. Formally:

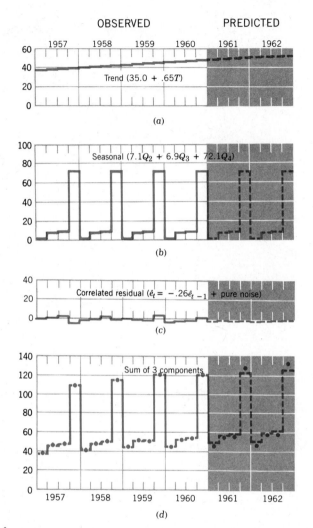

Figure 24-4

Decomposition of jewelry time series into its three components. On the right, each is projected with dashed lines into the gray future. (a) Trend component. (b) Seasonal component. (c) Residual. (d) Total (original) series of observed dots on the left is the sum of the three components above. On the right, compare the dashed-line projection with the actual sales (as they eventually turned out) shown as dots.

$$\text{Residual } \hat{e}_t = Y_t - \hat{Y}_t \tag{24-3}$$

For the time series we are analysing here, the residual in Figure 24-4c is so small that for many practical purposes it could be neglected. However, for most series the residual is substantial, and so we will examine in some detail how it should be treated. The model we use relates each e_t to its previous value e_{t-1} in a regression equation:

$$e_t = \rho e_{t-1} + v_t \qquad (24\text{-}4)$$

where ρ represents the strength of the serial correlation.

Note the slight differences from the ordinary regression model (12-3): there is no constant term; e_t is regressed on its own previous (lagged) value e_{t-1}, rather than on some other variable; and the regression coefficient is denoted by ρ instead of β. But we still assume that the v_t are independent and, hence, entirely unpredictable. Thus they are often called *pure noise* or *white noise* by time series analysts.

Because (24-4) is a regression of e_t on its own previous values, it is called an *autoregression* as well as a *serial regression*. To illustrate how it works, let us construct a simple case where $\rho = 1$, and the white noise term v_t is standard normal. Accordingly, let us draw a sample of 20 independent values of v_t from Appendix Table II, and graph them in panel (a) of Figure 24-5. Then starting with an initial value of $e_0 = 5$, for example, we use (24-4) to generate $e_1, e_2, e_3, \ldots, e_{20}$ as shown in panel (b). The existence of serial correlation is clear: e_t tends to be high whenever the previous value e_{t-1} is high, or low whenever e_{t-1} is low.

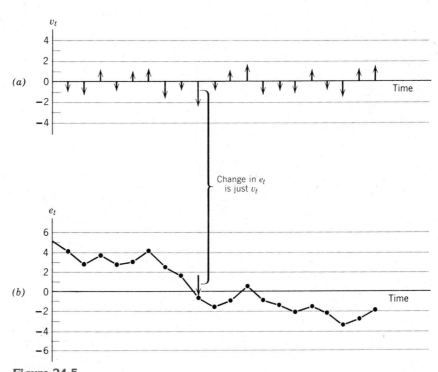

Figure 24-5
The construction of serially correlated e_t. (a) Independent v_t (white noise). (b) When $\rho = 1$, then $e_t = e_{t-1} + v_t$.

In practice, the parameter ρ is typically less than 1. Its estimate r can be calculated by applying the OLS formulas to the residuals $\hat{e}_t$. For example, the autoregression used to estimate the residuals of jewelry sales data in Figure 24-4c was

$$\hat{e}_t = r\hat{e}_{t-1} \qquad\qquad\qquad\qquad (24\text{-}5)$$

estimates (24-4)

$$\hat{e}_t = -.26\hat{e}_{t-1} \qquad\qquad\qquad (24\text{-}6)$$

The details of (24-6) are given in Appendix 24-2, which also shows that r is essentially the correlation coefficient between $\hat{e}_t$ and $\hat{e}_{t-1}$. This is why we use the symbol r (and ρ), and call it the *serial correlation*. The Durbin-Watson test for serial correlation (also detailed in the same appendix) can be viewed essentially as just a test of whether r is discernibly different from 0.

The estimate $r = -.26$ in (24-6) shows that autocorrelation can be negative, as well as positive. In this case, values tend to change sign (positive values tend to be followed by negative ones, rather than tracking themselves over time—as they did when positively correlated in Figure 24-5. Inventory series may provide another example of negative serial correlation, if particularly large purchases for inventory result in overstocking, hence smaller purchases in the following period.)

Finally, let us consider the generalization of (24-4):

$$\boxed{e_t = \rho_1 e_{t-1} + \rho_2 e_{t-2} + \cdots + \rho_k e_{t-k} + v_t} \qquad (24\text{-}7)$$

This could be called a multiple autoregression or, more commonly, a *k*th order *autoregression*.

D—FORECASTING

To review the decomposition of a time series such as jewelry sales into its components, we rewrite (24-3) as

$$Y = \hat{Y} + \hat{e}$$

When we substitute $\hat{Y}$ given in (24-2), the result is

$$\boxed{\begin{array}{l} Y = (35.0 + .65T) + (7.1Q_2 + 6.9Q_3 + 72.1Q_4) + (\hat{e}) \\ \text{Time series} = \text{trend} \quad + \quad \text{seasonal} \quad + \quad \text{residual} \end{array}} \qquad (24\text{-}8)$$

Of course, this is just the equation form of the graphical breakdown of a time series into its components shown in the first three panels of Figure 24-4—with panel (d) showing the way they all add up. To forecast the time series, we simply forecast each component in the first three panels, and add them up[3] in panel (d). These forecast values are shown in gray.

While forecasting the first two components in panels (a) and (b) involve no problem, how do we forecast the third component (the serially correlated residual) with its more erratic pattern? The answer is just to use the autoregression fit:

$$\hat{e}_t = r\hat{e}_{t-1} = -.26\hat{e}_{t-1} \qquad (24\text{-}9)$$
$$(24\text{-}6) \text{ repeated}$$

For example, the last $\hat{e}_t$ available was in the 16th quarter, and happened to be $\hat{e}_{16} = .5$. We can use this to forecast $\hat{e}_t$ in the next quarter:

$$\hat{e}_{17} = -.26\hat{e}_{16} = -.26(.5) = -.13 \qquad \text{like (24-9)}$$

With this in hand, we can forecast the next one again:

$$\hat{e}_{18} = -.26\hat{e}_{17} = -.26(-.13) = +.03$$

$$\hat{e}_{19} = -.26\hat{e}_{18} = -.26(.03) = -.01$$

Notice how this component, graphed in panel (c) of Figure 24-4, quickly dwindles away to zero in the future.

Turning finally to the forecast of the whole time series in panel (d), we wonder: With the benefit of hindsight, how well did it work out? Specifically, how close were the forecast values to the actual values we were later able to observe, which we continue to show as colored dots? We see that the forecast was very good for 1961, but not quite so good for 1962 because we were forecasting further into the future.

This is exactly what we would expect: The shorter the forecast, the better. Thus if funds and time are available, a forecast should be recomputed, to update it whenever new observations become available; specifically, the forecast for 1962 should be updated as soon as the 1961 figures become available.

While the regression analysis of this section has shown how a time series can be decomposed and used to forecast, more complex methods are often used in practice. Many of these have been developed *specifically for time series*: In Section 24-3, for example, we will look at a traditional

---

[3]For example, by evaluating each of the three components in (24-8) we predict the 17th and 18th quarters to be:

COMPONENT	T = 17	T = 18
Trend		
$35.0 + .65T$	$35.0 + .65(17) = 46.05$	$35.0 + .65(18) = 46.70$
Seasonal		
$7.1 Q_2 + 6.9 Q_3 + 72.1 Q_4$	Reference quarter = 0	Setting $Q_2 = 1$, $7.1(1) = 7.1$
Residual		
$-.26\hat{e}_{t-1}$	$-.26(.5) = -.13$	$-.26(-.13) = .03$
Total	45.92	53.83

decomposition that predates the computer, and in Sections 24-4 and
24-5 we will look at increasingly sophisticated forecasting techniques.

PROBLEMS

24-1   Given the following data on new car sales (in hundreds, *Statistics Canada*):

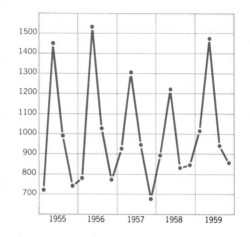

a.   Fit a linear trend by eye.

b.   Shift the linear trend in a parallel way so that it passes as close as
possible to the first quarter (reference quarter) points. Then roughly
estimate how much higher the second quarters rise, on average, above
this reference line. Repeat for the third and fourth quarters. This
produces an eyeball estimate of the seasonal component.

c.   Graph your eyeball fit as a jagged set of blocks, as in Figure 24-3*b*.

d.   If a trend-and-seasonal regression model like (24-2) was fitted, use
your eyeball fit to roughly estimate what the coefficients would be.

24-2   (Seasonal Adjustment) The car sales from Problem 24-1 above were fitted
with a regression equation like (24-2), and the computed seasonal component was 530 Q_2 + 80 Q_3 − 90 Q_4. If we tabulate this seasonal component S, we obtain the table at the end.

a.   Calculate the mean of the seasonal series $\bar{S}$. Then tabulate the seasonal
series in deviation form, $s = S − \bar{S}$. This gives us a seasonal series
that fluctuates up and down *around zero*.

b.   Subtract s from the original series of auto sales, Y. This new series
estimates how auto sales would move if there were no seasonal component, and so is called the *seasonally adjusted* series.

c.   Graph the seasonally adjusted series on the figure in Problem 24-1
for comparison.

d. Comparing the second quarter of 1957 with the first quarter, what was the increase in:

  i. Auto sales?

  ii. Seasonally adjusted auto sales?

Which of these figures is more meaningful, in terms of indicating how well the auto industry fared in that quarter?

AUTO SALES, Y		SEASONAL, S	(a) SEASONAL DEVIATIONS, s	(b) SEASONALLY ADJUSTED AUTO SALES, $Y - s$
1955	710	0		
	1440	530		
	980	80		
	740	−90		
1956	770	0		
	1530	530		
	1020	80		
	760	−90		
1957	920	0		
	1300	530		
	940	80		
	660	−90		
1958	890	0		
	1220	530		
	820	80		
	840	−90		
1959	1000	0		
	1470	530		
	930	80		
	850	−90		

24-3 **a.** Simulate a string of 10 serially correlated residuals e_t using Equation (24-4), with $\rho = .9$. Start with $e_o = 2$, say, and take v_t from Appendix Table II.

b. Graph e_t. Does it show random tracking like Figure 24-5?

24-4 **Average weekly carloadings (in thousands, from Moody, 1975)**

	QUARTER			
YEAR	1	2	3	4
1968	508	566	550	544
1969	507	562	545	557
1970	445	550	523	513
1971	486	526	477	457

The data above was fitted with a multiple regression (for $T = 1, 2, \ldots 16$):

$$\hat{Y}_t = 522 - 5.0T + 69Q_2 + 47Q_3 + 46Q_4$$

a. Calculate the residual $\hat{e}_t$.

b. Graph $\hat{e}_t$ against $\hat{e}_{t-1}$, and then fit by eye a straight line through the origin; that is, roughly estimate r in the simple autoregression:

$$\hat{e}_t = r\hat{e}_{t-1}$$

c. Forecast Y_t into the next 4 quarters. Compare to the four actual values: 468, 514, 500, 511.

d. Seasonally adjust the time series Y_t (by subtracting the zero-average seasonal component, as in Problem 24-2). Then, for example, how did carloadings in the second quarter of 1971 compare (on a seasonally adjusted basis) to carloadings in the first quarter?

e. On one graph, show:

i. Y_t and seasonally adjusted Y_t^a, for 1970 and 1971.

ii. Projected $\hat{Y}_t$ and actual Y_t, for 1972.

24-3 Traditional Decomposition: Ratio to Moving Average

In this section we examine a traditional method of decomposing a time series and making seasonal adjustments. Although it predates the computer, in one form or another it is still commonly used. (The U.S. Bureau of the Census uses a variation called X-11, for example, that provides short range forecasts as well as seasonal adjustment.)

This traditional decomposition of a time series Y assumes it has 4 components. These correspond *roughly* to the 3 components analyzed earlier in (24-8)—trend, seasonal, and residual—except that our earlier residual is now further broken down into a cyclical part C and a disturbance D. In other words, the components of a time series are:

Trend, Seasonal, Cyclical, and Disturbance (T, S, C, and D)   (24-10)

The strategy for actually carrying out this decomposition, and then using it to seasonally adjust the time series, is summarized at the end of this section in Table 24-3. You may find it helpful to check off the various steps in this analysis as you progress through this section.

A—CAPTURING *T* AND *C* WITH A MOVING AVERAGE

The first step is to smooth the time series by averaging. Specifically, we average the first four quarters, then move on to average the next four quarters and so on—thus forming the *moving average* shown in column 3 of Table 24-2. However, since we want to center each average on a particular quarter—rather than leaving it stranded half-way between two quarters—we average again (this time as simply as possible, two at a time). Thus we obtain the *centered moving average M* in column 4. (As Appendix 24-3 shows, we could express this with a formula, and generalize.)

TABLE **24-2** **Ratio-to-Moving-Average (Auto sales, from Figure 24-6)**

(1) TIME		(2) SALES Y	(3) MOVING AVERAGE	(4) CENTERED MOVING AVERAGE M	(5) Y/M	(6) SEASONAL S CALCULATED	(7) SEASONAL S REPEATED	(8) SEASONALLY ADJUSTED $Y^a = Y/S$
1955	1	710					.91	780
	2	1440					1.39	1040
	3	980	968	975	1.01		.96	1020
	4	740	982	994	.74		.75	990
			1005					
1956	1	770	1015	1010	.76	ave = .91	.91	850
	2	1530	1020	1018	1.50	1.39	1.39	1100
	3	1020	1058	1039	.98	.96	.96	1060
	4	760	1000	1029	.74	.75	.75	1010
1957	1	920	980	990	.93		.91	1010
	2	1300	955	968	1.34		1.39	940
	3	940	947	951	.99		.96	980
	4	660	928	938	.70		.75	880
1958	1	890	897	913	.97		.91	980
	2	1220	943	920	1.33		1.39	880
	3	820	970	956	.86		.96	850
	4	840	1032	1001	.84		.75	1120
1959	1	1000	1060	1046	.96		.91	1100
	2	1470	1063	1061	1.39		1.39	1060
	3	930					.96	970
	4	850					.75	1130

When M is graphed in Figure 24-6, we see that it is indeed much smoother than the original series, because it doesn't contain the large seasonal variation S. This seasonal fluctuation was removed because M averaged over all 4 quarters, so that each value of M contains in its average exactly one observation for each quarter (in particular, one of the very large second-quarter observations, and one of the very small fourth-quarter observations).

More generally, the M series smooths all the blips from the original series, whether they are generated by seasonal variation S or by the disturbance D. With these components removed, the moving average M consists of the remaining components—trend T and cyclical C. We confirm from its wave-like pattern in Figure 24-6 that M does indeed capture the cyclical component C; and we notice that M also captures the trend T, which could be isolated by fitting a least squares regression line.

B—CAPTURING S AND D USING THE RATIO TO MOVING AVERAGE

With the moving average M defining trend T and cyclical C, we now turn to the task of analysing the remaining two components—seasonal varia-

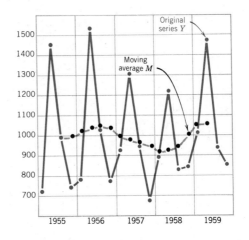

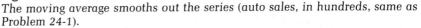

Figure 24-6
*The moving average smooths out the series (auto sales, in hundreds, same as
Problem 24-1).*

tion S and disturbance D. Since S and D are the components that are
included in the observed time series Y but are removed in calculating M,
we could capture them by using either:

i. $Y - M$, the difference between the observed Y and the moving
average M; or

ii. Y/M, the *ratio* of the observed Y to the moving average M, called
the *ratio to moving average*. Traditionally, this is the method
that is chosen; it is calculated in column 5 of Table 24-2, and is
graphed in Figure 24-7a.

We can confirm that the ratio Y/M does indeed capture the
seasonal effect: Every second quarter, this ratio substantially exceeds 1 because Y is substantially greater than M.

C—ISOLATING THE SEASONAL VARIATION S

The problem in Figure 24-7 is that Y/M in panel (a) gives us not only S,
but also the disturbance D. How do we isolate S? The answer is to average
all the first quarter values of Y/M; then similarly, average all the second
quarter values, and so on. To visualize this, imagine cutting panel (a) up
into 1-year periods, superimposing them in panel (b), and finally averaging
as shown in panel (c). (The actual numerical calculations are set out in
column 6 of Table 24-2, and are recorded in full in column 7.)

D—USING S TO SEASONALLY ADJUST THE ORIGINAL SERIES Y

Once the seasonal component S has been calculated, it can be used to
seasonally adjust (deseasonalize) the original time series Y. This is done
simply by dividing Y by the seasonal pattern in column 7 of Table 24-2,

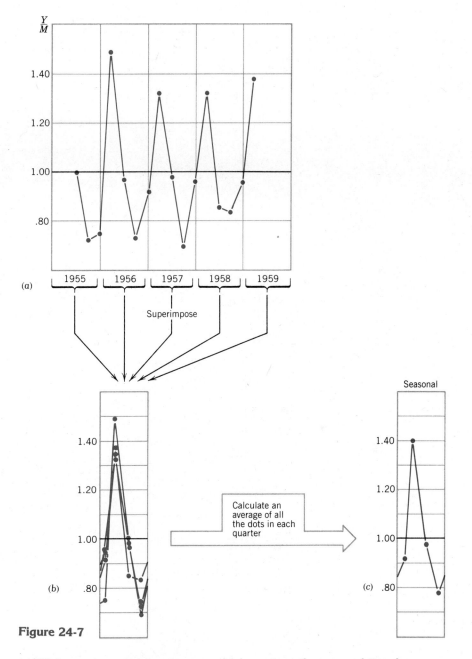

Figure 24-7

(a) Y/M = ratio to moving average, which captures the seasonal S and disturbance D. (b) Superimposing the yearly slices permits averaging to filter out the disturbance, leaving the seasonal in (c).

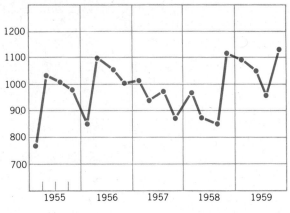

Figure 24-8
Seasonally adjusted series Y^a.

recording the result in column 8. When we graph this adjusted series Y^a in Figure 24-8, we see that the clear-cut seasonal pattern in the original data Y (graphed in Problem 24-1) has now been eliminated.

To sum up: This ratio-to-moving-average approach is primarily used to decompose a time series—and, in particular, to isolate S, which is then used to seasonally adjust the original series. See Table 24-3.

TABLE **24-3**   **Summary of the Ratio-to-Moving-Average Decomposition**

TIME SERIES Y IS MADE UP OF:	
TREND T AND CYCLICAL C	**SEASONAL S AND DISTURBANCE D**
(A) T and C are captured in the moving average M	(B) S and D are captured by the ratio Y/M. (C) S is then isolated by averaging the yearly slices. (D) S can then be used on Y to provide a seasonally adjusted time series Y^a. Specifically, $Y^a = Y/S$.

PROBLEMS

24-5   Analyze the jewelry sales data in Table 24-1, with the classical ratio-to-moving-average analysis. Specifically:

a. Calculate the moving average (4 quarters). Does it smooth out the large fourth quarter every year?

b. Calculate the ratio-to-moving average.

   c. Calculate the seasonal component S by averaging the series in **b** across all four years.

   d. Calculate the seasonally adjusted series, and graph it.

24-6 (Forecasting)

   a. Graph the moving average in Problem 24-5, part **a,** and extend it by eye another eight quarters. Roughly tabulate these extended values.

   b. Multiply the extended values of part **a** by the seasonal factor S found in Problem 24-5, part **c,** to obtain the *forecast* values. How do they compare to the values forecast with regression in Figure 24-4d?

24-7 Repeat Problem 24-5 for the carloading data of Problem 24-4.

24-4 Forecasting Using Exponential Smoothing

In this section we change gears. We concentrate on *forecasting* instead of decomposition, and accordingly we return to the approach of Section 24-2. In that context we assume—in the interests of simplifying a complex analysis—that trend and seasonal variation are not present. (For example, suppose the series is annual, and therefore cannot reveal any seasonal pattern.)

Nevertheless, the series will contain serial correlation, and we can forecast it using the approach already developed for a serially correlated residual. Specifically, we can use the *autoregressive model* that predicts Y_t as a linear combination of its previous values:

$$\hat{Y}_t = b_1 Y_{t-1} + b_2 Y_{t-2} + b_3 Y_{t-3} + \cdots \qquad (24\text{-}11)$$
$$\text{like (24-7)}$$

where the coefficients $b_1, b_2, \ldots$ are estimated by least squares multiple regression.

How far back into the past should the autoregression (24-11) go? One solution is to let it go back indefinitely, with smaller and smaller weights that decline exponentially:

$$b_k = b\lambda^k, \qquad k = 1, 2, \ldots \text{ and } 0 < \lambda < 1 \qquad (24\text{-}12)$$

Not only does this capture the sensible notion that more distant values should have less effect, it also turns out to be very convenient: As Appendix 24-4 shows, when we substitute (24-12) into (24-11), it reduces to a very simple form:

$$\boxed{\hat{Y}_t = (1 - \lambda)Y_{t-1} + \lambda\hat{Y}_{t-1}} \qquad (24\text{-}13)$$

where the weight λ can be estimated by least squares. (The details are more complicated than in Chapter 11, but the principle is the same: Choose λ to minimize the squared errors that result from this fitting procedure.)

It is easy to interpret (24-13): To predict this year's sales, we just take a weighted average of last year's sales (Y_{t-1}), and the *prediction* ($\hat{Y}_{t-1}$) of last year's sales we made a year earlier. This formula is so easy it can be calculated on the back of an envelope. An example will illustrate:

EXAMPLE **24-1**   A wholesale furniture salesman taking over a new territory was instructed to forecast each year's sales by the formula:

$$\text{Forecast sales this year} = .30 \text{ (last year's sales)}$$
$$+ .70 \text{ (previous year's forecast}$$
$$\text{of last year's sales)}$$

[In other words, it had been estimated that λ in (24-13) was .70.] Suppose the forecast of last year's sales was 3.70 (million dollars). But the actual sales last year turned out to be 3.92.

a.   What is his sales forecast for this year?

b.   If this year's sales turn out to be 3.64, what would the forecast be for next year?

Solution   a.   Substitute appropriately into (24-13):

$$\text{Forecast} = .30(3.92) + .70(3.70) = 3.77$$

b.   In (24-13) notice that t is now next year, and $t - 1$ is this year. Thus

$$\text{Forecast} = .30(3.64) + .70(3.77) = 3.73 \qquad (24\text{-}14)$$

Of course, this forecast has been kept relatively simple because we have assumed that trend and seasonal components are not present. If they are, each component should be forecast, using essentially the same exponential smoothing developed above. Then they can be added together (just as in Figure 24-4), to provide a forecast of the overall time series. This is called a *Holt-Winters* forecast (Wonnacott, 1984 IM).

PROBLEMS

24-8   To forecast next week's sales of its bread, a bakery used an exponential smoothing formula with $\lambda = .40$:

$$\hat{Y}_t = .60 \ Y_{t-1} + .40 \ \hat{Y}_{t-1} \qquad \text{like (24-13)}$$

a. Last week's forecast was 24,200 while actual sales were 24,700. What is the forecast of this week's sales?

b. If this week's sales turned out to be 26,000, what is the forecast of next week's sales?

c. Continue making weekly forecasts, if actual sales continued on from 26,000 as follows:

25,800   24,800   23,700   23,600

24-5 Forecasting Using Box-Jenkins Models

In order to understand this increasingly popular but complex method of forecasting, we again keep the argument simple by initially assuming the time series is free of trend and seasonal components. Then an autoregressive model of the following form is appropriate:

$$Y_t = \beta_1 Y_{t-1} + \beta_2 Y_{t-2} + \cdots + \beta_p Y_{t-p} + u_t \qquad (24\text{-}15)$$
$$\text{like (24-11)}$$

The potential number of lags p can be very large, of course. In the exponential smoothing technique of the previous section, we allowed an infinite number by locking in a rigid assumption of how the coefficients decreased exponentially into the past.

If we tried to avoid this restrictive assumption by a completely free fit of all the β parameters in (24-15), we would have too many parameters to estimate from a sample of relatively small size. The Box-Jenkins method is a compromise: It is a freer fit than exponential smoothing, yet it doesn't require all the parameters of a completely free fit of a long autoregression.

Specifically, the Box-Jenkins technique cuts Equation (24-15) off at a small number of lags—that is, includes only the most important recent lagged values. This means we cannot capture all of the serial correlation, so some remains in u_t. We could therefore model u_t with another auto-regression (that is, a regression of u_t on its own previous values). It is more effective, however, to model u_t as a *moving average* of serially independent terms v_t (white noise):

$$u_t = \alpha_0 v_t + \alpha_1 v_{t-1} + \cdots + \alpha_q v_{t-q} \qquad (24\text{-}16)$$

The moving average (24-16) might be more accurately called a "moving linear combination" since we don't require the coefficients to sum to 1, or even be positive. In fact, it is customarily written in a slightly different form:

$$u_t = v_t - \alpha_1 v_{t-1} - \alpha_2 v_{t-2} - \cdots - \alpha_q v_{t-q} \qquad (24\text{-}17)$$

When this moving average (24-17) is substituted into the autoregression (24-15), we obtain a combination of an autoregression (AR) and moving average (MA) whose lengths are p and q, respectively:

ARMA (*p*,*q*) model:

$$Y_t = \beta_1 Y_{t-1} + \cdots + \beta_p Y_{t-p}$$
$$+ v_t - \alpha_1 v_{t-1} - \cdots - \alpha_q v_{t-q} \qquad (24\text{-}18)$$

Although estimating the β and α parameters is complex, it can be done routinely with a computer package.

In practice, the Box-Jenkins model is even more complex. It also incorporates a seasonal component and a trend, and is known as a SARIMA model. Once its parameters are estimated, the model can be used for forecasting.

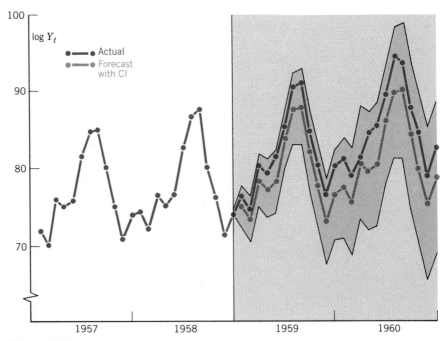

Figure 24-9
An example of a Box-Jenkins forecast, based on 120 months of data. The last 24 months are graphed, along with the gray forecast (and the 95% confidence interval around it) made in the last month, for the next 24 months into the future. Compare this forecast with the future series as it actually turned out, in blue. (Y_t* = total number of international airline passengers; series G of Box and Jenkins, 1976.)*

Although the details (Box and Jenkins, 1976; or Wonnacott, 1984 IM) would fill a book, we do have space here to show just how well a SARIMA model can forecast. Figure 24-9 shows a computer forecast of airline passengers up to 24 months ahead (McLeod, 1982). It includes a 95% confidence band that of course grows vaguer as the forecast reaches further into the future.

*24-6 Serial Correlation and Generalized Least Squares (GLS)

Again returning to our initial analysis of a time series in Section 24-2, recall that we used regression to estimate the trend and seasonal components, leaving a residual series to be analyzed for serial correlation. Unfortunately, a serially correlated residual mixed up in the original series will reduce the reliability of the regression. In this section we discuss this problem and suggest possible solutions.

A second objective is to extend our analysis. So far in this chapter we have concentrated on predicting *one* series, such as business sales, by examining its own seasonal, trend, and serial correlation characteristics. Now we wish to relate business sales to entirely different variables as well, such as national income. Thus our interest is not only in forecasting, but also in *understanding* how two or more time series variables may be related.

A—THE PROBLEM WITH SERIAL CORRELATION

Briefly, the problem with serial correlation is that successive observations are dependent to some extent. For example, positive serial correlation means that successive observations tend to resemble previous observations and hence give little new information. Thus n serially correlated observations give less information about trend (and seasonal) than n independent observations would. Consequently, our estimates will be less reliable, and our confidence intervals should reflect this.

B—THE PROBLEM ANALYZED

For simplicity, we take as our regression model,

$$Y_t = \alpha + \beta X_t + e_t \tag{24-19}$$

So far X_t has represented time itself, or seasonal dummies. Now let it represent *any* time series (or combination of series) deemed influential in determining Y_t. (Notice how this can take us, as we wish, beyond our earlier analysis of just one time series, to an analysis of how two or more time series are related.)

The residual e_t in (25-19) of course is assumed to be serially correlated, and for simplicity, we again take $\rho = 1$:

$$e_t = e_{t-1} + v_t \qquad (24\text{-}20)$$
$$\text{like (24-4)}$$

where v_t is white noise (serially uncorrelated) as usual.

Now suppose that the true regression line (defined by α and β) is the one shown in panel (a) of Figure 24-10. Suppose further that the residual series $e_1, e_2, \ldots$ is given in panel (b)—the same one generated earlier in Figure 24-5. This produces the pattern of observations shown in panel (a): Once the observations are above the true regression, they tend to stay above it.

We immediately can see the difficulties that the serial correlation causes by observing in panel (a) how badly the estimated regression through this scatter fits the true regression: β is seriously underestimated. But in another sample we might have observed precisely the opposite pattern of residuals, with e_t initially taking on negative values, followed by positive ones. In this case, we would overestimate β. Since we are as likely to get an overestimate as an underestimate, the problem is therefore not bias. Rather, the problem is that estimates tend to be wide of the target—they have large variance.

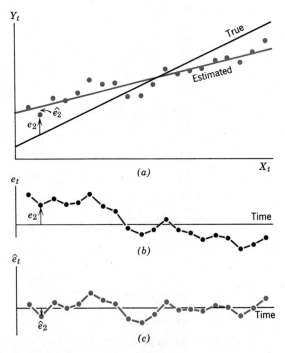

Figure 24-10
Regression with serially correlated residual. (a) True and estimated regression lines. (b) True residual (from Figure 24-5). (c) Estimated residual (around the estimated regression in panel (a)).

Panel (c) of Figure 24-10 illustrates a second problem: Because serial correlation tends to produce a smooth series, the *estimated* residuals $\hat{e}_t$ tend to be smaller than the true residuals e_t. Thus the residual variance s^2 will be smaller than σ^2, and the subsequent confidence interval for β will be too narrow.

To sum up: Because of serial correlation, the OLS estimate of β is less reliable. But it has the illusion of being *more* reliable because autocorrelated data is smooth and yields smaller observed residuals.

C—FIRST DIFFERENCES

The remedy for serial correlation is to transform the data to a form that does satisfy the assumptions of OLS. Since (24-19) holds true for any time t, it holds for time $(t - 1)$:

$$Y_{t-1} = \alpha + \beta X_{t-1} + e_{t-1} \tag{24-21}$$

We now can examine the change over time of our variables, by subtracting (24-21) from (24-19):

$$(Y_t - Y_{t-1}) = \beta(X_t - X_{t-1}) + (e_t - e_{t-1}) \tag{24-22}$$

Note from (24-20) that:

$$e_t - e_{t-1} = v_t \tag{24-23}$$

and define:

$$\left. \begin{aligned} \Delta Y_t &\equiv Y_t - Y_{t-1} \\ \Delta X_t &\equiv X_t - X_{t-1} \end{aligned} \right\} \qquad \begin{aligned} &(24\text{-}24) \\ &\text{like (24-23)} \end{aligned}$$

Then (24-22) can be written:

$$\boxed{\Delta Y_t = \beta \Delta X_t + v_t} \tag{24-25}$$

where v_t is a white noise term with all the properties required by OLS. Thus, β may be estimated validly by OLS regression of ΔY_t against ΔX_t.

D—GENERALIZED DIFFERENCES: GENERALIZED LEAST SQUARES (GLS)

A more useful model for the serially correlated residual is:

$$e_t = \rho e_{t-1} + v_t \qquad \begin{aligned} &(24\text{-}26) \\ &(24\text{-}4) \text{ repeated} \end{aligned}$$

where we no longer[4] insist that $\rho = 1^{\cdot}$ as in (24-20); instead,

$$|\rho| < 1 \tag{24-27}$$

We continue to assume the linear regression model:

$$Y_t = \alpha + \beta X_t + e_t \tag{24-28}$$
$$\text{(24-19) repeated}$$

To estimate β, we difference the data in much the same way as before. Equation (24-28) is reexpressed for time $(t - 1)$ and is then multiplied by ρ:

$$\rho Y_{t-1} = \rho\alpha + \rho\beta X_{t-1} + \rho e_{t-1} \tag{24-29}$$

Subtracting (24-29) from (24-28):

$$(Y_t - \rho Y_{t-1}) = \alpha(1 - \rho) + \beta(X_t - \rho X_{t-1}) + (e_t - \rho e_{t-1}) \tag{24-30}$$

Now define the *generalized differences*:

$$\left.\begin{array}{l} \Delta Y_t \equiv Y_t - \rho Y_{t-1} \\ \Delta X_t \equiv X_t - \rho X_{t-1} \end{array}\right\} \tag{24-31}$$
$$\text{like (24-24)}$$

Then (24-30) can be written:

$$\boxed{\Delta Y_t = \alpha(1 - \rho) + \beta\Delta X_t + v_t} \tag{24-32}$$

where v_t is again white noise, with all the properties required by OLS. Thus, after this transformation, we may regress ΔY_t against ΔX_t to estimate β. This technique is an example of *generalized least squares* (GLS).

However, prior to this regression, one additional adjustment must be made to the data. The *first* observed values Y_1 and X_1 cannot be transformed by (24-31) since their previous values are not available; instead, the appropriate transformation is[5]:

---

[4] If $|\rho| \geq 1$ as in (24-20), a major theoretical problem arises because the residual e_t becomes "explosive"; that is, the variance of e_t increases over time without limit.

[5] This transformation of the data is very important; without it, the regression (24-32) may be no better than OLS, and perhaps not even as good. And of course, the constant regressor (whose coefficient is α) also must be transformed. By analogy with (24-31) and (24-33), it becomes $(1 - \rho)$ for $t = 2,3, \ldots$ and $\sqrt{1 - \rho^2}$ for $t = 1$.

The transformation (24-33) must be used with some care. It is appropriate if and only if the process (24-28) generating the residual has been going on undisturbed for a long time previous to collecting the data. In practice, however, the first observation sometimes is taken just after a war or some other catastrophe, which seriously disturbs the residual.

$$\boxed{\begin{aligned} \Delta Y_1 &\equiv \sqrt{1 - \rho^2}\, Y_1 \\ \Delta X_1 &\equiv \sqrt{1 - \rho^2}\, X_1 \end{aligned}} \tag{24-33}$$

The problem is, of course, that all of these transformations require a value for ρ, which is unknown. We have already seen one way to obtain an estimate r for ρ: We use OLS to fit

$$\hat{e}_t = r\hat{e}_{t-1} \tag{24-34}$$
$$(24\text{-}5)\text{ repeated}$$

Unfortunately, r has a bias, which can be seen from Figure 24-10: The estimated residuals $\hat{e}_t$ in panel (c) fluctuate around zero more (i.e., bounce back and forth across the baseline more) and hence have a smaller serial correlation r than do the true residuals e_t in panel (b). Thus r tends to underestimate ρ.

E—AN ALTERNATIVE METHOD TO ESTIMATE ρ

An alternative that can be used in practice to estimate ρ is to rearrange (24-30) into the following regression equation:

$$Y_t = \alpha(1 - \rho) + \rho Y_{t-1} + \beta X_t - \beta\rho X_{t-1} + (e_t - \rho e_{t-1}) \tag{24-35}$$

where $e_t - \rho e_{t-1} = v_t$ is again the white noise term. Accordingly, run a regression of Y_t on Y_{t-1}, X_t, and X_{t-1}, retaining only the estimated coefficient of Y_{t-1}; this provides the estimate of ρ that can then be used in (24-31) and (24-33) to yield estimates of α and β.

CHAPTER 24 SUMMARY

24-1 Observations taken over time present special problems of serial correlation and seasonal variation.

24-2 Multiple regression can be used to decompose a time series. The seasonal component is easily estimated with dummies in regression—and at the same time, regression estimates trend. Any serially correlated residual that remains then requires special treatment as an autoregression.

Each of the time series components can be forecast, and then summed to provide a forecast of the original time series.

24-3 In the era before computers, decomposition of a time series started with a moving average to smooth the series. Then the ratio of the series to this moving average provided the seasonal component, plus a disturbance that could be averaged out.

When the series is divided by the seasonal pattern, we obtain a seasonally adjusted series that is often useful for its own sake.

24-4   An autoregression of Y_t on its own past values Y_{t-1}, Y_{t-2}, ... is very useful for prediction. If we assume the coefficients exponentially decrease further and further into the past, we greatly simplify the model.

*24-5   Box-Jenkins provides another, more flexible way to simplify the autoregression: After a few terms, the autoregression is chopped off, with the truncated part represented by a moving average.

*24-6   The OLS regression of Y_t against X_t is not fully efficient when the residuals are serially correlated. The remedy is to transform the regression equation by differencing, to produce *uncorrelated* residuals; then ordinary regression formulas can be used. This whole process is called Generalized Least Squares (GLS).

REVIEW PROBLEMS

24-9   Select the correct choice in each bracket:

   a.   Serial correlation in a time series means that the OLS fit is [particularly appropriate, inappropriate] as a forecast.

   b.   [Annual, Quarterly] economic series are often seasonally adjusted. Then the seasonally adjusted series [shows the actual historic record, may give a better idea of relative performance].

24-10   In which of the following series would you expect to find periodic patterns (such as the seasonal pattern in Figure 24-2)?

   a.   A sample of 100 monthly mean temperatures at Newark airport.

   b.   A sample of 500 hourly temperature readings at Atlanta airport.

   c.   Annual GNP growth from 1950 to 1985.

   d.   Quarterly GNP growth from 1950 to 1985.

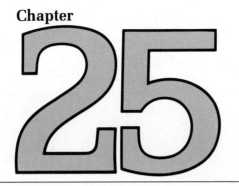

*Simultaneous Equations

(Requires Chapter 12)

Whatever moves is moved by another.　　　　ST. THOMAS AQUINAS

25-1　Introduction: The Bias in OLS

Economic relationships are often more complicated than the techniques that we have introduced so far suggest. Specifically, seldom is a variable determined by a single relationship (equation). Instead, it is usually determined simultaneously with many other variables in a whole system of simultaneous equations. For example, the price of corn is determined simultaneously with the price of rye, the price of hogs, and so on.

As another illustration, consider a very simple model of national income using just two equations:

$$Y = \alpha + \beta X + e \qquad (25\text{-}1)$$

$$X = Y + I \qquad (25\text{-}2)$$

Equation (25-1) is the standard form of the consumption function that relates consumer expenditure Y to income X. The parameters of this function that must be estimated are α (the intercept) and β (the slope, or marginal propensity to consume). We assume that the error term is well-behaved: Successive values of e are assumed to be independent and identically distributed, with mean 0 and variance σ^2. Equation (25-2) states that national income is defined as the sum of consumption and investment

I. (Since both sides of this equation are equal by economic definition, no error term appears in this equation—but the analysis remains the same in any case).

An important distinction must be made between two kinds of variables in our system. By assumption, *I* is determined *outside* the system of equations, so it is called *predetermined* (or *exogenous*). The essential point is that its values are determined elsewhere, and are not influenced by *Y*, *X*, or *e*. In particular, we emphasize that:

> **Because *I* is predetermined, *I* and *e* are statistically independent.**   (25-3)

On the other hand, *X* and *Y* are determined *within* the system and thus are influenced by *I* and *e*; they are often called mutually *dependent* (or *endogenous*). The two equations in this simultaneous system can be solved for the two dependent variables (the model is mathematically complete).

A diagram will be useful to illustrate the statistical difficulties encountered. To highlight the problems the statistician will face, let us suppose that we have some sort of omniscient knowledge, so that we know the true consumption function $Y = \alpha + \beta X$, as shown in Figure 25-1. And let us watch what happens to the statistician—a mere mortal—who does not have this knowledge but must try to estimate this function by observing only *Y* and *X*. Specifically, let us show how badly things will turn out if

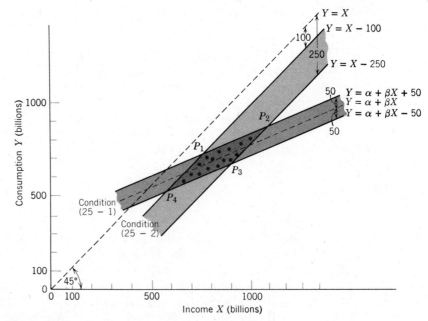

Figure 25-1
The consumption function, and the scatter of observed points around it.

he estimates α and β by fitting a line by ordinary least squares (OLS). To find the sort of scatter of Y and X that he will observe, we must remember that all observations must satisfy both Equations (25-1) and (25-2).

Consider (25-1) first. Whenever e takes on a zero value, the observation of Y and X must fall somewhere along the true consumption function $Y = \alpha + \beta X$, shown in Figure 25-1. If e takes on a value greater than zero (say, + \$50 billion), then consumption is greater as a consequence and the observation of Y and X must fall somewhere along $Y = \alpha + \beta X + 50$. Similarly, if e takes on a value of -50, the statistician will observe a point on the line $Y = \alpha + \beta X - 50$. According to the standard assumptions, e is distributed about a zero mean. To keep the geometry simple, we further assume that e is equally likely to take on any value between + 50 and -50. Thus the statistician will observe Y and X falling within this band around the consumption function, shaded in Figure 25-1.

Any observed combination of Y and X also must satisfy (25-2). What does this imply? This condition can be rewritten as:

$$Y = X - I \qquad (25\text{-}4)$$

If I were zero, then Y and X would be equal, and any observation would fall on the 45° line where $Y = X$. Let us suppose that when I is determined by outside factors, it is distributed uniformly through a range of 100 to 250. If $I = 100$, then from (25-4) any observation of Y and X must fall along $Y = X - 100$, which is simply the line lying 100 units below the 45° line. Similarly, when $I = 250$, an observation of Y and X would fall along the line $Y = X - 250$. These two lines define the steeper band within which observations must fall to satisfy (25-2).

Since any observed combination of Y and X must satisfy *both* conditions, all observations will fall within the parallelogram $P_1P_2P_3P_4$. To clarify, this parallelogram of observations is reproduced in Figure 25-2. When the statistician regresses Y on X using OLS, the result is shown as $Y = a + bX$. When this is compared with the true consumption function $Y = \alpha + \beta X$, it is clear that the statistician has come up with a bad fit; his estimate of the slope has an upward bias. What has gone wrong?

The observations around P_2 have "pulled" the estimating line above the true regression; similarly, the observations around P_4 have pulled the estimating line below the true regression. It is the pull on both ends that has tilted this estimated line. Moreover, increasing sample size will not help to reduce this bias. If the number of observations in this parallelogram is doubled, this bias will remain.[1] Hence, OLS is inconsistent; this technique that worked so well in a single-equation model clearly is less satisfactory when the problem is to estimate an equation that is embedded within a system of simultaneous equations.

_____

[1] With an increase in sample size, the reliability of b as an estimator will be increased somewhat, because its variance will decrease towards zero. However, its bias will not be reduced.

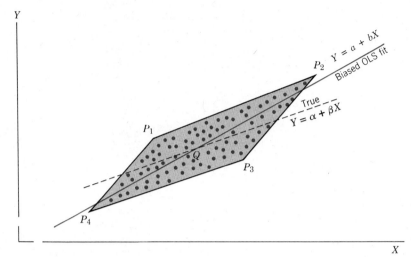

Figure 25-2
Inconsistent OLS fit of the consumption function.

The reason is evident: for the single-equation model, it was assumed in (12-3) that the expected value of e was 0 for every value of X (i.e., e was independent of X), and this is why OLS works so well. But in a simultaneous-equation model, this assumption no longer holds.

> **In an equation in a simultaneous system, regressors that are not predetermined are not independent of the error term e.**

(25-5)

To see why, note, for example, in Figure 25-2 how e is correlated positively with X: e tends to be positive (i.e., the observed point lies above the true regression $Y = \alpha + \beta X$) when X is large, and e tends to be negative (with the observed point below the true regression) when X is small.[2] Consequently, the OLS fit has too large a slope. The reason is intuitively clear: in explaining Y, OLS gives as little credit as possible to the error, and as much credit as possible to the regressor X. When the error and regressor are correlated, then some of the effect of the error is attributed wrongly to the regressor.

In conclusion, the problem is this: The single-equation technique of OLS is inconsistent in simultaneous equations. Perhaps a diagram will clarify just how simultaneous equations differ from single equations (even systems of single equations). In Figure 25-3, we see that the distinguishing feature of simultaneous equations is that the *mutually* dependent variables

---

[2]Note that the reason for this is precisely because of the second equation in the model: thus, in Figure 25-2, the observed points are not only determined by Equation (25-1), but they are also determined, or "pulled," by Equation (25-2). And it is this latter pull that results in the positive correlation of X and e.

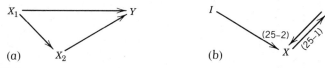

Figure 25-3
Contrast of single equations and simultaneous equations. (a) Single equations, easily analyzed by path analysis in Chapter 13. (b) Simultaneous equations (25-1) and (25-2).

influence each other, as indicated by the arrows going *both* ways between X and Y in panel (b).

PROBLEMS

25-1 In the consumption model shown in Figure 25-1, suppose the true consumption function is $Y = 10 + .6X$ and the following combinations of Y, X, and I have been observed:

Y	X	I
46	60	14
31	45	14
61	75	14
58	80	22
43	65	22
73	95	22
70	100	30
55	85	30
85	115	30

  a. Graph the true consumption function and the scatter of (X, Y) observations.

  b. Regress Y on X using OLS, and graph the estimated consumption function. Is it unbiased?

25-2 A sociologist observes that the number of children available for adoption (Y_1) depends on the number of births (Y_2). At the same time, births depend on income (X) and on the number of children available for adoption.

  a. Draw a schematic figure similar to Figure 25-3 illustrating these relationships.

  b. Set up a system of two simultaneous equations showing how Y_1 and Y_2 are jointly determined.

25-2 The Remedy: Instrumental Variables (IV)

Since OLS is inconsistent, we now begin the search for an estimator that *is* consistent. The first step is to develop an important new device that will have very broad application.

A—THE COVARIANCE OPERATOR

The *covariance* of any two variables, V and Y say, is defined much like the variance:

$$\text{Covariance } s_{VY} \equiv \frac{\Sigma(V - \overline{V})\,(Y - \overline{Y})}{n - 1} \qquad (25\text{-}6)$$

like (2-11), (5-10)

Just as the variance summarizes the variability of a variable, so the covariance is a single number that nicely summarizes the covariability or relation between two variables. To see how this concept of covariance can help us, consider the following familiar equation:

$$Y = \alpha + \beta X + e \qquad (25\text{-}7)$$

If we express all variables in deviation form, then the constant conveniently drops out[3]:

$$(Y - \overline{Y}) = \beta(X - \overline{X}) + (e - \overline{e}) \qquad (25\text{-}8)$$

We now select another variable V, and to both sides of (25-8) we do the following three things:

1. multiply by $(V - \overline{V})$,
2. sum from 1 to n,
3. divide by $n - 1$.

The result is the following equation in covariances:

$$s_{VY} = \beta\, s_{VX} + s_{Ve} \qquad (25\text{-}9)$$

Notice how we have transformed the equation relating variables (25-7) to the equation relating covariances (25-9). This transformation is called *applying the covariance operator* to (25-7), or *taking covariances of every variable in this equation with respect to V.*

So far, these are only mathematical manipulations. The question is: How can they help us to estimate the unknown target β?

B—THE INSTRUMENTAL VARIABLE ESTIMATE OF β

Let us solve (25-9) for the unknown parameter β by dividing through by s_{VX}:

---

[3]To prove this, we simply take averages in (25-7):
$$\overline{Y} = \alpha + \beta\overline{X} + \overline{e} \qquad \text{like (2-19)}$$
When we subtract this from (25-7), then (25-8) follows.

$$\frac{s_{VY}}{s_{VX}} = \beta + \frac{s_{Ve}}{s_{VX}} \tag{25-10}$$

If the last term is small enough to be negligible and drops out, we have an estimate of β—s_{VY}/s_{VX}. Since this estimate depends upon the variable V that was *instrumental* in obtaining it, we name it accordingly:

Instrumental Variable (IV) estimator of β:

$$b_V \equiv \frac{s_{VY}}{s_{VX}} \tag{25-11}$$

Finally, let us substitute (25-11) back into (25-10):

$$b_V = \beta + \left(\frac{s_{Ve}}{s_{VX}}\right) \tag{25-12}$$

The bracketed term represents the error in using b_V to estimate the target β. If this error is large, b_V will be a bad estimate; if it is small, b_V will be a good estimate. Thus we will look for an instrumental variable V that will make this bracketed error term small—that is, will make its numerator small and its denominator large.

Requirements of a good instrumental variable V

1. **To make s_{Ve} small, V should be uncorrelated with the error e ($s_{Ve} \to 0$).[4] If it is, the estimate is consistent.**

2. **To make s_{VX} large, V should be highly correlated to X.** (25-13)

To sum up. It doesn't matter where (25-7) comes from. That is, it doesn't matter whether it is one equation in a simultaneous system—like (25-1)—or if it is the one equation in a single equation model—like (12-3). In either case, we can take covariances of all variables in this equation with respect to V, to get the covariance equation (25-9). This in turn yields the IV estimator (25-11) which is a good estimator if conditions (25-13) are satisfied. To illustrate when this IV estimator (25-11) is a good estimator and when it is not, we next consider three cases.

C—APPLYING THE IV ESTIMATOR (25-11)

Case 1: On a single equation. Return to the simple situation in Chapter 12 where there was no problem of simultaneous equations: (25-7) was a

---

[4]For V and e to be uncorrelated means that the *population* covariance σ_{Ve} is 0; that is, the sample covariance s_{Ve} approaches 0 as $n \to \infty$.

single equation relating Y to X. If we try to estimate β by using this new technique, we have to select an instrumental variable. What could be better than X itself? [It perfectly satisfies the second condition in (25-13), because X is perfectly correlated with itself. And the first condition is discussed below.] When we thus use X as the instrumental variable V, (25-11) becomes:

$$b_X = \frac{s_{XY}}{s_{XX}} \qquad (25\text{-}14)$$

But this is just the OLS estimate[5] b. Thus:

> **For the equation $Y = \alpha + \beta X + e$, OLS is equivalent to using X as an instrumental variable (IV).**   (25-15)

Moreover, the first condition in (25-13) tells us that OLS—or equivalently, using X as an IV—requires that:

> **For consistency, X must be uncorrelated with the error e.**   (25-16)

This requirement was indeed guaranteed by our OLS model in Chapter 12. [In (12-3), we required the expected value of e to be zero for *every value* of X. Thus, when X was large, for example, there was no tendency for e to be positive (or negative). That is, the correlation of X and e was 0.]

Case 2:   In a simultaneous system, a mutually dependent variable gives an inconsistent estimate.   Now suppose the equation we are estimating is—like (25-1)—part of a simultaneous system of equations. If we calculate the estimator (25-14) using X as an instrument—that is, if we apply OLS to this equation—the result is an inconsistent estimator, because (25-16) is violated—specifically, by (25-5). This confirms our earlier conclusion that OLS was inconsistent in Figure 25-2.

Case 3:   In a simultaneous system, a predetermined variable gives a consistent estimate.   If using X as the instrumental variable (i.e., OLS) is not consistent in a simultaneous system, what *would* be? The second requirement in (25-13) is still that the instrumental variable be highly correlated with X. So we had better stick to a variable that is relevant, that is, one of the variables in the system.

   Why not use I instead of X as the instrumental variable? In fact, it passes with flying colors because, according to (25-3), it is uncorrelated with e.

---

[5]To prove this, we write:

$$b_X \equiv \frac{s_{XY}}{s_{XX}} = \frac{\Sigma(X - \overline{X})(Y - \overline{Y})/(n-1)}{\Sigma(X - \overline{X})^2/(n-1)} = \frac{\Sigma xy}{\Sigma x^2} \equiv b$$

Thus we obtain a consistent IV estimator of the slope β:

$$b_I = \frac{s_{IY}}{s_{IX}} \qquad (25\text{-}17)$$

In the next section, we will generalize this same issue—finding the best IV—to more complicated situations.

PROBLEMS

25-3 **a.** Use the data in Problem 25-1 to actually calculate the consistent IV estimate (25-17) of the consumption slope β. How far off the target $\beta = .6$ is it?

      b. Explain why this small sample was a lucky one. What would you expect if your small sample is less lucky?

25-4 For the consumption model in this chapter, suppose the covariance table (also called the covariance *matrix*) of X, Y, I has been computed for a sample of $n = 50$ observations:

$$\begin{bmatrix} s_{XX} & s_{YX} & s_{IX} \\ s_{XY} & s_{YY} & s_{IY} \\ s_{XI} & s_{YI} & s_{II} \end{bmatrix} = \begin{bmatrix} 130 & 100 & 30 \\ 100 & 80 & 20 \\ 30 & 20 & 10 \end{bmatrix}$$

      a. Calculate the estimate of the consumption slope β (i) using X as the IV, and (ii) using I as the IV.

      b. Which of the estimates in part (a) are biased? How much?

25-5 Consider the multiple regression model:

$$Y = \beta_0 + \beta_1 X_1 + \beta_2 X_2 + e \qquad (13\text{-}2)\text{ repeated}$$

      a. If you take covariances using the instrumental variable X_1, what equation do you get? If you take covariances using X_2 as an instrument, what equation do you get?

      b. Outline how you would use these two equations to estimate β_1 and β_2.

      c. To estimate β_1 and β_2, which is better—the IV estimates obtained from **b**, or the OLS estimates obtained from (13-5) and (13-6)? Or are they the same?

25-3 Two Stage Least Squares (2SLS)

A—INTRODUCTION

As we saw in our simple two-equation system, it is desirable, if possible, to use as instrumental variables only those predetermined variables that explicitly appear in our model—that is, which appear in at least one of the equations in the system. It is evident that neither this approach, nor any other, entirely overcomes our difficulties; the decision on which variables may be used as instrumental variables merely is pushed onto the researcher who specifies the model and, in particular, the predetermined variables that are included. Specification of the model remains arbitrary to a degree, and this gives rise to some arbitrariness in statistical estimation. But this cannot be avoided. We can only conclude that the first task of specifying the original structure of the model is a very important one, since it involves a prior judgment of which variables are "close to" X and Y, and which variables are relatively "far away."

Now let us consider more complicated and typical systems. We will denote the predetermined or exogenous variables (such as investment) by $X_1, X_2, \ldots$, and the mutually dependent or endogenous variables (such as income and consumption) by $Y_1, Y_2, \ldots$. For the system to be complete, there must be as many equations as there are dependent variables. For example, to determine 4 dependent Y variables, 4 equations would be required[6]:

$$Y_1 = \gamma_2 Y_2 + \gamma_3 Y_3 + \beta_1 X_1 + \beta_4 X_4 + \beta_6 X_6 + e \qquad (25\text{-}18)$$

$$Y_2 = \cdots$$

$$Y_3 = \cdots$$

$$Y_4 = \cdots$$

For simplicity, we explicitly showed only the first equation. Once we show how to estimate it, every other equation can be similarly estimated.

It is customary to specify that some of the coefficients are zero. For example, in (25-18) β_2, β_3, β_5, and γ_4 have been specified as zero, so that the terms in X_2, X_3, X_5, and Y_4 do not appear. This specification keeps each equation of manageable length (or *identified*, as we will show in Problem 25-8).

B—CHOOSING INSTRUMENTAL VARIABLES

In (25-18), what can we use as instrumental variables (IV)? To find variables that are highly correlated with the regressors—the second require-

---

[6]For simplicity, in (25-18) we have omitted the constant term, because it drops out when the covariance operator is applied. Then it can easily be estimated after everything else, using an equation like (13-7).

ment in (25-13)—we might try the regressors themselves (Y_2, Y_3, X_1, X_4, and X_6). No problem is involved in using X_1, X_4, and X_6; because they are predetermined, (25-3) means that they satisfy the first requirement in (25-13). But unfortunately, Y_1, Y_2, and Y_3 are mutually dependent (i.e., determined by the predetermined X variables and the errors) and so are correlated with the error e; thus Y_2 and Y_3 violate this first requirement. One of the most ingenious solutions involves two simple steps:

1. first purge Y_2 and Y_3 of their dependence on e, so that they will satisfy the first requirement in (25-13);
2. apply them, along with X_1, X_4, and X_6, as our instrumental variables.

These two stages—known as *Two Stage Least Squares* (2SLS)—in fact work well, as we will now show in detail.

C—FIRST STAGE

To purge Y_2 of its dependence on e, we regress it on all the exogenous X variables in the system,[7] obtaining the OLS fitted value:

$$\hat{Y}_2 = b_0 + b_1X_1 + b_2X_2 + \cdots \qquad (25\text{-}19)$$

Since each exogenous X variable is independent of e, their linear combination $\hat{Y}_2$ will also be independent[8] of e. Therefore, $\hat{Y}_2$ (and similarly $\hat{Y}_3$) can be used as instrumental variables for consistent estimation.

D—SECOND STAGE

For the second stage, we just apply all 5 instruments—$\hat{Y}_2$, $\hat{Y}_3$, and of course X_1, X_4, and X_5. For example, using $\hat{Y}_2$ as an IV on (25-18) yields

$$s_{\hat{Y}_2 Y_1} = \gamma_2 s_{\hat{Y}_2 Y_2} + \gamma_3 s_{\hat{Y}_2 Y_3} + \beta_1 s_{\hat{Y}_2 X_1} + \beta_4 s_{\hat{Y}_2 X_4} + \beta_6 s_{\hat{Y}_2 X_6} + s_{\hat{Y}_2 e}$$

while using $\hat{Y}_3$ as an IV yields

$$s_{\hat{Y}_3 Y_1} = \gamma_2 s_{\hat{Y}_3 Y_2} + \cdots \qquad\qquad\qquad + s_{\hat{Y}_3 e}$$

Thus, each such application of an IV to the equation (25-18) will give us this sort of estimating equation—for a total of 5 such equations, which

---

[7] Strictly speaking, when Theil (1957) introduced 2SLS, he specified the whole system of equations. However, the actual form of the other equations in the system doesn't matter. So statisticians applying 2SLS often just use whatever exogenous variables are reasonably relevant and available in the data bank, without specifying exactly what the system is.

[8] Not quite independent. Since the b coefficients in (25-19) are not absolute constants, but are estimates depending slightly on e, $\hat{Y}_2$ depends slightly on e too. But with larger and larger sample size n, this problem disappears. So we still obtain a consistent estimate.

can then be solved for estimates of the 5 unknown parameters γ_2, γ_3, ... (Since the last terms s_{Y_2e}, etc. approach zero, they can be disregarded.)

PROBLEMS

25-6 Suppose the following equations were set up as a simple macroeconomic model of the U.S. Altogether 2 mutually dependent Y variables were simultaneously determined by 3 predetermined X variables:

$$Y_1 = \gamma_2 Y_2 + \beta_1 X_1 + \beta_2 X_2 + e$$

$$Y_2 = \gamma_1 Y_1 + \beta_3 X_3 + e$$

Outline how you would estimate:

a. The first equation.

b. The second equation.

25-7 (The Identification Problem) Now consider a smaller simultaneous system than Problem 25-6, with 2 mutually dependent Y variables and just 1 predetermined X variable:

$$Y_1 = \gamma_2 Y_2 + e$$

$$Y_2 = \gamma_1 Y_1 + \beta_1 X_1 + e$$

	Y_1	Y_2	X_1
	38	8	21
	27	15	17
	31	10	20
	21	18	14
	20	15	12
	43	6	24
Averages	30	12	18

a. Using the data above, derive the 2SLS estimate of the coefficient γ_2 in the first equation.

b. Using 2SLS on the second equation, set up the appropriate estimating equations for γ_1 and β_1—in symbols and numbers. Why can't you solve these for γ_1 and β_1? (The original equation $Y_2 = \gamma_1 Y_1 + \beta_1 X_1 + e$ is therefore called *unidentified*.)

25-8 (The Identification Problem in General) In Problem 25-7 part **b**, the equation was unidentified because it had two parameters that had to be estimated (two variables on the right side), yet there was only one predetermined variable (X_1) in the system of equations. In a sense, the equation was "too long" for the available data to handle. And this may be proved to be more generally true: *If the number of parameters to be estimated in an equation[9] exceeds the number of exogenous variables (appearing anywhere in the system), then the equation cannot be estimated, and is therefore called unidentified.*

Using this principle, decide which equations are not identified in the following system containing 5 mutually dependent Y variables and 3 predetermined X variables: (For simplicity, we represent each coefficient with an asterisk, rather than a Greek symbol.)

$$Y_1 = * Y_3 + * X_1 + * X_3$$
$$Y_2 = * Y_1 + * Y_3 + * Y_4 + * X_1 + * X_3$$
$$Y_3 = * Y_4 + * X_2$$
$$Y_4 = * X_1 + * X_2 + * X_3$$
$$Y_5 = * Y_1 + * Y_2 + * Y_3 + * Y_4$$

CHAPTER 25 SUMMARY

25-1 OLS is biased when the error is correlated with the regressor; then some of the effect of the error term gets mistakenly attributed to the regressor. This bias occurs commonly in systems of simultaneous equations when mutually dependent variables appear as regressors.

25-2 A consistent estimate can be obtained using an instrumental variable (IV) that is highly correlated with the regressor, yet uncorrelated with the error term.

25-3 Two Stage Least Squares (2SLS) is a popular way to obtain consistent estimates in simultaneous equations: The first stage calculates the most effective IVs; the second stage applies them.

REVIEW PROBLEMS

25-9 Answer True or False; if False, correct it: An IV should be highly correlated with the error term in the equation, and uncorrelated with the regressor(s). Therefore, mutually dependent variables qualify as instrumental variables, but predetermined variables do not.

---

[9]We continue to ignore the constant term, which is recovered when all other estimation is complete.

25-10 For the consumption model in this chapter—equations (25-1) and (25-2)—suppose the covariance matrix of X, Y, and I has been computed for a sample of $n = 30$ observations:

$$\begin{bmatrix} s_{XX} & s_{YX} & s_{IX} \\ s_{XY} & s_{YY} & s_{IY} \\ s_{XI} & s_{YI} & s_{II} \end{bmatrix} = \begin{bmatrix} 340 & 260 & 80 \\ 260 & 200 & 60 \\ 80 & 60 & 20 \end{bmatrix}$$

Calculate the estimate of the consumption slope β:

a. Using OLS.

b. Using a consistent method.

APPENDIXES

APPENDIX TO SECTION **2-2**

Careful Approximation of the Median

To illustrate how a median can be approximated quite accurately from grouped data, consider the distribution of heights in Table 2-2. A graph will clarify where the first 50% of the observations lie.

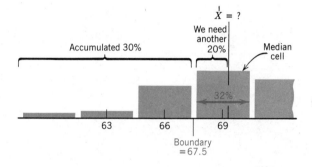

Recall that 30% of the observations were accumulated to the left of the median cell. To get the median, we must therefore pick up another 20% of the observations. Since the median cell includes 32% of the observations, we move 20/32 of the way through it. Starting at the cell boundary 67.5, and remembering that the cell width is 3, we therefore approximate the median as:

$$X \simeq 67.5 + \left(\frac{20}{32}\right)3$$

$$= 69.4 \text{ inches} \qquad (2\text{-}1) \text{ confirmed}$$

Any other percentile can be approximated the same way.

PROBLEM

2-1A   Verify that the lower quartile is about 66.8, and the upper quartile is about 71.9.

APPENDIX TO SECTION **2-5**

Effects of a Linear Transformation: Proofs

We will prove the two parts of Equation (2-19)—first the mean and then the variance. To establish the mean, we start with the definition for the new mean $\overline{Y}$:

$$\overline{Y} \equiv \frac{1}{n}[Y_1 + Y_2 + \cdots]$$

$$= \frac{1}{n}[(a + bX_1) + (a + bX_2) + \cdots]$$

$$= \frac{1}{n}[na + b(X_1 + X_2 + \cdots)]$$

$$= a + b\overline{X} \qquad \text{like (2-3)}$$

Next, to establish the variance, we start with the definition (2-11) for the new variance s_Y^2:

$$s_Y^2 \equiv \frac{1}{n-1}\{(Y_1 - \overline{Y})^2 + (Y_2 - \overline{Y})^2 + \cdots\}$$

$$= \frac{1}{n-1}\{[(a + bX_1) - (a + b\overline{X})]^2 + [(a + bX_2) - (a + b\overline{X})]^2 + \cdots\}$$

$$= \frac{1}{n-1}\{[b(X_1 - \overline{X})]^2 + [b(X_2 - \overline{X})^2] + \cdots\}$$

$$= \frac{b^2}{n-1}\{(X_1 - \overline{X})^2 + (X_2 - \overline{X})^2 + \cdots\}$$

$$= b^2 s_X^2$$

Thus $s_Y = |b|s_X$ (2-19) proved.

APPENDIX TO SECTION **3-7**

Probability As Axiomatic Mathematics

Historically, probability theory was set up on a proper axiomatic basis by the Russian mathematician Kolmogoroff in the 1930s. Of course, Kolmogoroff's axioms and theorems were much more rigorous than the extremely simple version that we now set out in this appendix, to illustrate the spirit of the axiomatic approach.

AXIOMS

A_1: $\Pr(e) \geq 0$

for every point e.

A_2: $\Pr(e_1) + \Pr(e_2) + \cdots + \Pr(e_N) = 1$

where N is the total number of points in the sample space.

A_3: $\Pr(E) = \Sigma \Pr(e)$

where the sum extends over all those points e that are in E.

Then the other properties are theorems that can be proved from these axioms. For example:

THEOREMS

$$T_1: 0 \leq \Pr(E)$$

$$T_2: \qquad \Pr(E) \leq 1$$

$$T_3: \qquad \Pr(\overline{E}) = 1 - \Pr(E)$$

Proofs   According to axioms A_1 and A_2, $\Pr(E)$ is the sum of terms that are positive or zero, and is therefore itself positive or zero; thus theorem T_1 is proved.

To prove T_3, we start with axiom A_2:

$$\underbrace{\Pr(e_1) + \Pr(e_2) +}_{\text{Terms for } E} \underbrace{\cdots + \Pr(e_N)}_{\text{Terms for } \overline{E}} = 1$$

According to A_3, this is just:

$$\Pr(E) + \Pr(\overline{E}) = 1$$

from which T_3 follows.

In T_1, we proved that every probability is positive or zero. In particular, $\Pr(\overline{E})$ is positive or zero; substituting this into the equation above finally ensures that:

$$\Pr(E) \leq 1 \qquad\qquad \text{T}_2 \text{ proved}$$

APPENDIX TO SECTION 4-2

Easier Formula for σ^2: Proof

To confirm (4-6), we begin with the definition of variance:

$$\sigma^2 \equiv \Sigma(x - \mu)^2 p(x) \qquad \text{(4-5) repeated}$$
$$= \Sigma(x^2 - 2\mu x + \mu^2)p(x)$$

and, noting that μ is a constant:

$$\sigma^2 = \Sigma x^2 p(x) - 2\mu \Sigma x p(x) + \mu^2 \Sigma p(x)$$

Since $\Sigma xp(x) = \mu$ and $\Sigma p(x) = 1$,

$$\sigma^2 = \Sigma x^2 p(x) - 2\mu(\mu) + \mu^2(1)$$
$$= \Sigma x^2 p(x) - \mu^2 \qquad \text{(4-6) proved}$$

PROBLEM (Requires all of Chapter 4)

4-1A   If the mean μ of a random variable X is approximated by a, prove that its variance can be approximated by $E(X - a)^2$—with a very small correction, as follows:

$$\sigma^2 = E(X - a)^2 - (\mu - a)^2 \quad \text{like (2-15)}$$

[*Hint*: Let us start out with the deviations of X from the approximate mean a, that is,

$$\text{let } Y = X - a$$

Since (4-6)—or its equivalent form (4-35)—is true for any random variable, it is true for Y:

$$\sigma_Y^2 = E(Y^2) - \mu_Y^2 \qquad \text{like (4-35)}$$

Then use the theory of linear transformations (4-29) to reexpress σ_Y^2 and μ_Y.]

APPENDIX TO SECTION **4-3**

Binomial Formula: Proof

The derivation of the binomial formula (4-8) will be a lot easier to follow if we split our page, showing on the left an example, and on the right the generalization. The example will be a familiar one—tossing a coin $n = 5$ times. We suppose the coin is somewhat biased, coming up heads (H) with probability $\pi = .60$, and tails (T) with probability $1 - \pi = .40$. Let us calculate the probability of the event shown in white in Figure 4-1A—that is, the probability that the number of heads $S = 3$.

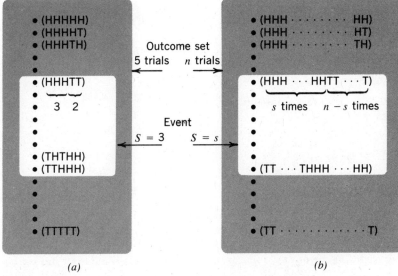

Figure 4-1A
Computing binomial probability. (a) Special case: 3 heads in 5 tosses of a coin. (b) General case: s successes in n trials.

We begin by looking at just one of the many ways we could get $X = 3$: We could get the 3 heads first, followed by 2 tails. In other words, we could get the sequence

or, in general, the sequence

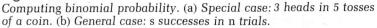

It has probability

It has probability

$$(.60)(.60)(.60)(.40)(.40) = (.60)^3(.40)^2$$

$$\pi \cdot \pi \cdots (1 - \pi) \cdot (1 - \pi) \cdots = \pi^s(1 - \pi)^{n-s}$$

where the simple multiplication is justified by the *independence* of the trials.

But there are many other ways we could get exactly 3 heads in 5 tosses. For example, we might get the sequence (*THTHH*), which has a probability

$$(.40)(.60)(.40)(.60)(.60) = (.60)^3(.40)^2$$

This is the same probability as before, because we are multiplying together exactly the same numbers as before. (The different order doesn't matter.) In fact, all sequences in the event $S = 3$ will have this same probability.

So the final question is, how many such sequences are there in the white region of Figure 4-1A? That is, how many ways are there to get exactly 3 heads? The answer: the number of different ways that three *H*'s and two *T*'s can be arranged. This number of ways is denoted by $\binom{5}{3}$, and can be calculated from

$$\binom{5}{3} = \frac{5!}{3!2!} = 10 \qquad (4\text{-}1A)$$

(as we will show at the end of the appendix)

To summarize:

The event

$$(S = 3)$$

includes

$$\binom{5}{3} = 10$$

outcomes, each with probability

$$(.60)^3(.40)^2 = .035$$

Hence its probability is:

$$p(3) = \binom{5}{3}(.60)^3(.40)^2$$

$$= 10(.035) = .35$$

To finish our proof, we must finally show where formula (4-1A) comes from. In how many ways can we arrange five distinct objects, designated H_1, H_2, H_3, T_1, T_2? We have a choice of 5 objects to fill the first position, 4 the second, and so on; thus the total number of arrangements is:

and, in general,

$$\binom{n}{s} = \frac{n!}{s!(n - s)!}$$

In general, the event

$$(S = s)$$

includes

$$\binom{n}{s}$$

outcomes, each with probability

$$\pi^s(1 - \pi)^{n-s}$$

Hence its probability is:

$$\boxed{p(s) = \binom{n}{s}\pi^s(1 - \pi)^{n-s}}$$

Thus (4-8) is proved.

$$5 \cdot 4 \cdot 3 \cdot 2 \cdot 1 = 5! \qquad \text{(4-2A)}$$

But this is not quite the problem at hand; in fact we cannot distinguish between H_1, H_2, and H_3—all of which appear as H. Thus many of the separate and distinct arrangements counted in (4-2A) cannot be distinguished, and appear as a single arrangement (e.g., $H_1\, H_2\, H_3\, T_1\, T_2$ and $H_2\, H_3\, H_1\, T_1\, T_2$ and many others appear as the single arrangement $HHHTT$). Thus (4-2A) involves serious overcounting. How much?

We overcounted $3 \cdot 2 \cdot 1 = 3!$ times because we assumed in (4-2A) that we could distinguish between H_1, H_2, and H_3, when in fact we could not. (3! is simply the number of distinct ways of rearranging H_1, H_2, and H_3). Similarly, we overcounted $2 \cdot 1 = 2!$ times because we assumed in (4-2A) that we could distinguish between T_1 and T_2, when in fact we could not. When (4-2A) is deflated for overcounting in both these ways, we have:

$$\frac{5!}{3!2!} \qquad \text{(4-1A) proved}$$

PROBLEM

4-2A   To better understand the binomial coefficient $\binom{n}{s}$, consider again the tossing of a coin. If we toss it $n = 5$ times, the possible ways we could get exactly $S = 3$ heads are:

$$HHHTT$$
$$HHTHT$$
$$\cdot$$
$$\cdot$$
$$\cdot$$

a.   Complete this list. How long is it? Does this agree with $\binom{5}{3}$ calculated from (4-1A)?

b.   Repeat **a** for $S = 2$ heads in $n = 4$ tosses.

APPENDIX TO SECTION **4-4**

Calculus for Continuous Distributions

For continuous random variables, sums are replaced by integrals—the limiting sum in calculus. Thus, for example, probability is the area under the curve $p(x)$, and is calculated as the integral:

$$\Pr(a \le X \le b) = \int_a^b p(x)dx \qquad \text{like (4-12)}$$

Similarly, the mean and variance are calculated as integrals:

$$\mu = \int_{\text{all } x} xp(x)dx \qquad \text{like (4-4)}$$

$$\sigma^2 = \int_{\text{all } x} (x - \mu)^2 p(x)dx \qquad \text{like (4-5)}$$

All the theorems that we state about discrete random variables then are equally valid for continuous random variables, with summations replaced by integrals.

PROBLEMS

4-3A   Suppose that a continuous random variable X has the probability distribution:

$$p(x) = 3x^2 \qquad 0 \le x \le 1$$
$$= 0 \qquad \text{otherwise}$$

a.   Calculate $\Pr(0 \le X \le .5)$, and show it on the graph of $p(x)$.

b.   Find the mean, median, and mode. Show them on the graph. Are they in the order you expect?

c.   Find σ, and show it on the graph. Is it a typical deviation in the sense of (2-14)?

4-4A   Repeat Problem 4-3A for each of the following distributions defined on the interval

0 to 1. [Called Beta distributions, they are all polynomials of the form $x^a(1 - x)^b$]:

i. $p(x) = 6x(1 - x)$

ii. $p(x) = 12x^2(1 - x)$

iii. $p(x) = 105x^4(1 - x)^2$

APPENDIX TO SECTION 5-3

Independent Implies Uncorrelated: Proof

To prove (5-12), we are given that X and Y are independent. According to (5-6), this means that for all x and y,

$$p(x,y) = p(x)p(y)$$

We substitute this into the definition of the covariance (5-10):

$$\sigma_{XY} \equiv \sum_x \sum_y (x - \mu_X)(y - \mu_Y)p(x,y)$$

$$= \sum_x \sum_y (x - \mu_X)(y - \mu_Y)p(x)p(y)$$

When we sum over y, we can take out as a common factor whatever does not depend on y (i.e., whatever depends on x alone):

$$\sigma_{XY} = \sum_x (x - \mu_X)p(x)\sum_y (y - \mu_Y)p(y)$$

The right-hand sum over y is zero—since every weighted sum of deviations is zero. [The proof would be like (2-9).] Thus

$$\sigma_{XY} = 0$$

That is, X and Y are uncorrelated, and (5-12) is proved.

APPENDIX TO SECTION 6-3

The Central Limit Theorem

Mathematicians call the Normal Approximation Rule by a different name—the *Central Limit Theorem*. Each of these three words deserves some explanation:

i.   It is a *theorem*, that is, it can be proved in general. The standard proof (Lindgren, 1976) requires two assumptions: *Random* sampling, from a population that has a *finite* variance (i.e., terribly extreme values are not too probable). In practice, the requirement of finite variance is nearly always met. Unfortunately, the requirement of randomness often is not.

ii. It is a *limit* theorem, that is, its conclusion is carefully stated in the form of a limit as follows: The probability that the standardized sample mean $(\overline{X} - \mu)/(\sigma/\sqrt{n})$ falls in a given interval converges in the limit to the probability that the standard normal variable Z falls in that interval.

iii. It is *central*, in the sense that it describes how $\overline{X}$ concentrates around its central value, the target μ.

APPENDIX TO SECTION 6-4

Continuity Correction: Graphical Explanation

In the figure below, we show the actual discrete distribution of P with bars, centered at the possible values of P: 0, 1/15, 2/15, ... The exact probability we want is the sum of all the shaded bars, from 11/15 on up ("more than 10 boys").

To approximate this sum with the normal curve as closely as possible, we should start the normal curve at the same place as those bars start. Since the first bar is *centered* at 11/15, it *starts* at $10\frac{1}{2}/15$, and this is precisely our continuity correction (c.c.):

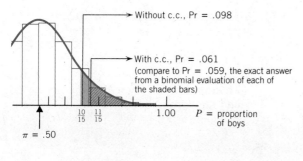

Without c.c., Pr = .098

With c.c., Pr = .061
(compare to Pr = .059, the exact answer
from a binomial evaluation of each of
the shaded bars)

$\frac{10}{15}$ $\frac{11}{15}$ 1.00 P = proportion of boys

$\pi = .50$

$$\Pr\left(P > \frac{10\frac{1}{2}}{15}\right) = .061 \qquad \text{(6-15) proved}$$

The probability of .061 obtained with c.c. compares very favorably with the exact answer of .059 (obtained from a tedious evaluation of all the shaded bars using the binomial formula).

APPENDIX TO SECTION **7-2**

Standard Error of $\overset{\shortmid}{X}$

A general large–sample formula for the standard error of the sample median for *any* continuous population can be derived:

$$\text{SE of } \overset{\shortmid}{X} \simeq \frac{1}{2\sqrt{n}\,p(\nu)} \qquad \text{(7-1A)}$$

where n is the sample size, and $p(\nu)$ is the height of the population probability density at the median ν.

For example, the symmetric Laplace distribution has density

$$p(x) = \frac{1}{\sqrt{2}\,\sigma}\, e^{-\sqrt{2}\left|\frac{x-\mu}{\sigma}\right|}$$

This differs from the normal curve (4-26) primarily because the negative exponent is not squared, and consequently the curve does not drop off so fast in the tails. To get the SE of $\overset{\shortmid}{X}$, symmetry assures us that $\nu = \mu$, and so $p(\nu)$ reduces to $p(\mu) = 1/\sqrt{2}\sigma$. Thus (7-1A) yields

$$\text{SE of } \overset{\shortmid}{X} \simeq \frac{1}{2\sqrt{n}\,(1/\sqrt{2}\sigma)}$$

$$= \frac{\sigma}{\sqrt{2}\sqrt{n}} = .71\,\frac{\sigma}{\sqrt{n}} \qquad \text{(7-7) confirmed}$$

By a similar argument, we can confirm the SE for the normal distribution given earlier: From (4-26), $p(\nu) = 1/\sqrt{2\pi}\,\sigma$, and thus

$$\text{SE of } \overset{\shortmid}{X} \simeq \frac{1}{2\sqrt{n}\,(1/\sqrt{2\pi}\,\sigma)}$$

$$= \sqrt{\frac{\pi}{2}}\,\frac{\sigma}{\sqrt{n}} = 1.25\,\frac{\sigma}{\sqrt{n}} \qquad \text{(7-5) confirmed}$$

APPENDIX TO SECTION **7-4**

Consistency: Careful Definition

The precise definition of consistency involves a limit statement:

> **V is defined to be a consistent estimator of θ, if for any positive δ (no matter how small),**
>
> $$\Pr(|V - \theta| < \delta) \to 1 \\ \text{as } n \to \infty \qquad \text{(7-2A)}$$

This is seen to be just a formal way of stating that in the limit as $n \to \infty$, it eventually becomes certain that V will be as close to θ as we please (within δ).

To be concrete, in Example 7-5 we show that the sample proportion P is a consistent estimator of π; for this case, (7-2A) therefore becomes:

$$\Pr(|P - \pi| < \delta) \to 1 \\ \text{as } n \to \infty \qquad \text{(7-3A)}$$

Stated informally, (7-3A) means that it eventually becomes certain that P will get as close to π as we please (within δ). In fact, (7-3A) is the precise statement of the law of large numbers (3-2).

As (7-21) states (and advanced texts prove), if the MSE approaches zero, then V is consistent. But the definition (7-2A) is broad enough that V can be consistent in other cases as well. An example will illustrate. Consider an estimator V with the following sampling distribution:

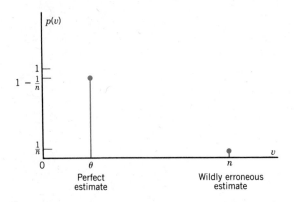

Perfect
estimate

Wildly erroneous
estimate

As n increases, note how the distribution of V becomes increasingly concentrated on its target θ while the probability of getting an erroneous estimate becomes smaller and smaller (as 1/n decreases). Thus, V is a consistent estimator according to the definition (7-2A).

But now let us consider the MSE of V:

$$
\begin{aligned}
\mathrm{MSE} &= E(V - \theta)^2 \\
&= \Sigma(v - \theta)^2 p(v) \\
&= 0^2\left(1 - \frac{1}{n}\right) + (n - \theta)^2\left(\frac{1}{n}\right) \\
&= n - 2\theta + \frac{\theta^2}{n}
\end{aligned}
$$

As $n \to \infty$, this MSE does not approach zero (in fact, it behaves much worse—it approaches infinity!) In conclusion, then, we have found an estimator V that is consistent, even though its MSE does not tend to zero.

PROBLEM

7-1A   Prove formula (7-11), that MSE = variance + bias².

[*Hint:* The first equation in Problem 4-1A can be rewritten as

$$
E(X - a)^2 = \sigma^2 + (a - \mu)^2
$$

Then put the estimator V in the role of X, and the target θ in the role of a.]

APPENDIX TO SECTION **8-3**

The Standard Error of $(\overline{X}_1 - \overline{X}_2)$: Proof

The difference $(\overline{X}_1 - \overline{X}_2)$ is a linear combination of two random variables that are independent (since the two samples are assumed independent). Its variance can therefore be found from the general formula (5-22):

$$
\mathrm{var}(aX + bY) = a^2 \,\mathrm{var}\, X + b^2 \,\mathrm{var}\, Y
$$

Substitute $X = \overline{X}_1$, $Y = \overline{X}_2$, $a = 1$ and $b = -1$.

$$
\begin{aligned}
\mathrm{var}(\overline{X}_1 - \overline{X}_2) &= 1^2 \,\mathrm{var}\, \overline{X}_1 + (-1)^2 \,\mathrm{var}\, \overline{X}_2 \\
&= \mathrm{var}\, \overline{X}_1 + \mathrm{var}\, \overline{X}_2
\end{aligned}
$$

For single samples, we already know the variances:

$$
\mathrm{var}\, \overline{X} = \mathrm{SE}^2 = \frac{\sigma^2}{n} \qquad \text{like (6-5)}
$$

Thus $$\mathrm{var}(\overline{X}_1 - \overline{X}_2) = \frac{\sigma_1^2}{n_1} + \frac{\sigma_2^2}{n_2}$$

$$
\mathrm{SE} = \sqrt{\frac{\sigma_1^2}{n_1} + \frac{\sigma_2^2}{n_2}}
$$

When this is substituted into (8-17), we obtain the desired confidence interval (8-18).

PROBLEM

8-1A   Calculate the SE of the unbiased estimate in Problem 7-8. Then find the 95% confidence interval for the population proportion.

APPENDIX TO SECTION **8-5**

Confidence Interval for π: Derivation of Graph

To see why Figure 8-4 works, let us return to first principles, and see how a deduction can be turned around into an induction.

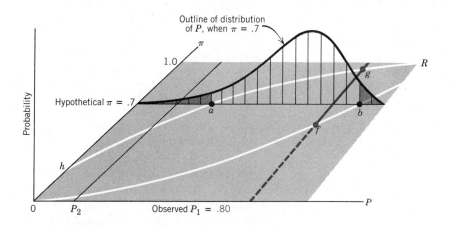

Our deduction of how the estimator P is distributed for a sample size $n = 20$ and $\pi = .7$ for example, is shown in the figure above. The range ab marks off the central chunk that contains 95% of the probability. Similarly, for every other possible value of π, the distribution of P determines the two critical points like a and b. When they are all joined, they make up the two white curves enclosing the 95% *probability band*.

This deduction of how the statistic P is related to the population π is now turned around to draw a statistical inference about π from a given sample P. For example, if we observe a sample proportion $P_1 = 16/20 = .80$, then the 95% confidence interval for π is defined by the interval fg contained within this probability band above P_1:

$$.55 < \pi < .95 \qquad \text{(8-28) proved}$$

Whereas the (deduced) probability interval is defined in the horizontal direction of the P axis, the (induced) confidence interval is defined in the vertical direction of the π axis.

To see why this works, suppose, for example, that the true value of π is .7. Then the probability is 95% that a sample P will fall between a and b. If and only if it does (e.g., P_1) will the confidence interval we construct bracket the true $\pi = .7$. We therefore are using an estimation procedure that is 95% likely to bracket the true value of π, and thus yield a correct statement. But we must recognize the 5% probability that the sample P will fall beyond a or b (e.g., P_2); in this case, our interval estimate will not bracket $\pi = .7$, and our conclusion will be wrong.

APPENDIX TO SECTION **10-1**

One-Way Analysis of Variation: Proof

Let us begin with a concrete example. In Table 10-1, the third observation in the second column (58) is greater than the grand mean (52). This total deviation can be broken down into two parts, noting that the second column mean is 56:

$(58 - 52) = (58 - 56) + (56 - 52)$
total   = deviation + deviation between
deviation   within   this column
              column   and others

That is,

$$6 = \quad 2 \quad + \quad 4$$

In words: the given observation is 6 better than the grand average—2 better than its column average, which in turn is 4 better than the grand average.

The generalization starts with the typical observation X_{it}:

$$(X_{it} - \overline{\overline{X}}) = (X_{it} - \overline{X}_i) + (\overline{X}_i - \overline{\overline{X}})$$

This equation must always be true, because the two occurrences of $\overline{X}_i$ cancel on the right side. Now let us interchange the order on the right-hand side, then square both sides, and finally sum over all i and t:

$$\sum_i \sum_t (X_{it} - \overline{\overline{X}})^2 = \sum_i \sum_t (\overline{X}_i - \overline{\overline{X}})^2 + 2 \sum_i \sum_t (\overline{X}_i - \overline{\overline{X}})(X_{it} - \overline{X}_i) + \sum_i \sum_t (X_{it} - \overline{X}_i)^2$$

On the right the first term can be reduced:

$$\sum_{t=1}^{n} \underbrace{\sum_{i=1}^{c} (\overline{X}_i - \overline{\overline{X}})^2}_{} = n \sum_{i=1}^{c} (\overline{X}_i - \overline{\overline{X}})^2$$

independent of t

Furthermore, the next term (cross product) vanishes:

$$2 \sum_{i=1}^{c} (\overline{X}_i - \overline{\overline{X}}) \underbrace{\sum_{t=1}^{n} (X_{it} - \overline{X}_i)}_{} = 0$$

This is 0, because the algebraic sum of deviations about the mean is always 0

With these two simplifications, we thus obtain our goal—Equation (10-14):

$$\sum_i \sum_t (X_{it} - \overline{\overline{X}})^2$$
$$= n \sum_i (\overline{X}_i - \overline{\overline{X}})^2 + \sum_i \sum_t (X_{it} - \overline{X}_i)^2$$
total   = between-   + within-
variation   column   column
             variation   variation

APPENDIX TO SECTION **10-2**

Two-Way Analysis of Variation: Proof

This may be proved just like one-factor ANOVA in Appendix 10-1. We begin with the similar identity:

$$(X_{ij} - \overline{\overline{X}}) = (X_{ij} - \hat{X}_{ij}) + (\overline{X}_{i\cdot} - \overline{\overline{X}}) + (\overline{X}_{\cdot j} - \overline{\overline{X}})$$

where

$$\hat{X}_{ij} = (\overline{X}_{i\cdot} - \overline{\overline{X}}) + (\overline{X}_{\cdot j} - \overline{\overline{X}}) + \overline{\overline{X}}$$
$$= \overline{X}_{i\cdot} + \overline{X}_{\cdot j} - \overline{\overline{X}}$$

By squaring and summing and noting how all the cross-product terms drop out, a little rearranging finally gives us our goal—Equation (10-18):

$$\sum_{i=1}^{c} \sum_{j=1}^{r} (X_{ij} - \overline{\overline{X}})^2$$
$$= r \sum_{i=1}^{c} (X_{i\cdot} - \overline{\overline{X}})^2 + c \sum_{j=1}^{r} (\overline{X}_{\cdot j} - \overline{\overline{X}})^2$$
$$+ \sum_{i=1}^{c} \sum_{j=1}^{r} (X_{ij} - \hat{X}_{ij})^2$$
total   = between-   + between- + residual
variation   column   row   variation
             variation   variation

APPENDIX TO SECTION **11-1**

Lines and Planes

LINES

The definitive characteristic of a straight line is that it continues forever in the same constant direction. In Figure 11-1A we make this idea precise. In moving from one point P_1 to another point P_2, we denote the horizontal distance by ΔX (where Δ is the Greek letter D, for difference) and the vertical distance by ΔY. Then the slope is defined as

$$\text{Slope} \equiv \frac{\Delta Y}{\Delta X} \qquad (11\text{-}1A)$$

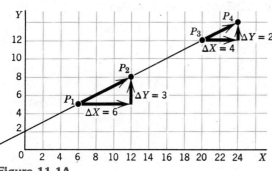

Figure 11-1A
A straight line is characterized by constant slope, $\Delta Y/\Delta X = b$.

Slope is a concept useful in engineering as well as mathematics. For example, if a highway rises 12 feet over a distance of 200 feet, its slope is $12/200 = 6\%$. The characteristic of a *straight* line is that the slope remains the same everywhere:

$$\boxed{\frac{\Delta Y}{\Delta X} = b \text{ (a constant)}} \qquad (11\text{-}2A)$$

For example, the slope between P_3 and P_4 is the same as between P_1 and P_2, as calculation will verify:

$$P_1 \text{ to } P_2: \quad \frac{\Delta Y}{\Delta X} = \frac{3}{6} = .50$$

$$P_3 \text{ to } P_4: \quad \frac{\Delta Y}{\Delta X} = \frac{2}{4} = .50 \qquad (11\text{-}3A)$$

A very instructive case occurs when X increases just one unit; then (11-2A) yields

$$\boxed{\text{When } \Delta X = 1, \quad \Delta Y = b} \qquad (11\text{-}4A)$$

In words, "b is the increase in Y that accompanies a unit increase in X," which agrees with the regression interpretation (11-8).

It is now very easy to derive the equation of a line if we know its slope b and any one point on the line. Suppose the one point we know is the Y intercept—the point where the line crosses the Y axis, at height a, let us say. As Figure 11-2A shows, in moving to any other point (X, Y) on the line, we may write

$$\text{Slope}, \frac{\Delta Y}{\Delta X} = \frac{Y - a}{X - 0} \qquad (11\text{-}5A)$$

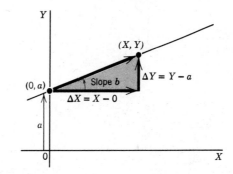

Figure 11-2A
Derivation of the equation of a straight line.

For the line to be straight, we insisted in (11-2A) that this slope must equal the constant b:

$$\frac{Y - a}{X - 0} = b$$

$$Y - a = bX$$

$$\boxed{Y = a + bX} \qquad (11\text{-}6A)$$

This is the required equation of a line (11-1).

PLANES

In Figure 11-3A we show a plane in the three dimensional (X_1, X_2, Y) space. Let L_1 denote the line where this plane cuts the X_1Y plane. Then we may think of the plane as a roof made of rafters parallel to the line L_1 and having the same slope b_1. That is, for every foot we move in the X_1 direction, the rafter rises vertically in the Y direction by b_1 feet.

Equally well, of course, we may think of the plane as a grid of rafters parallel to the line L_2 on the X_2Y plane, with slope b_2.

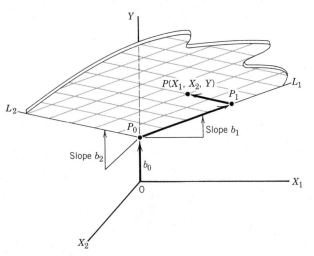

Figure 11-3A
Derivation of the equation of a plane.

It is now very easy to derive the equation of a plane. Referring to Figure 11-3A, let us start at the Y intercept, the point P_0 at height b_0. Let us move to the typical point P with coordinates (X_1, X_2, Y), in two steps: (1) we move along the line L_1 to the intermediate point P_1; and from there (2) we move parallel to L_2 to reach the final height Y at P.

(1) Just as (11-6A) gives the height as a function of X, likewise we find the height of P_1 on L_1 as a function of X_1:

$$b_0 + b_1X_1 \qquad \text{(11-7A)}$$

(2) For our final move from P_1 to P, note that our intercept has now become $b_0 + b_1X_1$ as given above. Note also that we will be moving along a grid line (rafter) parallel to L_2, with slope b_2. Applying (11-6A) to this case,

$$Y = (b_0 + b_1X_1) + b_2X_2$$

$$\boxed{Y = b_0 + b_1X_1 + b_2X_2} \qquad \text{(11-8A)}$$

This is the required equation of a plane (13-4).

APPENDIX TO SECTION 11-2

Least Squares Formulas: Proofs

REEXPRESSING THE CRITERION

We will find the proofs in the next two chapters much easier if we invest a little time now to express $\hat{Y}$ in a mathematically more convenient form. Recall that the straight line we are fitting is

$$\hat{Y} = a + bX \qquad \text{(11-6A) repeated}$$

This can be reexpressed as

$$\hat{Y} = (a + b\overline{X}) + b(X - \overline{X})$$

That is,

$$\boxed{\hat{Y} = a_* + bx} \qquad \text{(11-9A)}$$

where

$$a_* = a + b\overline{X} \qquad \text{(11-10A)}$$
$$x = X - \overline{X} \qquad \text{(11-11A)}$$
$$\text{(11-4) repeated}$$

This reexpression (11-9A) just uses the deviation form $(x = X - \overline{X})$, and as a consequence a must be replaced by a_*. Then the least squares criterion takes the form:

$$\text{minimize } \Sigma[Y - (a_* + bx)]^2 \qquad \text{like (11-2)}$$

This sum of squares will be denoted by $S(a_*, b)$, as a reminder that it is a function of a_* and b. Thus our problem is to select the values of a_* and b that minimize:

$$S(a_*, b) \equiv \Sigma(Y - a_* - bx)^2 \qquad \text{(11-12A)}$$

As a_* and b vary—that is, as various straight lines of the form (11-9A) are tried out—then $S(a_*, b)$ will vary too. Those values of a_* and b that make $S(a_*, b)$ a minimum give us the least squares line.

FINDING THE LEAST SQUARES VALUES OF a_* AND b

Minimizing $S(a_*, b)$ requires setting its partial derivatives equal to zero. So we first set the partial derivative with respect to a_* equal to zero:

$$\frac{\partial}{\partial a_*} \Sigma(Y - a_* - bx)^2 =$$

$$\Sigma \, 2(Y - a_* - bx)^1 \, (-1) = 0$$

Dividing through by -2:

$$\Sigma(Y - a_* - bx) = 0 \qquad \text{(11-13A)}$$

Rearranging:

$$\Sigma Y - na_* - b\Sigma x = 0 \qquad \text{(11-14A)}$$

Now we see why the deviation form x is so useful. Since the positive and negative deviations exactly cancel, their sum is zero [as shown in Table 11-2, for example, or Equation (2-9) in general]. In other words:

$$\Sigma x = 0 \qquad \text{(11-15A)}$$

Thus (11-14A) greatly simplifies: The term in b drops out, and we can solve for a_*:

$$a_* = \frac{\Sigma Y}{n} = \overline{Y} \qquad \text{(11-16A)}$$

Returning to $S(a_*, b)$ in (11-12A), we must also set its partial derivative with respect to b equal to zero:

$$\frac{\partial}{\partial b} \Sigma(Y - a_* - bx)^2 =$$

$$\Sigma 2(Y - a_* - bx)^1(-x) = 0$$

Dividing through by -2:

$$\Sigma x(Y - a_* - bx) = 0 \qquad \text{(11-17A)}$$

Rearranging:

$$\Sigma xY - a_*\Sigma x - b\Sigma x^2 = 0$$

Noting again that $\Sigma x = 0$, we can solve easily for b:

$$b = \frac{\Sigma xY}{\Sigma x^2} \qquad \text{(11-18A)}$$

REEXPRESSING THE SOLUTION

Let us solve (11-10A) for the original intercept a:

$$a = a^* - b\overline{X}$$

Substituting (11-16A),

$$a = \overline{Y} - b\overline{X} \qquad \text{(11-6) proved}$$

Finally, let us reexpress b. We shall use the deviation form for Y as well as X; in other words, let

$$y = Y - \overline{Y} \qquad \text{(11-4) repeated}$$

That is, $Y = y + \overline{Y}$, which we can substitute into equation (11-18A):

$$b = \frac{\Sigma x(y + \overline{Y})}{\Sigma x^2} \qquad \text{(11-19A)}$$

We now reexpress the numerator (noting that $\overline{Y}$ is a constant and can therefore be taken outside the summation):

$$\Sigma x(y + \overline{Y}) = \Sigma xy + \overline{Y}\Sigma x$$
$$= \Sigma xy$$

since $\Sigma x = 0$ according to (11-15A). Consequently, the solution for b simplifies to

$$b = \frac{\Sigma xy}{\Sigma x^2} \qquad \text{(11-5) proved}$$

PROBLEM

11-1A If the regression equation is lengthened to include another regressor, we obtain the *multiple* regression equation:

$$\hat{Y} = b_0 + b_1 x_1 + b_2 x_2 \quad \text{like (11-8A)}$$

Show that the least squares values of b_0, b_1, and b_2 must satisfy the three equations given in (13-5) to (13-7).

APPENDIX TO SECTION 12-3

A One-sided or Two-sided Test?

SIMPLEST CASE, IMPLIED BY WORDING OF THE QUESTION

In regression, we again encounter the issue first raised in Chapter 9: Should we use a one- or two-sided test (or p-value, or confidence interval)? Sometimes the answer may be found in the way the question is worded. For instance, in Example 12-3, the null hypothesis was that "yield *does not increase* with fertilizer." This implies the alternative hypothesis is that yield *does* increase with fertilizer—that is,

$$H_1: \quad \beta > 0$$

Since this is a one-sided hypothesis, the one-sided p-value was appropriate.

On the other hand, in Example 12-2, the null hypothesis was that "yield is unrelated to fertilizer." This implies the alternative hypothesis that yield is related to fertilizer in either a positive *or* negative way. Thus we may write this alternative hypothesis as:

$$H_1: \quad \beta > 0 \text{ or } \beta < 0$$

that is, $\qquad H_1: \quad \beta \neq 0$

Since this is a two-sided hypothesis, the two-sided test based on the two-sided confidence interval is appropriate.

Thus, just as in Chapter 9, the wording of the question typically provides the key. Phrases like: Y is "unrelated to" or "unaffected by" X suggest that a two-sided test is appropriate. On the other hand, phrases like: X "raises," "lowers," "improves," or "decreases" X, suggest that a one-sided test is required.

SUBTLE CASE, REQUIRING *A PRIORI* REASONING

The problem is that we are often *not* given the question, but have to phrase it ourselves. Should we use a one- or two-sided approach? The answer depends on prior theoretical reasoning. To illustrate, suppose we wish to investigate the effect of wages W on the national price level P, using the simple relationship,

$$P = \alpha + \beta W$$

On theoretical grounds, it may be concluded a priori that if wages affect prices at all, this relation will be a positive one. In this case, H_0 is tested against the one-sided alternative:

$$H_1: \quad \beta > 0$$

On the other hand, as an example of a case in which such clear prior guidelines do not exist, consider an equation explaining saving. How does it depend on the interest rate? It is not clear, on theoretical grounds, whether this effect is a positive or a negative one. Since interest is the reward for saving, a high interest rate should provide an incentive to save, leading us to expect the interest rate to affect saving positively. But if individuals save in order to accumulate some target sum (perhaps for their retirement, or to buy a house), then the higher the interest rate, the more rapidly any saving will accumulate, hence the less they need to save to reach this target. In this case, interest affects saving negatively. If it is not clear which of these two effects may dominate, it is appropriate to test H_0 against the two-sided alternative:

$$H_1: \quad \beta \neq 0$$

APPENDIX TO SECTION **12-4**

Confidence Intervals above X_0: Proofs

CONFIDENCE INTERVAL FOR THE MEAN μ_0

We refer once more to the sampling distribution of $\hat{Y}_0$ shown in Figure 12-6b. Sometimes $\hat{Y}_0$ is too high, sometimes too low, but on average, it seems just right. That is, $\hat{Y}_0$ appears to be an *unbiased* estimator. Let us now verify this; that is, show that $E(\hat{Y}_0) = \mu_0$.

In the theoretical approach that follows, it is convenient—as in Appendix 11-2—to use the deviation form $x = X - \overline{X}$. Then the fitted line has the form:

$$\hat{Y} = a_* + bx \qquad (12\text{-}1A)$$
$$\text{like } (11\text{-}9A)$$

In particular, when x is set at a specific level x_0

$$\hat{Y}_0 = a_* + bx_0 \qquad (12\text{-}2A)$$

For the true regression line, there is a corresponding form denoted with Greek symbols:

$$E(Y) = \alpha_* + \beta x \qquad (12\text{-}3A)$$

Then $\qquad \mu_0 = E(Y_0) = \alpha_* + \beta x_0$

We have already shown in equation (12-19) that b fluctuates around β with variance $\sigma^2/\Sigma x^2$. Similarly, we could show that a_* fluctuates around α_* with variance σ^2/n. [According to (11-16A), $a_* = \overline{Y}$, so this variance is just the formula (6-6) for sample means.] Furthermore, b and a_* are uncorrelated (as proved in advanced texts such as Wonnacott, 1981). This zero correlation in fact is an important reason for using the deviation form x.

The prediction $\hat{Y}_0$ in (12-2A) fortunately is a linear combination of these two variables (estimates) b and a_* (with coefficients 1 and x_0). Thus the theory of linear combinations can be applied. From (5-17),

$$E(\hat{Y}_0) = E(a_* + bx_0)$$
$$= E(a_*) + x_0 E(b)$$
$$= \alpha_* + x_0 \beta$$
$$= \mu_0$$

Thus $\hat{Y}_0$ is indeed an unbiased estimator of μ_0.

With the mean of $\hat{Y}_0$ thus determined, we now turn to its variance. First, note from (5-22) that

$$\text{var}(\hat{Y}_0) = \text{var}(a_* + bx_0)$$
$$= \text{var } a_* + x_0^2 \text{ var } b$$
$$= \frac{\sigma^2}{n} + x_0^2 \frac{\sigma^2}{\Sigma x^2}$$
$$= \sigma^2 \left[\frac{1}{n} + \frac{x_0^2}{\Sigma x^2} \right] \qquad (12\text{-}4A)$$
$$\text{SE of } \hat{Y}_0 = \sigma \sqrt{\frac{1}{n} + \frac{x_0^2}{\Sigma x^2}}$$

To make a 95% confidence interval, as usual we substitute s for σ, and use $t_{.025}$ to compensate:

$$\mu_0 = \hat{Y}_0 \pm t_{.025} \text{ SE}$$

That is,

$$\mu_0 = (a_* + bx_0) \pm t_{.025}\, s \sqrt{\frac{1}{n} + \frac{x_0^2}{\Sigma x^2}}$$

While this is the formula for the deviation form x, it can be easily seen that it translates back into the original X values, as follows:

$$\mu_0 = (a + bX_0) \pm t_{.025}\, s \sqrt{\frac{1}{n} + \frac{(X_0 - \overline{X})^2}{\Sigma(X - \overline{X})^2}}$$

This completes our derivation of (12-32).

PREDICTION INTERVAL FOR A SINGLE OBSERVATION Y_0

With x expressed in deviation form, the best prediction of a single observed Y_0 is again the point on the estimated regression line above x_0.

As already noted, an *interval* estimate for Y_0 requires (1) the variance (12-4A) reflecting the errors in estimating the fitted regression *plus* (2) the variance σ^2 reflecting the fluctuation in an individual Y observation. Adding these both together, we obtain:

$$\sigma^2 \left(\frac{1}{n} + \frac{x_0^2}{\Sigma x^2} \right) + \sigma^2 = \sigma^2 \left(\frac{1}{n} + \frac{x_0^2}{\Sigma x^2} + 1 \right)$$

Except for this larger variance, the prediction interval for Y_0 is the same as the confidence interval for μ_0 given earlier:

$$Y_0 = (a + bX_0) \pm t_{.025}\, s\, \sqrt{\frac{1}{n} + \frac{(X - X_0)^2}{\Sigma x^2} + 1}$$

This completes our derivation of (12-33).

APPENDIX TO SECTION 13-2

Solution of a Set of Simultaneous Linear Equations

To illustrate, let us take the pair of estimating equations in Table 13-2:

$$16,500 = 280,000b_1 + 7000b_2 \qquad (1)$$

$$600 = 7000b_1 + 400b_2 \qquad (2)$$

To solve a set of simultaneous equations, we eliminate one unknown at a time. To eliminate the first unknown b_1, we match its coefficient in equations (1) and (2), by multiplying equation (2) by 40:

$$24,000 = 280,000b_1 + 16,000b_2 \qquad (3)$$

When equation (1) is subtracted from (3), b_1 is indeed eliminated:

$$7500 = 9000b_2$$

$$b_2 = \frac{7500}{9000} = .833 \qquad (4)$$

With b_2 solved, we can use it to solve for the other unknown b_1. We substitute equation (4) back into (1):

$$16,500 = 280,000b_1 + 7000(.833)$$

$$b_1 = \frac{10,667}{280,000} = .0381 \qquad (5)$$

Thus equations (4) and (5) are the required solution.

APPENDIX TO SECTION 13-5

Direct Plus Indirect Relation: Proof

We will prove in general that the total relation of Y to X_1 (i.e., the simple regression coefficient of Y against X_1) is equal to the direct plus the indirect relation, as set out in (13-41). We begin with the first estimating equation (13-5) for the multiple regression of Y against X_1 and X_2:

$$\Sigma x_1 y = b_1 \Sigma x_1^2 + b_2 \Sigma x_1 x_2$$

If we divide this equation by Σx_1^2, we get:

$$\left(\frac{\Sigma x_1 y}{\Sigma x_1^2}\right) = b_1 + b_2 \left(\frac{\Sigma x_1 x_2}{\Sigma x_1^2}\right)$$

The bracketed expression on the left is the simple regression coefficient of Y against X_1—that is, the *total* relation. The bracketed expression on the right is the simple regression coefficient of X_2 against X_1, denoted b. Thus the whole equation may be written as the required equation (13-41):

$$\text{Total relation} = b_1 + b_2 b$$

APPENDIX TO SECTION 14-4

Log Regression Handles a Multiplicative Error Term

Earlier in Chapter 14, we said nothing about population models or error terms; we simply showed how various equations can be fitted. Standard assumptions about the error term would have justified the regression procedure, and there was no point in repeating the obvious.

However, an exception now occurs in log regression, where we must make a different assumption about the error term. For example, when we estimate exponential growth, we are assuming that the original population model has an error u that appears in the following *multiplicative* form:

$$P = \gamma e^{\beta X} u \qquad \text{like (14-23)}$$

Taking logs:

$$\log P = \log \gamma + \beta X + \log u$$

Letting $\log P \equiv Y$, $\log \gamma \equiv \alpha$, and $\log u \equiv e$, our population model takes the familiar form:

$$Y = \alpha + \beta X + e$$

Note how the log transformation has made the error term additive, as required. If we make the other usual assumptions about e (independence and constant variance), then this becomes the standard regression model (12-3).

PROBLEM

14-1A　In each model in Problem 14-22, what form should the error take in order for your transformation to reduce the model to the standard linear form?

APPENDIX TO SECTION 15-1

Correlation in Chapter 15 Agrees with Chapter 5

The correlation defined in Chapter 15 is the same correlation ρ that we met in Chapter 5. To see exactly why, let us begin with the equation that calculates the correlation for the whole population of N individuals:

$$\rho = \frac{\sum\limits^{N} xy}{\sqrt{\sum\limits^{N} x^2 \sum\limits^{N} y^2}} \qquad \text{like (15-2)}$$

To reexpress this, we can write out the deviations explicitly: $x = X - \mu_X$, and $y = Y - \mu_Y$, where, of course, μ_X and μ_Y are population (rather than sample) means. Let us also divide by N in the numerator, and again in the denominator to compensate:

$$\rho = \frac{\frac{1}{N} \sum\limits^{N} (X - \mu_X)(Y - \mu_Y)}{\sqrt{\frac{1}{N} \sum\limits^{N} (X - \mu_X)^2 \frac{1}{N} \sum\limits^{N} (Y - \mu_Y)^2}}$$

We recognize the numerator as the covariance σ_{XY} given in (5-10), and the denominator as $\sqrt{\sigma_X^2 \sigma_Y^2}$. Thus ρ takes on the familiar form we have met already:

$$\rho = \frac{\sigma_{XY}}{\sigma_X \sigma_Y} \qquad \text{(5-13) repeated}$$

APPENDIX TO SECTION 15-2

ANOVA and r^2: Proofs

THE ANALYSIS OF VARIATION

The proof of (15-10) is very similar to the proof in Appendix 10-1. We just square both sides of (15-8), and sum:

$$\Sigma(Y - \overline{Y})^2 = \Sigma[(\hat{Y} - \overline{Y}) + (Y - \hat{Y})]^2$$
$$= \Sigma(\hat{Y} - \overline{Y})^2 + 2\Sigma(\hat{Y} - \overline{Y})(Y - \hat{Y})$$
$$+ \Sigma(Y - \hat{Y})^2 \qquad \text{(15-1A)}$$

To prove the second-last sum is zero, once more the deviation form used in Appendix 11-2 will be useful:

$$\hat{Y} = a_\star + bx \qquad \text{(15-2A)}$$
$$\text{(11-9A) repeated}$$

from (11-16A),

$$= \overline{Y} + bx$$

Thus

$$\hat{Y} - \overline{Y} = bx \qquad \text{(15-3A)}$$

Substitute this into the second-last sum in (15-1A); this term then becomes

$$2\Sigma bx(Y - \hat{Y})$$

Ignoring the constant $2b$, we obtain:

$$\Sigma x(Y - \hat{Y}) = \Sigma x(Y - (a_\star + bx))$$

from (11-17A), $= 0$

With its second-last sum reduced to 0, (15-1A) becomes:

$$\Sigma(Y - \overline{Y})^2 = \Sigma(\hat{Y} - \overline{Y})^2 + \Sigma(Y - \hat{Y})^2 \quad (15\text{-}4A)$$
$$(15\text{-}9) \text{ proved}$$

If we substitute (15-3A) into the middle term of this, we obtain another useful form:

$$\Sigma(Y - \overline{Y})^2 = b^2\Sigma x^2 + \Sigma(Y - \hat{Y})^2 \quad (15\text{-}5A)$$
$$(15\text{-}10) \text{ proved}$$

r^2 IS THE COEFF. OF DETERMINATION

To prove (15-16), we begin by expressing b in terms of r. To do so, we divide equation (15-1) by (15-2), obtaining:

$$\frac{b}{r} = \frac{\sqrt{\Sigma y^2}}{\sqrt{\Sigma x^2}}$$

Solve for b:

$$b = r\frac{\sqrt{\Sigma y^2}}{\sqrt{\Sigma x^2}} \quad (15\text{-}6A)$$

Incidentally, if we divide numerator and denominator by $\sqrt{n-1}$, we obtain the equivalent form:

$$b = r\frac{\sqrt{\Sigma y^2/(n-1)}}{\sqrt{\Sigma x^2/(n-1)}} = r\frac{s_Y}{s_X} \quad (15\text{-}7A)$$
$$(15\text{-}7) \text{ proved}$$

Now we substitute (15-6A) into the fundamental ANOVA relation (15-5A):

$$\Sigma(Y - \overline{Y})^2 = r^2 \Sigma y^2 + \Sigma(Y - \hat{Y})^2$$

The solution for r^2 is:

$$r^2 = \frac{\Sigma(Y - \overline{Y})^2 - \Sigma(Y - \hat{Y})^2}{\Sigma y^2} \quad (15\text{-}8A)$$

We can reexpress the numerator using (15-4A). And the denominator, by definition, is $\Sigma y^2 \equiv \Sigma(Y - \overline{Y})^2$. Thus

$$r^2 = \frac{\Sigma(\hat{Y} - \overline{Y})^2}{\Sigma(Y - \overline{Y})^2}$$

$$= \frac{\text{explained variation of } Y}{\text{total variation of } Y} \quad (15\text{-}16) \text{ proved}$$

$$= \text{coefficient of determination}$$

RESIDUAL VARIANCE IS REDUCED BY r^2

To prove (15-17), note first that the *coefficient of indetermination* (the complement of r^2) is:

$$1 - r^2 = \frac{\text{unexplained variation of } Y}{\text{total variation of } Y}$$

To convert variations to variances, divide both numerator and denominator by $n-1$. Then in the denominator we get s_Y^2, the total variance of Y. In the numerator we almost get s^2, the residual variance (the slight difference is that s^2 should have a divisor of $n-2$ rather than $n-1$). Thus

$$1 - r^2 \simeq \frac{s^2}{s_Y^2}$$

That is, $s^2 \simeq (1-r^2)s_Y^2 \quad (15\text{-}17) \text{ proved}$

APPENDIX TO SECTION 18-2

MLE For Some Familiar Cases: Proofs

MLE FOR A PROPORTION π

Given a sample where S successes were observed in n trials, the likelihood function is given by the binomial formula,

$$L(\pi) = \binom{n}{S} \pi^S(1 - \pi)^{n-S}$$

The reader who has carefully learned that π is a fixed population parameter may wonder how it can appear in the likelihood function as a variable. This is simply a mathematical convenience. The true value of π is, in fact, fixed. But since it is unknown, in MLE we let it range through all of its possible or hypothetical values; in other words, we treat it like a variable.

To find the maximum of $L(\pi)$, we therefore set its derivative equal to zero:

$$\frac{dL(\pi)}{d\pi} = \binom{n}{S}[\pi^S(n - S)(1 - \pi)^{n-S-1}(-1)$$
$$+ S\pi^{S-1}(1 - \pi)^{n-S}] = 0$$

Divide by $\binom{n}{S}\pi^{S-1}(1 - \pi)^{n-S-1}$ to obtain:

$$-\pi(n - S) + S(1 - \pi) = 0$$
$$-n\pi + S = 0$$
$$\pi = \frac{S}{n} = P$$

We could easily confirm that this is a maximum (rather than a minimum or inflection point). Actually, this derivation assumed that $0 < S < n$. When $S = 0$, the proof does not even require calculus: When $S = 0$, then $L(\pi) = (1 - \pi)^n$, which is clearly a maximum at the end-point $\pi = 0$. Similarly, when $S = n$, then $L(\pi) = \pi^n$, which is clearly a maximum at the other end-point, $\pi = 1$. Thus we have proven (18-5) valid in all possible cases.

MLE IS IDENTICAL TO LEAST SQUARES IN A NORMAL REGRESSION MODEL

To simplify, the proof below is for simple regression; but the argument may be easily generalized to cover multiple regression as well. Suppose that we have a sample of n points, rather than just 3. Because Y_1, for example, is assumed normally distributed, its probability (density) is given by (4-26) as:

$$p(Y_1) = \frac{1}{\sqrt{2\pi}\sigma} \exp\left[-\frac{1}{2}\left(\frac{Y_1 - \mu}{\sigma}\right)^2\right]$$

where $\exp[\ ]$ is just a more convenient way of writing the exponential function $e^{[\ ]}$.

Now μ, the mean of Y_1, is given by (12-1) as:

$$\mu = \alpha + \beta X_1$$

Substituting this yields:

$$p(Y_1) = \frac{1}{\sqrt{2\pi}\sigma} \exp\left[-\frac{1}{2}\left(\frac{Y_1 - (\alpha + \beta X_1)}{\sigma}\right)^2\right]$$

In terms of the geometry of Figure 18-4, $p(Y_1)$ is the arrow above Y_1. The probability of Y_2 is similar, except that the subscript 2 replaces 1 throughout, and so on, for all the other observed Y values.

The independence of the Y values justifies multiplying all these probabilities together to find their joint probability (that is, the probability they would all occur, given α and β):

$$p(Y_1, Y_2, \ldots, Y_n \,/\, \alpha, \beta)$$
$$= \frac{1}{\sqrt{2\pi}\sigma} \exp\left[-\frac{1}{2}\left(\frac{Y_1 - (\alpha + \beta X_1)}{\sigma}\right)^2\right]$$
$$\frac{1}{\sqrt{2\pi}\sigma} \exp\left[-\frac{1}{2}\left(\frac{Y_2 - (\alpha + \beta X_2)}{\sigma}\right)^2\right] \ldots$$

Recall that the observed Y's are given. We are speculating on various values of α and β. To emphasize this, we call this the likelihood function $L(\alpha,\beta)$. At the same time, using the familiar rule for exponents (namely, $e^a e^b = e^{a+b}$), this product can be reexpressed by summing exponents:

$$L(\alpha,\beta) = \left(\frac{1}{\sqrt{2\pi}\sigma}\right)^n \exp\left(-\frac{1}{2\sigma^2}\sum_{}^{n}[Y_i - (\alpha + \beta X_i)]^2\right)$$

Now we ask, what values of α and β produce the largest L? The only place α and β appear is in the exponent; moreover, maximizing a function with a negative exponent involves minimizing the magnitude of the exponent. Hence the MLE estimates are obtained by choosing the α and β that

$$\text{Minimize } \Sigma[Y_i - (\alpha + \beta X_i)]^2$$

The selection of maximum likelihood estimates of α and β to minimize this is identical to the selection of least squares estimates α and β to minimize (11-2). That is, MLE and least squares yield identical estimates and (18-7) is proved.

COROLLARY: MLE FOR THE MEAN OF A NORMAL POPULATION

In the special case of regression where $\beta = 0$, then every Y comes from the *same* population centered at α. That is, we have a sample of n observations from one population. The MLE will still be the least squares estimate, of course, which we found in (11-9) was just $\overline{Y}$. Thus we have proved (18-6):

In a normal population,

$$\text{MLE of population mean} = \overline{Y}$$

PROBLEM

18-1A To prove the MLE is identical to least squares in a normal *multiple* regression model, how would the proof have to be changed?

APPENDIX TO SECTION **19-2**

Bayesian Confidence Interval for π: Proof

BASIC STATISTICAL ARGUMENT

Note that in Figure 19-5 or 19-6, the posterior distribution is approximately normal. This will be true in general, especially when S^* and F^* are each greater than 5. This normal approximation means that 95% of the probability lies within ± 1.96 standard deviations of the mean. To prove (19-19), therefore, we need only show:

$$\text{Posterior mean} = P^*$$

$$\text{Posterior variance} \simeq \frac{P^*(1 - P^*)}{n^*}$$

CALCULUS PREREQUISITE

The key result we need from integral calculus is:

$$\int_0^1 \pi^u (1 - \pi)^v \, d\pi = \frac{u! \, v!}{(u + v + 1)!} \quad (19\text{-}1A)$$

The proof could be carried out by mathematical induction, when u and v are positive integers. (If other u and v are of interest, see a text in advanced calculus.)

PROOF CARRIED OUT

Let us express the posterior distribution in terms of the convenient symbols S^* and F^*. According to (19-17), $a + S = S^* - 1$ and $b + F = F^* - 1$, so that the posterior can be written as:

$$p(\pi/S) \propto \pi^{S^* - 1}(1 - \pi)^{F^* - 1} \quad (19\text{-}2A)$$
$$\text{like (19-12)}$$

i. First, let us find out what the constant of proportionality is, in (19-2A). We substitute $u = S^* - 1$ and $v = F^* - 1$ into (19-1A):

$$\int_0^1 \pi^{S^* - 1}(1 - \pi)^{F^* - 1} \, d\pi = \frac{(S^* - 1)! \, (F^* - 1)!}{(S^* + F^* - 1)!}$$

We should divide (19-2A) by this, to make $p(\pi/S)$ a proper probability distribution with a total probability (integral) of 1:

$$p(\pi/S) = \frac{(S^* + F^* - 1)!}{(S^* - 1)! \, (F^* - 1)!} \, \pi^{S^* - 1} (1 - \pi)^{F^* - 1}$$

ii. The mean is easily found:

$$E(\pi) \equiv \int_0^1 \pi p(\pi/S) \, d\pi$$

Substitute $p(\pi/S)$:

$$E(\pi) = \frac{(S^* + F^* - 1)!}{(S^* - 1)! \, (F^* - 1)!} \int_0^1 \pi \pi^{S^* - 1} (1 - \pi)^{F^* - 1} \, d\pi$$

If we temporarily ignore the large expression outside the integral, we can concentrate on the integral itself:

$$\int_0^1 \pi^{S^*} (1 - \pi)^{F^* - 1} \, d\pi$$

This can be evaluated by substituting $u = S^*$ and $v = F^* - 1$ into (19-1A):

$$\int_0^1 \pi^{S^*} (1 - \pi)^{F^* - 1} \, d\pi = \frac{S^*! \, (F^* - 1)!}{(S^* + F^*)!}$$

When we substitute this into $E(\pi)$, many of the factorials cancel out, leaving the desired conclusion:

$$E(\pi) = \frac{S^*}{S^* + F^*} = P^*$$

iii. To calculate the variance, it would be easiest to first calculate $E(\pi^2)$, and then use

$$\sigma^2 = E(\pi^2) - [E(\pi)]^2 \qquad \text{like (4-35)}$$

The details are left as an exercise. They would finally establish the desired conclusion:

$$\sigma^2 = \frac{P^*(1 - P^*)}{n^* + 1} \simeq \frac{P^*(1 - P^*)}{n^*}$$

APPENDIX TO SECTION 19-3

Posterior Distribution of μ in a Normal Model: Proof

We will show that the posterior distribution of μ is normal, with the mean and standard error as claimed in (19-23) and (19-24). Denoting by $K_1, K_2, \ldots$ the unimportant constants that do not involve the unknown μ, we use the normal formula (4-26) to give us the two basic components:

Prior: $p(\mu) = K_1 \exp\left[-\frac{1}{2} \left(\frac{\mu - \mu_0}{\sigma_0} \right)^2 \right]$

Likelihood: $p(\overline{X}/\mu) = K_2 \exp\left[-\frac{1}{2} \left(\frac{\overline{X} - \mu}{\sigma/\sqrt{n}} \right)^2 \right]$

Recall that posterior = (prior) × (likelihood)

$$= K_1 K_2 \exp\left[-\frac{1}{2} \left[\left(\frac{\mu - \mu_0}{\sigma_0} \right)^2 + \left(\frac{\overline{X} - \mu}{\sigma/\sqrt{n}} \right)^2 \right] \right]$$

Let us consider in detail the exponent, which may be rearranged as a quadratic function of μ:

$$-\frac{1}{2} \left[\mu^2 \left(\frac{1}{\sigma_0^2} + \frac{n}{\sigma^2} \right) - 2\mu \left(\frac{\mu_0}{\sigma_0^2} + \frac{n\overline{X}}{\sigma^2} \right) + K_3 \right]$$

From (19-22), we substitute n_0/σ^2 for $1/\sigma_0^2$:

$$-\frac{1}{2} \left[\mu^2 \left(\frac{n_0 + n}{\sigma^2} \right) - 2\mu \left(\frac{n_0\mu_0 + n\overline{X}}{\sigma^2} \right) + K_3 \right]$$

$$= -\frac{1}{2} \left(\frac{n_0 + n}{\sigma^2} \right) \left[\mu^2 - 2\mu \left(\frac{n_0\mu_0 + n\overline{X}}{n_0 + n} \right) + K_4 \right]$$

$$= -\frac{1}{2} \left(\frac{1}{\sigma^2/(n_0 + n)} \right) \left[\mu - \left(\frac{n_0\mu_0 + n\overline{X}}{n_0 + n} \right) \right]^2 + K_5$$

Ignoring K_5, this exponent has the form

$$-\frac{1}{2} \frac{1}{b^2} [\mu - a]^2$$

in which—by analogy with the exponent in (4-26)—we recognize that μ has a normal distribution with mean a and standard deviation b. That is:

$$\text{Mean} = \frac{n_0\mu_0 + n\overline{X}}{n_0 + n} \qquad \text{(19-23) proved}$$

$$\text{SE} = \sigma/\sqrt{n_0 + n} \qquad \text{(19-24) proved}$$

APPENDIX TO SECTION 19-4

Posterior Distribution of β in Normal Regression: Proof

We assume a normal prior, and a normal likelihood (which is approximately true if n is large, or exactly true if the Y observations are distributed normally about the line). Then we can appeal to the normal model already derived in Section 19-3. Into (19-22), (19-23), (19-24) we substitute:

$$\beta \text{ for } \mu$$

$$b \text{ for } \overline{X}$$

$$\sigma/\sigma_x \text{ for } \sigma$$

and of course,

$$\beta_0 \text{ for } \mu_0$$

$$\sigma_0 \text{ for } \sigma_0$$

Then we obtain (19-29), (19-30), (19-31), as required.

APPENDIX TO SECTION **19-5**

Bayesian Shrinkage Confidence Intervals

We have seen how a Bayesian shrinkage estimate uses all the data—not just the estimate itself, but also its measure of reliability F. This has two advantages: (1) better point estimates, as already noted; (2) reduced variability of these estimates, which we will discuss now.

When F is reasonably large ($F \geq 3$, say), the variance of the estimate is reduced by the same shrinkage factor $1 - (1/F)$ that is used for the estimate itself. Consequently, the SE (standard error) of the estimate is reduced by the square root $\sqrt{1 - (1/F)}$:

> Bayesian shrinkage
> (if $F \geq 3$): Shrink the SE
> by the factor $\sqrt{1 - (1/F)}$

PROBLEMS

19-1A Contrast the classical and Bayesian shrinkage 95% confidence intervals, calculated for each of the following problems (in each case, calculate $F = t^2$ to determine the amount of shrinkage towards zero):
  a. Problem 19-15
  b. Problem 19-16
  c. Problem 19-17

19-2A In each case in Problem 19-1A, when is the Bayesian confidence interval better than the classical? And worse?

APPENDIX TO SECTION **20-3**

Bayesian Hypothesis Tests Proofs

PROOF OF THE GENERAL CASE (20-11)

We follow the same logic as in Table 20-6. Recall that the average loss was obtained by weighting each loss with its probability; for example, the first loss in Table 20-6 is

$$L(a_0) = (.75)10 + (.25)100 = 32.5$$

In the corresponding general case in Table 20-7,

$$L(a_0) = p(\theta_0/X)l_{00} + p(\theta_1/X)l_{01}$$

The other average loss is calculated similarly:

$$L(a_1) = p(\theta_0/X)l_{10} + p(\theta_1/X)l_{11}$$

Now we should choose a_0 whenever its average loss is less—that is, whenever

$$L(a_0) < L(a_1)$$

When we substitute $L(a_0)$ and $L(a_1)$, and simplify, we obtain the criterion: Choose a_0 whenever:

$$p(\theta_1/X)[l_{01} - l_{11}] < p(\theta_0/X)[l_{10} - l_{00}]$$

Noting the definition of regrets in (20-9) and (20-10), we can rewrite this as

$$p(\theta_1/X)r_1 < p(\theta_0/X)r_0$$

That is:

$$\frac{p(\theta_1/X)}{p(\theta_0/X)} < \frac{r_0}{r_1}$$

The posterior probabilities in this equation now can be expressed using (19-5), and noting that $p(X)$ cancels:

$$\frac{p(\theta_1)p(X/\theta_1)}{p(\theta_0)p(X/\theta_0)} < \frac{r_0}{r_1}$$

This can be easily reexpressed as:

$$\frac{p(X/\theta_1)}{p(X/\theta_0)} < \frac{r_0p(\theta_0)}{r_1p(\theta_1)} \qquad \text{(20-11) proved}$$

Remarks How does this relate to the approach of classical statistics, which considers only the data in the sample and ignores priors and losses? Two classical statisticians developed a similar criterion 50 years ago; it still remains a milestone in statistical theory:

> *Neyman-Pearson* **Criterion:**
> **For optimal decisions, accept**
> H_0 **if and only if**
>
> $$\frac{p(X/\theta_1)}{p(X/\theta_0)} < k_\alpha$$

where k_α is chosen to provide the desired error level of the test (usually $\alpha = 5\%$). The Bayesian criterion (20-11) is much more useful, however. Instead of invoking an arbitrary level α, it uses the prior probabilities and losses explicitly on the right-hand side.

PROOF FOR TESTING μ IN A NORMAL POPULATION (20-13)

To prove (20-13), we take logs of the general Bayesian criterion (20-11):

$$\log p(X/\theta_1) - \log p(X/\theta_0) < \log \left[\frac{r_0p(\theta_0)}{r_1p(\theta_1)} \right]$$

For the given problem of a normal population and its sample of n observations, we must substitute μ_0 for θ_0, μ_1 for θ_1, and the sample $(X_1, X_2 \ldots)$ for the data X:

$$\log p(X_1, X_2 \ldots /\mu_1)$$
$$- \log p(X_1, X_2, \ldots /\mu_0) < \log \left[\frac{r_0p(\mu_0)}{r_1p(\mu_1)} \right] \qquad \text{(20-1A)}$$

Now, for any normally distributed observation X_i, its probability distribution is given by (4-26); taking logs of this yields

$$\log p(X_i/\mu) = -\frac{1}{2\sigma^2}(X_i - \mu)^2 - \log \sqrt{2\pi}\,\sigma$$

Since the X_i are independent, we can multiply their probabilities, or add their logs:

$$\log p(X_1, X_2, \ldots /\mu)$$
$$= -\frac{1}{2\sigma^2} \Sigma(X_i - \mu)^2 - n \log \sqrt{2\pi}\,\sigma \quad \text{like (18-11)}$$

When this is substituted into the two places on the left-side of (20-1A), the two occurrences of $n \log \sqrt{2\pi}\,\sigma$ cancel, leaving:

$$-\frac{1}{2\sigma^2}[\Sigma(X_i - \mu_1)^2 - \Sigma(X_i - \mu_0)^2] =$$

$$-\frac{1}{2\sigma^2}[\Sigma(X_i^2 - 2\mu_1X_i + \mu_1^2) - \Sigma(X_i^2 - 2\mu_0X_i + \mu_0^2)]$$

$$= -\frac{1}{2\sigma^2}[-2(\mu_1 - \mu_0)\Sigma X_i + n(\mu_1^2 - \mu_0^2)]$$

$$= \frac{1}{\sigma^2}[(\mu_1 - \mu_0)\Sigma X_i - \frac{n}{2}(\mu_1 - \mu_0)(\mu_1 + \mu_0)]$$

$$= \frac{n(\mu_1 - \mu_0)}{\sigma^2}\left[\frac{\Sigma X_i}{n} - \frac{\mu_1 + \mu_0}{2}\right]$$

Substitute this into (20-1A), and cross multiply:

$$\frac{\Sigma X_i}{n} - \frac{\mu_1 + \mu_0}{2} < \frac{\sigma^2}{n(\mu_1 - \mu_0)} \log \left[\frac{r_0p(\mu_0)}{r_1p(\mu_1)} \right]$$

On the left, the first term is $\overline{X}$. The second term can be moved over to the right, which proves (20-13):

$$\overline{X} < \frac{\mu_1 + \mu_0}{2} + \frac{\sigma^2}{n(\mu_1 - \mu_0)} \log \left[\frac{r_0p(\mu_0)}{r_1p(\mu_1)} \right]$$

APPENDIX TO SECTION 24-2

Serial Correlation and the Durbin-Watson Test

ESTIMATING ρ

In the autoregression (24-5), an estimate r for ρ can be computed with the usual least squares regression formula. Since residuals have 0 mean, they are already in deviation form and can therefore be substituted into (11-5) directly, which gives:

$$r = \frac{\sum_{t=2}^{n} \hat{e}_{t-1}\, \hat{e}_t}{\sum_{t=2}^{n} \hat{e}_{t-1}^2} \qquad (24\text{-}1A)$$

Because of the shift of one time period, both sums had to start at $t = 2$, not $t = 1$. For the jewelry sales in Figure 24-4, these two sums can be calculated in the usual tableau:

	GIVEN		REGRESSION CALCULATIONS	
	$y =$ residual	$x =$ lagged residual		
t	$\hat{e}_t$	$\hat{e}_{t-1}$	$xy = \hat{e}_{t-1}\hat{e}_t$	$x^2 = \hat{e}_{t-1}^2$
1	.4			
2	.6	.4	.24	.16
3	1.2	.6	.72	.36
4	−3.7	1.2	−4.44	1.44
5	− .2	−3.7	.74	13.69
6	.	− .2	.	.
.	.	.	.	.
.	.	.	.	.
			−10.5	41.0

$$r = \frac{-10.5}{41.0} = -.26 \qquad (24\text{-}6)\ \text{proved}$$

r IS APPROXIMATELY THE CORRELATION COEFFICIENT

Let us calculate the correlation between $x = \hat{e}_{t-1}$ and $y = \hat{e}_t$:

$$\text{corr} = \frac{\sum_{t=2}^{n} \hat{e}_{t-1}\, \hat{e}_t}{\sqrt{\sum_{t=2}^{n} \hat{e}_{t-1}^2 \ \sum_{t=2}^{n} \hat{e}_t^2}} \qquad \begin{array}{c}(24\text{-}2A) \\ \text{like (15-2)}\end{array}$$

In the denominator, the second sum is the same as the first, except for the first and last term (because t and $t-1$ are one unit apart). That is:

$$\sum_{t=2}^{n} \hat{e}_t^2 \approx \sum_{t=2}^{n} \hat{e}_{t-1}^2$$

When this is substituted into (24-2A), we obtain (24-1A). That is, in an autoregressive equation the regression coefficient is essentially the correlation coefficient.

THE DURBIN-WATSON TEST

The Durbin-Watson statistic to test for $\rho = 0$ is defined as:

$$DW \equiv \frac{\sum_{t=2}^{n} (\hat{e}_t - \hat{e}_{t-1})^2}{\sum_{t=1}^{n} \hat{e}_t^2} \qquad (24\text{-}3A)$$

If there is serial correlation, then $\hat{e}_t$ will tend to track $\hat{e}_{t-1}$, and the numerator of DW—and hence DW itself—will be small. Hence small values of DW define the region for establishing serial correlation (i.e., rejecting H_0) as tabulated in Appendix Table IX.

Another useful view is obtained by expanding the numerator of (24-3A):

$$DW = \frac{\sum_{t=2}^{n} \hat{e}_t^2 - 2\left(\sum_{t=2}^{n} \hat{e}_t\, \hat{e}_{t-1}\right) + \sum_{t=2}^{n} \hat{e}_{t-1}^2}{\sum_{t=1}^{n} \hat{e}_t^2}$$

Note that the first and last terms are each approximately equal to the denominator. Moreover, when the middle term in brackets is matched with the denominator, it gives approximately the serial correlation r in (24-1A). Thus:

$$\boxed{DW \approx 2 - 2r}$$

This makes it clear that if the serial correlation increases (r rises towards 1), then DW falls towards zero.

i. If we average 4 at a time, and then again 4 more at a time.

ii. If we average 4 at a time once more (for the third time).

APPENDIX TO SECTION 24-3

Moving Averages in General

In Table 24-2, we first averaged 4 quarters at a time, and then in order to center, averaged 2 at a time. The net effect was to average over 5 time periods, with each end getting half weights. Thus, the moving average M can be written with the formula:

$$M(Y_t) = \frac{Y_t + Y_{t+1} + Y_{t-1} + .5Y_{t+2} + .5Y_{t-2}}{4}$$

That is, M is the average of the nearby values. We could easily generalize to any number of values and different weights:

$$M(Y_t) = a_0Y_t + a_1Y_{t+1} + a_{-1}Y_{t-1} + \cdots + a_kY_{t+k} + a_{-k}Y_{t-k}$$

where $\Sigma a_i = 1$. The weights a_i typically decrease so that more distant values count less—until finally, beyond a distance of k quarters, they don't count at all.

Although moving averages are most commonly used on seasonal data, they can be used for annual data without any seasonal component. They still accomplish the goal of every average: In averaging out the fluctuations, the average smooths the time series.

PROBLEM

24-1A **a.** If we average 4 at a time, and then 2 at a time, prove that the resulting moving average is the one given in the first equation above.

b. Similarly, calculate the formula for the overall moving average:

APPENDIX TO SECTION 24-4

Exponential Smoothing: Proof

To establish (24-13), we substitute (24-12) into (24-11):

$$\hat{Y}_t = b\lambda Y_{t-1} + b\lambda^2 Y_{t-2} + b\lambda^3 Y_{t-3} + \cdots$$

As an aside, notice that since this applies for any time period t, we can set $t = t - 1$, and thus get

$$\hat{Y}_{t-1} = b\lambda Y_{t-2} + b\lambda^2 Y_{t-3} + b\lambda^3 Y_{t-4} + \cdots$$

Returning now to the first equation, note that it appears to be impossible to estimate, because it is an infinite series. However, we can greatly simplify it by rewriting it:

$$\hat{Y}_t = b\lambda Y_{t-1} + \lambda(b\lambda Y_{t-2} + b\lambda^2 Y_{t-3} + \cdots)$$

The expression in brackets turns out to be $\hat{Y}_{t-1}$ given in the second equation. Thus:

$$\hat{Y}_t = b\lambda Y_{t-1} + \lambda\hat{Y}_{t-1}$$

If we require the weights to add to 1, then the first weight must be the complement of the second weight λ; that is:

$$\hat{Y}_t = (1 - \lambda)Y_{t-1} + \lambda\hat{Y}_{t-1} \quad \text{(24-13) proved}$$

APPENDIX TO SECTION 24-5

Forecasting Using Box-Jenkins Models

In this appendix, we will continue the ARMA model of Section 24-5, showing how it can be used to forecast, and how it can include trend and seasonal components.

FORECASTING

Since (24-18) is true for any t, it is true for $t + 1$:

$$Y_{t+1} = (\beta_1 Y_t + \beta_2 Y_{t-1} + \cdots) + \\ (v_{t+1} - \alpha_1 v_t - \alpha_2 v_{t-1} - \cdots)$$

To get the forecast value $\hat{Y}_{t+1}$, we substitute the estimated coefficients $(b_1, b_2, \ldots, a_1, a_2, \ldots)$ for their unknown targets $(\beta_1, \beta_2, \ldots, \alpha_1, \alpha_2, \ldots)$. We also substitute the residuals from the estimation process $(\hat{v}_t, \hat{v}_{t-1}, \ldots)$. Finally, we must estimate the future disturbance v_{t+1}; since it is pure noise and perfectly unpredictable, we can only estimate it with its mean value 0. Thus:

$$\hat{Y}_{t+1} = b_1 Y_t + b_2 Y_{t-1} + \cdots \\ + 0 - a_1 \hat{v}_t - a_2 \hat{v}_{t-1} - \cdots$$

To forecast another period ahead, of course it would be best to use the actual Y_{t+1} if it were available, and recalculate all the estimates $b_1, b_2, \ldots, a_1, a_2$, and so on. But if Y_{t+1} is not available, its estimate above can be used instead to obtain

$$\hat{Y}_{t+2} = b_1 \hat{Y}_{t+1} + b_2 Y_t + \cdots \\ + 0 - a_1(0) - a_2 \hat{v}_t - \cdots$$

and so on.

TREND HANDLED BY DIFFERENCING: ARIMA MODELS

For a series that has an increasing trend, it is natural to consider the change in the series:

$$\Delta Y_t \equiv Y_t - Y_{t-1}$$

If the trend is linear, this difference will remove it. Then we fit the appropriate ARMA (p,q) model to the new ΔY_t series.

The original series Y_t may be called the *integrated* series (in order to distinguish it from the differenced series ΔY_t). Thus the whole model is called an ARIMA model (Auto Regression, Integrated, Moving Average). If differencing is applied d times successively (to remove a polynomial trend of degree d), then the model is called an ARIMA (p,d,q) model.

SEASONAL INCLUDED TOO: SARIMA MODELS

To handle seasonal variation, let us consider what all the winter (first) quarters would look like in a series like Figure 24-2. Being all winter values, they would form a subseries without the ups and downs of seasonal variation (which occur only as we pass from season to season). Thus we could handle them with an ARIMA model, with parameters (p_s, d_s, q_s), say.

Similarly, all the spring (second) quarters could be fitted with a seasonal ARIMA model, and so on. [To keep the number of parameters small, this estimation is often constrained so that it will generate only one set of coefficients $(b_1, b_2, \ldots, a_1, a_2, \ldots)$ for all four quarters.]

From this seasonal fitting, there will be estimated pure noise terms $\hat{v}_t$. Strung together as a new time series, they can be fitted with another ARIMA model (p,d,q). The whole model then is called a *Seasonal ARIMA*, or more formally, a SARIMA $(p,d,q) \times (p_s, d_s, q_s)_4$ model.

In practice, to keep the complexity down to a reasonable level, the SARIMA model is often specified to include very few parameters. For example, a $(0,1,1) \times (0,1,1)_4$ model is common—or, for monthly data, a $(0,1,1) \times (0,1,1)_{12}$ model. Such a model is closely related to the exponential smoothing discussed in Section 24-4.

TABLES

TABLE **I** **Random Digits (Blocked merely for convenience)**

39 65 76 45 45	19 90 69 64 61	20 26 36 31 62	58 24 97 14 97	95 06 70 99 00
73 71 23 70 90	65 97 60 12 11	31 56 34 19 19	47 83 75 51 33	30 62 38 20 46
72 20 47 33 84	51 67 47 97 19	98 40 07 17 66	23 05 09 51 80	59 78 11 52 49
75 17 25 69 17	17 95 21 78 58	24 33 45 77 48	69 81 84 09 29	93 22 70 45 80
37 48 79 88 74	63 52 06 34 30	01 31 60 10 27	35 07 79 71 53	28 99 52 01 41
02 89 08 16 94	85 53 83 29 95	56 27 09 24 43	21 78 55 09 82	72 61 88 73 61
87 18 15 70 07	37 79 49 12 38	48 13 93 55 96	41 92 45 71 51	09 18 25 58 94
98 83 71 70 15	89 09 39 59 24	00 06 41 41 20	14 36 59 25 47	54 45 17 24 89
10 08 58 07 04	76 62 16 48 68	58 76 17 14 86	59 53 11 52 21	66 04 18 72 87
47 90 56 37 31	71 82 13 50 41	27 55 10 24 92	28 04 67 53 44	95 23 00 84 47
93 05 31 03 07	34 18 04 52 35	74 13 39 35 22	68 95 23 92 35	36 63 70 35 33
21 89 11 47 99	11 20 99 45 18	76 51 94 84 86	13 79 93 37 55	98 16 04 41 67
95 18 94 06 97	27 37 83 28 71	79 57 95 13 91	09 61 87 25 21	56 20 11 32 44
97 08 31 55 73	10 65 81 92 59	77 31 61 95 46	20 44 90 32 64	26 99 76 75 63
69 26 88 86 13	59 71 74 17 32	48 38 75 93 29	73 37 32 04 05	60 82 29 20 25
41 47 10 25 03	87 63 93 95 17	81 83 83 04 49	77 45 85 50 51	79 88 01 97 30
91 94 14 63 62	08 61 74 51 69	92 79 43 89 79	29 18 94 51 23	14 85 11 47 23
80 06 54 18 47	08 52 85 08 40	48 40 35 94 22	72 65 71 08 86	50 03 42 99 36
67 72 77 63 99	89 85 84 46 06	64 71 06 21 66	89 37 20 70 01	61 65 70 22 12
59 40 24 13 75	42 29 72 23 19	06 94 76 10 08	81 30 15 39 14	81 83 17 16 33
63 62 06 34 41	79 53 36 02 95	94 61 09 43 62	20 21 14 68 86	94 95 48 46 45
78 47 23 53 90	79 93 96 38 63	34 85 52 05 09	85 43 01 72 73	14 93 87 81 40
87 68 62 15 43	97 48 72 66 48	53 16 71 13 81	59 97 50 99 52	24 62 20 42 31
47 60 92 10 77	26 97 05 73 51	88 46 38 03 58	72 68 49 29 31	75 70 16 08 24
56 88 87 59 41	06 87 37 78 48	65 88 69 58 39	88 02 84 27 83	85 81 56 39 38
22 17 68 65 84	87 02 22 57 51	68 69 80 95 44	11 29 01 95 80	49 34 35 86 47
19 36 27 59 46	39 77 32 77 09	79 57 92 36 59	89 74 39 82 15	08 58 94 34 74
16 77 23 02 77	28 06 24 25 93	22 45 44 84 11	87 80 61 65 31	09 71 91 74 25
78 43 76 71 61	97 67 63 99 61	80 45 67 93 82	59 73 19 85 23	53 33 65 97 21
03 28 28 26 08	69 30 16 09 05	53 58 47 70 93	66 56 45 65 79	45 56 20 19 47
04 31 17 21 56	33 73 99 19 87	26 72 39 27 67	53 77 57 68 93	60 61 97 22 61
61 06 98 03 91	87 14 77 43 96	43 00 65 98 50	45 60 33 01 07	98 99 46 50 47
23 68 35 26 00	99 53 93 61 28	52 70 05 48 34	56 65 05 61 86	90 92 10 70 80
15 39 25 70 99	93 86 52 77 65	15 33 59 05 28	22 87 26 07 47	86 96 98 29 06
58 71 96 30 24	18 46 23 34 27	85 13 99 24 44	49 18 09 79 49	74 16 32 23 02
93 22 53 64 39	07 10 63 76 35	87 03 04 79 88	08 13 13 85 51	55 34 57 72 69
78 76 58 54 74	92 38 70 96 92	52 06 79 79 45	82 63 18 27 44	69 66 92 19 09
61 81 31 96 82	00 57 25 60 59	46 72 60 18 77	55 66 12 62 11	08 99 55 64 57
42 88 07 10 05	24 98 65 63 21	47 21 61 88 32	27 80 30 21 60	10 92 35 36 12
77 94 30 05 39	28 10 99 00 27	12 73 73 99 12	49 99 57 94 82	96 88 57 17 91

TABLE **II** Random Normal Numbers, $\mu = 0$, $\sigma = 1$
(Rounded to 1 Decimal Place)

.5	.1	2.5	-.3	-.1	.3	-.3	1.3	.2	-1.0
.1	-2.5	-5	-.2	.5	-1.6	.2	-1.2	.0	.5
1.5	-.4	-.6	.7	.9	1.4	.8	-1.0	-.9	-1.9
1.0	-.5	1.3	3.5	.6	-1.9	.2	1.2	-.5	-.3
1.4	-.6	.0	.3	2.9	2.0	-.3	.4	.4	.0
.9	-.5	-.5	.6	.9	-.9	1.6	.2	-1.9	.4
1.2	-1.1	.0	.8	1.0	.7	1.1	-.6	-.3	-.7
-1.5	-.5	-.2	-.1	1.0	.2	.4	.7	-.4	-.4
-.7	.8	-1.6	-.3	-.5	-2.1	-.5	-.2	.9	-.5
1.4	.2	.4	.8	.2	-.7	1.0	-1.5	-.3	.1
-.5	1.7	-.1	-1.2	-.5	.9	-.5	-2.0	-2.8	-.2
-1.4	-.2	1.4	-.6	-.3	-.2	.2	.8	1.0	-.9
-1.0	.6	-.9	1.6	.1	.4	-.2	.3	-1.0	-1.0
.0	-.9	.0	-.7	1.1	-.1	1.1	.5	-1.7	.4
1.4	-1.2	-.9	1.2	-.2	-.2	1.2	-2.6	-.6	.1
-1.8	-.3	1.2	1.0	-.5	-1.6	-.1	-.4	-.6	.6
-.1	-.4	-1.4	.4	-1.0	-.1	-1.7	-2.8	-1.1	-2.4
-1.3	1.8	-1.0	.4	1.0	-1.1	-1.0	.4	-1.7	2.0
1.0	.5	.7	1.4	1.0	-1.3	1.6	-1.0	.5	-.3
.3	-2.1	.7	-.9	-1.1	-1.4	1.0	.1	-.6	.9
-1.8	-2.0	-1.6	.5	.2	-.2	.0	.0	.5	-1.0
-1.2	1.2	1.1	.9	1.3	-.2	.2	-.4	-.3	.5
.7	-1.1	1.2	-1.2	-.9	.4	.3	-.9	.6	1.7
-.4	.4	-1.9	.9	-.2	.6	.9	-.4	-.2	-.1
-1.4	-.2	.4	-.6	-.6	.2	-.3	.5	.7	-.3
.2	.2	-1.1	-.2	-.3	1.2	1.1	.0	-2.0	-.6
.2	.3	-.3	.1	-2.8	-.4	-.8	-1.3	-.6	-1.0
2.3	.6	.6	-.7	.2	1.3	.1	-1.8	-.7	-1.3
.0	-.3	.1	.8	-.6	.5	.5	-1.0	.5	1.0
-1.1	-2.1	.9	.1	.4	-1.7	1.0	-1.4	-.6	-1.0
.8	.1	-1.5	.0	-2.1	.7	.1	-.9	-.6	.6
.4	-1.7	-.9	.2	-.7	.3	-.1	-.2	-.1	.4
-.5	-.3	.2	-.7	1.0	.0	.4	-.8	.2	.1
.3	-.5	1.3	-1.2	-.9	.1	-.5	-.8	.0	.5
1.0	3.0	-.6	-.5	-1.1	1.3	-1.4	-1.3	-3.0	.5
-1.3	1.3	-.6	-.1	-.5	-.6	2.9	.5	.4	.3
-.3	-.1	-.3	.6	-.5	-1.2	-1.2	-.3	-.1	1.1
.2	-.9	-.9	-.5	1.4	-.5	.2	-.4	1.5	1.1
-1.3	.2	-1.2	.4	-1.0	.8	.9	1.0	.0	.8
-1.2	-.2	-.3	1.8	1.4	.6	1.2	.7	.4	.2
.6	-.5	.8	.1	.5	-.4	1.7	1.2	.9	-.3
.4	-1.9	.2	-.5	.7	-.1	-.1	-.5	.5	1.1
-1.4	.5	-1.7	-1.2	.8	-.7	-.1	1.0	-.8	.2
-.2	-.2	-.4	-.8	.3	1.0	1.8	2.9	-.8	-.1
-.3	.5	.4	-1.5	1.5	2.0	-.1	.2	.0	-1.2
.4	-.4	.6	1.0	-.1	.1	.5	-1.3	1.1	1.1
.6	.7	-1.1	-1.4	-1.6	-1.6	1.5	1.3	.7	-.9
.9	-.9	-.1	-.5	.5	1.4	.0	-.3	-.3	1.2
.2	-.6	.0	-.5	-.9	-.4	-.5	1.7	-.2	-1.2
-.9	.4	.8	.8	.4	-.3	-1.1	.6	1.4	1.3

TABLE III(a) Binomial Coefficients $\binom{n}{s}$

n	0	1	2	3	4	5	6	7	8	9	10
0	1										
1	1	1									
2	1	2	1								
3	1	3	3	1							
4	1	4	6	4	1						
5	1	5	10	10	5	1					
6	1	6	15	20	15	6	1				
7	1	7	21	35	35	21	7	1			
8	1	8	28	56	70	56	28	8	1		
9	1	9	36	84	126	126	84	36	9	1	
10	1	10	45	120	210	252	210	120	45	10	1
11	1	11	55	165	330	462	462	330	165	55	11
12	1	12	66	220	495	792	924	792	495	220	66
13	1	13	78	286	715	1287	1716	1716	1287	715	286
14	1	14	91	364	1001	2002	3003	3432	3003	2002	1001
15	1	15	105	455	1365	3003	5005	6435	6435	5005	3003
16	1	16	120	560	1820	4368	8008	11440	12870	11440	8008
17	1	17	136	680	2380	6188	12376	19448	24310	24310	19448
18	1	18	153	816	3060	8568	18564	31824	43758	48620	43758
19	1	19	171	969	3876	11628	27132	50388	75582	92378	92378
20	1	20	190	1140	4845	15504	38760	77520	125970	167960	184756

Note. $\binom{n}{s} = \dfrac{n(n-1)(n-2)\cdots(n-s+1)}{s(s-1)(s-2)\cdots3\cdot2\cdot1}$; $\binom{n}{0} = 1$; $\binom{n}{1} = n$.

For coefficients missing from the above table, use the relation:

$\binom{n}{s} = \binom{n}{n-s}$, e.g., $\binom{20}{11} = \binom{20}{9} = 167960$.

TABLE III(b) Individual Binomial Probabilities $p(s)$

n	s	.10	.20	.30	.40	.50	.60	.70	.80	.90
1	0	.900	.800	.700	.600	.500	.400	.300	.200	.100
	1	.100	.200	.300	.400	.500	.600	.700	.800	.900
2	0	.810	.640	.490	.360	.250	.160	.090	.040	.010
	1	.180	.320	.420	.480	.500	.480	.420	.320	.180
	2	.010	.040	.090	.160	.250	.360	.490	.640	.810
3	0	.729	.512	.343	.216	.125	.064	.027	.008	.001
	1	.243	.384	.441	.432	.375	.288	.189	.096	.027
	2	.027	.096	.189	.288	.375	.432	.441	.384	.243
	3	.001	.008	.027	.064	.125	.216	.343	.512	.729
4	0	.656	.410	.240	.130	.063	.026	.008	.002	.000
	1	.292	.410	.412	.346	.250	.154	.076	.026	.004
	2	.049	.154	.265	.346	.375	.346	.265	.154	.049
	3	.004	.026	.076	.154	.250	.346	.412	.410	.292
	4	.000	.002	.008	.026	.063	.130	.240	.410	.656
5	0	.590	.328	.168	.078	.031	.010	.002	.000	.000
	1	.328	.410	.360	.259	.156	.077	.028	.006	.000
	2	.073	.205	.309	.346	.313	.230	.132	.051	.008
	3	.008	.051	.132	.230	.313	.346	.309	.205	.073
	4	.000	.006	.028	.077	.156	.259	.360	.410	.328
	5	.000	.000	.002	.010	.031	.078	.168	.328	.590
6	0	.531	.262	.118	.047	.016	.004	.001	.000	.000
	1	.354	.393	.303	.187	.094	.037	.010	.002	.000
	2	.098	.246	.324	.311	.234	.138	.060	.015	.001
	3	.015	.082	.185	.276	.313	.276	.185	.082	.015
	4	.001	.015	.060	.138	.234	.311	.324	.246	.098
	5	.000	.002	.010	.037	.094	.187	.303	.393	.354
	6	.000	.000	.001	.004	.016	.047	.118	.262	.531
7	0	.478	.210	.082	.028	.008	.002	.000	.000	.000
	1	.372	.367	.247	.131	.055	.017	.004	.000	.000
	2	.124	.275	.318	.261	.164	.077	.025	.004	.000
	3	.023	.115	.227	.290	.273	.194	.097	.029	.003
	4	.003	.029	.097	.194	.273	.290	.227	.115	.023
	5	.000	.004	.025	.077	.164	.261	.318	.275	.124
	6	.000	.000	.004	.017	.055	.131	.247	.367	.372
	7	.000	.000	.000	.002	.008	.028	.082	.210	.478
8	0	.430	.168	.058	.017	.004	.001	.000	.000	.000
	1	.383	.336	.198	.090	.031	.008	.001	.000	.000
	2	.149	.294	.296	.209	.109	.041	.010	.001	.000
	3	.033	.147	.254	.279	.219	.124	.047	.009	.000
	4	.005	.046	.136	.232	.273	.232	.136	.046	.005
	5	.000	.009	.047	.124	.219	.279	.254	.147	.033
	6	.000	.001	.010	.041	.109	.209	.296	.294	.149
	7	.000	.000	.001	.008	.031	.090	.198	.336	.383
	8	.000	.000	.000	.001	.004	.017	.058	.168	.430

TABLE **III(b)** (continued)

					π					
n	s	.10	.20	.30	.40	.50	.60	.70	.80	.90
9	0	.387	.134	.040	.010	.002	.000	.000	.000	.000
	1	.387	.302	.156	.060	.018	.004	.000	.000	.000
	2	.172	.302	.267	.161	.070	.021	.004	.000	.000
	3	.045	.176	.267	.251	.164	.074	.021	.003	.000
	4	.007	.066	.172	.251	.246	.167	.074	.017	.001
	5	.001	.017	.074	.167	.246	.251	.172	.066	.007
	6	.000	.003	.021	.074	.164	.251	.267	.176	.045
	7	.000	.000	.004	.021	.070	.161	.267	.302	.172
	8	.000	.000	.000	.004	.018	.060	.156	.302	.387
	9	.000	.000	.000	.000	.002	.010	.040	.134	.387
10	0	.349	.107	.028	.006	.001	.000	.000	.000	.000
	1	.387	.268	.121	.040	.010	.002	.000	.000	.000
	2	.194	.302	.233	.121	.044	.011	.001	.000	.000
	3	.057	.201	.267	.215	.117	.042	.009	.001	.000
	4	.011	.088	.200	.251	.205	.111	.037	.006	.000
	5	.001	.026	.103	.201	.246	.201	.103	.026	.001
	6	.000	.006	.037	.111	.205	.251	.200	.088	.011
	7	.000	.001	.009	.042	.117	.215	.267	.201	.057
	8	.000	.000	.001	.011	.044	.121	.233	.302	.194
	9	.000	.000	.000	.002	.010	.040	.121	.268	.387
	10	.000	.000	.000	.000	.001	.006	.028	.107	.349
11	0	.314	.086	.020	.004	.000	.000	.000	.000	.000
	1	.384	.236	.093	.027	.005	.001	.000	.000	.000
	2	.213	.295	.200	.089	.027	.005	.001	.000	.000
	3	.071	.221	.257	.177	.081	.023	.004	.000	.000
	4	.016	.111	.220	.236	.161	.070	.017	.002	.000
	5	.002	.039	.132	.221	.226	.147	.057	.010	.000
	6	.000	.010	.057	.147	.226	.221	.132	.039	.002
	7	.000	.002	.017	.070	.161	.236	.220	.111	.016
	8	.000	.000	.004	.023	.081	.177	.257	.221	.071
	9	.000	.000	.001	.005	.027	.089	.200	.295	.213
	10	.000	.000	.000	.001	.005	.027	.093	.236	.384
	11	.000	.000	.000	.000	.000	.004	.020	.086	.314
12	0	.282	.069	.014	.002	.000	.000	.000	.000	.000
	1	.377	.206	.071	.017	.003	.000	.000	.000	.000
	2	.230	.283	.168	.064	.016	.002	.000	.000	.000
	3	.085	.236	.240	.142	.054	.012	.001	.000	.000
	4	.021	.133	.231	.213	.121	.042	.008	.001	.000
	5	.004	.053	.158	.227	.193	.101	.029	.003	.000
	6	.000	.016	.079	.177	.226	.177	.079	.016	.000
	7	.000	.003	.029	.101	.193	.227	.158	.053	.004
	8	.000	.001	.008	.042	.121	.213	.231	.133	.021
	9	.000	.000	.001	.012	.054	.142	.240	.236	.085
	10	.000	.000	.000	.002	.016	.064	.168	.283	.230
	11	.000	.000	.000	.000	.003	.017	.071	.206	.377
	12	.000	.000	.000	.000	.000	.002	.014	.069	.282

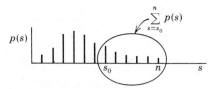

TABLE III(c) Cumulative Binomial Probability in Right-Hand Tail

n	s_0	.10	.20	.30	.40	.50	.60	.70	.80	.90
2	1	.190	.360	.510	.640	.750	.840	.910	.960	.990
	2	.010	.040	.090	.160	.250	.360	.490	.640	.810
3	1	.271	.488	.657	.784	.875	.936	.973	.992	.999
	2	.028	.104	.216	.352	.500	.648	.784	.896	.972
	3	.001	.008	.027	.064	.125	.216	.343	.512	.729
4	1	.344	.590	.760	.870	.937	.974	.992	.998	1.000
	2	.052	.181	.348	.525	.687	.821	.916	.973	.996
	3	.004	.027	.084	.179	.312	.475	.652	.819	.948
	4	.000	.002	.008	.026	.062	.130	.240	.410	.656
5	1	.410	.672	.832	.922	.969	.990	.998	1.000	1.000
	2	.081	.263	.472	.663	.812	.913	.969	.993	1.000
	3	.009	.058	.163	.317	.500	.683	.837	.942	.991
	4	.000	.007	.031	.087	.187	.337	.528	.737	.919
	5	.000	.000	.002	.010	.031	.078	.168	.328	.590
6	1	.469	.738	.882	.953	.984	.996	.999	1.000	1.000
	2	.114	.345	.580	.767	.891	.959	.989	.998	1.000
	3	.016	.099	.256	.456	.656	.821	.930	.983	.999
	4	.001	.017	.070	.179	.344	.544	.744	.901	.984
	5	.000	.002	.011	.041	.109	.233	.420	.655	.886
	6	.000	.000	.001	.004	.016	.047	.118	.262	.531
7	1	.522	.790	.918	.972	.992	.998	1.000	1.000	1.000
	2	.150	.423	.671	.841	.937	.981	.996	1.000	1.000
	3	.026	.148	.353	.580	.773	.904	.971	.955	1.000
	4	.003	.033	.126	.290	.500	.710	.874	.967	.997
	5	.000	.005	.029	.096	.227	.420	.647	.852	.974
	6	.000	.000	.004	.019	.062	.159	.329	.577	.850
	7	.000	.000	.000	.002	.008	.028	.082	.210	.478
8	1	.570	.832	.942	.983	.996	.999	1.000	1.000	1.000
	2	.187	.497	.745	.894	.965	.991	.999	1.000	1.000
	3	.038	.203	.448	.685	.855	.950	.989	.999	1.000
	4	.005	.056	.194	.406	.637	.826	.942	.990	1.000
	5	.000	.010	.058	.174	.363	.594	.806	.944	.995
	6	.000	.001	.011	.050	.145	.315	.552	.797	.962
	7	.000	.000	.001	.009	.035	.106	.255	.503	.813
	8	.000	.000	.000	.001	.004	.017	.058	.168	.430

TABLE **III(c)**   (continued)

n	s_0	.10	.20	.30	.40	.50	.60	.70	.80	.90
						π				
9	1	.613	.866	.960	.990	.998	1.000	1.000	1.000	1.000
	2	.225	.564	.804	.929	.980	.996	1.000	1.000	1.000
	3	.053	.262	.537	.768	.910	.975	.996	1.000	1.000
	4	.008	.086	.270	.517	.746	.901	.975	.997	1.000
	5	.001	.020	.099	.267	.500	.733	.901	.980	.999
	6	.000	.003	.025	.099	.254	.483	.730	.914	.992
	7	.000	.000	.004	.025	.090	.232	.463	.738	.947
	8	.000	.000	.000	.004	.020	.071	.196	.436	.775
	9	.000	.000	.000	.000	.002	.010	.040	.134	.387
10	1	.651	.893	.972	.994	.999	1.000	1.000	1.000	1.000
	2	.264	.624	.851	.954	.989	.998	1.000	1.000	1.000
	3	.070	.322	.617	.833	.945	.988	.998	1.000	1.000
	4	.013	.121	.350	.618	.828	.945	.989	.999	1.000
	5	.002	.033	.150	.367	.623	.834	.953	.994	1.000
	6	.000	.006	.047	.166	.377	.633	.850	.967	.998
	7	.000	.001	.011	.055	.172	.382	.650	.879	.987
	8	.000	.000	.002	.012	.055	.167	.383	.678	.930
	9	.000	.000	.000	.002	.011	.046	.149	.376	.736
	10	.000	.000	.000	.000	.001	.006	.028	.107	.349
11	1	.686	.914	.980	.996	1.000	1.000	1.000	1.000	1.000
	2	.303	.678	.887	.970	.994	.999	1.000	1.000	1.000
	3	.090	.383	.687	.881	.967	.994	.999	1.000	1.000
	4	.019	.161	.430	.704	.887	.971	.996	1.000	1.000
	5	.003	.050	.210	.467	.726	.901	.978	.998	1.000
	6	.000	.012	.078	.247	.500	.753	.922	.988	1.000
	7	.000	.002	.022	.099	.274	.533	.790	.950	.997
	8	.000	.000	.004	.029	.113	.296	.570	.839	.981
	9	.000	.000	.001	.006	.033	.119	.313	.617	.910
	10	.000	.000	.000	.001	.006	.030	.113	.322	.697
	11	.000	.000	.000	.000	.000	.004	.020	.086	.314
12	1	.718	.931	.986	.998	1.000	1.000	1.000	1.000	1.000
	2	.341	.725	.915	.980	.997	1.000	1.000	1.000	1.000
	3	.111	.442	.747	.917	.981	.997	1.000	1.000	1.000
	4	.026	.205	.507	.775	.927	.985	.998	1.000	1.000
	5	.004	.073	.276	.562	.806	.943	.991	.999	1.000
	6	.001	.019	.118	.335	.613	.842	.961	.996	1.000
	7	.000	.004	.039	.158	.387	.665	.882	.981	.999
	8	.000	.001	.009	.057	.194	.438	.724	.927	.996
	9	.000	.000	.002	.015	.073	.225	.493	.795	.974
	10	.000	.000	.000	.003	.019	.083	.253	.558	.889
	11	.000	.000	.000	.000	.003	.020	.085	.275	.659
	12	.000	.000	.000	.000	.000	.002	.014	.069	.282

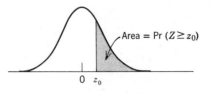

TABLE IV Standard Normal, Cumulative Probability in Right-Hand Tail (For Negative Values of z, Areas are Found by Symmetry)

z_0	NEXT DECIMAL PLACE OF z_0									
	0	1	2	3	4	5	6	7	8	9
0.0	.500	.496	.492	.488	.484	.480	.476	.472	.468	.464
0.1	.460	.456	.452	.448	.444	.440	.436	.433	.429	.425
0.2	.421	.417	.413	.409	.405	.401	.397	.394	.390	.386
0.3	.382	.378	.374	.371	.367	.363	.359	.356	.352	.348
0.4	.345	.341	.337	.334	.330	.326	.323	.319	.316	.312
0.5	.309	.305	.302	.298	.295	.291	.288	.284	.281	.278
0.6	.274	.271	.268	.264	.261	.258	.255	.251	.248	.245
0.7	.242	.239	.236	.233	.230	.227	.224	.221	.218	.215
0.8	.212	.209	.206	.203	.200	.198	.195	.192	.189	.187
0.9	.184	.181	.179	.176	.174	.171	.169	.166	.164	.161
1.0	.159	.156	.154	.152	.149	.147	.145	.142	.140	.138
1.1	.136	.133	.131	.129	.127	.125	.123	.121	.119	.117
1.2	.115	.113	.111	.109	.107	.106	.104	.102	.100	.099
1.3	.097	.095	.093	.092	.090	.089	.087	.085	.084	.082
1.4	.081	.079	.078	.076	.075	.074	.072	.071	.069	.068
1.5	.067	.066	.064	.063	.062	.061	.059	.058	.057	.056
1.6	.055	.054	.053	.052	.051	.049	.048	.047	.046	.046
1.7	.045	.044	.043	.042	.041	.040	.039	.038	.038	.037
1.8	.036	.035	.034	.034	.033	.032	.031	.031	.030	.029
1.9	.029	.028	.027	.027	.026	.026	.025	.024	.024	.023
2.0	.023	.022	.022	.021	.021	.020	.020	.019	.019	.018
2.1	.018	.017	.017	.017	.016	.016	.015	.015	.015	.014
2.2	.014	.014	.013	.013	.013	.012	.012	.012	.011	.011
2.3	.011	.010	.010	.010	.010	.009	.009	.009	.009	.008
2.4	.008	.008	.008	.008	.007	.007	.007	.007	.007	.006
2.5	.006	.006	.006	.006	.006	.005	.005	.005	.005	.005
2.6	.005	.005	.004	.004	.004	.004	.004	.004	.004	.004
2.7	.003	.003	.003	.003	.003	.003	.003	.003	.003	.003
2.8	.003	.002	.002	.002	.002	.002	.002	.002	.002	.002
2.9	.002	.002	.002	.002	.002	.002	.002	.001	.001	.001

z_0	DETAIL OF TAIL ($._2135$, FOR EXAMPLE, MEANS .00135)									
2.	$._2228$	$._1179$	$._1139$	$._1107$	$._2820$	$._2621$	$._2466$	$._2347$	$._2256$	$._2187$
3.	$._2135$	$._3968$	$._3687$	$._3483$	$._3337$	$._3233$	$._3159$	$._3108$	$._4723$	$._4481$
4.	$._4317$	$._4207$	$._4133$	$._5854$	$._5541$	$._5340$	$._5211$	$._5130$	$._6793$	$._6479$
5.	$._6287$	$._6170$	$._7996$	$._7579$	$._7333$	$._7190$	$._7107$	$._8599$	$._8332$	$._8182$
	0	1	2	3	4	5	6	7	8	9

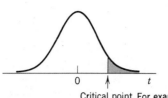

Critical point. For example:
$t_{.025}$ leaves .025 probability
in the tail.

TABLE **V** *t* **Critical Points**

d.f.	$t_{.25}$	$t_{.10}$	$t_{.05}$	$t_{.025}$	$t_{.010}$	$t_{.005}$	$t_{.0025}$	$t_{.0010}$	$t_{.0005}$
1	1.00	3.08	6.31	12.7	31.8	63.7	127	318	637
2	.82	1.89	2.92	4.30	6.96	9.92	14.1	22.3	31.6
3	.76	1.64	2.35	3.18	4.54	5.84	7.45	10.2	12.9
4	.74	1.53	2.13	2.78	3.75	4.60	5.60	7.17	8.61
5	.73	1.48	2.02	2.57	3.36	4.03	4.77	5.89	6.87
6	.72	1.44	1.94	2.45	3.14	3.71	4.32	5.21	5.96
7	.71	1.41	1.89	2.36	3.00	3.50	4.03	4.79	5.41
8	.71	1.40	1.86	2.31	2.90	3.36	3.83	4.50	5.04
9	.70	1.38	1.83	2.26	2.82	3.25	3.69	4.30	4.78
10	.70	1.37	1.81	2.23	2.76	3.17	3.58	4.14	4.59
11	.70	1.36	1.80	2.20	2.72	3.11	3.50	4.02	4.44
12	.70	1.36	1.78	2.18	2.68	3.05	3.43	3.93	4.32
13	.69	1.35	1.77	2.16	2.65	3.01	3.37	3.85	4.22
14	.69	1.35	1.76	2.14	2.62	2.98	3.33	3.79	4.14
15	.69	1.34	1.75	2.13	2.60	2.95	3.29	3.73	4.07
16	.69	1.34	1.75	2.12	2.58	2.92	3.25	3.69	4.01
17	.69	1.33	1.74	2.11	2.57	2.90	3.22	3.65	3.97
18	.69	1.33	1.73	2.10	2.55	2.88	3.20	3.61	3.92
19	.69	1.33	1.73	2.09	2.54	2.86	3.17	3.58	3.88
20	.69	1.33	1.72	2.09	2.53	2.85	3.15	3.55	3.85
21	.69	1.32	1.72	2.08	2.52	2.83	3.14	3.53	3.82
22	.69	1.32	1.72	2.07	2.51	2.82	3.12	3.50	3.79
23	.69	1.32	1.71	2.07	2.50	2.81	3.10	3.48	3.77
24	.68	1.32	1.71	2.06	2.49	2.80	3.09	3.47	3.75
25	.68	1.32	1.71	2.06	2.49	2.79	3.08	3.45	3.73
26	.68	1.31	1.71	2.06	2.48	2.78	3.07	3.43	3.71
27	.68	1.31	1.70	2.05	2.47	2.77	3.06	3.42	3.69
28	.68	1.31	1.70	2.05	2.47	2.76	3.05	3.41	3.67
29	.68	1.31	1.70	2.05	2.46	2.76	3.04	3.40	3.66
30	.68	1.31	1.70	2.04	2.46	2.75	3.03	3.39	3.65
40	.68	1.30	1.68	2.02	2.42	2.70	2.97	3.31	3.55
60	.68	1.30	1.67	2.00	2.39	2.66	2.92	3.23	3.46
120	.68	1.29	1.66	1.98	2.36	2.62	2.86	3.16	3.37
∞	.67	1.28	1.64	1.96	2.33	2.58	2.81	3.09	3.29
	$= z_{.25}$	$= z_{.10}$	$= z_{.05}$	$= z_{.025}$	$= z_{.010}$	$= z_{.005}$	$= z_{.0025}$	$= z_{.0010}$	$= z_{.0005}$

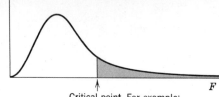

Critical point. For example:
$F_{.05}$ leaves 5% probability in the tail.

TABLE **VI** *F* **Critical Points**

(c-1)

DEGREES OF FREEDOM FOR NUMERATOR

c(N-1)

		1	2	3	4	5	6	8	10	20	40	∞
1	$F_{.25}$	5.83	7.50	8.20	8.58	8.82	8.98	9.19	9.32	9.58	9.71	9.85
	$F_{.10}$	39.9	49.5	53.6	55.8	57.2	58.2	59.4	60.2	61.7	62.5	63.3
	$F_{.05}$	161	200	216	225	230	234	239	242	248	251	254
2	$F_{.25}$	2.57	3.00	3.15	3.23	3.28	3.31	3.35	3.38	3.43	3.45	3.48
	$F_{.10}$	8.53	9.00	9.16	9.24	9.29	9.33	9.37	9.39	9.44	9.47	9.49
	$F_{.05}$	18.5	19.0	19.2	19.2	19.3	19.3	19.4	19.4	19.4	19.5	19.5
	$F_{.01}$	98.5	99.0	99.2	99.2	99.3	99.3	99.4	99.4	99.4	99.5	99.5
	$F_{.001}$	998	999	999	999	999	999	999	999	999	999	999
3	$F_{.25}$	2.02	2.28	2.36	2.39	2.41	2.42	2.44	2.44	2.46	2.47	2.47
	$F_{.10}$	5.54	5.46	5.39	5.34	5.31	5.28	5.25	5.23	5.18	5.16	5.13
	$F_{.05}$	10.1	9.55	9.28	9.12	9.10	8.94	8.85	8.79	8.66	8.59	8.53
	$F_{.01}$	34.1	30.8	29.5	28.7	28.2	27.9	27.5	27.2	26.7	26.4	26.1
	$F_{.001}$	167	149	141	137	135	133	131	129	126	125	124
4	$F_{.25}$	1.81	2.00	2.05	2.06	2.07	2.08	2.08	2.08	2.08	2.08	2.08
	$F_{.10}$	4.54	4.32	4.19	4.11	4.05	4.01	3.95	3.92	3.84	3.80	3.76
	$F_{.05}$	7.71	6.94	6.59	6.39	6.26	6.16	6.04	5.96	5.80	5.72	5.63
	$F_{.01}$	21.2	18.0	16.7	16.0	15.5	15.2	14.8	14.5	14.0	13.7	13.5
	$F_{.001}$	74.1	61.3	56.2	53.4	51.7	50.5	49.0	48.1	46.1	45.1	44.1
5	$F_{.25}$	1.69	1.85	1.88	1.89	1.89	1.89	1.89	1.89	1.88	1.88	1.87
	$F_{.10}$	4.06	3.78	3.62	3.52	3.45	3.40	3.34	3.30	3.21	3.16	3.10
	$F_{.05}$	6.61	5.79	5.41	5.19	5.05	4.95	4.82	4.74	4.56	4.46	4.36
	$F_{.01}$	16.3	13.3	12.1	11.4	11.0	10.7	10.3	10.1	9.55	9.29	9.02
	$F_{.001}$	47.2	37.1	33.2	31.1	29.8	28.8	27.6	26.9	25.4	24.6	23.8
6	$F_{.25}$	1.62	1.76	1.78	1.79	1.79	1.78	1.77	1.77	1.76	1.75	1.74
	$F_{.10}$	3.78	3.46	3.29	3.18	3.11	3.05	2.98	2.94	2.84	2.78	2.72
	$F_{.05}$	5.99	5.14	4.76	4.53	4.39	4.28	4.15	4.06	3.87	3.77	3.67
	$F_{.01}$	13.7	10.9	9.78	9.15	8.75	8.47	8.10	7.87	7.40	7.14	6.88
	$F_{.001}$	35.5	27.0	23.7	21.9	20.8	20.0	19.0	18.4	17.1	16.4	15.8
7	$F_{.25}$	1.57	1.70	1.72	1.72	1.71	1.71	1.70	1.69	1.67	1.66	1.65
	$F_{.10}$	3.59	3.26	3.07	2.96	2.88	2.83	2.75	2.70	2.59	2.54	2.47
	$F_{.05}$	5.59	4.74	4.35	4.12	3.97	3.87	3.73	3.64	3.44	3.34	3.23
	$F_{.01}$	12.2	9.55	8.45	7.85	7.46	7.19	6.84	6.62	6.16	5.91	5.65
	$F_{.001}$	29.3	21.7	18.8	17.2	16.2	15.5	14.6	14.1	12.9	12.3	11.7
8	$F_{.25}$	1.54	1.66	1.67	1.66	1.66	1.65	1.64	1.63	1.61	1.59	1.58
	$F_{.10}$	3.46	3.11	2.92	2.81	2.73	2.67	2.59	2.54	2.42	2.36	2.29
	$F_{.05}$	5.32	4.46	4.07	3.84	3.69	3.58	3.44	3.35	3.15	3.04	2.93
	$F_{.01}$	11.3	8.65	7.59	7.01	6.63	6.37	6.03	5.81	5.36	5.12	4.86
	$F_{.001}$	25.4	18.5	15.8	14.4	13.5	12.9	12.0	11.5	10.5	9.92	9.33
9	$F_{.25}$	1.51	1.62	1.63	1.63	1.62	1.61	1.60	1.59	1.56	1.55	1.53
	$F_{.10}$	3.36	3.01	2.81	2.69	2.61	2.55	2.47	2.42	2.30	2.23	2.16
	$F_{.05}$	5.12	4.26	3.86	3.63	3.48	3.37	3.23	3.14	2.94	2.83	2.71
	$F_{.01}$	10.6	8.02	6.99	6.42	6.06	5.80	5.47	5.26	4.81	4.57	4.31
	$F_{.001}$	22.9	16.4	13.9	12.6	11.7	11.1	10.4	9.89	8.90	8.37	7.81

DEGREES OF FREEDOM FOR DENOMINATOR

		DEGREES OF FREEDOM FOR NUMERATOR

		1	2	3	4	5	6	8	10	20	40	∞
10	$F_{.25}$	1.49	1.60	1.60	1.59	1.59	1.58	1.56	1.55	1.52	1.51	1.48
	$F_{.10}$	3.28	2.92	2.73	2.61	2.52	2.46	2.38	2.32	2.20	2.13	2.06
	$F_{.05}$	4.96	4.10	3.71	3.48	3.33	3.22	3.07	2.98	2.77	2.66	2.54
	$F_{.01}$	10.0	7.56	6.55	5.99	5.64	5.39	5.06	4.85	4.41	4.17	3.91
	$F_{.001}$	21.0	14.9	12.6	11.3	10.5	9.92	9.20	8.75	7.80	7.30	6.76
12	$F_{.25}$	1.56	1.56	1.56	1.55	1.54	1.53	1.51	1.50	1.47	1.45	1.42
	$F_{.10}$	3.18	2.81	2.61	2.48	2.39	2.33	2.24	2.19	2.06	1.99	1.90
	$F_{.05}$	4.75	3.89	3.49	3.26	3.11	3.00	2.85	2.75	2.54	2.43	2.30
	$F_{.01}$	9.33	6.93	5.95	5.41	5.06	4.82	4.50	4.30	3.86	3.62	3.36
	$F_{.001}$	18.6	13.0	10.8	9.63	8.89	8.38	7.71	7.29	6.40	5.93	5.42
14	$F_{.25}$	1.44	1.53	1.53	1.52	1.51	1.50	1.48	1.46	1.43	1.41	1.38
	$F_{.10}$	3.10	2.73	2.52	2.39	2.31	2.24	2.15	2.10	1.96	1.89	1.80
	$F_{.05}$	4.60	3.74	3.34	3.11	2.96	2.85	2.70	2.60	2.39	2.27	2.13
	$F_{.01}$	8.86	5.51	5.56	5.04	4.69	4.46	4.14	3.94	3.51	3.27	3.00
	$F_{.001}$	17.1	11.8	9.73	8.62	7.92	7.43	6.80	6.40	5.56	5.10	4.60
16	$F_{.25}$	1.42	1.51	1.51	1.50	1.48	1.48	1.46	1.45	1.40	1.37	1.34
	$F_{.10}$	3.05	2.67	2.46	2.33	2.24	2.18	2.09	2.03	1.89	1.81	1.72
	$F_{.05}$	4.49	3.63	3.24	3.01	2.85	2.74	2.59	2.49	2.28	2.15	2.01
	$F_{.01}$	8.53	6.23	5.29	4.77	4.44	4.20	3.89	3.69	3.26	3.02	2.75
	$F_{.001}$	16.1	11.0	9.00	7.94	7.27	6.81	6.19	5.81	4.99	4.54	4.06
20	$F_{.25}$	1.40	1.49	1.48	1.46	1.45	1.44	1.42	1.40	1.36	1.33	1.29
	$F_{.10}$	2.97	2.59	2.38	2.25	2.16	2.09	2.00	1.94	1.79	1.71	1.61
	$F_{.05}$	4.35	3.49	3.10	2.87	2.71	2.60	2.45	2.35	2.12	1.99	1.84
	$F_{.01}$	8.10	5.85	4.94	4.43	4.10	3.87	3.56	3.37	2.94	2.69	2.42
	$F_{.001}$	14.8	9.95	8.10	7.10	6.46	6.02	5.44	5.08	4.29	3.86	3.38
30	$F_{.25}$	1.38	1.45	1.44	1.42	1.41	1.39	1.37	1.35	1.30	1.27	1.23
	$F_{.10}$	2.88	2.49	2.28	2.14	2.05	1.98	1.88	1.82	1.67	1.57	1.46
	$F_{.05}$	4.17	3.32	2.92	2.69	2.53	2.42	2.27	2.16	1.93	1.79	1.62
	$F_{.01}$	7.56	5.39	4.51	4.02	3.70	3.47	3.17	2.98	2.55	2.30	2.01
	$F_{.001}$	13.3	8.77	7.05	6.12	5.53	5.12	4.58	4.24	3.49	3.07	2.59
40	$F_{.25}$	1.36	1.44	1.42	1.40	1.39	1.37	1.35	1.33	1.28	1.24	1.19
	$F_{.10}$	2.84	2.44	2.23	2.09	2.00	1.93	1.83	1.76	1.61	1.51	1.38
	$F_{.05}$	4.08	3.23	2.84	2.61	2.45	2.34	2.18	2.08	1.84	1.69	1.51
	$F_{.01}$	7.31	5.18	4.31	3.83	3.51	3.29	2.99	2.80	2.37	2.11	1.80
	$F_{.001}$	12.6	8.25	6.60	5.70	5.13	4.73	4.21	3.87	3.15	2.73	2.23
60	$F_{.25}$	1.35	1.42	1.41	1.38	1.37	1.35	1.32	1.30	1.25	1.21	1.15
	$F_{.10}$	2.79	2.39	2.18	2.04	1.95	1.87	1.77	1.71	1.54	1.44	1.29
	$F_{.05}$	4.00	3.15	2.76	2.53	2.37	2.25	2.10	1.99	1.75	1.59	1.39
	$F_{.01}$	7.08	4.98	4.13	3.65	3.34	3.12	2.82	2.63	2.20	1.94	1.60
	$F_{.001}$	12.0	7.76	6.17	5.31	4.76	4.37	3.87	3.54	2.83	2.41	1.89
120	$F_{.25}$	1.34	1.40	1.39	1.37	1.35	1.33	1.30	1.28	1.22	1.18	1.10
	$F_{.10}$	2.75	2.35	2.13	1.99	1.90	1.82	1.72	1.65	1.48	1.37	1.19
	$F_{.05}$	3.92	3.07	2.68	2.45	2.29	2.17	2.02	1.91	1.66	1.50	1.25
	$F_{.01}$	6.85	4.79	3.95	3.48	3.17	2.96	2.66	2.47	2.03	1.76	1.38
	$F_{.001}$	11.4	7.32	5.79	4.95	4.42	4.04	3.55	3.24	2.53	2.11	1.54
∞	$F_{.25}$	1.32	1.39	1.37	1.35	1.33	1.31	1.28	1.25	1.19	1.14	1.00
	$F_{.10}$	2.71	2.30	2.08	1.94	1.85	1.77	1.67	1.60	1.42	1.30	1.00
	$F_{.05}$	3.84	3.00	2.60	2.37	2.21	2.10	1.94	1.83	1.57	1.39	1.00
	$F_{.01}$	6.63	4.61	3.78	3.32	3.02	2.80	2.51	2.32	1.88	1.59	1.00
	$F_{.001}$	10.8	6.91	5.42	4.62	4.10	3.74	3.27	2.96	2.27	1.84	1.00

DEGREES OF FREEDOM FOR DENOMINATOR

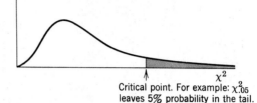

Critical point. For example: $\chi^2_{.05}$
leaves 5% probability in the tail.

TABLE **VII** χ^2 Critical Points

d.f.	$\chi^2_{.25}$	$\chi^2_{.10}$	$\chi^2_{.05}$	$\chi^2_{.025}$	$\chi^2_{.010}$	$\chi^2_{.005}$	$\chi^2_{.001}$
1	1.32	2.71	3.84	5.02	6.63	7.88	10.8
2	2.77	4.61	5.99	7.38	9.21	10.6	13.8
3	4.11	6.25	7.81	9.35	11.3	12.8	16.3
4	5.39	7.78	9.49	11.1	13.3	14.9	18.5
5	6.63	9.24	11.1	12.8	15.1	16.7	20.5
6	7.84	10.6	12.6	14.4	16.8	18.5	22.5
7	9.04	12.0	14.1	16.0	18.5	20.3	24.3
8	10.2	13.4	15.5	17.5	20.1	22.0	26.1
9	11.4	14.7	16.9	19.0	21.7	23.6	27.9
10	12.5	16.0	18.3	20.5	23.2	25.2	29.6
11	13.7	17.3	19.7	21.9	24.7	26.8	31.3
12	14.8	18.5	21.0	23.3	26.2	28.3	32.9
13	16.0	19.8	22.4	24.7	27.7	29.8	34.5
14	17.1	21.1	23.7	26.1	29.1	31.3	36.1
15	18.2	22.3	25.0	27.5	30.6	32.8	37.7
16	19.4	23.5	26.3	28.8	32.0	34.3	39.3
17	20.5	24.8	27.6	30.2	33.4	35.7	40.8
18	21.6	26.0	28.9	31.5	34.8	37.2	42.3
19	22.7	27.2	30.1	32.9	36.2	38.6	32.8
20	23.8	28.4	31.4	34.2	37.6	40.0	45.3
21	24.9	29.6	32.7	35.5	38.9	41.4	46.8
22	26.0	30.8	33.9	36.8	40.3	42.8	48.3
23	27.1	32.0	35.2	38.1	41.6	44.2	49.7
24	28.2	33.2	36.4	39.4	32.0	45.6	51.2
25	29.3	34.4	37.7	40.6	44.3	46.9	52.6
26	30.4	35.6	38.9	41.9	45.6	48.3	54.1
27	31.5	36.7	40.1	43.2	47.0	49.6	55.5
28	32.6	37.9	41.3	44.5	48.3	51.0	56.9
29	33.7	39.1	42.6	45.7	49.6	52.3	58.3
30	34.8	40.3	43.8	47.0	50.9	53.7	59.7
40	45.6	51.8	55.8	59.3	63.7	66.8	73.4
50	56.3	63.2	67.5	71.4	76.2	79.5	86.7
60	67.0	74.4	79.1	83.3	88.4	92.0	99.6
70	77.6	85.5	90.5	95.0	100	104	112
80	88.1	96.6	102	107	112	116	125
90	98.6	108	113	118	124	128	137
100	109	118	124	130	136	140	149

TABLE VIII Wilcoxon Rank Test (Two Independent Samples)

The one-sided p-value (p) corresponding to the rank sum W of the smaller sample, ranking from the end where this smaller sample is concentrated. For $n_2 > 6$ or $p > .25$, see Equation (16-10).

$n_2 = 2$

$n_1 = 1$		$n_1 = 2$	
W	p	W	p
1	.333	3	.167
		4	.333

$n_2 = 3$ — SMALLER SAMPLE SIZE, n_1

$n_1 = 1$		$n_1 = 2$		$n_1 = 3$	
W	p	W	p	W	p
1	.250	3	.100	6	.050
2	.500	4	.200	7	.100
		5	.400	8	.200
				9	.350
				10	.500

LARGER SAMPLE SIZE, $n_2 = 4$ — SMALLER SAMPLE SIZE, n_1

$n_1 = 1$		$n_1 = 2$		$n_1 = 3$		$n_1 = 4$	
W	p	W	p	W	p	W	p
1	.200	3	.067	6	.029	10	.014
2	.400	4	.133	7	.057	11	.029
		5	.267	8	.114	12	.057
		6	.400	9	.200	13	.100
				10	.314	14	.171
				11	.429	15	.243
						16	.343

LARGER SAMPLE SIZE, $n_2 = 5$ — SMALLER SAMPLE SIZE, n_1

$n_1 = 1$		$n_1 = 2$		$n_1 = 3$		$n_1 = 4$		$n_1 = 5$	
W	p	W	p	W	p	W	p	W	p
1	.167	3	.048	6	.018	10	.008	15	.004
2	.333	4	.095	7	.036	11	.016	16	.008
3	.500	5	.190	8	.071	12	.032	17	.016
		6	.286	9	.125	13	.056	18	.028
		7	.429	10	.196	14	.095	19	.048
				11	.286	15	.143	20	.075
				12	.393	16	.206	21	.111
				13	.500	17	.278	22	.155
						18	.365	23	.210
						19	.452	24	.274
								25	.345
								26	.421
								27	.500

LARGER SAMPLE SIZE, $n_2 = 6$ — SMALLER SAMPLE SIZE, n_1

$n_1 = 1$		$n_1 = 2$		$n_1 = 3$		$n_1 = 4$		$n_1 = 5$		$n_1 = 6$	
W	p	W	p	W	p	W	p	W	p	W	p
1	.143	3	.036	6	.012	10	.005	15	.002	21	.001
2	.286	4	.071	7	.024	11	.010	16	.004	22	.002
3	.429	5	.143	8	.048	12	.019	17	.009	23	.004
		6	.214	9	.083	13	.033	18	.015	24	.008
		7	.321	10	.131	14	.057	19	.026	25	.013
		8	.429	11	.190	15	.086	20	.041	26	.021
				12	.274	16	.129	21	.063	27	.032
				13	.357	17	.176	22	.089	28	.047
				14	.452	18	.238	23	.123	29	.066
						19	.305	24	.165	30	.090
						20	.381	25	.214	31	.120
						21	.457	26	.268	32	.155
								27	.331	33	.197
								28	.396	34	.242
								29	.465	35	.294

TABLE **IX** **Critical Points of the Durbin-Watson Test for Autocorrelation**

This table gives two limiting values of critical D (D_L and D_U), corresponding to the two most extreme configurations of the regressors; thus, for every possible configuration, the critical value of D will be somewhere between D_L and D_U:

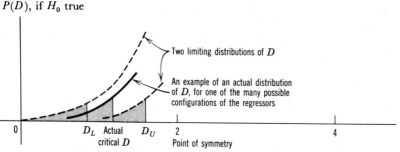

$P(D)$, if H_0 true

Two limiting distributions of D

An example of an actual distribution of D, for one of the many possible configurations of the regressors

D_L Actual D_U 2 4
critical D Point of symmetry

As an example of a test for positive serial correlation, suppose that there are n = 15 observations and k = 3 regressors (excluding the constant) and we wish to test $\rho = 0$ versus $\rho > 0$ at the level $\alpha = .05$. Then if D falls below $D_L = .82$, reject H_0. If D falls above $D_U = 1.75$, do not reject H_0. If D falls between D_L and D_U, this test is indecisive.

To test for negative serial correlation ($\rho = 0$ versus $\rho < 0$), the right-hand tail of the distribution defines the critical region. The symmetry of the distribution permits us to calculate these values very easily. With the same sample size, number of regressors, and level α as before, our new critical values would be $4 - D_L = 4 - .82 = 3.18$, and $4 - D_U = 4 - 1.75 = 2.25$. Accordingly, if D falls beyond 3.18, reject H_0. If D falls short of 2.25, do not reject H_0. If D falls between 2.25 and 3.18, this test is indecisive.

SAMPLE SIZE = n	$p =$ PROBABILITY IN LOWER TAIL (LEVEL, α)	$k =$ NUMBER OF REGRESSORS (EXCLUDING THE CONSTANT)									
		1		2		3		4		5	
		D_L	D_U	D_L	D_U	D_L	D_U	D_L	D_U	D_L	D_U
15	.01	.81	1.07	.70	1.25	.59	1.46	.49	1.70	.39	1.96
	.025	.95	1.23	.83	1.40	.71	1.61	.59	1.84	.48	2.09
	.05	1.08	1.36	.95	1.54	.82	1.75	.69	1.97	.56	2.21
20	.01	.95	1.15	.86	1.27	.77	1.41	.68	1.57	.60	1.74
	.025	1.08	1.28	.99	1.41	.89	1.55	.79	1.70	.70	1.87
	.05	1.20	1.41	1.10	1.54	1.00	1.68	.90	1.83	.79	1.99
30	.01	1.13	1.26	1.07	1.34	1.01	1.42	.94	1.51	.88	1.61
	.025	1.25	1.38	1.18	1.46	1.12	1.54	1.05	1.63	.98	1.73
	.05	1.35	1.49	1.28	1.57	1.21	1.65	1.14	1.74	1.07	1.83
40	.01	1.25	1.34	1.20	1.40	1.15	1.46	1.10	1.52	1.05	1.58
	.025	1.35	1.45	1.30	1.51	1.25	1.57	1.20	1.63	1.15	1.69
	.05	1.44	1.54	1.39	1.60	1.34	1.66	1.29	1.72	1.23	1.79
60	.01	1.38	1.45	1.35	1.48	1.32	1.52	1.28	1.56	1.25	1.60
	.025	1.47	1.54	1.44	1.57	1.40	1.61	1.37	1.65	1.33	1.69
	.05	1.55	1.62	1.51	1.65	1.48	1.69	1.44	1.73	1.41	1.77
80	.01	1.47	1.52	1.44	1.54	1.42	1.57	1.39	1.60	1.36	1.62
	.025	1.54	1.59	1.52	1.62	1.49	1.65	1.47	1.67	1.44	1.70
	.05	1.61	1.66	1.59	1.69	1.56	1.72	1.53	1.74	1.51	1.77
100	.01	1.52	1.56	1.50	1.58	1.48	1.60	1.46	1.63	1.44	1.65
	.025	1.59	1.63	1.57	1.65	1.55	1.67	1.53	1.70	1.51	1.72
	.05	1.65	1.69	1.63	1.72	1.61	1.74	1.59	1.76	1.57	1.78

REFERENCES

Anderson, T. W. (1958), *An Introduction to Multivariate Analysis*. New York: Wiley.

Anderson, T. W., an S. L. Sclove (1978), *Statistical Analysis of Data*. Boston: Houghton-Mifflin.

Bailey, Martin (1980), *Reducing Risks to Life*. Washington: American Enterprise Institute.

Bickel, P. J., and J. W. O'Connell (1975), "Sex Bias in Graduate Admissions: Data from Berkeley," *Science* 187, February, pp. 398–404.

Blau, P. M., and O. D. Duncan (1967), *The American Occupational Structure*. New York: Wiley.

Bostwick, Burdette E. (1977), *Finding the Job You've Always Wanted*. New York: Wiley.

Box. G. E. P., and G. C. Tiao (1973), *Bayesian Inference in Statistical Analysis*. Reading, Mass.: Addison-Wesley.

Breiman, L., and D. Freedman (1983), "How Many Variables Should be Entered in a Regression Equation?," *Journal of the American Statistical Association*, Vol. 78, March, pp. 131–136.

Brownlee, K. A. (1965), *Statistical Theory and Methodology in Science and Engineering*. New York: Wiley.

Chatfield, C. (1980), *The Analysis of Time Series: An Introduction*. London: Chapman and Hall.

Choldin, H. M. (1978), "Urban Density and Pathology," *Annual Review of Sociology*, Vol. 4, pp. 91–113.

Clark, K. B., and M. P. Clark (1958), "Racial Identification and Preference in Negro Children," in *Readings in Social Psychology*, edited by E. E. Maccoby et al. New York: Holt, Rinehart and Winston.

Clopper, C. J., and E. S. Pearson (1934), "The Use of Confidence or Fiducial Limits Illustrated in the Case of the Binomial," *Biometrika* 26: p. 404.

Cochrane, W. G. (1977), *Sampling Techniques*, 3rd ed. New York: Wiley.

Conner, J. R., and K. C. Gibbs, and J. E. Reynolds (1973), "The Effects of Water Frontage on Recreational Property Values," *Journal of Leisure Research* 5, pp. 26–38.

713

Conover, W. J. (1980), *Practical Nonparametric Statistics*, 2nd ed. New York: Wiley.

Consumer Reports (1976), "Is Vitamin C Really Good for Colds?" *Consumer Reports*, February, 1976, Vol. 41, No. 2, pp. 68–70.

Cramer, H. (1946), *Mathematical Methods of Statistics*. Princeton, N.J.: Princeton University Press.

Daniel, Cuthbert, and F. S. Wood (1971), *Fitting Equations to Data*. New York: Wiley.

Dawes, R. M. (1971), "A Case Study of Graduate Admissions: Application of Three Principles of Human Decision Making," *America Psychologist*, Vol. 26, No. 2, pp. 180–188.

Dean, Joel (1941), "Statistical Cost Functions of a Hosiery Mill," *Journal of Business*.

Fadeley, R. C. (1965), "Oregon Malignancy Pattern Physiographically Related to Hanford, Washington, Radioistopic Storage," *Journal of Environmental Health* 27, pp. 883–897.

Fairley, W. B., and F. Mosteller (1977) (eds.), *Statistics and Public Policy*. Reading, Mass.: Addison-Wesley.

Fraumini, J. F., Jr. (1968), "Cigarette Smoking and Cancers of the Urinary Tract: Geographic Variation in the Uniited States," *Journal of National Cancer Institute*, pp. 1205–1211.

Freedman, D., R. Purvis, and R. Pisani (1978), *Statistics*. New York: Norton.

Freedman, J. L. (1975), *Crowding and Behavior*. New York: Viking Press.

Gallup, George H. (1976), "What Mankind Thinks About Itself," *Readers Digest*, October 1976, pp. 25–31.

Gilbert, J. P., B. McPeek, and F. Mosteller (1977), "Statistics and Ethics in Surgery and Anesthesia," *Science* (18 November), pp. 684–689.

Glick, N. (1970), "Hijacking Planes to Cuba: An Updated Version of the Birthday Problem," *The American Statistician*, February, pp. 41–44.

Glover, D. and J. L. Simon (1975), "The Effects of Population Density Upon Infrastructure: The Case of Road Building," *Economic Development and Cultural Change*, Vol. 23, pp. 453–468.

Gourou, Pierre (1966), *The Tropical World, Its Social and Economic Conditions and its Future Status*. New York: Wiley.

Griliches, Zvi, and W. M. Mason (1973), "Education, Income, and Ability," in A. S. Goldberger and O. D. Duncan (eds.), *Structural Equation Models in the Social Sciences*. New York: Seminar Press.

Haupt, A., and T. T. Kane (1978), *Population Handbook*. Washington, D.C.: Population Reference Bureau.

Hooker, R. H. (1907), "The Correlation of the Weather and Crops," *Journal of the Royal Statistical Society* 70, pp. 1–42.

Howard, R. A. (1972), "The Decision to Seed Hurricanes," *Science* 176, June.

Huff, Darrell (1957), *How to Take a Chance*. New York: Norton (paperback).

Huff, Darrell (1954), *How to Lie with Statistics*. New York: Norton (paperback).

Jones, K. L., D. W. Smith, A. P. Streissgrath, and N. C. Myrianthopoulos (1974), "Outcome in Offspring of Chronic Alcoholic Women," *Lancet* (1 June), pp. 1076–1078.

Katz, D. A. (1973), "Faculty Salaries, Promotions, and Productivity at a Large University," *American Economic Review* 63, pp. 469–477.

Kendall, M. G. and C. A. O'Muircheartaigh (1977), *Path Analysis and Model Building* (Tech. Bulletin 414), World Fertility Survey.

Kennel, J. H., D. K. Voos, and M. H. Klaus (1979), "Parent-Infant Bonding," Chapter 23 in J. Oxofsky, *Handbook of Infant Development*. New York: Wiley.

Klaus, M. H., R. Jerauld, N. C. Kreger, W. McAlpine, M. Steffa, and J. H. Kennel (1972), "Maternal Attachment," *New England Journal of Medicine*, 2 March.

Klitgaard, R. E., and G. R. Hall (1977), "A Statistical Search for Unusually Effective Schools," in Fairley and Mosteller (1977), pp. 51–86.

Knodel, John (1977), "Breast-feeding and Population Growth," *Science* (16 December), pp. 1111–1115.

Langer, W. L. (1976), "Immunization Against Smallpox Before Jenner," *Scientific American*, January, pp. 112–117.

Lave, L. B., and E. P. Seskin (1977), "Does Air Pollution Shorten Lives?" in Fairly and Mosteller (1977), pp. 143–160.

Lefcoe, N. M., and T. H. Wonnacott, "The Prevalence of Chronic Respiratory Disease in Four Occupational Groups," *Archives of Environmental Health* 29, September, pp. 143–146.

Lindgren, B. W. (1976), *Statistical Theory*. New York: Macmillan.

Malcolm, J. G. (1983), "Practical Application of Bayes Formulas," *1983 Proceedings, Annual Reliability and Maintainability Symposium*.

McLeod, A. I. (1982), personal communication.

McNeil, B. J., R. Weichselbaum, and S. G. Pareker (1978), "Fallacy of the Five-Year Survival in Lung Cancer," *New England Journal of Medicine* 229, December, pp. 1397–1401.

Medical Research Council (1948), "Streptomycin treatment of pulmonary tuberculosis." *British Medical Journal*, vol. 2, p. 769.

Meier, Paul (1977), "The Biggest Public Health Experiment Ever: The 1954 Field Trial of the Salk Poliomyelitis Vaccine," pp. 88–100 in *Statistics: A Guide to the Study of the Biological and Health Sciences*. San Francisco: Holden Day.

Miksch, W. F. (1950), "The Average Statistician," *Colliers* (17 June).

Moody, (1975), *Moody's Transportation Manual*.

Moore, David S. (1979), *Statistics: Concepts and Controversies*. San Francisco: Freeman.

Mosteller, F., and J. W. Tukey (1977), *Data Analysis and Regression: A Second Course in Statistics*. Reading, Mass.: Addison-Wesley.

Nemenyi, P., S. K. Dixon, N. B. White Jr., and M. L. Hedstrom (1977), *Statistics from Scratch* (Pilot Edition). San Francisco: Holden Day.

Neter, John (1976), "How Accountants Save Money by Sampling," Chapter 9 of *Statistics: A Guide to Business and Economics*, ed. by J. M. Tanur et al. San Francisco: Holden-Day, pp. 83–92.

Peacock, E. E. (1972), quoted in *Medical World News*, September 1, 1972, p. 45.

Pindyck, R. S., and D. L. Rubinfeld (1976), *Econometric Models and Economic Forecasts*. New York: McGraw-Hill.

Press, S. James (1972), *Applied Multivariate Analysis*. New York: Holt, Rinehart and Winston.

Raiffa, H. (1968). Decision Analysis. Reading, Mass.: Addison-Wesley.

Raiffa, H., and R. L. Keeney (1972), *Decisions with Multiple Objectives*. New York: Wiley.

Rosenzweig, M., E. L. Bennett, and M. C. Diamond (1964), "Brain Changes in Response to Experience," *Scientific American* (February), pp. 22–29.

Ryan, T. A., B. L. Joiner, and B. F. Ryan (1976), *Minitab Student Handbook*. North Scituate, Mass.: Duxbury Press.

Scheffe, H. (1959), *The Analysis of Variance*. New York: Wiley.

Schunk, G. J., and W. L. Lehman (1954), "Mongolism and Congenital Leukemia," *Journal of the American Medical Society* (May).

Simon, J. L. (1981), *The Ultimate Resource*. Princeton, N.J.: Princeton University Press.

Slonim, M. J. (1966), *Sampling in a Nutshell*. New York: Simon and Shuster (paperback).

Snedecor, G. W., and W. G. Cochran (1967), *Statistical Methods*, 6th ed. Ames, Iowa: Iowa State University Press.

SAS Institute Inc. (1979), *SAS User's Guide, 1979 Edition*. Cary, North Carolina.

Tanur, J. M., F. Mosteller, W. H. Kruskal, R. F. Link, R. S. Pieters, G. R. Rising, and E. L. Lehman (1977), *Statistics: A Guide to the Unknown*. Also revised as three smaller paperbacks; (1) *Statistics: A Guide to Business and Economics*; (2) *Statistics: A Guide to Biological and Health Sciences*; (3) *Statistics: A Guide to Political and Social Issues*. San Francisco: Holden-Day.

Theil, H. (1975), "Specification Errors and the Estimation of Economic Relationships," *Review of the International Statistical Institute*, Vol. 25, pp. 41–51.

Tufte, E. R. (1974), *Data Analysis for Politics and Policy*. Englewood Cliffs, N.J.: Prentice-Hall.

Tukey, J. W. (1977), *Exploratory Data Analysis*. Reading, Mass: Addison-Wesley.

U.S. Department of Health, Education, and Welfare (1964), *Smoking and Health: Report of the Advisory Committee to the Surgeon General of the Public Health Service*.

U.S. Department of Transportation (1981), National Highway Traffic Safety Administration, *National Accident Sampling System Report on Traffic Accidents and Injuries for 1979*.

U.S. Surgeon-General (1979), *Smoking and Health, a Report of the Surgeon-General*. U.S. Dept. of Health, Education and Welfare.

Van de Geer, J. P. (1971), *Introduction to Multivariate Analysis for the Social Sciences*. San Francisco: Freeman.

Wallis, W. A., and H. V. Roberts (1956), Statistics: A New Approach. New York: The Free Press. Abbreviated Paperback appears as *The Nature of Statistics*, 1962.

Wonnacott, T. H., and R. J. Wonnacott (1977), *Introductory Statistics for Business and Economics*, 2nd ed. New York: Wiley.

Wonnacott, T. H., and R. J. Wonnacott (1981), *Regression: A Second Course in Statistics*. New York: Wiley.

Wonnacott, T. H., and R. J. Wonnacott (1984 IM), *Instructor's Manual for Introductory Statistics for Business and Economics*. New York: Wiley.

Zeisel, H., and H. Kalven (1972), "Parking Tickets and Missing Women," in *Statistics, A Guide to the Unknown*, edited by J. Tanur, F. Mosteller, et al. San Francisco: Holden-Day.

Answers, Odd-Numbered Problems

1-1   **a.** .51 ± .0253 ≈ 51% ± 3%
.64 ± .0243 ≈ 64% ± 2%
.50 ± .0253 ≈ 50% ± 3%
.38 ± .0246 ≈ 38% ± 2%
.51 ± .0253 ≈ 51% ± 3%
.48 ± .0253 ≈ 48% ± 3%
   b. The intervals for 1964 and 1980 are wrong.

1-3   In each case, bias can be eliminated by taking a *random* sample.
   a. People with moderate views may be vastly under-represented.
   b. Working couples would be completely left out.
   c. Alumni who are most distant, most busy, and most reluctant to reveal their incomes, would be vastly under-represented.

1-7   **b.** Pairing achieves perfect balance for the factor of marital status. The randomization also protects us on average against all other extraneous factors, even the ones we didn't think of.

1-9   **a.** Poverty, racial tension, social upheaval, youth, etc.
   b. (i) observational studies
       (ii) extraneous factors
       (iii) confounded with

1-13   This statement claims too much. Causation is never proved by a naive observational study. For example, the MBA may be associated with high ambition, and this may be what increases income, rather than the MBA itself. To reduce this bias, analyze the data by multiple regression.

1-15   **a.** .005 ± .001, .023 ± .003
   b. .004 ± .001, .013 ± .002
   c. It is foolhardy not to buckle up. (It seems to reduce the chance of injury to less than half.)

1-17   6.0% ± 0.2%

2-1

2-3

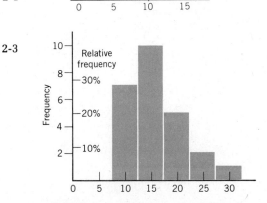

2-5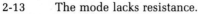

2-7   **a.** 11.0
      b. 16.0

2-9   **a.** The cell frequencies are 5, 7, 8.
      b. mean = $71.50, mode = $80

2-11 **a.**

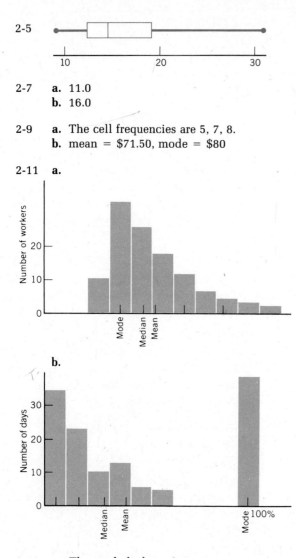

b.

2-13 The mode lacks resistance.

2-15 **a.** $\overline{X} = 9$
      b. average deviation = 0
      c. average deviation = 0
      d. See footnote on p. 36. The average deviation from the median is *not* zero.

2-17 **a.** 8,   3
      b. 2.4,   8.0,   10,   3.2

2-19 **a.** The central value is becoming smaller. So is the spread.
      b. means: 74.8,   71.3,   70.3
             st. dev: 4.84,   4.14,   2.75
             These calculations confirm part **a.**

c. The overall $\overline{X}$ (72.1) equals the average of the three component $\overline{X}$ values. The overall s (4.425) is greater than the average of the three component s values.

2-21 **a.** 40
      b. 2400
      c. 12,000

2-23 $\overline{X} = 24{,}510{,}000$, $s_x = 46{,}000$

2-25 **a.**

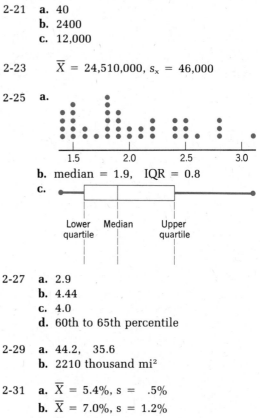

      b. median = 1.9,   IQR = 0.8
      c.

2-27 **a.** 2.9
      b. 4.44
      c. 4.0
      d. 60th to 65th percentile

2-29 **a.** 44.2,   35.6
      b. 2210 thousand mi²

2-31 **a.** $\overline{X} = 5.4\%$, $s = .5\%$
      b. $\overline{X} = 7.0\%$, $s = 1.2\%$
      c. $\overline{X} = 6.2\%$, $s = 1.2\%$
         The overall $\overline{X}$ equals the average of the two component $\overline{X}$ values. The overall s is greater than the average of the two component s values.

2-33 **a.** 47, 64
      b. 104, 38.5

3-1   See Figure 3-1, for example.

3-3   **a.** .50
      b. relative frequency
      c. relative frequency, or by reasoning from the symmetry of the dice, 8/36 = 22%

3-5   **a.** .216 ≈ 22%
      b. .432 ≈ 43%

3-7 **a.** $.255 \simeq 25\%$
    b. $.091 \simeq 9\%$

3-9 **a.** .50, .30, .65, .15
    b. $.65 = .50 + .30 - .15$
    c. .50, .70, .85, .35
    d. $.85 = .50 + .70 - .35$

3-11 **a.** $(1/2)^{10} = .001$
    b. .998

3-13 **a.** .52   **d.** .33
    b. .17   **e.** .77
    c. .40

3-15

	For example, rel. freq.	Long-run rel. freq.
a.	$27/50 = .54$	$Pr(G) = .50$
b.	$10/50 = .20$	$Pr(G \cap H) = .125$
c.	$10/27 = .37$	$Pr(H/G) = .25$

3-17 **a.** .875   **c.** .57
    b. .50   **d.** .75

3-19 **a.** Were these 15 the *only* material witnesses, or 15 out of many?
    b. Absurd, of course. The *conditional* probability of another passenger carrying a live bomb is one in a million, no matter what he himself carries on with him.

3-21 **a.** $1/221 = .0045$
    b. $1/1326 = .00075$
    c. .143

3-23 **a.** .26
    b. .26
    c. yes

3-25 **a.** yes, independent
    b. no, dependent

3-27 **a.** yes, independent
    b. yes, independent
    c. no, dependent

3-29 .64, .36

3-31 .40, .44, .16

3-33 **a.** .84
    b. .96

3-37 all valid

3-39 **a.** True
    b. True
    c. False: A and B are *independent* . . .
    d. False: . . . will be 1/2.

3-41 **a.** 0
    b. 0
    c. .8

3-43 **a.** .33
    b. .50
    c. 0
    d. .17

3-45 $1 - \dfrac{120}{120} \times \dfrac{119}{120} \times \ldots \dfrac{99}{120} = 87\%$

3-47 Only **(b)** and **(d)** are true.

3-49 Even the richest man has finite wealth, and risks losing it all. Why should he take even a tiny risk, just to win peanuts?

4-1

	a.			**b.**	
	x	$p(x)$		y	$p(y)$
	0	1/8		1	2/8
	1	3/8		2	4/8
	2	3/8		3	2/8
	3	1/8			

4-3 **a.**

x	$p(x)$
0	1/8
1	3/8
2	3/8
3	1/8

    b. 1.50
    c. .75

4-5 **a.**

x	$p(x)$
0	.072
1	.104
2	.176
3	.648

    b. .824

4-7 **a.** 62.0, 9.5
    b. 80%, 20%
    c. 70

4-9   If he doesn't mind taking risks, he should
not cancel. This would minimize his
expected loss at $24,500.

4-11   **b.** .980 ≈ 98%

4-13   **a.** 3.00,   1.22
b. 7.20,   1.20

4-15   **a.** $(.98)^{50} = 36\%$
b. Independence is not at all valid because
learning, for example, improves the
chance of survival in later raids.

4-17   **a.** .055   **e.** .475
b. .044   **f.** .950
c. .949   **g.** .682
d. .103   **h.** .006

4-19   **a.** .001
b. .159
c. .840

4-21   **a.** .055
b. $72

4-23   **a.** E(X) = 3.5
b. E(Y) = 15
c. True

4-25   **a.** μ = 4.6,   σ² = 2.04
b. 1.50
c. 23.2

4-27   **a.** 9%
b. 91st percentile
c. 99th percentile

4-29   **a.** 2.55, 2, 2, 2.55, 1.95. The mean
(2.55) is what determines population
growth.
c. 2, 24%, 7.44 million
d. 27%, 73%

4-31   **a.** 71%
b. $.99^7 = 93\%$
c. 4.20,   1.30

4-33   .9999

4-35   **a.** 12.040, .094
b. .050

4-37   **a.** .035
b. Yes, we would question the hypothesis
of "no effect."

4-39   **a.** $\bar{X} = 3.00$,   s = .86
b.

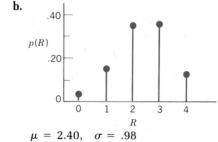

μ = 2.40,   σ = .98

5-1   **a.**

x	70	80	90
p(x)	.20	.60	.20

Y has the same distribution
b. no
c. 80, 80
d. 40, 40

5-3

y	70	80	90
p(y/x)	.50	.50	0

5-5

x	1	2	3
p(x/ Y= 1)	.17	.66	.17

Since this conditional distribution does
not equal the unconditional (marginal)
distribution, X and Y are not
independent.

5-7   **b.**

x	p(x)	y	p(y)
0	.008	0	.008
1	.096	100	.064
2	.384	200	.160
3	.512	300	.256
		400	.512

And p(x) is the same as in Problem 4-4,
of course.
c. 2.40, $320
d. no

5-9   **a.** 16.17
b. 6.6
c. 6145
d. 1.288,   23.3,   no

5-11 **a.** 13/6, 29/36, .898
b. 13/6, 17/36, .687
c. 26/6, 46/36, 1.130
d. For means and variances, yes. For standard deviations, no.

5-13 **a.** For part **c**, $\sigma_{XY} = 0$, and X and Y are independent.
For part **d**, $\sigma_{XY} = 0$, yet X and Y are not independent.
b. (1) True
(2) False

5-15 **a.** 1.20, 1.42, 1.50
b. .0585, .153
c. If self-reported happiness is believable, (i) and (ii) are true. (iii) Last sentence is false: Education does not necessarily *cause* the greater happiness, (In an observational study, some extraneous third factor may be the cause.)

5-17 **b.** $E(X) = 30$, var $(X) = 60$
$E(Y) = 25$, var $(Y) = 40$
c. $\sigma_{XY} = 20$
d. $E(S) = 55$, var $(S) = 140$
e. $E(W) = 38$, var $(W) = 66.4$
f. $E(D) = 5$, var $(D) = 60$
g. Not very good. In an observational study like this, some extraneous factor may cause the difference in income.

5-19 (i) b (200), then c (118), and finally a (110)
(ii) b (5.9), then c (6.7), and finally a (9.5)

5-21 **a.** .532
b. .841
c. .898

5-23 (i)

	mean	S.D.
	65	10.0
	70	11.5
	30	34.6
(ii)	65	14.1
	70	14.9
	30	28.3

5-25 In each of these observational studies, the *cause* is likely an extraneous factor instead.

5-27 **b.** $\mu = 9.65$, $\sigma = 7.5$
c. mode = 2.5 (or 0, if finer detail were given). Median is in between the mode and mean.

5-29 The ABC poll gave a biased view of the whole country, since it relied only on volunteer calls. So of course the polls did not agree.

5-31

a.	**b.**
1000	.026
974	.009
965	.019
947	.022
926	.053
877	.133
760	.288
541	.521
259	.803
51	1.000

b. (i) Correction: ... the *second* decade
(ii) Improvement: ... *from age 30 to age 90.*
c. (i) About $19 (ii) about $54
d. (i) .049 (ii) .926 (iii) .053

5-33 **a.**

x	0	1	2	3
p(x)	.512	.384	.096	.008

b. $\mu = .60$, $\sigma^2 = .48$
c. $13.88

5-35 **a.** about 5 people
b. about 293 people
c. only 32%
d. There are so many non-diabetics that even the small error rate of 2% produces relatively many "false alarms."

5-37 **a.** 2, 1, 4, 700
b. no
c. yes

5-39 **a.** $\mu = 1.21$, $\sigma = 1.34$
b. $\mu = 2.7$, $\sigma = 1.1$

5-41 **a.** 55 tickets
b. The boys came from families typical of the subdivision, which in turn is typical of the whole U.S. in 1972. Also assume both parents living, etc.

6-1 **b.** The small sample with the high response rate generally has a mean closer to the target μ.

6-3 **c.** $\mu = 4.5$
d. $\overline{X}$ is indeed closer to μ than many of the individual observations.

6-5 **a.** True
b. Correction: SE of only $\sigma/\sqrt{n}$

6-7 **a.** .010 thousand = $10
b. .031 thousand = $31

6-9 **a.** $20,000, $1,200, normal
b. .047

6-11 **a.** .047
b. $< .00048$
c. $\simeq 0\ (z = -6.67)$

6-13 **a.** 34, 8
b. .000048

6-15 **a.** $\mu = .30$, which exactly equals the population proportion of 1's.
b. (i) .164   (ii) Can't say

6-17 $.018 \simeq 2\%$ (wcc, .036)

6-19 $.038 \simeq 4\%$ (wcc, .056)

6-21 **a.** $.264 \simeq 26\%$
b. Between 13,000 and 14,000

6-25 **a.** True
b. True
c. False: . . . then the reduction factor is between .995 and 1, and so changes the SE less than 1/2%.

6-27 **a.** $\mu = .769,\quad \sigma = 1.31$
b. $.233 \simeq 23\%$ (wcc, .271)

6-31 **b.** The results should indeed be like Figure 6-11.

6-33   equally

6-35 **a.** 1/5%
b. Yes, random selection ensures that it is indeed unbiased and representative.

6-37 **a.** $.251 \simeq 25\%$
b. $.00048 \simeq .05\%$

6-39   .001 (wcc, .003)

6-41 **a.** 5.26 cents per dollar bet
b. (i) $.548 \simeq 55\%$
(ii) $.603 \simeq 60\%$
(iii) $.722 \simeq 72\%$

6-43 **a.** $.00023 \simeq .02\%$
b. If, for example, all passengers travelled and cancelled in pairs, the probability of overload would increase (to .007).

7-1 **a.** False: $\overline{X}$ is a . . . . the parameter μ.
b. True
c. False: . . . are both *unbiased* . . . the sample median is *less efficient than* the sample mean.

7-3   64%

7-5 **a.** True
b. Are you sure you have analyzed it efficiently? If not, maybe it would be wiser to spend $99,000 collecting the data and $1,100 analyzing it.

7-7 **a.** bias of 18 percentage points toward R
b. Yes.

7-9 **a.** .000308
b. .000182, 69% more efficient
c. .000156, 97% more efficient
d. True, as part (**a**) showed.

7-11 **a.** For a normal distribution, the best estimate is $\overline{X} = 8.1$.
b. Because of possible outliers, the best estimate is $\overline{X}_{.25} = 7.9\%$ or perhaps $\overset{\shortmid}{X} = 8.0\%$.
c. Because of possible outliers, the best estimate is $\overline{X}_{.25} = 15.18$.

7-13 **a.** 104,   109.4,   105.3
c. $\overset{\shortmid}{X} = 104$

7-15   105.13

7-17 **a.** When the population is asymmetric, only $\overline{X}$ estimates μ consistently.
b. none.

7-19   **a.** True
  b. False: If we *quadruple* the sample size, we halve . . .
  c. True
  d. False . . . when the population shape is unknown *and may be any shape, even very long-tailed.*
  e. False: $\overset{\backprime}{X}$ or $\overline{X}_{.25}$ is a robust . . . , *while the IQR is a robust* . . .

7-21   **a.** scale A, with MSE = .0001
  b. True

7-23   **a.** Only the high quality (100% response) is unbiased
  b. MSE are .00199, .00172, .00216 (or more). Thus the "high quality" survey (ii) is most efficient.

7-25   **a.** They are slightly different
  b. average first (method ii)
  c. zero bias

7-27   **b.** $\overline{X}$ is best for a normal population

8-1   **a.** $\overline{X}, \mu$
  b. $\sigma/\sqrt{n}$, standard error SE
  c. 2, μ
  d. about 50 times. But unfortunately he would not know . . .
  e. wider

8-3   **a.** $\mu = 3.3 \pm .392 \simeq 3.3 \pm .4$
   Note: In all confidence intervals, we should round the confidence allowance (.4) to the same number of decimal places as the estimate (3.3). We give the unrounded value (.392) merely as a check on your methodology.
  b. $\mu = 3.3 \pm .516 \simeq 3.3 \pm .5$

8-5 $T = 9,990,000 \pm 2,120,000$

8-7   **a.** $\mu = 2.20 \pm .81$
  b. $\mu = 2.20 \pm 1.34$

8-9 $\mu = 16.18 \pm .023$
   $\simeq 16.18 \pm .02$

8-11   **c.** .95
  d. False: . . . are wider *on average* than . . .

8-13   **a.** $\mu_M - \mu_W = 5.0 \pm 5.8$
   That is, the men at the university earn 5.0 (± 5.8) thousand dollars more than the women, on average. Or, equivalently, $\mu_W - \mu_M = -5.0 \pm 5.8$
   That is, women earn 5.0 (± 5.8) *less* than men.
  b. It fails to show discrimination on two counts: (1) This is an observational study, and whatever differences exist may be due to extraneous factors such as men having better qualifications, more experience, etc. (2) We're not even sure a difference exists in the population; the confidence interval includes $\mu_1 - \mu_2 = 0$ (i.e., no difference).

8-15   **a.** $\mu_T - \mu_C = 1.93 \pm 1.02$
  b. $\mu_T - \mu_C = 3.64 \pm 1.88$

8-17 $\Delta = 40 \pm 57.6 \simeq 40 \pm 58$

8-19   **a.** $\Delta = 3.00 \pm 1.35$
  b. There is good evidence that an interesting environment stimulates the growth of the cortex in rats, increasing its weight by about 5% ($\pm 2\%$).

8-21   **a.** $0 < \pi < .45$
  b. $.01 < \pi < .30$
  c. $.04 < \pi < .22$, or $\pi = .10 \pm .08$
  d. $\pi = .10 \pm .042 \simeq .10 \pm .04$

8-23   **a.** $\pi_1 - \pi_2 = .07 \pm .077 \simeq 7\% \pm 8\%$
  b. $\pi_1 - \pi_3 = -.17 \pm .079 \simeq -17\% \pm 8\%$

8-25   **a.** 28.5, 71.0, 46.2 (per 100,000)
  b. We estimate the vaccine reduces the polio rate from 71 to 28 cases per 100,000, a reduction of 43 cases per 100,000 (with 95% confidence, a reduction of 43 $\pm$ 14).
  c. **i** An observational study like that would have confounded volunteering with the vaccine. **ii** And since volunteering is a substantial effect (an estimated 25 cases per 100,000) it would seriously bias the estimated effect of the vaccine (downwards by about 25 cases per 100,000).

8-27 $\mu_M - \mu_W > 5.0 - 3.32 \approx 1.7$
That is, the average male earns more than the average female professor by an amount we estimate to be $5,000; with 95% confidence, it is at least $1,700.

8-29 For 8-15 **a**: $\mu_T - \mu_c > 1.09$
For 8-17 : $\Delta > 40 - 44.2 \approx -4$
For 8-25 **b**: The reduction is at least 31 cases per 100,000 with 95% confidence.

8-31 **a.** .634
b. .635 ± .083
c. The kill rate for the whole county.

8-33 **a.** $s^2 = 3528$
b. $\sigma^2 = 3528 \pm 2516$
c. MSD = 2646. Yes, the MSD is biased, since it underestimates σ^2, on average.
d. $\sigma^2 = 3528 \pm 1677$
This confidence interval is narrower.
e. False: . . . the jacknifed *point* estimate coincides with the unbiased *point* estimate. (For a general proof of this bonus, see Wonnacott, 1984 IM).

8-35 **a.** $\mu = 44.21 \pm 2.68$
b. $\mu = 16,800 \pm 1,020$

8-37 $\mu_M - \mu_W = 9.0 \pm 8.0$
Or perhaps the one-sided claim is better: $\mu_M - \mu_W > 2.6$

8-39 **a.** Alcoholism in the mothers causes a drop of 21 ± 14 IQ points, on average—*if* this had been a randomized study (which of course would be unethical).
b. The extraneous factors that were not matched remain as a possible source of bias.

8-41 **a.** $\pi = .523 \pm .148 \approx 52\% \pm 15\%$
b. We agree.

8-43 **a.** less than 95%
b. 95%
c. much less than 95%
d. very likely more than 95%

9-1 The difference is statistically dicernible at the 5% level in all four cases.

9-3 **a.** H_0: Judge is fair, i.e., $\pi = .29$
b. $Z = 8.16$, $p \approx 0$

9-5 **a.** mean increase > 9.9
b. $t = 5.28$, $p < .0025$

9-7 **a.** In the population as a whole, men outsmoke women by at least 5 percentage points (4.86, exactly).
b. $t = z = 3.6$, $p = .00016$
c. Yes, is discernible.

9-9 **a.** H_0: the proportion of defectives is the old value of 10%, that is, $\pi = .10$.
H_1: the proportion of defectives is worse than 10%, that is, $\pi > .10$.
b. Reject H_0 if $P > 14\%$ (.140, exactly).
c. Reject all but the first and third shipments.

9-11 I, α. II, β. α, β.

9-13 **a.** No, do not reject H_0.
b. We would follow common sense, because the classical test is narrow-minded, and arbitrarily sets $\alpha = .05$.
c. Since $\overline{X} = 1245$ exceeds the new critical value of 1235, we reject H_0. So the problem in (**a**) was indeed inadequate sample size.
d. Since $\overline{X} = 1201$ exceeds the new and very stringent critical value of 1200.5, we reject H_0; therefore, we find the increase of 1 unit "statistically significant." Conclusion is true—and shows another weakness of classical tests.

9-15 α would likely decrease, and consequently β would increase.

9-17 **a.** H_1: $\pi = .30$
b. $.149 = 15\%$

9-19 **a.** $\overline{X}_c = 8.664$
b. yes, reject H_0
c. $\beta = .74, .09, .00$ (for $\mu_1 = 8.6, 8.8, 9.0$)
d. .01

9-21 **a.** $\alpha = .13$ (WCC, .11)
b. $\beta = $ .500, .176, .043, .008
(WCC, .540, .203, .052, .010)
d. Power is just OCC flipped upside down.

9-23 **a.** $\pi_{75} - \pi_{72} = .03 \pm .0355 \approx .03 \pm .04$
    b. .096
    c. No, not discernible.

9-25 **b.** $\overline{X} = 4.0, 3.2$
    $s = .76, 1.22$

9-27 **a.** False. See (**c**) for correct version
    b. True
    c. True
    d. True

9-29 **a.** 16.2%
    b. $p \approx .76$ or $p = .73$ (one-sided!)
    c. Given the eligibility rule, there is no evidence whatsoever that the grand jury discriminated against blacks.
    d. The major problem may be the racial discrimination in the past, which has left so many blacks illiterate.

10-1

Source	SS	df	MS	F	p
Fert.	312	2	156	9.0	<.01
Residual	156	9	17.3		
Total	468	11			

10-3

Source	SS	df	MS	F	p
Regions	1530	3	510	1.01	>.25
Residual	8046	16	503		
Total	9576	19			

10-5 **b.** To the extent that the class is large, the frequency distribution of F approximates the probability distribution in Figure 10-2, and approximately 5% of the F values exceed 3.89.

10-9

Source	SS	df	MS	F	p
Hours	18	2	9	3.6	<.25
Men	78	2	39	15.6	<.05
Residual	10	4	2.5		
Total	106	8			

10-11

Source	SS	df	MS	F	p
Fert.	608	2	304	51.68	<.001
Seeds	183	1	183	31.11	<.001
Blocks	261	3	87	14.79	<.001
Residual	100	17	5.88		
Total	1152	23			

10-13 **a.** $\mu_{NE} - \mu_S = 15.0 \pm 30.1$ ($000)
    b. $\mu_{NE} - \mu_S = 15.0 \pm 44.2$ ($000)

10-15 **a.**

Source	SS	df	MS	F	p
Varieties	54.0	2	27.0	4.5	< .10
Soils	186.0	2	93.0	15.5	< .05
Residual	24.0	4	6.0		
Total	264.0	8			

    b. $\mu_A - \mu_B = -3.00 \pm 7.45$
        $\mu_A - \mu_C = 3.00 \pm 7.45$
        $\mu_B - \mu_C = 6.00 \pm 7.45$
    c. He seems to have succeeded—B's yield in loam (31) is much better than any other yield in the table. This means the assumption of additivity is suspect, and so is the analysis in parts a and b.

10-17 **a.** i. $\pi_M - \pi_W = .023 \pm .024$
        ii. $\pi_M - \pi_W = -.045 \pm .032$
        iii. $\pi_M - \pi_W = .097 \pm .018$
    b. The hypothetical population of all those students with the same sorts of qualifications, who might have applied.
    c. The only faculty with discernibly different rates is Science, where women have a *higher* admission rate. The reason women have a discernibly lower admission rate in the whole school is that they tend to apply to the tougher faculty—not sex discrimination. For more detail, see Problem 1-8**d** on p. 17.

10-19 **a.** False: ... an *acceptable* hypothesis, while ... a *rejected* hypothesis.
    b. False: The *sample* mean ... with expectation μ and approximate standard deviation 10.
    c. False: ... the true but *unknown* population mean μ.

10-21 With 95% confidence:
$$\pi_H - \pi_L = .10 \pm .18$$
We estimate the 3-month survival rate is better for high concentration of oxygen than for low—better by 10 ± 18 percentage points. That is, the survival rate may be as much as 28 percentage points higher, or 8 lower.

10-23 $.047 \simeq 5\%$

10-25 a. $p \simeq 22\%$ (one-sided. $t \simeq z = .76$)
b. $p \simeq 34\%$ (one-sided. $t \simeq z = .41$)

10-27 a. 2 sided $p < .10$ ($t = 2.17$)
b. $p < .10$ ($F = 4.70$)
c. The answers agree.
In both **a** and **b** we assumed *independent* random samples. (And in calculating p, as always we assumed H_0 was true, i.e., the two populations were identical).

10-29 a. True
b. True

10-31 Not fair at all. Some extraneous factor such as ambition may be confounded with education to produce the high incomes of college graduates.

11-1 a. $\hat{Y} = 59 + 7.0X$
b.
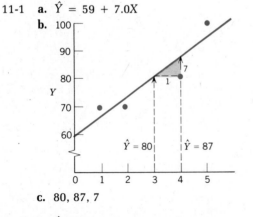
c. 80, 87, 7

11-3 a. $\hat{Y} = 119 + 9.0X$
b. 164, 119

c.
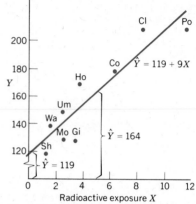

d. Since this is an uncontrolled observational study, it does not provide proof. (Other evidence, however, shows radiation is harmful).

11-5 a. $P = 30 + .50R$
c. It only estimates roughly how they are *related*. Causation cannot be inferred from an observational study.

11-7 a. $\hat{Y} = 156 + 2.35X$
Oberb: $\hat{Y} = 304$, error $= 14$
Pfalz: $\hat{Y} = 191$, error $= 23$
b. The positive slope indicates higher bottle feeding is associated with higher mortality—hardly a benefit. Since this is an uncontrolled observational study, however, it cannot prove that bottle feeding is harmful. (Other evidence, however, shows it is not as good as breastfeeding, at least in poor countries.)

12-3 a. .067
b. $(Y_7 - Y_1) / (X_7 - X_1)$
c. yes
d. yes
e. By the Gauss-Markov theorem, var $b_1 >$ var b.
f. $SE(b_1) = .0024\sigma$
$SE(b) = .0019\sigma$

12-5 $\beta = 7.0 \pm 10.1$

       $\beta = .144 \pm .148$

12-7   Only $\beta = .50$ in part (**d**) is rejected

12-9   **a.** $H_0: \beta = 0$

         $H_1: \beta > 0$

     b. $p < .05$

     c. $\beta > .035$

     d. reject H_0 on both counts

12-11 **b.** $15.0 \pm 3.9, 39.0 \pm 2.5$

        $63.0 \pm 1.7, 87.0 \pm 2.5$

     c. This band contains far fewer than 95% of the points.

12-13 **a.** $\beta = 800 \pm 490$

     b. Is discernible, since 800 exceeds $\pm$ 490.

     c. \$9,200 $\pm$ \$14,800 (very roughly, since income is not normally distributed)

     d. No, since this is just an observational study. Perhaps men with higher education also tend to have more ambition, and this may be what produces the higher income.

12-15 $p = .003$ $(t = z = 2.76)$

12-17 **a.** $\hat{Y} = 0.7 + 1.266X$

     c. $\mu_Y = 229 \pm 45$

     d. $\mu_Y = 140 \pm 117$

     e. The confidence interval is centered better, and narrower

12-19 **b.** $\alpha = 0.7 \pm 57.4$

13-1   **a.** The graph shows two lines, nearly parallel, with slope of about .18

     b. $+.18$ approximately

     c. $-.7$ approximately

     d. $\hat{Y} \approx 2.5$

13-3   soil fertility, temperature, hail and wind damage, insect and disease damage, etc.

13-5   **a.** Four equally spaced and parallel lines with slope of $-.13$.

     b. Decrease of .13 percentage points.

     c. decrease of .034 percentage points.

     d. 1.4 percentage points.

13-7 $b = \dfrac{\Sigma x_1 y}{\Sigma x_1^2}$, which is the simple regression formula

13-9   **a.**

SE	(.046)	(.010)
95% CI	($\pm$.15)	($\pm$.032)
t ratio	(-2.8)	(-3.4)
p	($<.05$)	($<.025$)

     b. We are assuming the 6 years represent a random sample from a large hypothetical population (which is pretty far fetched).

     c. change in real income, X_2

     d. keep it, on both prior and statistical grounds.

13-11 **a.** \$3.30 less, per front foot

     b. \$ .67 higher, per front foot

     c. *For the same kind of lot*, the price trend was up by \$1.50 per front foot, per year.

13-13 **a.** 10.14   **d.** 9.66

     b. .36   **e.** 9.66

     c. 10.50   **f.** -2.08

13-15 **a.** False: ... a year with *above*-average rainfall would *tend to* produce above average yield.

     b. Could be improved: In view of the posterior multiple regression coefficient, it would improve the crop to irrigate, *assuming* (i) irrigation acts like rain water, (ii) the level of water is not near the saturation point where further water would hinder rather than help, (iii) temperature is the only important extraneous factor that needs to be allowed for in the multiple regression etc.

13-17 **a.** 3.38

     b. 3.22

     c. increase yield by 3.38 (making the same assumptions as in problem 13-15b).

     d. $3.22, -.208$

13-19 a.

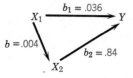

b. .039

c. The simple regression fits the data as well as any single line possibly can (though not as well as the 3 separate lines).

d. Yes, yes—and that is the great virtue of randomization.

13-21 a. .26 children fewer (on average)
b. .07 children fewer
c. .16 children fewer
d. .069 children more

13-23 b. .490, .303, .304
c. (i)   decrease of .030
(ii) increase of .912

13-25 Five years is a biased figure because of omitted extraneous factors. The bias could be reduced by a multiple regression, or eliminated (in theory) by randomized control. We would *guess* the unbiased figure was about 3 instead of 5. The major defect is not a formal statistical problem, however. It is a question of accurate reporting. Smoking does not cut 3 years of senility off the *end* of your life. It more likely cuts 3 years of vigorous living from the *best* of your life. (See Problem 13-24d for an explanation).

13-27 a. 100 (deaths per 10,000). This is a very good prediction.
b. 91 (deaths per 10,000)
c. (i)   decrease by 29 (deaths per 10,000)
(ii) decrease by 0.4 (deaths per 10,000)
(iii) decrease by 4 (deaths per 10,000)
d. (i)   .041 ± .032
(ii) $p < .010$ (t = 2.5)
(iii) Yes, because of (i) or (ii).

13-29 a. (i)   .68 st. dev. more
(ii) .41 st. dev. more
(iii) 1.26 st. dev. more
b. Disagree. When other variables are held constant, the number of police has a

slightly negative relation $(-.004)$ to the number of riots.

14-1 a.

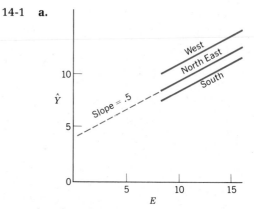

b. The 3 lines are no longer parallel.

14-3 a. -39 ± 3.5, -9.0 ± 4.3
-350 ± 90, -380 ± 104
-180 ± 106
b. age and cigarette smoking, 350 ml, *lower*
c. 30 ml, *higher*
d. 39 ml, *lower*
e. 180 ml, *lower*
f. 4.6 years, *important variables being omitted from the regression*

14-5 a. $.0021 \pm .0018$, $.0047 \pm .0022$
$.065 \pm .047$, $.002 \pm .053$
$-.032 \pm .053$
b. age and cigarette smoking, 6.5 percentage points (pp), *higher*
c. 6.3 pp, *higher*
d. .21 pp, *higher*
e. 9.4 pp, *higher*
f. 45 years, *important variables omitted*

14-7 a. 15.6 ± 6.1
b. 15.6 ± 6.1
c. They are identical. A much better CI is obtainable from the multiple regression coefficient, which allows for the extraneous effect of weight: 10.6 ± 6.5

14-9 a. 61, 70, 73
b. 61, 70, 73, identical
c. one factor, with response being yield and the factor being fertilizer

14-11 a.

additive		interactive	
70	80	74	77
55	65	53	67
65	75	68	73

    b. *additive, both sexes*

14-13 a. For large I, the expected yield would surely curve down (negative term in I^2)
    b. $p < .001$ (t = −3.75), so that H_0 has little credibility and (**a**) is confirmed
    c. maximum at I = 4.0, but irrigate much less if expensive.
    d. 4.5

14-15 a. grain production seems to have risen more slowly than population.
    b. grain is up 28%, while population is up 18%, so grain increased more.

14-17 a. 4.6%
    b. 17%
    c. 0.1%
    d. −14% (i.e., 14% *less*)

14-19 b. Four parallel but unequally spaced lines

14-21 a. race, dummy variables
    b. just categorical factors, suitable
    c. X and X^2 and X^3 . . . as the regressors
    d. logarithms
    e. can, measured; unit, constant
    f. cannot, some omitted extraneous factors . . .

14-23 a. 5.8¢ higher, 8.2¢ lower, 0.5¢ higher
    b. suburban, young, and indoor

14-25 a. b.

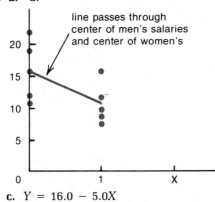

line passes through center of men's salaries and center of women's

    c. Y = 16.0 − 5.0X
    d. β = −5.0 ± 5.8. Women earn on average $5,000 (±$5,800) less than men.

e. Same
f. No. To measure discrimination, men and women *with the same qualifications* should be compared.

14-27 False: After holding constant *many* (*but not all*) other variables, women made $2,400 less than men. Therefore, $2,400 is our *best feasible* measure . . .
In conclusion, the $2,400 figure may be due to discrimination, but it also may be partly due to some more subtle difference between men and women *not measured in this study.*

14-29 Such severe rounding makes the coefficient easy to state and interpret, but it also loses some information and hence probably weakens the t ratio.

14-31 $\log Q = \log \beta_0 + \beta_1 \log K + \beta_2 \log L$
Regress Y = $\log Q$ on $X_1 = \log K$ and $X_2 = \log L$

15-1 a. $.787 \approx 79\%$
    b. $-.05 < \rho < .96$, approx.
    c. $H_0{:}\rho = 0$ is acceptable, that is, no proven relation.

15-3 a. .450
    b. $-.62 < \rho < .91$, approx.
    c. $\hat{Y} = .165 + .405X$
      $\beta = .405 \pm 1.476$
    e. i. No, because of (**c**)
      ii. No, because of (**b**)

15-5 a. $\hat{Y} = 1.15 + .0232X$
    b.

Source	SS	df	MS	F	p
Regression	5.73	1	5.73	44.6	<.001
Residual	6.17	48	.129		
Total	11.9	49			

Reject H_0 ($\beta = 0$) at the 5% error level, since its p-value is less than 5%.
    c. $\beta = .0232 \pm .0070$
    Yes, H_0 can be rejected.
    d. $.50 < \rho < .81$, approx.
    Yes, H_0 can be rejected.
    e. Yes
    f. 48%, yes; 52%, yes

15-7   **a.** True

       b. False: $\sigma_Y^2 = \sigma_X^2$

       c. True

       d. True

       e. False: ... the predicted X score is 75

15-9   **c.** $r = \sqrt{1(.50)} = .71$

       d. $\hat{Y} = 70$

       e. $\hat{X} = 65$

15-11   **a.** $R = .92$

       b. $r = .92$, the same

15-13 a with e; b with c and d.

15-15   **a.**

X_1	X_2
-2	2
-1	-1
0	-2
1	-1
2	2

           $r = 0$

       b. No.

       c. $SE_1 = 1.25$, $SE_2 = 1.06$

       d. You can simply drop off the X_2 term.

15-17 $r = 1.00$

15-19   **a.** $r = .57$

       b. $\hat{X}_2 = 23 + .61 X_1$

       c. 78, 47

       d. mean = 73, about the same as 78 in (c)

       e. mean = 43, about the same as 47 in (c)

       f. True with a minor qualification: The pilots who scored very well or very badly on the first text X_1 tended to be closer to the average on the second text X_2.

15-21   **a.** In this observational study, the effect of criticism (or praise) cannot be distinguished from the natural regression toward the mean.

       b. This design is still questionable. Now the effect of criticism cannot be distinguished from the effect of practice. The proper way to compare criticism with praise (or nothing), is with a *randomized* experimental design.

15-23   **a.** True

       b. False: Profits altogether are as extreme in 1985 as in 1984.

       c. True

       d. False: ... the best single prediction of its 1985 profit would be 9.5% (on the *regression* line).

15-25   **a.** $\hat{Y} = -120 + 921 X$

       b. $\beta \approx 920 \pm 560$

       c. $9,000 \pm 14,000$ (roughly—since income is not normally distributed).

15-27   **a.**

CI	$.09 \pm .02$	$.24 \pm .10$	2.61 ± 1.61
t	9.0	4.8	3.18
p	$<.0005$	$<.0005$	$<.001$
r	.54	.32	.22

       b. 32 minutes, 27.4 minutes

       c. The second prediction error was smaller— only 54% as big.

       d. Yes. [The prediction errors with and without regression are in the ratio (squared) of $(Y - \hat{Y})^2/(Y - \overline{Y})^2 = 5.4^2/10.0^2 = .29$. For all 200 patients, we likewise would obtain $\Sigma(Y - \hat{Y})^2/\Sigma(Y - \overline{Y})^2 = 1 - R^2 = 1 - .72 = .28$. So this one patient in part (c) was typical.]

16-1   **a.** $.109 \approx 11\%$

       b. $.344 \approx 34\%$

16-3   **a.** $.035 \approx 4\%$

       b. $.062 \approx 6\%$

       c. $.004 \approx 4/10\%$

16-5 $.968 \approx 97\%$

16-7   **a.** $67 \le \nu \le 76$

            $63 \le \nu \le 69$

              $3 \le \nu \le 7$

       b. $.930 = 93\%$

16-9   **a.** Yes, approximately normal.

       b. The nonparametric CI is likely less precise.

       c. $67.8 \le \mu \le 74.2$, which is indeed more precise (narrower).

16-11   **a.** $\overset{\star}{X} = 15$

       b. 99.8%

       c. $12 \le \nu \le 19$

16-13 $.047$

16-15 $.025$ (wcc, .033)

16-17 **b.** $t = -.84, p < .25$

16-19 **b.** $t = 2.73, p < .025$
    c. p-value is sharper (smaller) than earlier, because more information (ranking) was used.

16-21

Wilcoxon two-sample test	Kruskal-Wallis 1-factor ANOVA
Wilcoxon paired-sample test	Friedman 2-factor ANOVA

16-23 $r = .70$, not discernibly different from 0.

16-25 $p = .006$ (wcc, .011)

16-27 **a.** $p = .145$ (sign test), or $t = 1.84, p < .10$ (Wilcoxon)
    b. $p = .035$ (sign test), or $t = 3.43, p < .01$ (Wilcoxon)
    c. $p = .147 \cong 15\%$ (wcc, .159)

16-29 H_0 is that programs A and B are equally effective.
    $p = .032 \approx 3\%$ (wcc, .037)

16-31 **a.** $p = .152 \approx 15\%$ (wcc, .221)
    b. We cannot *a priori* expect a time series like this to be a random sample. And the test in part (**a**) does nothing to allay our fears. We therefore can't assume it is valid to proceed with the sign test.

17-1   **a.** $\chi^2 = 12.65, p < .01$
    b. yes, reject H_0

17-3   **c.** About 5% (the error level α)

17-5   **a.** Instead of χ^2, use the proportion of aces P. To calculate the p-value, standardize:

$$Z = \frac{P - 1/6}{\sqrt{\dfrac{(1/6)(5/6)}{30}}}$$

    c. much greater than 5% (the *power* of the test)

17-7   **a.** education and occupation are independent
    b. $\chi^2 = 53, p \ll .001$

17-9 $\chi^2 = 13.9, p < .001$

17-11 **a.** newspapers and class are independent
    b. $\chi^2 = 20.6, p < .005$, not discernible at level .001

17-13 **a.** degree and sex are independent
    b. $\chi^2 = 7.4, p \approx .025$
    c. $\pi_1 = 45\% \pm 3\%$
       $\pi_2 = 47\% \pm 6\%$
       $\pi_3 = 23\% \pm 14\%$

17-15 $\chi^2 = 36, p < .001$

18-1   MLE = .25, .50, .50, 0

18-3   **a.** $(1/90)^{10}$, the probability that a population of 90 rockets would produce the given sample of 10 serial numbers.
    b. $(1/77)^{10}$

18-5   **a.** MME
    b. MLE, in the sense of MSE (mean squared error), for example.

18-7   **a.** yes
    b. no, yes, yes

18-9 $\sqrt{\Sigma(X_i - \mu)^2/n}$

18-11 **a.** MME = MLE = $\bar{X}$ = \$21,000
    b. MME = 64, MLE = 44 is better (in MSE sense)
    c. MME = MLE = 6634

19-1   **a.** .10, .40, .50
    b. .28, .44, .28

19-3   40%

19-5   **a.** a lot narrower than in Problem 19-4, and a little narrower than the classical CI. Also centered better.
    b. Bayesian: $\pi = .56 \pm .13$
       Classical: $\pi = .60 \pm .14$

19-7   **a.** normal, with mean = 70 and st. dev. = 2.5
    b. at $\bar{X}$ = 70, the MLE
    c. the posterior is between, but 16 times closer to the likelihood, which has 16 times as much information.

d. Classical: $\mu = 70 \pm 4.9$
Bayesian: $\mu = 69.4 \pm 4.7$ (almost the same)

19-9 False. . . . n is *large*. It is only when n is small that . . .

19-11 a. $m = 13.97 \pm .36$
  b. 56%

19-13 b. The Bayesian estimates are shrunk 12% of the way to H_0 (52).
  c. (i) Figure 19-8
    (ii) the same, Figure 19-8

19-15 Shrinkage slope = .020
We assumed *a priori* that the population slope β was as likely to be $-$ as $+$.

19-17 Classical estimate = 6.0
Shrinkage estimate = 5.95. And for this, we assumed *a priori* that the population improvement was as likely to be $-$ as $+$.

19-19 a. 3500, assuming the shipment is like a *random* draw from past shipments, and then the 5 observations drawn from it are a *random* sample (VSRS)
  b. $\mu = 3500 \pm 160$

19-21 a. $\mu = 115 \pm 4.2$
  b. Each is making a correct inference; it's just that Dr. B knows more.
  c. $\mu = 117.0 \pm 3.4$
  d. $\mu = 117.0 \pm 3.4$
  e. practically the same

20-1 a. $L(a_1) = 22$ is min.
  b. if "rain," $L(a_3) = 15.9$ is min.
    if "shine," $L(a_1) = 17.7$ is min.
  c. False. Action a_1 is best when "shine" is predicted, and *also when no prediction is possible*. Action a_3 is best when "rain" is predicted. Action a_2 is never best.
  d. 4.7

20-3 a. $a_1 = 210$
    $a_2 = 220$
    $a_3 = 222$
  b. a_2
  c. a_3
  d. a_1

20-5 The mode, median, mean, and midrange, respectively.

20-7 30

20-9 a. 238.4
  b. 238.4
  c. 238.4

20-11 a. accept H_0
  b. accept H_1
  c. accept H_0
  d. accept H_1
  e. accept H_1
  f. accept H_1
  In all parts: accept H_0 if $p(X/S_0) > 6p(X/S_1)$, i.e., if $X < 34$

20-13

	20-10	20-11
(i)	accept H_1	accept H_0 if $X < 34$
(ii)	indifferent	accept H_0 if $X < 37$

	20-12 **(c)**	20-12 **(d)**
(i)	36.07	37.21
(ii)	37.5	37.5

20-15 a. 113.12
  b. .195
  c. Average loss increases over 3 times (by a factor of 3.15)

20-17 a.

150	200
191	200
199	200

  b.

150	200
191	200
199	200

20-19 a. about 5
  b. just these 5 statistically significant ones would get published, misleading the reader into thinking the Brainstorm effect really did exist.

20-21 the second kind (reliable)

20-23 pass if $X < 62.3$

21-1 **a.** choose the top option, with expected value of 20K
b. To avoid risk, she might be wise to choose the bottom option (the "sure thing")

21-3 **a.** seed, at an expected cost of 94.3 million
b. still seed, at an expected cost of 110.8 million

21-5 **a.** the pilot plant would be best, to net 64 million
b. continue the pilot plant, to keep losses down to 8.8 million

21-7 **a.** slightly worse
b. This half-price test is useless. So use the better seismic test again.

21-9 **a.** for homeowners who are risk averse, some form of insurance is wise to protect them from severe losses.
b. 52%
c. 38%, $43.60

21-11 **a.** We need to know the proportion of dissatisfied customers.
b. Assuming Hunter Creations is risk neutral, they should use the expensive spring, which on average costs $1.25 extra (compared to $2.10 extra for the cheap spring).

21-13 **a.** He should buy elsewhere and sue, without consulting the legal expert. Then expected cost is $45,000.
b. To $3,000 or less (EVPI = $3,000)
c. He should get expert legal advice; if it's encouraging, buy elsewhere and sue; if not, buy from Regal. Then expected disutility is about .13

21-15 **a.** Yes, the new can would increase expected profit by 7.5K
b. EVPI = 19K
c. (i) not worth it.
(ii) not worth it
(iii) 9.4K

22-1 **a.** 120.8, 120.0, 120.4
same order: PPI < FPI < LPI
b. 88.0, 87.4, 87.7
same order: PQI < FQI < LQI
c. cost index = 105.6
only Fisher's indexes pass
d. no difference

22-3

a.

133.3	183.3	133.3
133.3	175.3	133.3
133.3	179.3	133.3

b. LPI is simplest to calculate, and more important, it is simplest to measure: It does not require new quantity measurements every year (Q_t), but just initially (Q_0).

c.

137.5	74.6	102.6
134.0	72.7	97.4
135.7	73.7	100.0

The time reversal test is passed only by Fisher's Index.

22-9 **a.** 120.1, 115.3, 117.7
b. 114.6, 110.0, 112.3
c. 132.1

22-11 **a.** quantity, price
b. 1973–76 (29%).

23-1 **a.** $5,320,000 ± $300,000
b. $5,230,000 ± $220,000

23-3 **a.** 16 hours
b. 36 hours, which may be worse than useless.

23-5 **a.** $337
b. 220, 322, 260
c. SE = .39, a reduction of 13%
d. Earlier steps save more. Thus stratification reduces the SE by 24%, and sampling itself helps enormously.

23-7 **a.** USU, cluster
b. heterogeneous
c. sequential, overwhelming, random variable, less
d. may be disasterous, more biased

23-11 a. T
b. (i) False. . . regression *can be used for categorical stratification too, using dummy variables.* (ii) True.

23-13 a. T
b. T

24-1 d. See Problem 24-2.

24-3 b. Yes, it does show random tracking

24-5 If you use a computer package (which we highly recommend), you will get slightly different answers for Problems 24-5 and 24-7, due to the minor refinements the computer can easily make.

Starting at the third quarter of 1957:
a. Centered M = 58.0, 58.5, 59.0, 60.0, . . . , which does indeed smooth out the large fourth quarter
b. .776, 1.812, .644, .767, . . .
c. .763, 1.816, .654, .768, repeated
d. Starting now at the first quarter: 55, 57, 59, 58, . . .

24-7 Starting at the third quarter of 1968:
a. Centered M = 542, 541, 540, 541, . . . , which is indeed smooth
b. 1.015, 1.006, .939, 1.039, . . .
c. 1.018, 1.020, .917, 1.059, and repeated
d. Starting now at the first quarter: 554, 534, 540, 533, . . .

24-9 a. inappropriate
b. quarterly, may give a better idea of relative performance.

25-1 b. $\hat{Y} = -1.5 + .744\,X$, which is biased.

25-3 a. $\hat{Y} = 10 + .60X$, which is perfectly on target
b. This is a lucky sample, because the errors e_i exactly cancel.

25-5 a.
$$s_{X_1 Y} = \beta_1 s_{X_1 X_1} + \beta_2 s_{X_1 X_2}$$

$$s_{X_2 Y} = \beta_1 s_{X_2 X_1} + \beta_2 s_{X_2 X_2}$$

b. Solve the two equations for the two unknowns, β_1 and β_2.
b. They are the same, since the equations in part (**a**) are just the estimating equations (13-5) and (13-6)—except for the constant divisor $n - 1$.

25-7 a. $\hat{Y}_1 = -2.09\,Y_2$
b. Using $\hat{Y}_1$ as an IV (where $\hat{Y}_1 = 1.99\,X_1$):

$$s_{\hat{Y}_1 Y_2} = s_{\hat{Y}_1 Y_1}\,\gamma_1 + s_{\hat{Y}_1 X_1}\,\beta_1$$

i.e., $-38.6 = 80.8\,\gamma_1 + 40.6\,\beta_1$

Using X_1 as an IV:

$$s_{X_1 Y_2} = s_{X_1 Y_1}\,\gamma_1 + s_{X_1 X_1}\,\beta_1$$

i.e., $-19.4 = 40.6\,\gamma_1 + 20.4\,\beta_1$
But this is just the first equation restated (multiplied by 1/1.99).

25-9 False: An IV should be highly correlated with the *regressors*, and uncorrelated with the *error term*. Therefore, *predetermined variables in the equation* qualify as good instrumental variables, but *mutually dependent variables* do not.

Glossary Of Common Symbols

SYMBOL	MEANING	REFERENCE
	ENGLISH LETTERS	
a	regression intercept	Figure 11-4, (11-6)
ANOVA	analysis of variance	(10-14), (10-18)
b	regression slope	Figure 11-4, (11-5)
c	number of columns in a table	(10-7), (17-11)
cov	covariance, $= \sigma_{XY}$	(5-10), (5-18)
d.f.	degrees of freedom	(2-16), (8-12)
D	difference in two paired observations	(8-24), Table 8-3
e	error term in regression	(12-3)
E	event (also F, G, etc.)	(3-6)
$\bar{\bar{E}}$	not E	(3-14), (3-15)
E( )	expected value, $= \mu$	(4-32), (17-1)
F	variance ratio	(10-15)
H_0	null hypothesis	(9-9), (19-33)
H_1	alternative hypothesis	(9-10), (19-33)
L( )	likelihood function, or	(18-3), (18-4)
	average loss	(20-4)
MLE	maximum likelihood estimate	(18-4), (18-11)
MSD	mean squared deviation	(2-10), (2-23)
MS	mean square	Tables 10-4a, 10-8a
n	sample size	(6-7), (6-8)

SYMBOL	MEANING	REFERENCE
n_0	quasi-sample size	(19-22), (19-29)
N	population size	(6-24)
O	observed value	Table 17-1
OLS	ordinary least squares	(11-2)
$p(x)$	probability distribution	(4-2), (5-4)
$p(x,y)$	joint probability distribution	(5-2)
P	sample proportion	(6-12), (8-27)
$Pr(E)$	probability of event E	(3-2), (3-7)
$Pr(E/F)$	conditional probability	(3-17)
r	simple correlation, or	(15-2), (15-6)
	number of rows	Table 10-8 (17-11)
r^2	coefficient of determination	(15-16), (15-17)
r_0, r_1	regrets in Bayesian testing	(20-9), (20-11)
R	number of runs	(16-14)
R^2	coeff. of multiple determination	(15-28)
$\overline{R}^2$	R^2 corrected for d.f.	(15-29), (15-30)
s	standard deviation	(2-13), (2-14)
s^2	variance of sample, or	(2-11), (2-12)
	residual variance	Table 10-8, (12-20)
s_p^2	pooled variance of samples	(8-21), (10-5)
s_X^2	variance of X values in regression	(12-6), (15-7)
s_Y^2	variance of Y values in regression	(15-7), (15-17)
S	number of successes	(4-8), Table 4-3
SE	standard error	(6-7), (8-15), (12-22)
SS	sum of squares (variation)	(10-14), (10-18)
t	t variable	(8-15), (9-18)
var	variance, $= \sigma^2$	(4-5), (5-18)
w	bisquare weights	(7-15), (7-18), (11-13)
W	Wilcoxon test statistic	(16-9)
X	random variable, or	(4-2), Figure 4-1
	regressor in original form	(11-4)
x	realized value of X, or	Figure 4-1
	regressor in deviation form	(11-4)
$\overline{X}$	sample mean	(2-4), (6-9)
$\dot{X}$	sample median	(2-2)
Y	response in regression	(12-2)
$\hat{Y}$	fitted value of Y	(11-1), (12-20)
Z	standard normal variable	(4-16), (4-23), (6-11)

MATHEMATICAL SYMBOLS

$G \cup H$	G or H, or both	(3-8), (3-12)
$G \cap H$	G and H	(3-9), (3-18)
$\propto$	is proportional to	(19-6)
$\equiv$	equals, by definition	

SYMBOL	MEANING	REFERENCE
$\simeq$	approximately equals	
$>$	is greater than	
*	optional section or problem	

GREEK LETTERS ARE RESERVED FOR POPULATION PARAMETERS

α	probability of type I error, or	Figures 9-4, 9-6
	population regression intercept	Figure 12-2, (12-25)
β	probability of type II error, or	Figures 9-6, 9-9
	population regression slope	Figure 12-2, (12-23)
Δ	population mean difference	(8-24)
θ	any population parameter	(7-1), (18-10), (19-5)
μ	population mean	(4-4), (4-28), (8-11)
μ_0	regression mean at X_0, or	(12-32), Figure 12-7
	mean of prior distribution, or	(19-23), Figure 19-7
	null population mean	(9-9), (20-13)
μ_1	alternative population mean	(9-10), (20-13)
ν	population median	(16-2)
π	population proportion	(8-27), (19-19)
ρ	population correlation	(5-13), Figure 15-4
σ^2	population variance	(4-5), (5-18)
σ_{XY}	population covariance	(5-10), (5-18)
Σ	sum of	(2-4)
χ^2	chi-square variable	(17-2), (17-10)

GREEK ALPHABET

Letter	Name	English Equivalent	Letter	Name	English Equivalent
$A\alpha$	Alpha	a	$N\nu$	Nu	n
$B\beta$	Beta	b	$\Xi\xi$	Xi	x
$\Gamma\gamma$	Gamma	g	Oo	Omicron	o
$\Delta\delta$	Delta	d	$\Pi\pi$	Pi	p
$E\epsilon$	Epsilon	e	$P\rho$	Rho	r
$Z\zeta$	Zeta	z	$\Sigma\sigma$	Sigma	s
$H\eta$	Eta	—	$T\tau$	Tau	t
$\Theta\theta$	Theta	—	$\Upsilon\nu$	Upsilon	u or y
$I\iota$	Iota	i	$\Phi\phi$	Phi	—
$K\kappa$	Kappa	k	$X\chi$	Chi	—
$\Lambda\lambda$	Lambda	l	$\Psi\psi$	Psi	—
$M\mu$	Mu	m	$\Omega\omega$	Omega	—

Index